Principles of Population Genetics

Principles of

POPULATION GENETICS

DANIEL L. HARTL PURDUE UNIVERSITY

SINAUER ASSOCIATES, INC. • PUBLISHERS
Sunderland, Massachusetts

PRINCIPLES OF POPULATION GENETICS

Printed in U.S.A.

Library of Congress Cataloging in Publication Data

Hartl, Daniel L
Principles of population genetics.

Bibliography: p.
Includes index.
1. Population genetics. 2. Quantitative genetics.
I. Title.
QH455.H37 575.1 79-28384
ISBN 0-87893-272-0

9 8 7 6 5 4 3 2

For Christine

Contents

CONTENTS

Chapter 5. Population Genetics, Ecology, and Evolution 325

Preface

This is a textbook of population genetics designed for junior-senior undergraduates and graduate students. Population genetics is an extraordinarily diverse field. It cuts across molecular biology, genetics, ecology, evolutionary biology, systematics, natural history, plant breeding, animal breeding, many areas of conservation and wildlife management, human genetics, sociology, anthropology, mathematics, and statistics. Because population genetics plays a special role in so many areas of biology, I have tried to summarize the principles of the field for the widest possible audience. No prerequisites in genetics or advanced mathematics are necessary for an understanding of the material in the book. The relevant background is provided in Chapter 1. The first half of Chapter 1 deals with the necessary background in genetics and molecular biology; it is intended as a summary for students whose prior training has been in statistics or mathematics, or for students in biology who have not yet studied genetics. This material also serves as a memory refresher for students who have studied genetics. The second half of Chapter 1 provides the relevant statistical and mathematical background; it is designed primarily for students in any of the biological or social sciences.

Chapters 2 through 4 concern what is usually regarded as the core of population genetics. The theme of these chapters is the origin, maintenance, and significance of genetic variation. Chapter 2 focuses on the detection and measurement of genetic variation, and on the organization of genetic variation as influenced by particular mating systems such as random mating or inbreeding. Chapter 3 deals with the evolutionary effects of mutation, migration, selection, and small population size. Chapter 4 considers traits that are influenced by alleles at many loci as well as by the effects of environment. Although particular attention is devoted to methods of artificial selection that are used for the genetic improvement of crop plants and domesticated animals, Chapter

4 also discusses how the principles of quantitative genetics can be applied to humans and natural populations of other species.

Chapter 5 is a synthesis of the elements of population genetics as they apply to actual, evolving populations in an ecological setting. The first part of the chapter is devoted to Sewall Wright's shifting balance theory of evolution and its alternatives. The second part deals with modes of selection that do not act directly on the individual but rather on the individual's relatives: group selection, interdeme selection, and kin selection. The third part illustrates the principles of population genetics in action by means of four particularly well studied examples — the evolution of antibiotic resistance in bacteria, of heavy-metal tolerance in plants, of industrial melanism in moths, and of Batesian mimicry in butterflies. Finally, Chapter 5 deals with the role of population genetics in understanding the origin of species.

Throughout the book, I have tried to emphasize the interplay between experimental observations and theoretical deductions. The experimental side of population genetics is wonderfully rich and diverse. But the literature of experimental population genetics is scattered among a variety of journals extending from those in plant and animal breeding to those in ecology and evolution. One motivation for writing the book was to bring this disparate literature together and emphasize its underlying unity. The theoretical side of population genetics is less imposing than might be expected. Especially at this introductory level, the mathematics involved in theoretical population genetics is mainly simple and straightforward algebra. In addition, the theoretical arguments in the book are developed step by step and are illustrated with numerous actual examples.

A number of special features of the book deserve mention. Throughout the book are a number of boxes that are set off from the main text in smaller type. These boxes serve several functions. A few provide background material in experimental or statistical methods. A few others show how certain equations stated in the text can be derived. Most of the boxes, however, take up where the text leaves off; they extend the material in the text and briefly discuss more specialized or advanced topics. Although a few boxes make use of elementary calculus, emphasis is on the meaning and application of the results, not on the mathematics. Moreover, all required mathematical formulas are provided as needed.

Use of the boxes is at the instructor's discretion and depends on the level and orientation of the course. None of the boxes is required for an understanding of the text. In this sense, the boxes are optional. However, an instructor may wish to assign certain boxes as required reading; the boxes in Chapter 4 would be particularly relevant in courses oriented toward plant or animal breeding, for example. Many of the boxes contain sufficiently important material to justify their requirement in courses at the graduate level.

At the end of each chapter is a set of fifteen problems designed to help students verify their understanding of the text. Solutions to the problems, which include the methods of solution as well as final answers, can be found in the back of the book. Problems are also provided as a test of mastery in each box; these, too, are solved in full at the back of the book. In writing the problems, no special effort has been made to avoid the need for numerical calculations, as most students these days have access to electronic calculators.

At the end of each chapter is a set of further readings. These are general sources, mainly books and review articles. All references cited in the text are listed in the bibliography at the end of the book. Although the bibliography contains over 600 references to the primary literature of population genetics, the number could easily have grown much larger. In order to keep the bibliography in manageable proportions, I have emphasized review articles (wherever possible) and recent literature. The bibliography is designed mainly as an entry point to the literature. Moreover, I have made no serious attempt to cite the first or "priority" paper on any subject, on the grounds that such distinctions are best left to historians of the science. It should be noted, however, that most of the ideas of modern population genetics are derived from the works of Sewall Wright, R. A. Fisher, and J. B. S. Haldane. While I have tried to be fair to original investigators in acknowledging their contributions, there are no doubt a few unintentional oversights. For these I apologize.

Writing a book is always grueling, at least it is for me. This task was made much easier by reviewers who offered expert guidance in matters of emphasis and balance and in correcting wrong or misleading statements. I thank James F. Crow, Franklin D. Enfield, Joseph Felsenstein, Bruce R. Levin, Rollin Richmond, Michael J. Simmons, and Edward O. Wilson, each of whom read all or major parts of the book. Carl W. May did a splendid job of collating all the reviewers' suggestions and offering numerous helpful comments of his own.

PREFACE

The material in the book formed the basis of a course in experimental population genetics that was offered while I was Visiting Professor at the University of Zürich in Switzerland. I am grateful to Professor Hans Burla for arranging the course, to the students for their advice and criticism, and especially to Hans Jungen and Elizabeth Hauschteck-Jungen for their generous hospitality during my stay in Zürich.

Rollin Richmond used a preliminary draft of the book for a course in population genetics at Indiana University, and I am gratified by the favorable student response. Stephen Rich of Purdue University read the final draft and rechecked the answers to all the chapter-end and box problems.

The final manuscript was expertly typed by Betty Gick, Pat Oswalt, and Carol Wigg. Natalie Brown and Stephen Rich helped in proofreading. Credit for the final appearance of the book should go to Sinauer Associates and to Joseph Vesely Production Services. A special thanks goes to Andy Sinauer for his advice and encouragement.

The people in my laboratory deserve plaudits for their continuing interest and patience while I struggled with the manuscript. My thanks to Daniel Dykhuizen, Doris Freeman, and Luanne Wolfgram for keeping the house in order, and to John Dunne, David Haymer, Paul Kaytes, Nizam Kettaneh, and Susan Wurster for their forebearance.

I want also to thank Dana Margaret and Theodore James Hartl for their unflagging curiosity and enthusiasm for the project. Their constant "When will it be finished?" and "When can we see it?" served as a needed prod.

DANIEL L. HARTL

·1·

Genetic and Statistical Background

Population genetics is the study of how Mendel's laws and other genetic principles apply to entire populations. Such a study is essential to a proper understanding of evolution because, fundamentally, evolution is the result of progressive change in the genetic composition of a population. Population genetics thus seeks to understand and to predict the effects of such genetic phenomena as segregation, recombination, and mutation; at the same time, population genetics must take into account such ecological and evolutionary factors as population size, patterns of mating, geographic distribution of individuals, migration, and natural selection. The many genetic, ecological, and evolutionary factors that influence populations also interact and feed back, one upon the other. As might well be expected, to gain an understanding of such complex interactions is a formidable task.

Ideally, one would wish to know how to describe the types and frequencies of genes in a population, to explain how the population's genetic composition came to be the way it is, and to predict how the population would change as a result of natural selection or as a result of artificial selection applied by a plant or animal breeder. Population geneticists are a long way from being able to do these things as well as they might be done, but in the 80 or so years since the formal study of population genetics was inaugurated, considerable progress has been made. Indeed, a balanced view of the recent history of biology would not permit the brilliant and relatively rapid success of molecular biologists to completely overshadow the slow, steady progress of population geneticists against their own problems, problems that can be particularly stubborn because they are often so subtle and sometimes maddeningly complex.

Nevertheless, it is appropriate to point out that certain central problems in population genetics have been solved by use of relatively straightforward approaches and have yielded relatively simple principles; these principles have applications and implications for topics that range all the way from medicine to ecology, from anthropology to plant and animal breeding. This book is my attempt to summarize these principles.

The Genetic Background of Population Genetics

The first half of this chapter is intended as a memory refresher for students in any of the biological sciences and as a brief summary of genetics for students in mathematics, statistics, or the social sciences. Because major aspects of population genetics deal with the implications of Mendel's laws as they affect whole populations, we may do well to begin with Mendel — Gregor Johann Mendel (1822-1884), amateur plant breeder, who spent most of his adult life as a priest and later abbot in a monastery in the city of Brünn, Moravia, now Brno, Czechoslovakia. A few key terms must first be defined, however.

GENE is a general term meaning, loosely, that physical entity transmitted during the reproductive process that influences hereditary traits among the offspring. Genes influence such human traits as hair color, eye color, skin color, height, weight, and intelligence, although most of these traits are also influenced more or less strongly by environment. Genes also determine the nature of such proteins as hemoglobin, which carries oxygen in the red blood cells, or insulin, which is important in maintaining glucose balance in the blood. Genes can exist in different forms or states. For example, a gene for hemoglobin may exist in a normal form or in any one of a number of forms that produce hemoglobin molecules that are more or less abnormal. These alternative forms of a gene are called ALLELES.

From a biochemical point of view, a gene corresponds to a specific sequence of constituents (called nucleotides) along a molecule of DNA (deoxyribonucleic acid) — DNA is the genetic material. Different sequences of nucleotides that may occur in a gene, therefore, represent alleles. (See Figure 1 and the section on molecular genetics later in this chapter for further detail.)

Mendel studied seven traits in the garden pea *Pisum sativum.* He self-fertilized plants for several years until he obtained strains that "bred true"; that is, until he obtained strains that were genetically stable in the sense that the offspring in any generation strongly resembled each other and their parents with respect to each trait considered. For example, Mendel obtained a strain that bred true for *round* seeds; crosses between individuals of this strain always yielded plants with *round* seeds. He also obtained a strain that bred true for *wrinkled* seeds, and crosses within this strain always yielded plants with *wrinkled* seeds. Among the seven traits for which Mendel obtained true-breeding strains that exhibited contrasting appearances were, as mentioned, seed shape (*round* versus *wrinkled*), color of the seed (actually the color of the cotyledon, the first leaf of the embryo; *yellow* versus *green*), and seedpod shape (*smooth* versus *constricted*).

With the true breeding strains that he had developed, Mendel began his monumental series of breeding experiments that culminated in 1866 with publication of "Experiments on plant hybrids" in the *Proceedings of the Brünn Natural History Society.* (The original paper is in German, but an excellent English translation is available in Stern and Sherwood, 1966.) Mendel's paper was almost completely ignored until 1900, when Hugo de Vries in Amsterdam, Carl Correns in Tübingen, and Erich von Seysenegg Tschermak in Vienna independently rediscovered Mendel's findings and finally uncovered his original publication.

DOMINANCE

Now, when Mendel crossed plants from the true-breeding strain having *round* seeds with plants from the true-breeding strain having *wrinkled* seeds, he obtained a startling result: the offspring or HYBRID seeds were *round.* Similarly, crosses between the *yellow*-seed strain and the *green*-seed strain yielded *yellow* seeds; crosses between the *smooth*-pod and *constricted*-pod strains gave rise to plants with *smooth* pods, and so on for all seven traits. These results can be summa-

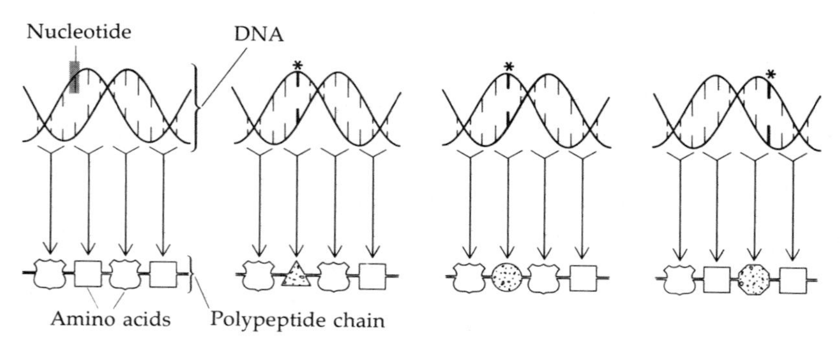

FIGURE 1. Alleles are alternative forms of a gene. DNA is composed of two intertwined strands, each consisting of a long linear sequence of nucleotides. Genes are fundamental units of genetic information that correspond chemically to the sequence of nucleotides in a segment of DNA. A typical gene consists of hundreds or thousands of nucleotides, only a few of which are shown here. The arrows show how the genetic information in a portion of the nucleotide sequence of DNA specifies the amino acid sequence in a portion of a polypeptide (or protein), each group of three adjacent nucleotides corresponding to one amino acid. Substitutions of one nucleotide for another (at positions indicated by the asterisks and heavy lines) in the DNA can lead to substitutions of one amino acid for another in the polypeptide (indicated by stippling).

rized in modern terminology by saying that the allele determining *round* seeds is DOMINANT to the allele determining *wrinkled* seeds; the allele for *yellow* seeds is dominant to that for *green* seeds. Conversely, one could say that the allele for *wrinkled* seeds is RECESSIVE to that for *round* seeds. The point is that the hybrids, which receive a different allele from each parental strain, express only the trait corresponding to the dominant allele and fail to express the trait corresponding to the recessive allele.

Mendel's LAW OF DOMINANCE states that, in any mating in which the hybrid receives different alleles from each parent, one or the other allele will be dominant and the trait corresponding to that allele will be expressed; the trait corresponding to the other, recessive, allele will remain unexpressed.

The occurrence of the phenomenon of dominance is by no means universal, however. For many traits, the hybrid will strikingly resemble neither parent, but will have an appearance more or less intermediate between the parents. Moreover, many alleles that are superficially dominant are imperfectly so, because the hybrid may be somewhat, though often subtly, different from the parent it resembles. The absence of "perfect" dominance is particularly true at the biochemical level when one examines, for example, the proteins determined by the alleles in question. Absence of dominance at the molecular level may occur even though one allele may appear to be dominant at the superficial morphological level. Indeed, even Mendel's *round*-seed allele is not completely dominant to the *wrinkled* one, as microscopic examination of the size and number of starch granules reveals the hybrid to be intermediate between the parents.

SEGREGATION

The heart of Mendelian genetics is segregation, which Mendel observed when he crossed the hybrids resulting from initial crosses of the true-breeding strains. SEGREGATION, which occurs during the formation of reproductive cells, refers to the separation of alleles of a gene and their distribution into the reproductive cells.

The offspring of crosses between the true-breeding strains constitute the so-called F_1 GENERATION; it is in this generation that dominance, or the lack of it, appears. The offspring of $F_1 \times F_1$ crosses constitute the so-called F_2 GENERATION; it is in this generation that segregation is observed.

When Mendel crossed his F_1 hybrid plants, he observed in the F_2 generation an almost perfect 3/4:1/4 (or 3:1) ratio of plants showing the trait associated with the dominant allele to plants showing the trait associated with the recessive allele. For example, in the cross involving alleles for *round* versus *wrinkled* seeds, he obtained 5474 *round* seeds and 1850 *wrinkled* seeds in the F_2, a ratio of 2.96:1, in similar crosses, he obtained 6022 *yellow* seeds and 2001 *green* seeds (a ratio of 3.01:1), and also 882 plants with *smooth* pods and 299 plants with *constricted* pods (a ratio of 2.95:1).

Mendel accounted for these 3:1 ratios in the F_2 by reasoning as follows: let *A* represent the dominant allele for, say, *round* seeds, and let *a* represent the recessive allele. Let us suppose that each GAMETE (a term for a mature reproductive cell such as an egg or sperm) carries one such allele. A gamete from a true-breeding *round*-seed strain must therefore carry *A*, and a gamete from a true-breeding *wrinkled*-seed strain must carry *a*. The hybrid, or F_1, must therefore be genetically *Aa* (Figure 2A). The genetic composition *Aa*

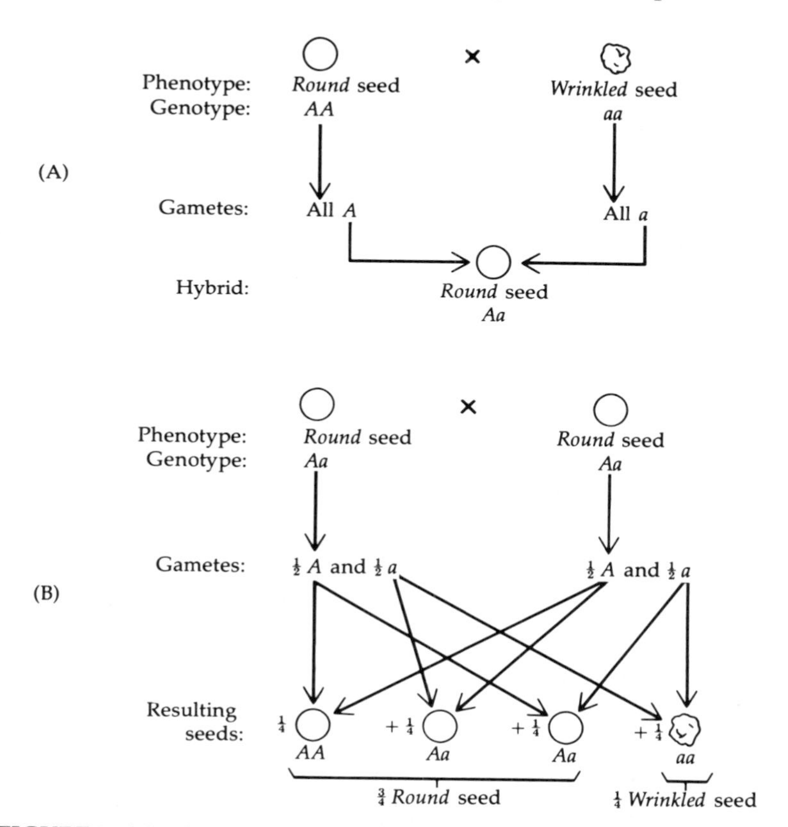

FIGURE 2. Mendel's results involving the inheritance of *round* or *wrinkled* seeds. (A) Mating a homozygous *AA* plant with a homozygous *aa* plant gives rise to heterozygous *Aa* seeds. The *Aa* seeds are *round,* like those in the *AA* parent, indicating that *A* is dominant to *a.* (B) Mating between two heterozygous *Aa* plants shows the result of segregation. Among the seeds that result from the mating, 1/4 are *AA,* 1/2 are *Aa,* and 1/4 are *aa.* Because *A* is dominant to *a,* both *AA* and *Aa* have *round* seeds; therefore the observed ratio of *round* to *wrinkled* seeds is 3/4 to 1/4.

must characterize body (SOMATIC) cells of the F_1 plants; when these plants form gametes, the gametes must carry only a single allele, *A* or *a*. Mendel assumed, first, that in the F_1 plants, the *A* and *a* alleles maintained their integrity and did not blend together or contaminate or change each other in any way. This assumption departed markedly from the prevailing views of heredity in Mendel's time.

Mendel then assumed — and this assumption is the LAW OF SEGREGATION — that when the F_1 hybrids form gametes, half the gametes carry *A* and half carry *a*. Combining gametes at random from two such plants, as shown in Figure 2B, yields a genetic ratio in the F_2 of 1/4 *AA*:1/2 *Aa*:1/4 *aa*. However, both *AA* and *Aa* have *round* seeds, the trait corresponding to the dominant allele *A*, so the observed F_2 ratio is 3/4 *round*:1/4 *wrinkled*, or 3:1.

To express Mendel's result in modern terminology, we would say that the ratio of genotypes in the F_2 is 1/4 *AA*:1/2 *Aa*:1/4 *aa* (GENOTYPE refers to the genetic constitution of an individual); because of dominance, the observed ratio of phenotypes is 3/4 *round*:1/4 *wrinkled* (PHENOTYPE refers to the physical appearance of an individual). It is important to note that *AA* and *Aa* have the same phenotype (*round* seeds), but different genotypes. Genotypes *AA* and *aa* are called HOMOZYGOUS, whereas the genotype having two different alleles, *Aa*, is called HETEROZYGOUS. In modern terminology, we would say that Mendel's original true-breeding strains were homozygous and that they bred true precisely because they were homozygous.

Results for these and other kinds of crosses can be predicted using the sort of "checkerboards" shown in Figure 3. These checkerboards, called PUNNETT SQUARES after their inventor Reginald C. Punnett (1875–1967), show that a cross of *Aa* × *aa* leads to a genotypic ratio in the next generation of 1/2 *Aa*:1/2 *aa*. Predictions of the results of crosses follow smoothly from Mendel's law of segregation. One has only to remember that, because of segregation, half the gametes from a heterozygote will carry one allele and the other half will carry the other allele.

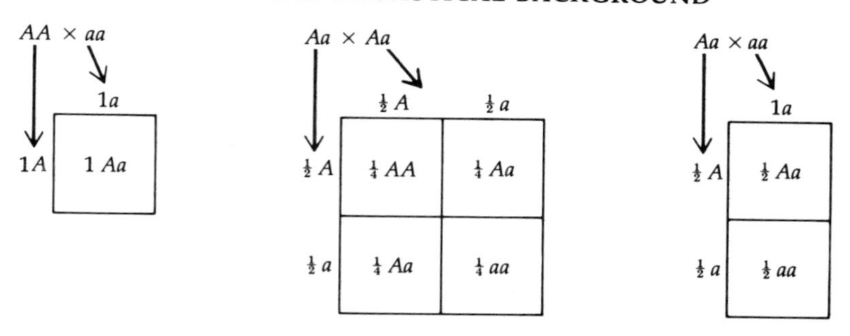

FIGURE 3. Punnett squares showing the genotypic ratios produced in various kinds of matings involving a single locus.

INDEPENDENT ASSORTMENT AND LINKAGE

Although one of Mendel's brilliant insights was to focus attention on the inheritance of single, easily identified, contrasting traits such as *round* versus *wrinkled* seeds, he also carried out experiments to determine how the inheritance of one trait might influence that of another. That is to say, he studied the simultaneous transmission of, for example, *round* versus *wrinkled* seeds and *yellow* versus *green* seeds. In this endeavor Mendel was extremely lucky, for he found that the inheritance of each of his seven traits was completely independent of any of the others that he examined. Such independence in genetic transmission is known as INDEPENDENT ASSORTMENT.

To understand the phenomenon of independent assortment, consider what happened when Mendel crossed a homozygous *round, yellow*-seed strain, the genotype of which may be represented as *AABB*, with another homozygous strain, this one having *wrinkled, green* seeds, genotypically *aabb*. [In modern pea breeding, the symbols *R* and *r* are used for the alleles associated with *round* and *wrinkled* seeds, respectively, and *I* and *i* are used for the alleles associated with *yellow* and *green* seeds, respectively (Blixt, 1974).] The F_1 seeds were then of genotype *AaBb* and were *round* and *yellow*, because *A* is dominant to *a* and *B* to *b* (Figure 4A). When he carried out the $F_1 \times F_1$ crosses, Mendel

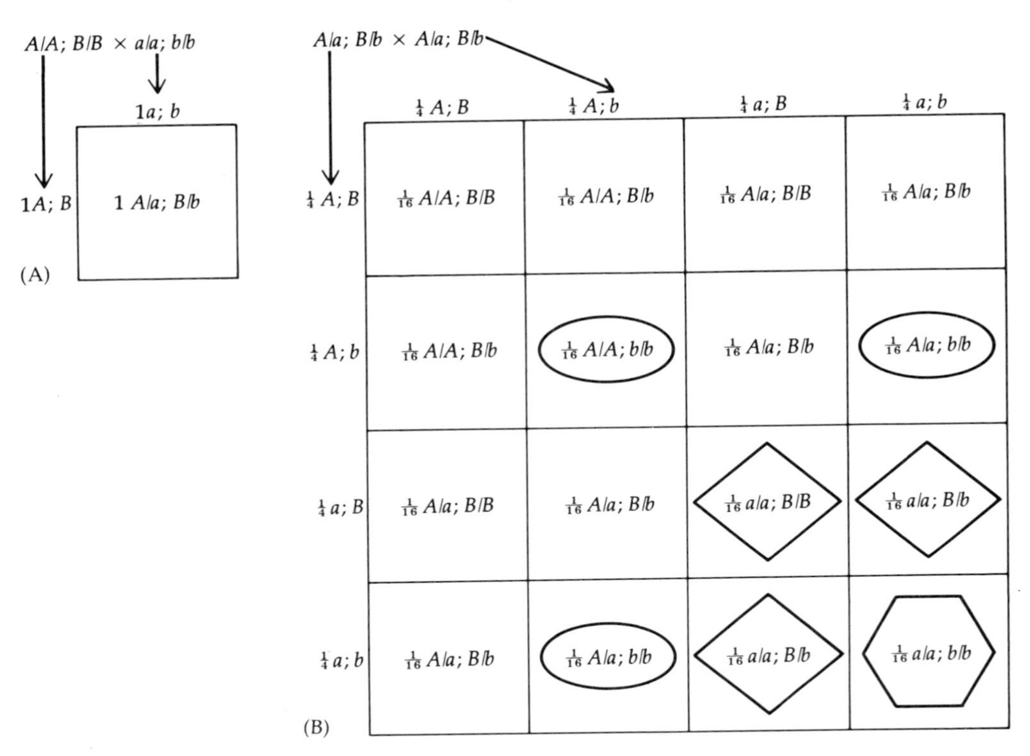

FIGURE 4. Punnett squares showing the results of matings involving two unlinked loci. (A) Mating between double homozygotes yields offspring that are heterozygous for both loci. If *A* represents the dominant allele for *round* seeds and *B* represents the dominant allele for *yellow* seeds, then the offspring of the mating (called the F_1 generation) will all have *round, yellow* seeds. (B) Mating between two F_1's. The offspring with *round, green* seeds are inscribed in ovals, those with *wrinkled, yellow* seeds are inscribed in diamonds, and the class with *wrinkled, green* seeds is inscribed in a hexagon. The overall phenotypic ratio of *round, yellow*: *round, green*: *wrinkled, yellow*: *wrinkled, green* is 9/16: 3/16: 3/16: 1/16.

found in the F_2 315 seeds that were *round, yellow*; 108 that were *round, green*; 101 *wrinkled, yellow*; and 32 *wrinkled, green* (Figure 4B). This ratio of 315:108:101:32 is 9.84:3.16:3.38:1, or nearly 9:3:3:1, which is to say 9/16:3/16:3/16:1/16. From Mendel's law of segregation one expects the F_2 ratio for each trait separately to be 3:1, that is to say, 3/4 *round*:1/4 *wrinkled* and

3/4 *yellow*:1/4 *green*. Were the traits inherited independently, we could multiply these proportions and obtain (3/4 *round* + 1/4 *wrinkled*) × (3/4 *yellow* + 1/4 *green*) = 9/16 *round, yellow* + 3/16 *round, green* + 3/16 *wrinkled, yellow* + 1/16 *wrinkled, green,* which is, of course, exactly the 9:3:3:1 ratio observed in the F_2 (Figure 4B). (The statistical method for showing that Mendel's ratio of 315:108:101:32 does not differ significantly from 9:3:3:1 is outlined later in this chapter.)

LINKAGE AND RECOMBINATION

Although all pairs of traits that Mendel studied undergo independent assortment, the phenomenon is not a general one. Independent assortment will occur if the alleles for the two traits are on different CHROMOSOMES: these are microscopic structures in the cell nucleus that contain (with other constituents) a single, long molecule of DNA and therefore have a linear arrangement of genes. The position of a gene on a chromosome is called a LOCUS; if the loci controlling two traits are close enough together on a single chromosome, they will not exhibit independent assortment. Lack of independent assortment due to loci being on the same chromosome is known as LINKAGE.

Linkage of loci on the same chromosome occurs because chromosomes behave as units during the process of cell division that occurs in the formation of gametes (called MEIOSIS). In most animals and plants, chromosomes occur in pairs; one of each pair is contributed by the fertilizing sperm and the other by the egg. Each chromosome of such a pair is said to be HOMOLOGOUS to the other member of the pair. Early in meiosis, each chromosome undergoes replication and also becomes aligned side-by-side with its homologue. At this stage, because of the previous replication, each chromosome actually consists of two CHROMATIDS, so the paired homologous chromosomes comprise four chromatids altogether (see Figure 5). In the FIRST MEIOTIC DIVISION, the homologous chromosomes separate from one another as the cell itself pinches into two; in the SECOND MEIOTIC DIVISION, the individual chromatids of each chromosome separate as the daughter cells produced by the first

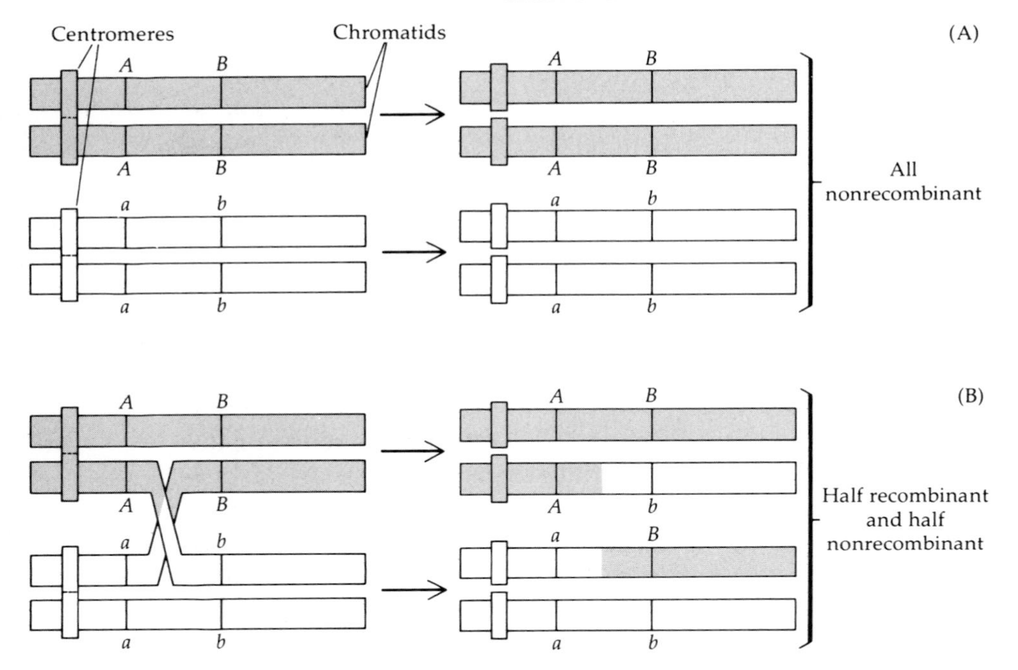

FIGURE 5. Crossing over and recombination. (A) A chromosome pair in a double heterozygote as it would appear early in the first meiotic division, with no crossover between the *A* and *B* loci of the two chromosomes. The centromeres are the sites of attachment of certain filaments that contract and so move the centromeres and their attached chromosome arms to opposite poles of a cell during cell division. In the first meiotic division, the two homologous centromeres shown at the left separate from one another; by this stage, the chromosomes have replicated, and each centromere carries two chromatids. In the second meiotic division, each centromere splits (as indicated by the dashed lines), and the chromatids (each now a chromosome in its own right) proceed to separate. In males, each chromosome resulting from the four chromatids of a pair at the beginning of meiosis ends up in a different sperm; in females, one chromosome ends up in the functional egg nucleus and the other three are extruded from the egg and do not participate in reproduction. When there is no crossover between the *A* and *B* loci, all of the resulting chromosomes are nonrecombinant. (B) When a crossover (breakage and reunion) does occur between the *A* and *B* loci, then half of the chromosomes resulting from meiosis will be nonrecombinant and half will be recombinant. If the probability of a crossover is c, then the sequence shown in (A) will occur with probability $1 - c$, and the sequence shown in (B) will occur with probability c. Consequently, the overall recombination fraction will be $r = (1 - c)(0) + c(1/2)$, so $r = c/2$.

meiotic division again divide. (See the legend of Figure 5 for more detail.)

Because of the meiotic process, alleles on the same chromosome would always segregate together as a unit were it not for the phenomenon of RECOMBINATION that results from a CROSSOVER (this event is hypothesized to be a breakage and reunion of chromatids of homologous chromosomes that occurs during the formation of gametes). The probability of a crossover occurring somewhere within a specified region of chromosome generally increases as the length of the region increases, so loci that are physically close together are more tightly linked and loci that are farther apart are more loosely linked. The degree of linkage is most easily quantified in terms of the RECOMBINATION FRACTION, which is simply the proportion of gametes produced by a double heterozygote (an individual heterozygous for two loci) that have undergone recombination. In Figure 6, it can be seen that when loci are so close together on a chromosome that no more than one crossover can occur between them, the recombination fraction r is then equal to one-half of the probability of a crossover.

The limiting values for the recombination fraction r are $r = 0$, which represents ABSOLUTE LINKAGE (the case in which the alleles on a particular chromosome are always transmitted together), and $r = 1/2$, which represents independent assortment. The value $r = 1/2$ occurs when the loci are on different chromosomes or when they are so far apart on the same chromosome that one or more crossovers always occurs between them. In general, a double heterozygote will produce the following types of gametes: two nonrecombinant types, each in the fraction $(1 - r)/2$; and two recombinant types, each in the fraction $r/2$, as shown in Figure 6. (Box A demonstrates estimation of r from actual data.)

As an interesting historical aside, it may be noted that Mendel's seven loci reside on only four of *Pisum's* seven chromosomes (Figure 7). Most pairs of Mendel's loci that are on the same chromosome are so far apart that they assort independently, but in one case, involving a locus that influences pod shape and another locus that influences plant

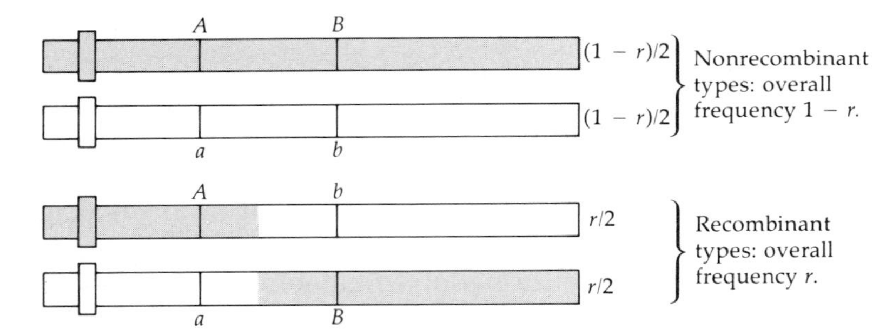

FIGURE 6. Kinds of gametes produced by a double heterozygote when the recombination fraction between the two loci is r.

height, the recombination fraction is $r = 0.12$. Mendel failed to discover linkage because he apparently did not study the simultaneous transmission of these two traits.

A Estimating Proportions and Their Standard Errors

In population genetics (and, indeed, in all experimental sciences), it is rarely possible to measure the exact value of any parameter. Usually, the best that can be accomplished is to estimate the value of the parameter based on some limited number of observations, though the more numerous the observations, the more reliable the estimate. It then becomes important to measure the reliability of the estimate, and the measure of reliability is a number called the STANDARD ERROR (s.e.) of the estimate. Here we illustrate the calculation and interpretation of the standard error of an estimate of a frequency (i.e., proportion), because such estimates are of great importance in population genetics.

For concreteness, consider the problem of estimating the recombination fraction. Suppose that among 140 gametes produced by an individual heterozygous at each of two linked loci, 42 were recombinants. Between the two loci there is some "true" recombination fraction, which may be symbolized as, say, ρ, the value of which is unknown. We do know, however, that $r = 42/140 =$

.300 is the recombination fraction observed in the data, and statistical theory assures us that the best estimate of ρ is, in fact, r. A

The standard error of the estimate $r = .300$ is calculated as $\sqrt{r(1-r)/n}$, where n is the sample size; in this case $n = 140$, so the standard error is $\sqrt{.3 \times .7/140} = .03873$, or, rounding off, .039. (The convention for representing the standard error of an estimate is to write $r = .300 \pm .039$.)

To interpret the standard error as a measure of reliability, imagine that someone else were to repeat the experiment with the same sample size, n. We will also assume that n is reasonably large and that r is not too close to 0 (or too close to 1, which may arise in other situations). The repetition of the experiment would yield another estimate of ρ; call this one r', and r' will not necessarily equal r. However, the probability that r' will lie within two standard errors of r is .96, and the probability that r' will lie within three standard errors of r is .997. Thus, with 96 percent confidence, we may assert that r' will be in the interval $.300 - 2(.039) = .222$ to $.300 + 2(.039) = .378$, and with 99.7 percent confidence that r' will lie between the values $.300 - 3(.039) = .183$ and $.300 + 3(.039) = .417$. The smaller the standard error, of course, the more likely it is that a subsequent estimate of the same parameter will be close to the estimate already obtained. Note that the standard error of a proportion decreases as $\sqrt{n}$, so an increase of a factor of 10 in sample size decreases the standard error by a factor of only $\sqrt{10} \approx 3.2$.

a. In a cross involving the genes *f* (*fine* stripe) and *an* (*anther* ear) on chromosome 1 of maize, R. A. Emerson (cited in Goodenough, 1978) obtained 147 recombinants and 732 nonrecombinants. What is the best estimate of the recombination fraction between *f* and *an* and its standard error? Calculate the 96 percent and 99.7 percent intervals as in the preceding paragraph, and interpret them.

b. Among 248 chromosomes studied in a population of American eels (*Anguilla rostrata*) off the coast of Halifax, Nova Scotia, 138 were found to carry a certain allele (call it *Adh*-1) at a locus coding for the enzyme alcohol dehydrogenase. What is the best estimate of the frequency of the *Adh*-1 allele among all chromosomes in the population, and what is its standard error? Calculate the 96 and 99.7 percent intervals as in part (a) and interpret. (Data from Williams et al., 1973.)

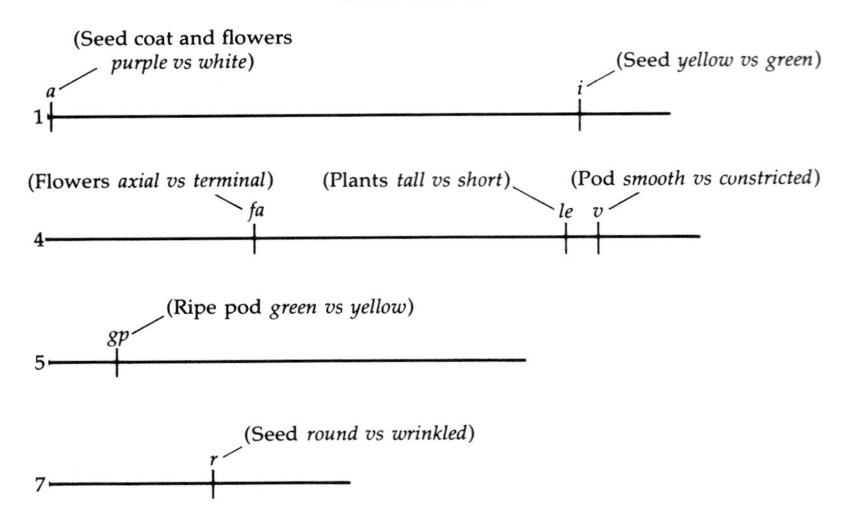

FIGURE 7. Genetic map of *Pisum* showing the locations of Mendel's seven loci and the symbols used for the loci in modern pea breeding. The numbers on the left are the chromosome numbers.

SEX LINKAGE

Notice that the Punnett squares in Figures 2, 3, and 4 take no account of sex; it does not matter in such a cross which parent is the male and which the female. In most cases, RECIPROCAL CROSSES (crosses in which the genotypes of the parents are the same but the sexes have been interchanged) are equivalent because, in both sexes, the chromosomes occur in pairs and, therefore, both sexes carry two alleles at each locus.

In most groups of bisexual animals, sex is determined by a special pair of chromosomes called the X chromosome and the Y chromosome. (The X and Y are called SEX CHROMOSOMES in contrast to the other chromosomes, which are called AUTOSOMES.) Females are chromosomally XX, males are chromosomally XY. Although all eggs carry a single X chromosome, half the sperm carry an X and the other half carry the Y. The genetic consequences of this situation are outlined in Figure 8A, which illustrates why a male's X chromosome must derive from his mother and can be transmitted only to his daughters. [In birds, moths, and butterflies, and in a few Diptera (flies), Crustacea, fish, amphibians, and reptiles, the situation is reversed — females are chromosomally XY and males are XX (White, 1973).]

In humans, the house mouse, and the fruit fly *Drosophila melanogaster,* where the matter has been particularly well studied, the Y chromosome carries very few genes beyond those involved in sex determination (in mammals) or male fertility. Because the Y chromosome does not carry alleles of loci present on the X chromosome, recessive alleles on the X chromosome become expressed in males. The pattern of inheritance of loci on the X chromosome (called SEX-LINKED or X-LINKED loci) is illustrated in Figures 8B and 8C. Note in Figure 8B that a male who carries a sex-linked recessive will have phenotypically normal sons and daughters (although all of his daughters will be heterozygous for the allele); in Figure 8C note that a heterozygous female will have half her sons receive and express the recessive allele, though all her daughters will be phenotypically normal (in this case, half the daughters will be heterozygous). For example, the *a* allele in the male in Figure 8B could represent the allele for

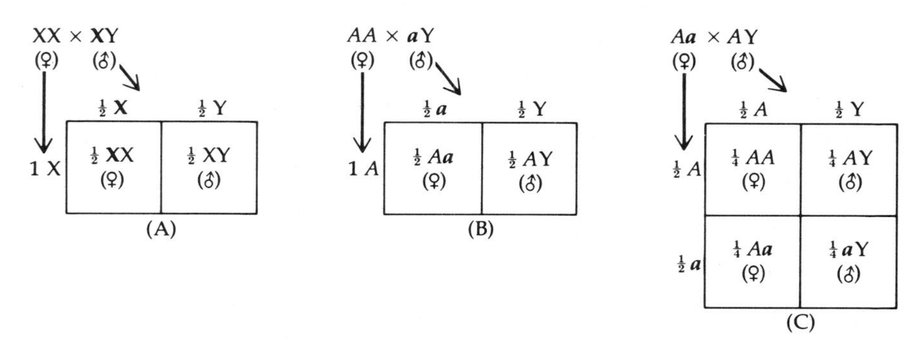

FIGURE 8. Sex-linked (or X-linked) inheritance of alleles (*A* and *a*) of a locus on the X chromosome. (A) A male's X chromosome is inherited from his mother and transmitted to his daughters. (B) A male affected with a condition due to a rare sex-linked recessive (*a*) has children that are all phenotypically normal, but all of his daughters are heterozygous. (C) If a female is heterozygous for a rare, sex-linked recessive, half of her sons will be affected and half of her daughters will be carriers.

the most common form of color blindness (green blindness, or deuteranopia); the male himself would be color blind, but all of his offspring would have normal color vision. The female in Figure 8C could represent a daughter of the male in Figure 8B, and half of her sons will be color blind.

VARIABLE EXPRESSIVITY AND PENETRANCE

The phenotype corresponding to a particular genotype may be expressed to different degrees in different individuals. For example, the dominant gene for polydactyly (extra fingers) in humans may, in one individual, lead to a perfectly formed extra finger, whereas in another individual, the gene may produce a mere nub. When such variation in the expression of a gene occurs, the gene is said to have VARIABLE EXPRESSIVITY. The reasons for variable expressivity are rarely known for certain, but they are usually thought to relate to the particular environment of the individual in question (the uterine environment, in the case of polydactyly), or to interactions of the gene with others in the genotype.

In certain cases, the phenotype corresponding to a particular genotype may not be expressed to any detectable extent in some individuals. In such cases, the gene is said to have reduced or incomplete penetrance, where the word PENETRANCE refers to the proportion of individuals of a particular genotype who actually express the corresponding phenotype. The dominant gene for polydactyly exhibits both incomplete penetrance and variable expressivity.

CHROMOSOMAL ABNORMALITIES

Although segregation and linkage are the soul of Mendelian genetics, certain kinds of chromosomal abnormalities are important in population genetics and evolution. While the detailed genetic consequences of chromosomal abnormalities can be rather complex, for present purposes a brief discussion will be sufficient.

Abnormalities in Chromosome Number. Abnormal chromosome numbers are conveniently classified into two types: those that involve only one of the chromosomes in a set (POLYSOMY) and those that involve an entire set of chromosomes (POLYPLOIDY). Of these, polyploidy is by far the more important in evolution.

Polysomic individuals have a chromosome represented too many or too few times. For example, in humans, individuals with Down's syndrome have chromosome number 21 represented three times instead of twice and are therefore TRISOMIC for chromosome 21; females with Turner's syndrome have the X chromosome represented only once (and have no Y chromosome) and are therefore MONOSOMIC for the X (written as XO). In general, polysomic individuals have a reduced chance of survival or a lowered fertility as compared with normal individuals, and this is probably why polysomy has apparently played no significant role in the chromosomal evolution of animals and plants.

Polyploid individuals have an excess or deficiency of entire sets of chromosomes. The normal condition for most species is the DIPLOID condition, in which each chromosome in the set is represented twice — a homologous pair. (The X and Y sex chromosomes in animals are something of an exception, of course, although the X and Y are both sex chromosomes and therefore constitute a sort of "pair.") An extra set of chromosomes added to a diploid creates a TRIPLOID, two extra sets a TETRAPLOID, four extra sets a HEXAPLOID, and so on. In humans, for example, the basic set of chromosomes has 23 members, and normal, diploid individuals therefore have 46 chromosomes; triploid (69 chromosomes) and tetraploid (92 chromosomes) human embryos have been found, but they undergo spontaneous abortion and do not survive.

Polyploidy has not played a significant role in the chromosomal evolution of animals, presumably because in many animals, polyploidy would disrupt the chromosomal mechanism of sex determination. In plants, however, which generally lack sex chromosomes and are often MONOECIOUS (that is, male and female elements coexist on the same plant), polyploidy has been of major importance in evolution. Many of our most valuable crop plants, including wheat, oats, cotton, potatoes, sugar cane, and bananas, are polyploid. Indeed, Stebbins (1950, 1977) has estimated that 25 to 35 percent of the species of flowering plants are of polyploid origin.

Two types of polyploidy among plants may be distinguished. In the first, called AUTOPOLYPLOIDY, all chromosome sets derive from a single ancestral species; in the second type of polyploidy, known as ALLOPOLYPLOIDY, the chromosome sets derive from different ancestral species, which is to say that hybridization between species, sometimes between widely different species, must have occurred somewhere in the ancestry of the polyploid. Common bread wheat (*Triticum aestivum*), for example, is an allohexaploid, having two sets of chromosomes from each of three distinct ancestral species (Figure 9). Stebbins (1950) estimates that half or more of all naturally occurring polyploids are, in fact, allopolyploids.

Abnormalities in Chromosome Structure. Four major types of abnormality in chromosome structure are illustrated in Figure 10. Each type warrants a brief discussion.

1. Chromosomes that have one or more loci represented twice on a chromosome, instead of only once, are said to have DUPLICATIONS (Figure 10B). Duplications are thought to provide the "raw material" for the evolution of new genes; when a gene is duplicated, one of the copies is then "free" to evolve in a novel direction because the other copy can retain its original function, which may be indispensable to the organism. Recent studies of families of related genes have made it quite clear that gene duplication and subsequent evolutionary divergence is indeed a key process in evolution. (For further discussion, see Chapter 3; reviews can be found in Bryson and Vogel, 1965; Ohno, 1970; Markert et al., 1975; MacIntyre, 1976.)

Duplications can give rise to chromosomes carrying three, four, or more copies of the originally duplicated locus. Such "repeats" of a locus (or parts of a locus; see Barker et al., 1978) are generated by mispairing of chromosomes and re-

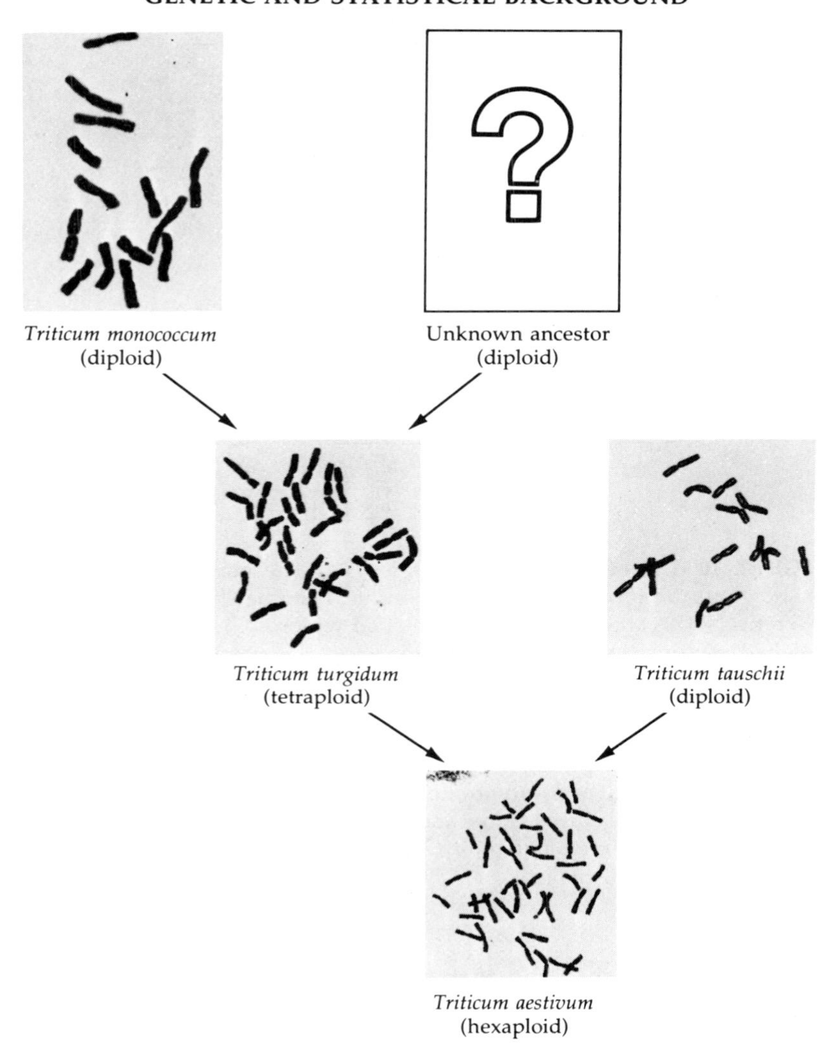

FIGURE 9. The chromosome constitution (karyotype) of various species of wheat. *Triticum monococcum,* a diploid species with 14 chromosomes, hybridized with an as yet unknown diploid species also having 14 chromosomes. Doubling of the chromosomes in the hybrid created *T. turgidum,* an allotetraploid having 28 chromosomes. *T. turgidum* then hybridized with the diploid *T. tauschii* (14 chromosomes); chromosome doubling in this hybrid created modern wheat, *T. aestivum,* an allohexaploid having 42 chromosomes (i.e., 14 from each of the three diploid ancestral species). (Photographs courtesy of G. Kimber.)

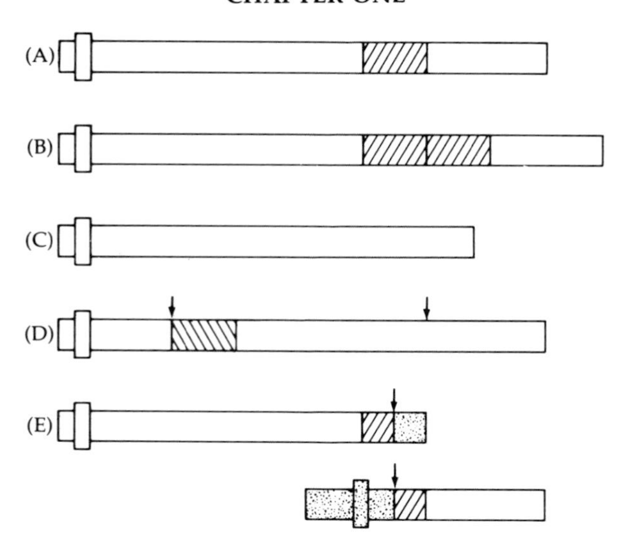

FIGURE 10. Chromosomal abnormalities. (A) The normal chromosome is shown at the top, and a small region has been hatched for an area of reference. (B) A duplication of the hatched region. (C) A deletion of the hatched region. (D) An inversion involving the hatched region, with breakpoints where indicated by the arrows; an inversion such as this one, which does not include the centromere, is known as a paracentric inversion; an inversion that does include the centromere is called a pericentric inversion. (E) A reciprocal translocation, which results from the interchange of parts between two nonhomologous chromosomes. The breakpoints of the normal chromosome and another one (stippled) are indicated by arrows, and an interchange of terminal segments has occurred prior to chromosome restitution.

combination during the formation of gametes, as shown in Figure 11, and they can give rise to multigene "families" of related proteins such as those involved in formation of antibodies in the immune system (Hood et al., 1975). Some theoretical aspects of gene duplication and multigene families are discussed in Spofford (1969) and Ohta (1978). Certain genes in higher organisms are known to be highly "redundant" — they are repeated many times in tandem along a chromosome. Examples include genes for ribosomal RNA (discussed later in this chapter; Tartof, 1975; Ritossa, 1976) and histones (important protein constituents of chromosomes; Reeck et al., 1978). It seems likely that such redundant loci were produced originally by a process similar to that outlined in Figure 11. (See also Chapter 3.)

2. Chromosomes that are lacking one or more loci have undergone DELETIONS (Figure 10C). Examination of related proteins in different organisms reveals that very small deletions can be important in evolution; these deletions are so small that they remove only one or a few subunits from a protein molecule. For example, in the divergent evolutionary lines leading to yeast, on the one hand, and horses, on the other, the gene that codes for the respiratory protein cytochrome *c* has, in the yeast line, sustained a deletion that eliminates one of the 104 subunits in the molecule, namely amino acid number 102 (Jukes, 1966). A major evolutionary role for large deletions is as yet unknown, however.

3. Chromosomes that have a sequence of genes in reverse of the normal order are said to have INVERSIONS (Figure 10D). Although not frequent in most natural populations, inversions are found at relatively high frequency in *Drosophila* species such as *D. pseudoobscura* and *D. persimilis* (Dobzhansky, 1970; Kastritsis and Crumpacker, 1966; Crumpacker et al., 1977), *D. subobscura* (Prevosti, 1966; Lakovaara and Saura, 1971; Sperlich et al., 1977; Pinsker et al., 1978), *D. melanogaster* (Stalker, 1976; Watanabe and Watanabe, 1977), and many others (examples: Burla et al., 1949; Brncic, 1970). (*Drosophila* and many other dipterans are especially convenient for studies of inversions because the inversions

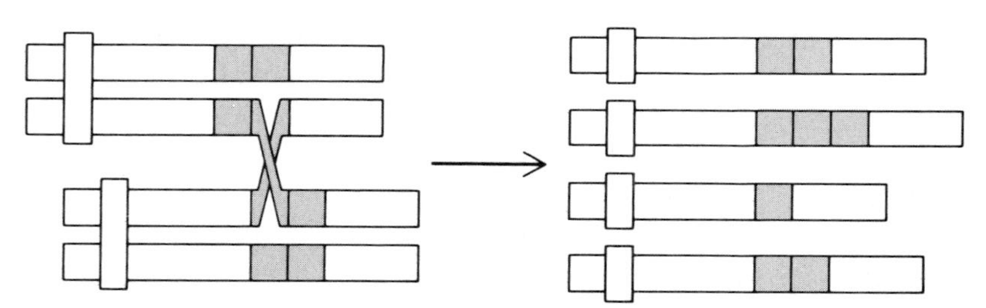

FIGURE 11. Unequal crossing over. Mispairing and crossing over of chromosomes that already carry duplicated regions can lead to chromosomes that have multiple copies of the region. Further mispairing and crossing over of chromosomes that carry multiple copies of the region can, of course, generate chromosomes that have even more copies.

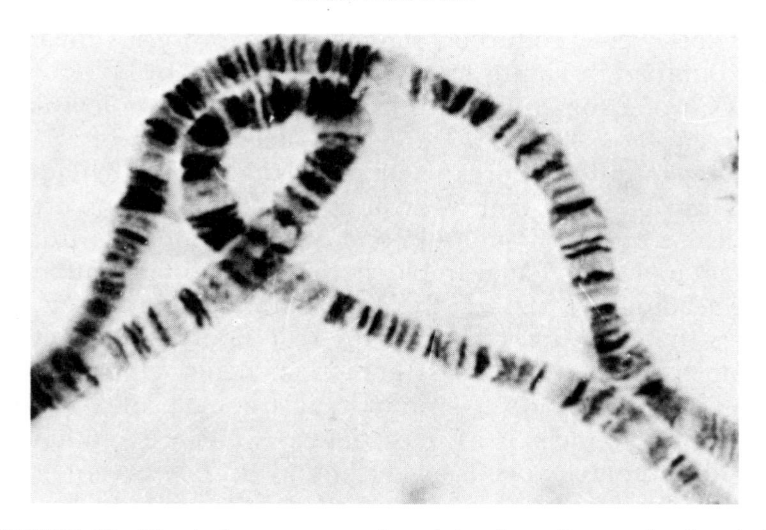

FIGURE 12. Giant chromosomes found in the salivary glands of the dipteran black fly *Simulium vittatum*. Such chromosomes arise from about 10 chromosome replications without intervening cell division. In this photograph, the individual is heterozygous for a paracentric inversion (i.e., an inversion that does not include the centromere). Note the loop that results from gene-for-gene pairing of the chromosomes. Note also that, in this cell, part of the inverted region in the vicinity of the breakpoints has remained unpaired. (Courtesy of K. H. Rothfels.)

form characteristic loops in the paired "giant" chromosomes of the larval salivary glands, as shown in Figure 12.) In most organisms, heterozygosity for a large inversion reduces fertility because crossing over somewhere in the inverted segment during meiosis produces abnormal gametes that carry chromosomes having large deletions or duplications, and these abnormal chromosomes generally lead to death of the gamete (usual in plants) or death of the embryo (usual in animals).

The genetic consequences of a heterozygous inversion are outlined in Figure 13, which illustrates a PARACENTRIC INVERSION, so-called because it does not include the centromere; paracentric inversions are the only kind normally found in *Drosophila* populations. (Inversions that do include the centromere are called PERICENTRIC INVERSIONS.)

The legend of Figure 13 explains how the reduction in fertility due to heterozygosity for an inversion is circumvented in *Drosophila* by the absence of crossing over in males and by the peculiar geometry involved in the formation of the *Drosophila* egg. In *Drosophila* species, therefore, heterozygous inversions cause no decline in fertility. Because chro-

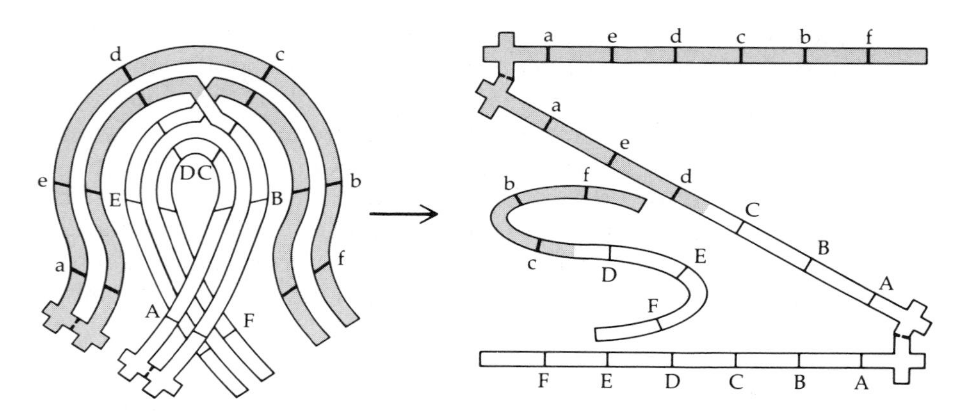

FIGURE 13. Consequences of crossing over in inversion heterozygotes. On the left is shown the pairing configuration of chromosomes during meiosis in an individual heterozygous for a paracentric inversion. One of the chromosomes forms a loop to allow gene-for-gene pairing. Crossing over within the inverted segment generates, after the first meiotic division, the products shown on the right. Note the dicentric (two-centromere) chromosome, and the acentric (without a centromere) chromosome. After the second meiotic division, during which the centromeres attached to the dicentric chromosome split (dotted lines), gametes that carry the inverted chromosome at the right top are genetically normal (except that they carry the inversion), those gametes carrying the chromosome at the right bottom are genetically normal, but any gametes carrying either the remaining acentric or the dicentric chromosome will have duplications for some genes and deficiencies for others. In *Drosophila* species, abnormal gametes resulting from crossing over within inversion loops do not arise. In males, there is no crossing over, so the abnormal gametes cannot be formed. In females, the products of the first meiotic division are linearly arranged, as shown here on the right, and the functional egg nucleus always includes a chromosome at the end of the linear arrangement, in this case either the chromosome at the right top or bottom. Thus, the dicentric chromosome bridge (the slanting portion here) ensures that the abnormal chromosomes will not be included in the functional egg nucleus.

mosomes that have undergone a crossover within the inverted segment are not included in functional gametes, recombination is effectively prevented. The upshot is that the alleles present in the inversion must segregate together as a unit — making a sort of "supergene" — and interesting evolutionary consequences of inversions arise from this fact (see Chapter 2).

4. TRANSLOCATIONS (Figure 10E), more properly called RECIPROCAL TRANSLOCATIONS, involve an interchange of terminal segments between two nonhomologous chromosomes. Segregation of chromosomes in a translocation heterozygote often results in a high frequency of chromosomally abnormal gametes and consequent semisterility (see Figure 14 and Endrizzi, 1974), so translocations are extremely rare in most natural populations. On the other hand, a very few species are known that are permanently heterozygous for translocations, the most notable instance being

FIGURE 14. Segregation from a translocation heterozygote. On the left is shown the pairing configuration in early meiosis in a translocation heterozygote. (Each chromosome is already replicated into two chromatids at this stage, which is indicated by the dashed lines.) Just prior to chromosome separation in the first meiotic division, the translocation configuration can align in four distinct ways, as shown on the right. The possibility of crossing over is ignored here, and the plane of cell division in the first meiotic division is indicated by thin solid lines. The two modes of alignment shown on the top lead to what is called "adjacent" segregation, and all gametes formed from adjacent segregation carry duplications of some chromosomal material and deficiencies of others. The other two modes of alignment, shown at the bottom, lead to "alternate" segregation, and in this case, half of the gametes carry the structurally normal chromosomes and half carry both parts of the reciprocal translocation, so none of the gametes carry duplications or deficiencies. In certain plants, such as some species of *Oenothera,* there is permanent heterozygosity for translocations, and segregation always occurs according to a particular one of the alternate modes. The translocation heterozygosity in *Oenothera* is permanent because one of the chromosomal constitutions of the alternate mode of segregation causes abortion of the pollen nucleus and the reciprocal chromosomal constitution of the alternate mode causes abortion of the egg nucleus. Thus the functional pollen nucleus and the functional egg nucleus contribute precisely complementary chromosomes, thereby reconstituting the translocation heterozygosity.

certain species of *Oenothera,* the evening primrose; in such cases, chromosomal segregation occurs in such a highly ordered and regular fashion that chromosomally abnormal gametes are not formed (see the legend of Figure 14 and Carson, 1967).

Although translocations are generally rare in natural populations, their potential importance in the evolutionary ori-

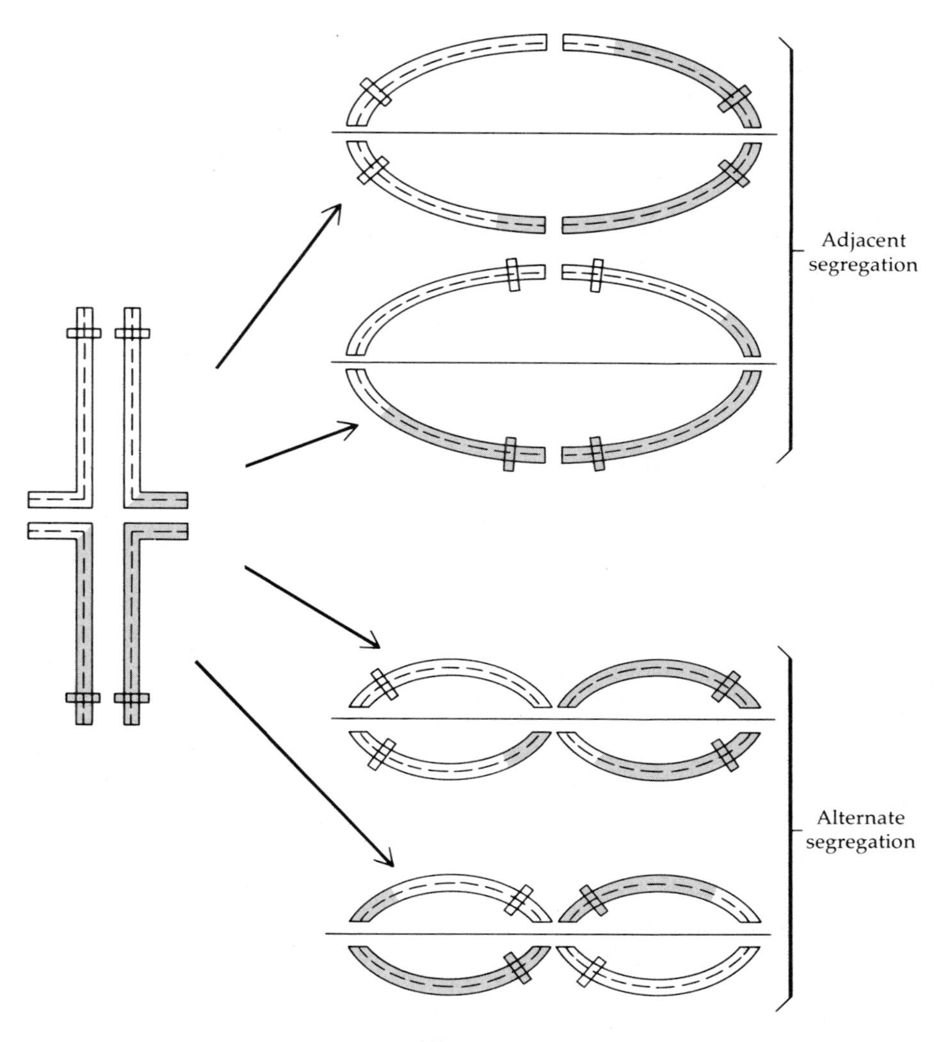

gin of new species has been emphasized by White (1978) and will be discussed later in Chapter 5.

The Molecular Basis of Molecular Population Genetics

In recent decades, population genetics has profited from the phenomenal successes in the field of molecular biology. Using new knowledge and techniques derived from molecular biology, certain fundamental problems of evolutionary biology have been attacked directly at the molecular level. In this section, we provide but a brief review of those aspects of molecular genetics that impinge directly on population genetics.

DEOXYRIBONUCLEIC ACID (DNA)

As mentioned earlier, DNA is the genetic material in almost all organisms. (Exceptions are some groups of viruses in which the genetic material is the related molecule, ribonucleic acid — RNA.) Structurally, a DNA molecule (Figure 15) consists of two long strands intertwined together in the famous double helix first proposed by Watson and Crick (1953). (See Watson, 1968; Olby, 1974; and Sayre, 1975 for fascinating accounts of the discovery of the structure.)

Each DNA strand in the molecule consists of a sequence of nucleotides (Figure 15). Four nucleotides are found in

FIGURE 15. A diagrammatic representation of a DNA molecule. The top and bottom "loops" show a representation similar to that used in Figure 1; here the base pairs are represented by horizontal lines and the "backbones" of the two intertwined strands are indicated as ribbons. The second loop from the bottom shows the alternating sugar (shaded pentagon) — phosphate (P) unit that constitutes the actual backbone of the strands. The third loop from the bottom shows how the bases (adenine, thymine, guanine, or cytosine) are attached to the sugars in the interior of the molecule. The several short lines between atoms in the base pairs denote hydrogen bonds, which are weak chemical attractions between nucleotides at corresponding positions along the backbone that serve to stabilize the double helical structure of the molecule. Note that A pairs only with T, and G pairs only with C. The unit composed of a base, a sugar, and a phosphate group is called a NUCLEOTIDE.

DNA: deoxyadenosine phosphate (A), thymidine phosphate (T), deoxyguanosine phosphate (G), and deoxycytidine phosphate (C). The nucleotides in the two strands of a DNA double helix have a very special relationship brought about by weak chemical bonds, called hydrogen bonds, that form between the nucleotides: where one strand carries an A, the other strand must carry a T; and where one strand

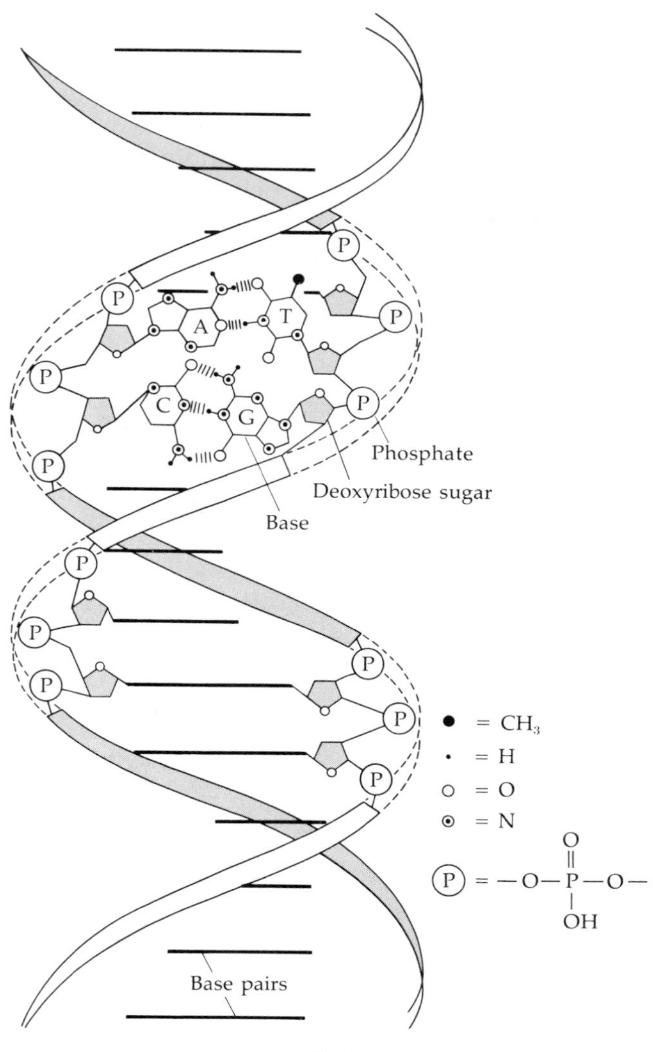

carries a G, the other must carry a C. (Figure 15 somewhat distorts the double helix in order to show the chemical structures of the nucleotides. In the actual molecule, the nucleotide pairs are flat and stacked one on top of the other like pennies in a roll, with the sugar-phosphate backbones winding around the outside of the roll.) This pairing of nucleotides is of fundamental importance, because when DNA replicates (Figure 16), the two strands separate and each strand serves as a template for the synthesis of a new strand; the overall consequence of replication is that each of the two daughter molecules produced are identical in nucleotide sequence to the parental molecule.

The genetic information in a DNA molecule is contained in its sequence of nucleotides, much as the intellectual information in this sentence is contained in its sequence of letters. Therefore, a change in one or more nucleotides in a

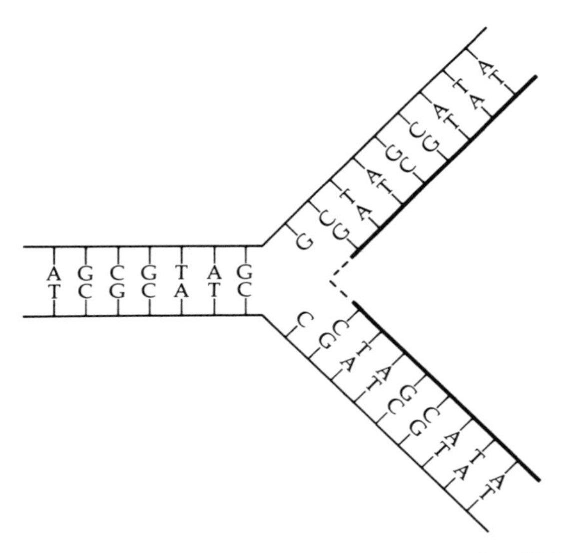

FIGURE 16. DNA replication. During replication, the DNA molecule unwinds to form a "replication fork," and enzymes synthesize new DNA strands (heavy lines) according to the base-pairing rules A-T and G-C. Note that each daughter molecule being formed has one parental DNA strand (lighter line) and one newly synthesized strand (heavy line). The dashed lines indicate continued DNA synthesis.

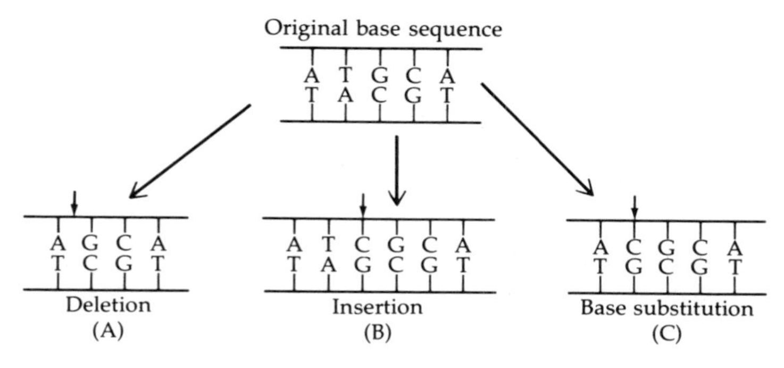

FIGURE 17. Mutation. Any heritable alteration in a DNA molecule is a mutation. Shown here (at positions indicated by arrows) are three kinds of mutations. (A) A single-base deletion. (B) A single-base insertion. (C) A substitution of one base pair for another, in this case a substitution of C-G for T-A.

DNA molecule alters the genetic information in the corresponding gene (Figure 17). Any heritable change in the genetic material, including changes at the nucleotide level, is called MUTATION. The RATE OF MUTATION — that is, the probability that a gene will undergo a mutation in a specified period of time — varies somewhat from gene to gene, but it is typically of the order of 10^{-4} to 10^{-6} per generation of organisms. That is to say, the probability that a locus in a gamete carries a newly arising mutation (one not present in the parental organism) is usually between 0.0001 and 0.000001. Mutation rates can, however, be increased an order of magnitude or more by ionizing radiation or mutagenic chemicals.

GENE ACTION

Transcription. Most genes code for the production of particular protein molecules with amino acid sequences that correspond to the sequences of nucleotides in the genes. The overall process of protein synthesis occurs in two distinct stages. In the first stage, called TRANSCRIPTION (Figure 18), the DNA double helix opens out and one of the strands

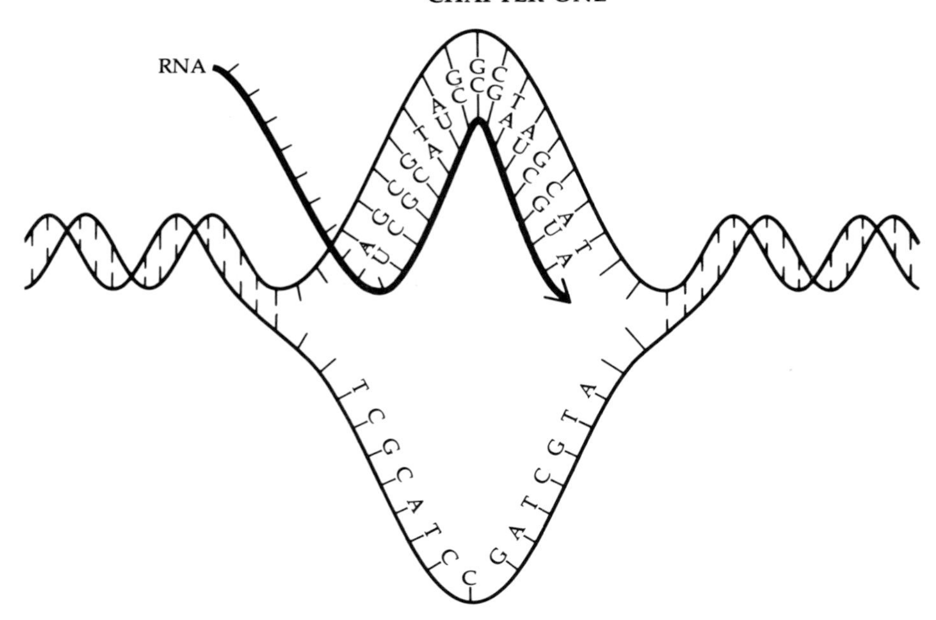

FIGURE 18. Transcription. In the process of transcription, a region of the DNA molecule "unwinds," and from one DNA strand (the "template"), a molecule of RNA (heavy line) complementary in base sequence to the template strand is synthesized. Note that the base thymine (T) in DNA is replaced with uracil (U) in RNA. For genes that specify the amino acid sequence of proteins, the RNA transcript undergoes a chemical processing in the nucleus during which, among other things, any intervening sequences (introns) are removed; after processing, the RNA is transported to the cytoplasm, where it functions as a messenger RNA. RNA transcripts from some genes are not converted to messenger RNA but function directly in cellular activities; examples include transfer RNA and ribosomal RNA, which are involved in the process of translation (Figure 19).

serves as a template for pairing of nucleotides in the synthesis of a molecule of RNA. RNA is chemically similar to DNA, except that (1) it is almost always single stranded, (2) the sugar deoxyribose in DNA is replaced with ribose in RNA, and (3) the nucleotide constituents in RNA are adenosine phosphate (A), uridine phosphate (U) (which replaces the T found in DNA), guanosine phosphate (G), and cytidine phosphate (C).

By a complex and little understood mechanism of chemical "processing" or modification, the RNA molecule produced by transcription of most genes is transformed into MESSENGER RNA, which is the molecule actually used by the cell in protein synthesis.

One of the steps involved in the processing of RNA transcripts in higher organisms is known as "splicing." In higher organisms, many, perhaps most, genes are "split" — the nucleotide sequences coding for amino acids in different parts of the protein molecule are interrupted by other sequences hundreds or thousands of nucleotides long. The functions of these interruptions, which are called INTERVENING SEQUENCES or INTRONS, are unknown, but a gene may contain up to 20 *different* introns. In the splicing process, the intervening sequences are excised from the transcript and the actual amino acid coding sequences (called EXONS) are fused into juxtaposition (see Crick, 1979, for a review). Processing of RNA transcripts into messenger RNA also involves "trimming" (removal of stretches of nucleotides at each end of the transcript), "capping" (addition of an unusual nucleotide in an unusual orientation at one end of the trimmed transcript), and usually "polyadenylation" (addition of a sequence of tens or hundreds of consecutive A's at the other end of the trimmed transcript).

Translation. In the final part of the process of protein synthesis [called TRANSLATION (Figure 19)], the nucleotides in the messenger RNA are "read" from one end to the other in groups of three (called CODONS). Each codon specifies a particular amino acid subunit to be inserted into the growing protein molecule (more accurately, a growing polypeptide). The correspondence between codons in messenger RNA and the amino acid residues in the protein is known as the GENETIC CODE and is shown in Table I. One noteworthy feature of the genetic code is the property called DEGENERACY, whereby most amino acids can be specified by two or more codons. Note particularly the degeneracy involving the third position of the codon, where A and G are usually translated identically, as are U and C.

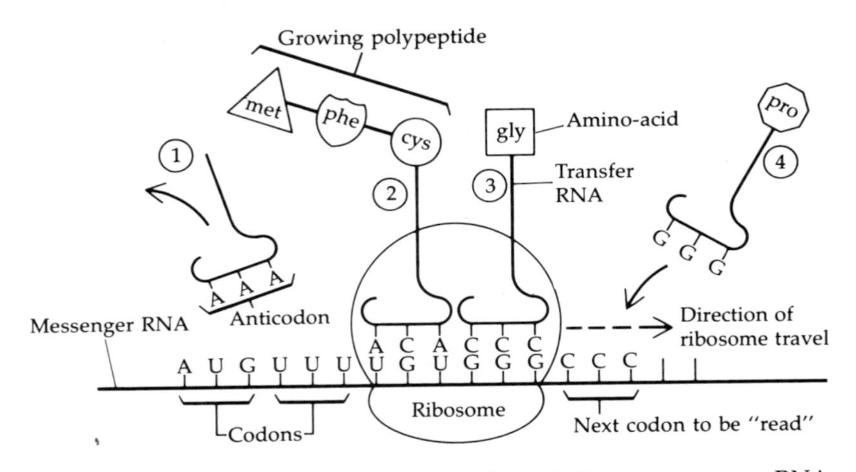

FIGURE 19. Translation. In the process of translation, a messenger RNA binds to a ribosome, and as the ribosome moves along the RNA (from left to right in the figure) each nonoverlapping group of three nucleotides (a codon) in the RNA is used to specify the amino acid to be placed at the corresponding position in a polypeptide according to the codon–amino acid correspondence in the genetic code. The figure shows a messenger RNA being translated. The hairpin-shaped structures are molecules of transfer RNA; each of these consists of about 75 nucleotides, of which three (which constitute the ANTICODON) are particularly important in translation, as each anticodon corresponds to a specific amino acid that is attached to the transfer RNA by certain enzymes. As shown near the ribosome, the transfer RNA for phenylalanine (1) has contributed its amino acid to the polypeptide chain, and the growing polypeptide chain is now attached to the transfer RNA for cysteine (2) at the left-hand position on the ribosome. (Note the base pairing of codon and anticodon.) The anticodon of the transfer RNA for glycine (3), the next amino acid in line, has undergone base pairing with the codon for glycine at the right-hand position on the ribosome. At the next step, three things happen almost simultaneously: (a) the cysteine of the growing polypeptide chain becomes detached from its transfer RNA and is attached to the glycine; (b) the transfer RNA corresponding to cysteine is released from the ribosome as the ribosome shifts one codon to the right, which places the polypeptide-bearing transfer RNA (3) in its left-hand position; (c) the transfer RNA corresponding to the codon CCC (number 4, which carries proline) comes to occupy the vacant rightmost position on the ribosome. The same three steps occur repeatedly as the messenger RNA is translated codon by codon. (The figure takes some liberties with anticodons because each is shown as the exact base-pair complement of the codon. In practice, anticodons often contain certain unusual bases other than A, U, G, or C.)

TABLE I. The genetic code.[a]

First nucleotide in codon	Second nucleotide in codon and translation product							
	U	Translation product	C	Translation product	A	Translation product	G	Translation product
U	UUU	phe(F)	UCU	ser(S)	UAU	tyr(Y)	UGU	cys(C)
	UUC	phe(F)	UCC	ser(S)	UAC	tyr(Y)	UGC	cys(C)
	UUA	leu(L)	UCA	ser(S)	UAA	*ochre*(stop)	UGA	*opal*(stop)
	UUG	leu(L)	UCG	ser(S)	UAG	*amber*(stop)	UGG	trp(W)
C	CUU	leu(L)	CCU	pro(P)	CAU	his(H)	CGU	arg(R)
	CUC	leu(L)	CCC	pro(P)	CAC	his(H)	CGC	arg(R)
	CUA	leu(L)	CCA	pro(P)	CAA	gln(Q)	CGA	arg(R)
	CUG	leu(L)	CCG	pro(P)	CAG	gln(Q)	CGG	arg(R)
A	AUU	ile(I)	ACU	thr(T)	AAU	asn(N)	AGU	ser(S)
	AUC	ile(I)	ACC	thr(T)	AAC	asn(N)	AGC	ser(S)
	AUA	ile(I)	ACA	thr(T)	AAA	lys(K)	AGA	arg(R)
	AUG*	met(M)	ACG	thr(T)	AAG	lys(K)	AGG	arg(R)
G	GUU	val(V)	GCU	ala(A)	GAU	asp(D)	GGU	gly(G)
	GUC	val(V)	GCC	ala(A)	GAC	asp(D)	GGC	gly(G)
	GUA	val(V)	GCA	ala(A)	GAA	glu(E)	GGA	gly(G)
	GUG	val(V)	GCG	ala(A)	GAG	glu(E)	GGG	gly(G)

[a] The asterisk indicates the "start" codon for protein synthesis; although methionine (met) is always the first amino acid in a polypeptide chain, it is frequently cleaved off the completed chain by enzymes. There are three "stop" codons, which signal the termination of translation; these are conventionally denoted *ochre, amber,* and *opal.* Two forms of abbreviation of amino acids are given in the table: one, a three-letter abbreviation and the other, a single-letter abbreviation (the latter is particularly useful for giving the amino acid sequence of long polypeptide chains). Three-letter abbreviations: ala, alanine; arg, arginine; asn, asparagine; asp, aspartic acid; cys, cysteine; glu, glutamic acid; gln, glutamine; gly, glycine; his, histidine; ile, isoleucine; leu, leucine; lys, lysine; met, methionine; phe, phenylalanine; pro, proline; ser, serine; thr, threonine; trp, tryptophan; tyr, tyrosine; val, valine.

Gene Effects. Protein products of genes affect cells in myriad ways. Some proteins, such as the oxygen-carrying protein hemoglobin, the contractile protein actin, or the support protein tubulin, participate directly in cellular activities. Other proteins, called enzymes, participate indirectly in cellular activities by catalyzing (changing the rate of) particular biochemical reactions. In cells, enzyme catalysts always act to accelerate reactions.

As an example of enzymatic effects on cells, consider the familiar ABO blood groups in humans. The genetic locus responsible for this system has basically three alleles, called I^A, I^B, and I^O. Now, associated with the membrane of red blood cells is a carbohydrate "precursor" molecule that can

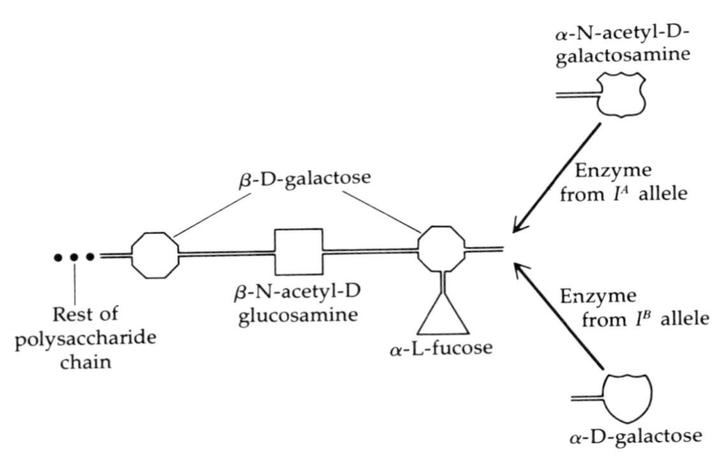

FIGURE 20. Chemistry of the ABO blood groups. The ABO blood groups are polysaccharides found on the cytoplasmic membrane surface of red blood cells. All red blood cells carry many copies of the precursor molecule, the terminal portion of which is shown at the left. The I^A allele specifies an enzyme that adds a particular sugar to the terminus of the precursor, whereas the I^B allele specifies an enzyme that adds a different sugar to the terminus. (The I^O allele, by contrast, is associated with no addition to the terminus.) It is the terminal sugar of the polysaccharide that confers the antigenic specificity of the ABO blood groups. Individuals of genotype I^A/I^B have some polysaccharides with the I^A-associated terminus and others with the I^B-associated terminus, so both the A and B antigens are present on the red blood cells.

be chemically modified by the enzyme product of the *I* locus (Figure 20); the I^A-coded enzyme adds α-*N*-acetyl-D-galactosamine to the precursor, the I^B-coded enzyme adds α-D-galactose, and the I^O gene product adds nothing (Giblett, 1977).

The two possible types of membrane-associated carbohydrates produced by the products of the I^A and I^B alleles are ANTIGENIC; that is to say, each of these carbohydrate antigens stimulates the immune system to produce a specific kind of ANTIBODY, a protein capable of combining chemically with a particular antigen. Thus, the A antigen associated with the I^A allele stimulates production of anti-A antibody, and the B antigen associated with the I^B allele stimulates production of anti-B antibody. When red blood cells carrying, for example, the A antigen are mixed with blood serum containing anti-A antibody, an agglutination reaction takes place in which the red blood cells become clumped together by the antibody. This agglutination reaction is the basis for "blood typing" individuals with respect to the ABO blood groups.

The relationship between genotype and phenotype for the ABO blood groups is shown in Table II. Note that individuals do not produce antibody against antigens that they themselves possess, a phenomenon known as immune tolerance.

TABLE II. Important properties of the ABO blood groups.

Blood group genotype	Antigens on red blood cells	Blood type	Antibodies present in blood	Antibodies present in saliva and other body fluids?
I^A/I^A or I^A/I^O	A	A	Anti-B	Yes, if secretor[a]
I^B/I^B or I^B/I^O	B	B	Anti-A	Yes, if secretor
I^A/I^B	A and B	AB	None	No
I^O/I^O	None	O	Anti-A and anti-B	Yes, if secretor

[a] Although all individuals have ABO antibodies present in the blood, a second locus determines whether these antibodies will be secreted into the saliva, sweat, and other body fluids. The second locus is the secretor locus; secretion is due to a dominant allele, so *Se*/*Se* and *Se*/*se* genotypes are secretors, whereas *se*/*se* genotypes are nonsecretors.

REPETITIVE DNA AND THE GENETIC MATERIAL

Now we come to one of the great riddles of modern molecular biology: a significant fraction of DNA in higher organisms is of unknown function. Three broad categories of DNA may be distinguished. The first category consists of UNIQUE SEQUENCES: DNA sequences roughly 1000 to 1500 nucleotides long (although some are much longer) that are present only once in each set of chromosomes in a gamete; the nucleotide sequences that code for most proteins are in this category. As noted earlier, certain important genes whose products are required in large amounts are duplicated. For example, genes coding for histone proteins are duplicated 250 to 500 times; genes for the RNA constituents of ribosomes are also duplicated hundreds of times. Table III shows that the percentage of the DNA that consists of unique sequences varies widely among organisms.

TABLE III. Percentage of unique-sequence DNA in various organisms.

Organism	Common name	Unique-sequence DNA[a] (percent)
Paramecium aurelia	Paramecium	85
Neurospora crassa	Bread mold	80
Triticum aestivum	Bread wheat	25
Glycine max	Soybean	40
Strongylocentrotus purpuratus	Sea urchin	38
Bombyx mori	Silk worm	55
Drosophila melanogaster	Fruit fly	78
Xenopus laevis	African clawed toad	54
Rana clamitans	Green frog	22
Mus musculus	House mouse	49
Bos taurus	Cattle	55
Homo sapiens	Human	64

[a] Data for wheat from Flavell and Smith (1976), soybean from Goldberg (1978), others from Fasman (1976).

A second category of DNA comprises so-called MIDDLE REPETITIVE SEQUENCES: sequences 300 to 500 nucleotides long, each repeated hundreds or thousands of times and scattered throughout the chromosomes. The function of middle repetitive DNA is unknown, but most molecular biologists now believe that at least some of these sequences are involved in the regulation of gene activity (Davidson and Britten, 1979). Indeed, a significant fraction of middle repetitive DNA sequences may consist of introns.

The third category of DNA consists of HIGHLY REPETITIVE SEQUENCES. The proportion of highly repetitive DNA varies extremely among organisms, averaging roughly 10 to 20 percent of the DNA. Because highly repetitive DNA is often associated with the centromeric regions of chromosomes, it is widely believed to be involved somehow in the proper movement of chromosomes during cell division. Highly repetitive DNA usually consists of simple nucleotide sequences that are repeated in tandem tens of thousands or hundreds of thousands of times. In the fruit fly *Drosophila virilis,* for example, some 40 to 50 percent of the DNA is highly repetitive, and it consists of tandem repeats of each of the three simple nucleotide sequences, ACAAACT, ATAAACT, and ACAAATT.

Unfortunately, at the present time, we know almost nothing about the tempo and mode of evolution of repetitive DNA sequences. If middle repetitive DNA sequences are involved in gene regulation, as they seem to be, and if, as Wilson et al. (1977) suggest, major evolutionary advances often come about through changes in gene regulation rather than through changes in those genes that code directly for proteins, then a really comprehensive view of molecular evolution must await close study of middle repetitive DNA sequences. Conversely, as Crick (1979) has pointed out, "A molecular biologist who wishes to discuss the evolution of the eukaryotic genome will need not only to know a lot about the way DNA and its transcripts can behave but also something about modern ideas on population genetics."

CHAPTER ONE

The Statistical Basis of Population Genetics

The basic mathematics and probability involved in introductory population genetics is quite straightforward. Although the mathematics is used to gain insight into the biology of a complex situation, it is best first to illustrate the mathematics with a simple nonbiological example.

PROBABILITY

Consider the deck of cards represented in Figure 21. Suppose the deck is well-shuffled and one card is drawn at random. Because there are 52 cards, there are exactly 52 possible outcomes for this experiment, one being, for example, that the card drawn is the six of clubs. The 52 possible outcomes of the card-drawing experiment are called ELEMENTARY OUTCOMES, because one and only one of these outcomes *must* occur.

To proceed further, we assign to each elementary outcome a PROBABILITY: a number between 0 and 1 that, in a sense, represents how much confidence we have that the outcome will occur. The probabilities actually assigned to the elementary outcomes are mathematically arbitrary, but in practice, they are based on simple genetic reasoning, intuition, or experience. One requirement that the assigned probabilities must meet is that the sum of the probabilities of all the

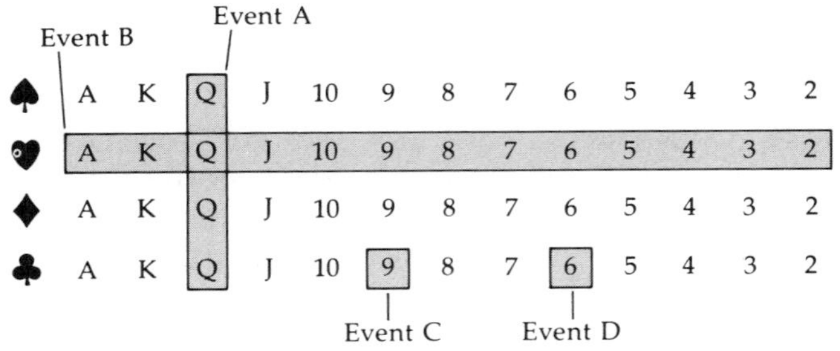

Event B

Event A

♠	A	K	Q	J	10	9	8	7	6	5	4	3	2
♥	A	K	Q	J	10	9	8	7	6	5	4	3	2
♦	A	K	Q	J	10	9	8	7	6	5	4	3	2
♣	A	K	Q	J	10	9	8	7	6	5	4	3	2

Event C

Event D

FIGURE 21. A standard deck of cards, showing several events.

elementary outcomes should add up to exactly 1; this is just the mathematical way of saying that one of the elementary outcomes *must* occur. If the deck in Figure 21 is well-shuffled and the card is really drawn at random, then it is intuitively reasonable to suppose that the 52 elementary outcomes are equally likely, so each of the elementary outcomes may be assigned the probability 1/52.

Events. An EVENT is simply a collection or a set of elementary outcomes. In Figure 21, for example, event A ("card drawn is a queen") consists of four elementary outcomes, namely, that the card drawn is the (1) queen of spades, (2) queen of hearts, (3) queen of diamonds, or (4) queen of clubs. Likewise event B ("card drawn is a heart") consists of 13 elementary outcomes. Elementary outcomes are themselves events, as exemplified in Figure 21 by events C ("card drawn is the nine of clubs") and D ("card drawn is the six of clubs").

The central statement of probability theory is this: *the probability of any event is the sum of the probabilities of the elementary outcomes that make up that event.* Thus, for example, the probability of event A in Figure 21 is $\Pr\{A\} = 1/52 + 1/52 + 1/52 + 1/52 = 4/52$. What this number means in practice is that, were the card-drawing experiment to be repeated many times, the proportion of times one would expect to draw a queen is $4/52 = 7.69$ percent. Because event B in Figure 21 consists of 13 elementary outcomes, $\Pr\{B\} = 13/52$. Similarly, $\Pr\{C\} = 1/52$, and $\Pr\{D\} = 1/52$.

Compound Events. Events that are combinations of other events are often important. There are three main categories of such compound events, symbolized as A + B, AB, and A|B.

The event A + B — read as A *or* B — consists of all elementary outcomes belonging to A or B or both. In Figure 21, for example, the symbol A + B refers to the event "card drawn is a queen or a heart or both"; there are 16 elementary outcomes in the event A + B, so $\Pr\{A + B\} = 16/52$.

The event AB — read as A *and* B — consists of all elementary outcomes belonging simultaneously to both A and B. In Figure 21, the symbol AB refers to the event "card drawn is a queen and a heart," the event AB consists of only one elementary outcome ("card drawn is the queen of hearts"), so Pr{AB} = 1/52.

The event A|B — read as A *given* B — consists of all elementary outcomes belonging to B that also belong to A. In Figure 21, the symbol A|B refers to the event "card drawn is a queen after one is already certain that the card drawn is a heart." Events like A|B arise whenever one has partial information on the outcome of an experiment. I may, for example, draw a card and say, "The card I drew is a heart (i.e., the event B has occurred with certainty). What is the probability that the card is a queen?"

Because of the partial information available in events like A|B, one must be very careful in calculating probabilities. Since one already knows that event B has occurred, there are no longer 52 elementary outcomes, but only 13; these 13 elementary outcomes are demarcated as event B in Figure 21, and it is reasonable to assign to each of the 13 elementary outcomes the probability 1/13. Now, among the elementary outcomes in event B, only one ("card drawn is the queen of hearts") is also in event A, so Pr{A|B} = 1/13.

Probabilities of Compound Events. Having defined the events A + B, AB, and A|B, we are in a position to discuss two fundamental formulas that link their probabilities together.

The first formula is the so-called "addition rule":

$$\Pr\{A + B\} = \Pr\{A\} + \Pr\{B\} - \Pr\{AB\}$$

This formula can easily be confirmed for the example in Figure 21 because we have already calculated that Pr{A + B} = 16/52, Pr{A} = 4/52, Pr{B} = 13/52, and Pr{AB} = 1/52.

If two events cannot occur simultaneously because the occurrence of one event precludes the occurrence of the other, then Pr{AB} = 0 and the events are said to be MUTUALLY EXCLUSIVE. Therefore, in the special case of mutually exclusive events, Pr{A + B} = Pr{A} + Pr{B}.

The second fundamental formula for combining probabilities is the so-called "multiplication rule":

$$\Pr\{AB\} = \Pr\{A|B\}\Pr\{B\}$$

Again the formula can be confirmed for Figure 21 because we have already calculated that $\Pr\{AB\} = 1/52$, $\Pr\{A|B\} = 1/13$, and $\Pr\{B\} = 13/52$. (Note: $\Pr\{A|B\}$ is often called the CONDITIONAL PROBABILITY of A given B.)

If the occurrence of event B provides no information on the likelihood of occurrence of event A, then $\Pr\{A|B\} = \Pr\{A\}$ and the events are said to be INDEPENDENT. Therefore, in the special case of independent events, $\Pr\{AB\} = \Pr\{A\}\Pr\{B\}$. In Figure 21, for example, the face value of a card and its suit are independent events.

REPEATED TRIALS

Often it is useful to think of a biological problem in terms of repeated trials. For example, whether a particular birth results in a boy or girl may be considered a "trial," and a sequence of births would then represent repeated "trials." The importance of repeated trials is that, very often, events pertaining to one trial are independent of events pertaining to different trials. For example, the sex of a first-born child (the first "trial") is independent of the sex of a second-born child (the second "trial").

Suppose, to return to the card-drawing example of Figure 21, that the experiment of drawing a card is carried out repeatedly, after each drawing the card is returned and the deck is reshuffled. Let B_i represent the event "the card drawn in the ith trial is a heart"; let the event "the card drawn in the ith trial is not a heart" be denoted $\bar{B}_i$ (read "not B_i"). We have already calculated that $\Pr\{B_i\} = 13/52 = 1/4$, and therefore $\Pr\{\bar{B}_i\} = 1 - 1/4 = 3/4$. For two trials, there are only four possible outcomes:

(a) first a heart, then another heart (event B_1B_2)
(b) first a heart, then a non-heart (event $B_1\bar{B}_2$)
(c) first a non-heart, then a heart (event $\bar{B}_1B_2$)
(d) first a non-heart, then another non-heart (event $\bar{B}_1\bar{B}_2$)

Now, any event pertaining to trial number 1 is independent of any event pertaining to trial number 2, so the probabilities associated with the above events can be multiplied to yield:

(a) $\Pr\{B_1B_2\} = \Pr\{B_1\}\Pr\{B_2\} = 1/4 \times 1/4 = 1/16$
(b) $\Pr\{B_1\bar{B}_2\} = \Pr\{B_1\}\Pr\{\bar{B}_2\} = 1/4 \times 3/4 = 3/16$
(c) $\Pr\{\bar{B}_1B_2\} = \Pr\{\bar{B}_1\}\Pr\{B_2\} = 3/4 \times 1/4 = 3/16$
(d) $\Pr\{\bar{B}_1\bar{B}_2\} = \Pr\{\bar{B}_1\}\Pr\{\bar{B}_2\} = 3/4 \times 3/4 = 9/16$

Moreover, the four events above are mutually exclusive, so, for example, to obtain the probability of the event $B_1\bar{B}_2 + \bar{B}_1B_2$ (which is the event "exactly one heart is drawn in two trials"), the probabilities can be added to yield:

$$\Pr\{B_1\bar{B}_2 + \bar{B}_1B_2\} = \Pr\{B_1\bar{B}_2\} + \Pr\{\bar{B}_1B_2\} = 3/16 + 3/16 = 6/16$$

Obviously, to make a list of all possible outcomes of a large number of repeated trials would be laborious. When there are only two possible outcomes for each trial, as is the case for "heart" and "non-heart" above, then there are 2^n possible outcomes for n trials. For $n = 20$, $2^n = 1{,}048{,}576$ — a formidable list indeed!

Luckily, the simple notions involved in two trials easily extend to many trials. Take the case of 20 trials, and suppose we're interested in the probability that the event B ("card drawn is a heart") occurs exactly five times and the event $\bar{B}$ ("card drawn is a non-heart") occurs exactly 15 times. (This is an interesting example because the event in question is the "expected" outcome and, indeed, it is the most probable outcome.)

Among the sequences containing exactly 5 hearts and 15 non-hearts, there is one sequence in which the 5 hearts are the first five cards drawn. Because the repeated trials are independent, the probability of this specific sequence is calculated by multiplying together five 1/4's and fifteen 3/4's, which is

$$(1/4)^5(3/4)^{15} = 1.30503 \times 10^{-5}$$

Every other sequence of 5 hearts and 15 non-hearts must have this same probability, and since each sequence is mutually exclusive with all the others, the total probability of the event in question is obtained by adding together 1.30503×10^{-5} as many times as there are such sequences.

The number of such sequences can be calculated from the so-called BINOMIAL COEFFICIENT:

$$\frac{n!}{r!(n - r)!}$$

where n represents the number of trials, r represents the number of occurrences of event B, and $n - r$ represents the number of occurrences of event $\bar{B}$. The symbol $n!$ (read "n factorial") means $1 \times 2 \times 3 \times \ldots \times (n - 1) \times n$. Most mathematical and statistical handbooks contain tabulations of binomial coefficients and factorials. 0! is defined to equal 1. In the case at hand, $n = 20$, $r = 5$, and $n - r = 15$. Thus, the overall probability of drawing exactly 5 hearts in a total of 20 trials is:

$$\frac{20!}{5!15!}(1/4)^5(3/4)^{15} = 15{,}504 \times 1.30503 \times 10^{-5} = .20233$$

That is to say, the most probable outcome of this experiment will occur a little over 20 percent of the time.

It is well to generalize this discussion of repeated trials. If an event B has probability p (say), then the probability of exactly r occurrences of event B in n repeated and independent trials is:

$$\frac{n!}{r!(n - r)!} p^r(1 - p)^{n-r}$$

For a generalization of this expression to more than two possible outcomes see Box B.

THE CHI-SQUARED TEST

I have actually performed the card-drawing experiment 20 times and obtained 8 hearts and 12 non-hearts. Of course, the "expected" numbers are 5 hearts and 15 non-hearts. What one would like to know in this case, and in many cases involving real biological data, is whether the observed result of the experiment represents a satisfactory fit to the theoretical expectations. If eight hearts is too many, perhaps I was not shuffling the deck well enough or perhaps had an incomplete deck containing too many hearts. It is in testing

B Multinomial Probabilities

The purpose of this box is to show how the binomial probability can be extended to include cases in which more than two outcomes are possible. For instance, the mating *AB/ab* × *ab/ab* can produce four genotypes of offspring (*AB/ab, Ab/ab, aB/ab,* and *ab/ab*), and we may be interested in the probability of obtaining a specified number of each type of offspring.

Generalization of the binomial probability to more than two events is quite straightforward. If $B_1, B_2, \ldots, B_r$ represent r mutually exclusive events with respective probabilities $p_1, p_2, \ldots, p_r$ $(p_1 + p_2 + \ldots + p_r = 1)$, then, in n repeated and independent trials, the probability of obtaining n_1 occurrences of B_1, n_2 occurrences of $B_2, \ldots, n_r$ occurrences of B_r (where $n_1 + n_2 + \ldots + n_r = n$), is given by:

$$\frac{n!}{n_1!n_2!\ldots n_r!} p_1^{n_1}p_2^{n_2}\ldots p_r^{n_r}$$

The first term of this expression, involving factorials, is usually called the MULTINOMIAL COEFFICIENT.

a. A mating of *AaBb* × *AaBb* produces exactly 16 offspring. Assuming *A* and *B* are dominant to their respective alleles and that the loci are unlinked, what is the probability that the observed phenotypic ratio is exactly 9:3:3:1?

b. A mating of *Ab/aB* × *ab/ab* produces 10 offspring. If the recombination fraction between the *A* and *B* loci is $r = .20$, what are the expected proportions of *Ab/ab, aB/ab, AB/ab,* and *ab/ab* offspring? What is the probability of finding exactly these proportions among the 10 offspring?

GOODNESS OF FIT between observed data and expected values that the chi-squared test is particularly useful.

To assess closeness (goodness) of fit, we first need some measure of "closeness." The measure conventionally used is called chi-squared and symbolized χ^2; it is calculated by adding together quantities of the form (obs − exp)^2/exp, where obs = the observed number and where exp = the corresponding expected number. That is to say,

$$\chi^2 = \sum \frac{(\text{obs} - \text{exp})^2}{\text{exp}}$$

where the $\sum$ sign means summation.

In the case at hand, I observed 8 hearts and 12 non-hearts, expecting 5 and 15. So

$$\chi^2 = \frac{(8 - 5)^2}{5} + \frac{(12 - 15)^2}{15} = 2.40$$

Now, the best way to assess goodness of fit of my observed result to the expected result is to calculate the probability of getting a fit as bad or worse than I did, under the assumptions that the proportion of hearts in my deck was indeed 1/4 and that I did indeed draw successive cards at random.

Table IV shows the 21 possible outcomes I could have obtained in the experiment, the associated probability for

TABLE IV. Probabilities of various outcomes in the card-drawing experiment.

Number of hearts	Probability	Associated χ^2
0	3.171×10^{-3}	6.67
1	2.114×10^{-2}	4.27
2	6.695×10^{-2}	2.40
3	1.339×10^{-1}	1.07
4	1.897×10^{-1}	.27
5	2.023×10^{-1}	0
6	1.686×10^{-1}	.27
7	1.124×10^{-1}	1.07
8	6.089×10^{-2}	2.40
9	2.706×10^{-2}	4.27
10	9.922×10^{-3}	6.67
11	3.007×10^{-3}	9.60
12	7.157×10^{-4}	13.07
13	1.542×10^{-4}	17.07
14	2.570×10^{-5}	21.60
15	3.426×10^{-6}	26.67
16	3.569×10^{-7}	32.27
17	2.799×10^{-8}	38.40
18	1.555×10^{-9}	45.07
19	5.457×10^{-11}	52.27
20	9.095×10^{-13}	60.00

each result [calculated as $20!(1/4)^r(3/4)^{20-r}/r!(20 - r)!$ for $r = 0, 1, \ldots, 20$], and the corresponding χ^2 values. Note that those outcomes containing two or fewer hearts, or eight or more, fit the expectations as badly or worse than the observed result because their associated χ^2 values are as large or larger than the one observed. When the probabilities associated with outcomes having two or fewer hearts, or eight or more hearts, are added, we obtain $p = .19$. This value is called the "probability associated with $\chi^2 = 2.4$," and it represents the proportion of times I would get a fit as bad or worse than I actually observed, were I to perform the same experiment many times.

Is a goodness of fit corresponding to $p = .19$ acceptable? Well, since $p = .19$, 19 percent of the time a result as bad or worse would have been obtained; whether this is good or bad is a matter for scientific judgment. There are some guidelines to go by, however. Conventionally, a value of p less than .05 is said to be "significant" and is taken as evidence of an unacceptable fit. In our case $p = .19 > .05$, so the goodness of fit *is* acceptable. Had it been otherwise, one would have had to reject the hypothesis that the probability of drawing a heart was 1/4 and that successive draws were random or independent. With rejection based on this criterion, there is always, of course, the possibility that the hypothesis is indeed true and that the poor fit results from the experiment having had an unusual or improbable outcome. Thus, a correct hypothesis can sometimes be rejected (this is called "type I error"); also, a false hypothesis can sometimes fail to be rejected ("type II error").

To make a complete list of outcomes, as in Table IV, is virtually impossible when the number of observations is large. The probability values corresponding to χ^2 have been extensively tabulated however (Figure 22); thus, once the χ^2 value has been calculated for observed data, the associated probability value can immediately be read from Figure 22. The probability values in the figure are approximations, and the larger the expected number in each class of data, the better is the approximation. As a rough rule, Figure 22 should not be used when the smallest expected number is

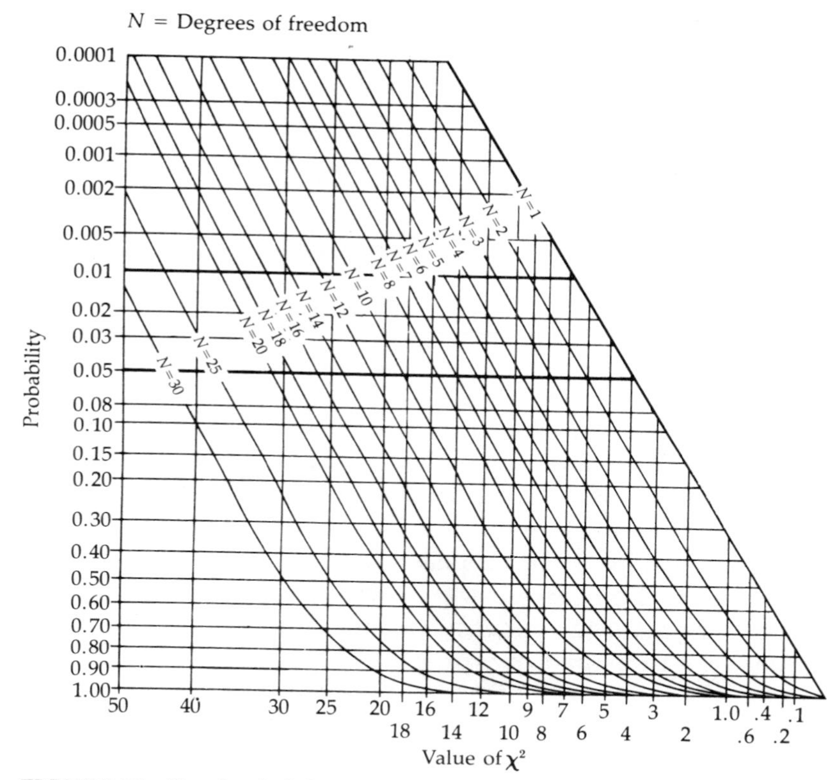

FIGURE 22. Graph of χ^2. To use the graph, find the value of χ^2 along the abscissa (the horizontal axis), then read the associated value of p corresponding to the appropriate number of degrees of freedom. (From Crow, 1945.)

less than five. In the present card-drawing case, the approximation turns out to be acceptable, even though the numbers involved are quite small. We had $\chi^2 = 2.4$, and reading off Figure 22 for one degree of freedom (see following), the p-value corresponding to $\chi^2 = 2.4$ is $p = .12$. (Recall that the exact probability calculated from Table IV is .19.)

The number of DEGREES OF FREEDOM in a chi-squared test of the type under discussion is one less than the number of classes of data. In the card-drawing experiment, the number of classes is two ("heart" versus "non-heart"), so the num-

TABLE V. Chi-squared test of Mendel's data involving segregation.

	Round	*Wrinkled*	**Total**
Observed	5474	1850	7324
Expected	(3/4)(7324) = 5493	(1/4)(7324) = 1831	7324

$$\chi^2 = \frac{(5474 - 5493)^2}{5493} + \frac{(1850 - 1831)^2}{1831} = 0.263$$

With $2 - 1 = 1$ degree of freedom; $p \approx 0.64$

ber of degrees of freedom is one. (Discussion of degrees of freedom will be taken up again in Chapter 2.)

In Table V, the χ^2 test for goodness of fit is applied to Mendel's previously mentioned data on segregation involving *round* versus *wrinkled* peas. The probability value (read from Figure 22) is about .64, which means, were Mendel to have carried out the experiment many times, he would have obtained a fit as bad or worse some 64 percent of the time. Because $p > .05$, we have no reason to reject the hypothesis that the ratio of *round* to *wrinkled* seeds is 3:1.

A χ^2 test of Mendel's experimental result on the independent assortment of *round* versus *wrinkled* and *yellow* versus *green* seeds is carried out in Table VI. Here there are three degrees of freedom (because there are four classes of data), and the p value estimated from Figure 22 is about .93; that is, Mendel would have gotten a fit as bad or worse 93 percent of the time. Here, because $p > .05$, there is no reason to reject the hypothesis of independent assortment. It is of interest to note that famed statistician R. A. Fisher examined Mendel's data quite carefully and concluded that the fit was a little *too* good — a little too close to be easily accounted for by chance. (See Fisher, 1936; the issue is also discussed in Sturtevant, 1965.)

The Scope of Population Genetics

As noted earlier, evolution occurs as a result of progressive change in the kinds and frequencies of genes that occur in

TABLE VI. Chi-squared test of Mendel's data involving independent assortment.

	Round, yellow	*Round, green*	*Wrinkled, yellow*	*Wrinkled, green*	**Total**
Observed	315	108	101	32	556
Expected	(9/16)(556) = 312.75	(3/16)(556) = 104.25	(3/16)(556) = 104.25	(1/16)(556) = 34.75	556

$$\chi^2 = \frac{(315 - 312.75)^2}{312.75} + \frac{(108 - 104.25)^2}{104.25} + \frac{(101 - 104.25)^2}{104.25} + \frac{(32 - 34.75)^2}{34.75} = 0.470$$

With $4 - 1 = 3$ degrees of freedom; $p \approx 0.93$

populations, and it results in progressive increase in the adaptation of organisms to their environment. No organism on earth could suddenly alter the codon-amino acid correspondences in its genetic code, even though the function of a particular protein might be improved by the change, for the reason that a change in the genetic code would automatically alter every protein produced by the organism, and most of these alterations would be detrimental. Again, in normal sexual organisms, evolution must work within the limits set by Mendelian segregation and recombination. Limits to evolution are also set by the patterns of control of gene activity during development; developmental pathways have evolved to function in a coordinated manner and involve so many components that they cannot undergo radical and instantaneous change. For example, in the course of human evolution there has been a dramatic increase in the size of the brain and cranium, with one result that human females sometimes have trouble giving birth because of the relatively large size of the newborn's head. This reproductive problem cannot be circumvented completely, however, because of the absolute necessity for intrauterine development. The purpose of population genetics, therefore, is to determine how evolution occurs within the framework of biological constraints.

MODELS

A primary tool in population genetics, as well as in such sciences as physics and chemistry, is a model. A MODEL is an intentional simplification of a complex situation designed to eliminate extraneous detail in order to focus attention on the essentials of the situation. In population genetics, we must contend with such factors as population size, patterns of mating, geographical distribution of individuals, mutation, migration, and natural selection. Although we wish ultimately to understand the combined effects of all these factors and more, the factors are so numerous and interact in such complex ways that they cannot usually be grasped all at once. Simpler situations are therefore devised, situations in which a few identifiable factors are the most important ones and others can be neglected. A typical experiment is a kind of model, because in an experiment, one strives to control or eliminate all sources of variation other than the variable or variables under scrutiny.

Perhaps the most important kind of model in population genetics is the MATHEMATICAL MODEL, which is a set of hypotheses that specifies the mathematical relationships between measured or measurable quantities (PARAMETERS) in a system or process. Mathematical models can be extremely useful: they express concisely the hypothesized quantitative relationships between parameters; they reveal which parameters are the most important ones in a system and thereby suggest critical experiments or observations; they serve as guides to the collection, organization, and interpretation of observed data; and they make quantitative predictions about the behavior of a system that can, within limits, be confirmed or shown to be false. The validity of a model must, of course, be tested by determining whether the hypotheses that it is based on and the predictions that grow out of it are consistent with observations.

Mathematical models are always simpler than the actual situation they are designed to elucidate. Many features of the actual system are intentionally left out of the model, because to include *every* aspect of the system would make the model too complex and unwieldy. Construction of a model always involves a compromise between realism and complexity; a completely realistic model is likely to be too complex to handle mathematically, and a model that is mathematically simple may be so unrealistic as to be useless. Ideally, a model should include all essential features of the system and exclude all nonessential ones. How good or useful a model is often depends on how closely this ideal is approximated. In short, a model is a sort of metaphor or analogy. Like all analogies, it is valid only within certain limits and when pushed beyond these limits becomes misleading or even absurd. (For a further discussion of models, see Levins, 1966.)

MODELS OF POPULATION GROWTH

To illustrate the nature of mathematical models, we will consider the dynamics of population growth, a subject of considerable interest in population genetics and ecology. Figure 23A (solid dots) shows the increase in a population of yeast cells in a defined quantity of medium. To obtain a mathematical model of the growth process, we may assume, as a first approximation, that population size increases by a constant fraction in each generation; thus, we may write $N_t = N_{t-1} + rN_{t-1}$, where N_t and N_{t-1} represent population size in generations t and $t - 1$, and where r is a constant called the INTRINSIC RATE OF INCREASE. The solution of this equation is:

$$N_t = N_0(1 + r)^t$$

where N_0 represents the initial population size. (For the derivation of this solution, see Box C.) Since, from the above

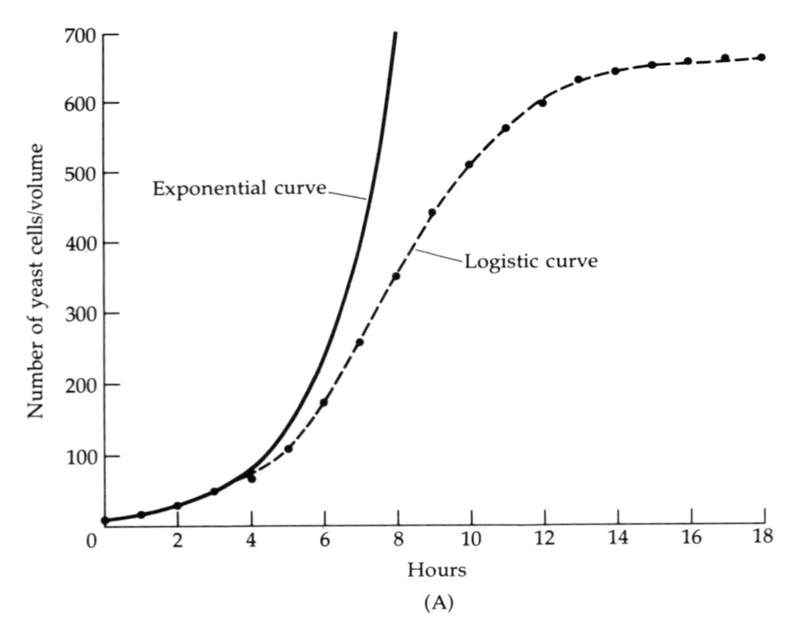

FIGURE 23. (A) Population growth of yeast in closed culture, showing exponential and logistic approximations (data from Pearl, 1927). (B) Erratic fluctuations in the population size of the German forest insect pest, *Bupalus piniarius*. Note that the ordinate is logarithmic and that the population fluctuates in size by a factor of 1000 or more. Such fluctuations are characteristic of this and many other insects. [Data for (B) from Schwerdtfeger, 1941.]

equation, $\ln(N_t/N_0) = t \ln(1 + r)$, where ln represents the natural logarithm, a plot of $\ln(N_t/N_0)$ against t yields a straight line with slope equal to $\ln(1 + r)$. (How straight lines are best fitted to data is described in Box D.)

If r is not too large, $(1 + r)^t$ is very close to e^{rt}, where e = 2.71828... is the base of natural logarithms; thus, in this case,

$$N_t = N_0 e^{rt}$$

(Introduction of the number e into the equation may seem artificial at this point, but students of elementary calculus will see in Box C that e intrudes itself quite naturally and unavoidably.)

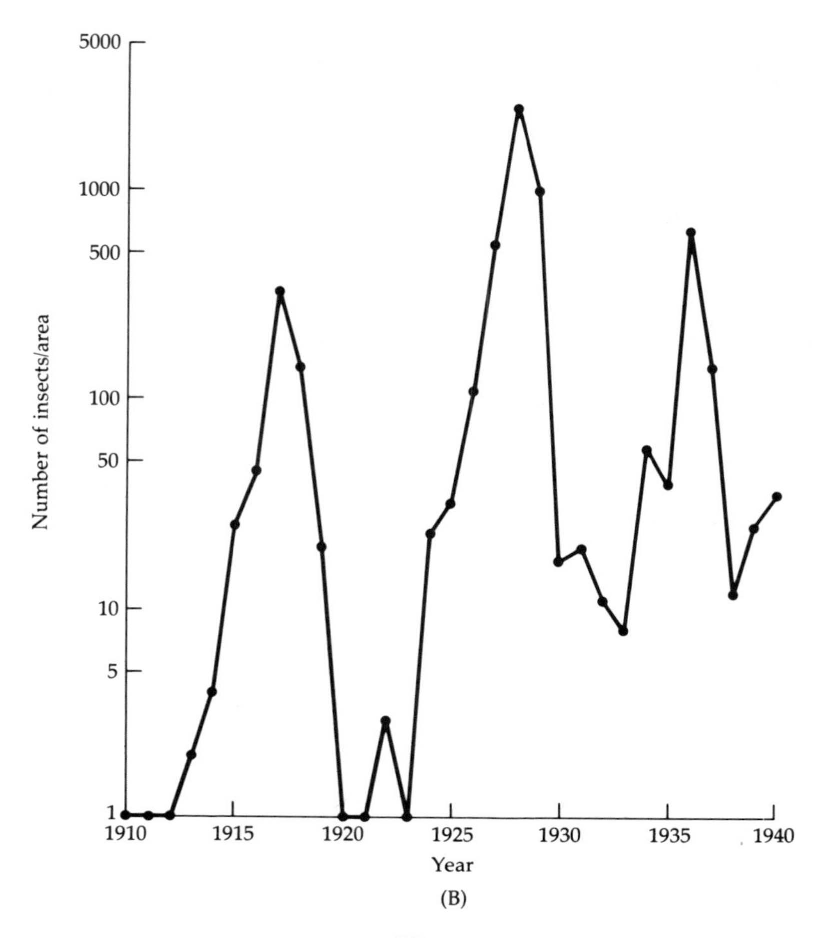

(B)

C Relationships between Population-Growth Models

For the models of population growth discussed in the text, the population size at any time, N_t, was expressed in terms of the population size in the previous generation, N_{t-1}. In this box we show that somewhat more realistic models can be devised using simple calculus, but first we consider situations in which population growth is governed by the equation $N_t = N_{t-1} + rN_{t-1}$. When population size changes according to this equation, then, for $t = 1$, $N_1 = N_0(1 + r)$, and $t = 2$, $N_2 = N_1(1 + r)$; substituting the expression for N_1 into that for N_2, we obtain $N_2 = [N_0(1 + r)](1 + r) = N_0(1 + r)^2$. Similarly, $N_3 = N_0(1 + r)^3$ and, in general, $N_t = N_0(1 + r)^t$. However, N_t is defined only for t equal to positive integers; that is to say, the population does not change gradually in size, but "jumps" instantaneously from one size to the next.

a. For $N_0 = 100$ and $r = .02$, calculate N_t for $t = 50, 100, 150, 200$, and 250 generations.

As noted above, a model in which population size changes smoothly can be devised using simple calculus. By analogy with $N_t - N_{t-1} = rN_{t-1}$, we write $dN/dt = r_0N$, where dN/dt is the first derivative of $N(t)$; that is to say, dN/dt is the rate of change in $N(t)$ at an instant in time. (The subscript 0 is attached to r_0 for later use.) A little manipulation of $dN/dt = r_0N$ yields $dN/N = r_0dt$. We can now use the integration formula $\int(1/x)dx = \ln x$ on the left-hand side, yielding $\int_{N_0}^{N_t}(1/N)dN = \ln(N_t) - \ln(N_0) = \ln(N_t/N_0)$; application of $\int dx = x$ to the right-hand side yields $\int_0^t r_0dt = r_0t - 0 = r_0t$. Setting the two sides equal: $\ln(N_t/N_0) = r_0t$, or, $e^{\ln(N_t/N_0)} = N_t/N_0 = e^{r_0t}$. Thus $N_t = N_0e^{r_0t}$, where now t can take on *any* value, not merely integers; so N_t changes smoothly.

For the two models to give the same population size at integer values of t, we would need $(1 + r)^t = e^{r_0t}$, or $r_0 = \ln(1 + r)$. But for $-1 < r < 1$, $\ln(1 + r) = r - r^2/2 + r^3/3 - r^4/4 + \ldots$, so if r is small enough that r^2 and higher powers can be neglected, then $\ln(1 + r) \approx r$, and $r_0 \approx r$ also.

b. With $N_0 = 100$ and $r = .02$ as in (a), use $N_t = N_0e^{r_0t}$ to calculate N_t at $t = 50, 100, 150, 200$, and 250, for $r_0 = \ln(1 + r)$, for $r_0 = r$, and for $r_0 = r - r^2/2$.

The logistic model of population growth is technically defined as $dN/dt = rN(K - N)/K$, where K is the carrying capacity. To solve the equation we separate variables as above, obtaining $dN/[N(1 - N/K)] = rdt$.

c. Use the formula $\int[1/x(1 + ax)]dx = \ln[x/(1 + ax)]$ to show that $N_t/(K - N_t) = N_0e^{rt}/(K - N_0)$, and use this expression to derive the logistic growth curve $N_t = K/(1 + Ce^{-rt})$, where $C = (K - N_0)/N_0$.

As is the case with exponential growth, the logistic growth curve solves $dN/dt = rN(K - N)/K$ and not $N_t - N_{t-1} = rN_{t-1}(K - N_{t-1})/K$. However, if population size changes slowly in any one generation, then $N_t - N_{t-1}$ approximates dN/dt, and therefore the logistic growth curve approximates the solution of $N_t - N_{t-1} = rN_{t-1}(K - N_{t-1})/K$.

d. For $N_0 = 1000$, $r = .10$, and $K = 2000$, calculate N_1, N_2, N_3, N_4, and N_5 exactly using $N_t - N_{t-1} = rN_{t-1}(K - N_{t-1})/K$. Then calculate the approximate values using the logistic growth curve.

Fitting Straight Lines to Data

For a set of points such as those in the figure in this box, it is often necessary to find the best fitting straight line. The appropriate procedure is called a LINEAR REGRESSION, and it will be exemplified using the data in the figure. There are four points (x, y) — namely (1,3), (2,7), (3,6), and (4,9). For the sake of generality, let n denote the number of points (here $n = 4$), and let a bar over a symbol

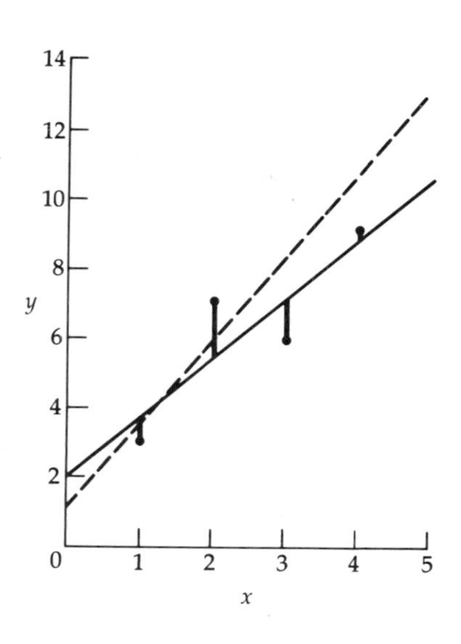

D denote an average [$\bar{x}$, for example, denotes the average of the x values, so $\bar{x} = (1 + 2 + 3 + 4)/4 = 2.500$]. (Note: the word MEAN is often used as a synonym for average.) The best fitting straight line will have the equation $y = a + bx$, where a and b are to be determined. The reasoning behind the calculations will be discussed in Chapter 4, but here we need only a recipe. To find a and b, prepare a list of the following quantities:

(1) the number, n, of points (here $n = 4$);
(2) the mean, $\bar{x}$, of x (in this case, as calculated above, $\bar{x} = 2.500$);
(3) the mean, $\overline{x^2}$, of the squares of x — i.e., $\overline{x^2} = (1^2 + 2^2 + 3^2 + 4^2)/4 = (1 + 4 + 9 + 16)/4 = 7.500$;
(4) the mean, $\bar{y}$, of y [here $\bar{y} = (3 + 7 + 6 + 9)/4 = 6.250$]; and
(5) the mean, $\overline{xy}$, of the products of x and y [in this example $\overline{xy} = (1 \times 3 + 2 \times 7 + 3 \times 6 + 4 \times 9)/4 = (3 + 14 + 18 + 36)/4 = 17.750$].

Then calculate:

(6) $\overline{xy} - \bar{x}\bar{y}$. From (2), (3), and (5), we have $\overline{xy} - \bar{x}\bar{y} = 17.75 - (2.500)(6.250) = 2.125$. This quantity is usually called the COVARIANCE of x and y and is symbolized as $Cov(x,y)$.
(7) $\overline{x^2} - \bar{x}^2$. From (2) and (3), we obtain $\overline{x^2} - \bar{x}^2 = 7.500 - (2.500)^2 = 7.500 - 6.250 = 1.250$. This quantity is usually called the VARIANCE of x and is symbolized as $Var(x)$.

Finally calculate:

(8) $b = Cov(x,y)/Var(x)$. From (6) and (7), $b = 2.125/1.250 = 1.700$.
(9) $a = \bar{y} - b\bar{x}$. From (2), (4), and (8), $a = 6.250 - (1.700)(2.500) = 6.250 - 4.250 = 2.000$.

Thus, the best fitting straight line has the equation $y = 2.000 + 1.700x$. This line is the unbroken line in the figure here, and the heavy line segments connect each actual point with a corresponding point on the line having the same value of x. The calculated line is the "best" line in the sense that the sum of the squares of the lengths of the heavy segments is a minimum; for this reason, the calculated line is often called the "least squares regression line."

Sometimes it is desirable to fit a line when the points have different degrees of importance, or WEIGHTS. For the above data, for example, the points (1,3), (2,7), (3,6), and (4,9) may have weights 10, 10, 1, and 1, meaning that each of the first two points is to be given 10 times as much importance as each of the last two. The appropriate method for fitting the equation $y = a + bx$ in this case is called a WEIGHTED REGRESSION, and the calculations are much the same as the unweighted regression calculations carried out earlier. Specifically, one requires: D

(1) the sum of the weights, symbolized Σw_i, which in this case is $\Sigma w_i = 10 + 10 + 1 + 1 = 22$;
(2) the weighted average of x, symbolized $\bar{x}_w$, which is the average of the x's, each multiplied by its weight; in the present example $\bar{x}_w = (10 \times 1 + 10 \times 2 + 1 \times 3 + 1 \times 4)/\Sigma w_i = 37/22 = 1.68182$;
(3) the weighted average of the squares of x, denoted $\overline{x_w^2}$, calculated as in (2); for the present case $\overline{x_w^2} = (10 \times 1^2 + 10 \times 2^2 + 1 \times 3^2 + 1 \times 4^2)/\Sigma w_i = (10 + 40 + 9 + 16)/22 = 75/22 = 3.40909$;
(4) the weighted average of y, called $\bar{y}_w$, which equals $(10 \times 3 + 10 \times 7 + 1 \times 6 + 1 \times 9)/\Sigma w_i = 115/22 = 5.22727$;
(5) the weighted mean of the products of x and y, symbolized $\overline{xy}_w$, which for this example is $(10 \times 1 \times 3 + 10 \times 2 \times 7 + 1 \times 3 \times 6 + 1 \times 4 \times 9)/\Sigma w_i = 224/22 = 10.18182$.

Then calculate:

(6) $Cov(x,y) = \overline{xy}_w - \bar{x}_w\bar{y}_w = 10.18182 - (1.68182)(5.22727) = 1.39049$, using the values in (2), (4), and (5), and
(7) $Var(x) = \overline{x_w^2} - \bar{x}_w^2 = 3.40909 - (1.68182)^2 = .58057$, using values in (2) and (3).

Finally, as before,

(8) $b = Cov(x,y)/Var(x) = 1.39049/.58057 = 2.39504$, from (6) and (7),
(9) $a = \bar{y}_w - b\bar{x}_w = 5.22727 - (2.39504)(1.68182) = 1.19924$.

For the weighted least squares regression, then, the best-fitting line is given by $y = 1.19924 + 2.39504x$, which is the broken line in the figure.

a. Obtain quantities (1) through (9) for an unweighted linear regression of y on x for the points (2,11), (4,21), (6,22), and (8,32).

b. Obtain quantities (1) through (9) for the same points in a weighted regression with weights (given in the same order as the points) 1, 1, 20, 20.

In Figure 23A, the solid line depicts the prediction of this so-called EXPONENTIAL GROWTH curve, $N_t = N_0e^{rt}$, when $N_0 = 9.6$ (the observed value) and $r = .5355$. As can be seen, the fit to the actual data is quite good for a period of three or four hours, but then it becomes increasingly poor. If one is interested in short-term population growth, therefore, the exponential growth model is reasonably good. For longer term yeast cultures, the model is clearly inadequate.

To obtain a somewhat better model of yeast population growth, we may assume that the increase in population size in each generation decreases as a linear function of N; that is to say,

$$N_t = N_{t-1} + rN_{t-1}\left(\frac{K - N_{t-1}}{K}\right)$$

where K is a constant known as the CARRYING CAPACITY. Observe that when N is very small compared to K, then $N_t \approx N_{t-1} + rN_{t-1}$, so population growth will be nearly exponential; on the other hand, when N is close to K, $N_t \approx N_{t-1}$, so population growth will cease. If population size changes slowly, then an approximate solution to this mathematical model of population growth is:

$$N_t = \frac{K}{1 + Ce^{-rt}}$$

where $C = (K - N_0)/N_0$; this equation is called the LOGISTIC GROWTH CURVE and the model that leads to it is called the LOGISTIC GROWTH MODEL. (For derivation, see Box C.) When population growth is logistic, population size increases according to a sort of S-shaped curve. The dashed line in Figure 23A is the prediction calculated from the logistic equation when $K = 665$, $r = .5355$, and $N_0 = 9.9$. The fit is remarkably good.

Thus, in two steps, we have achieved a rather good mathematical model of the growth of yeast populations in closed cultures. For more complex growth situations, such as that in Figure 23B, even the logistic model is inadequate. In such cases, more realistic models might have to take into account individuals that enter the population from outside (IMMIGRATION) or that leave the population (EMIGRATION); or take into account other populations, such as predators or competitors, whose size influences the size of the population in question (Box E); or take into account the fact that individuals of different ages have different chances of dying or giving birth (see Chapter 2); or take into account such random and unpredictable influences on population size as disease epidemics or bad weather. Such complications are more properly in the purview of ecology or demography, however, not population genetics (see Wilson and Bossert, 1971; Poole, 1978).

Lotka-Volterra Competition Model

The purpose of this box is to show how the logistic model of population growth can be extended to encompass certain kinds of competitive interactions between species. The model was developed by the mathematicians Alfred J. Lotka (1925) and Vito Volterra (1926). While by no means universally valid, the Lotka-Volterra model is nevertheless simple and useful (Wangersky, 1978). As noted above, the model grows out of the logistic equation $dN/dt = rN\,(1 - N/K)$ (see Box C), with the additional assumption that each of the two competing species "uses up" some of the other's carrying capacity. To be specific, the Lotka-Volterra competition model is defined by:

$$\frac{dN_1}{dt} = r_1N_1\left[1 - \frac{(N_1 + \alpha_{21}N_2)}{K_1}\right]$$

and

$$\frac{dN_2}{dt} = r_2N_2\left[1 - \frac{(N_2 + \alpha_{12}N_1)}{K_2}\right]$$

where the subscripts 1 and 2 denote those species to which the parameters refer. The quantity $1/K_1$ measures the overall effect of an individual of species 1 on the growth of species 1, and α_{21}/K_1 measures the overall effect of an individual of species 2 on the growth of species 1; similarly, $1/K_2$ and α_{12}/K_2 measure the overall effects of an individual of species 2 or of species 1, respectively, on the growth of species 2. (The quantities α_{21} and α_{12} are often called "competition coefficients.") The calculus symbols dN_1/dt and dN_2/dt represent the rate of change in N_1 and N_2 in an infinitesimally small interval of time, but if N_1 and N_2 are changing slowly, dN_1/dt approximates $N_1(t + 1) - N_1(t)$ — that is, the change in N_1 in one generation — and dN_2/dt approximates $N_2(t + 1) - N_2(t)$.

Complete analysis of the Lotka-Volterra equations is beyond the

E scope of this book; the equations are introduced here for later use in Chapter 5. However, the most important case arises when $\alpha_{12}/K_2 < 1/K_1$ and $\alpha_{21}/K_1 < 1/K_2$ — that is, when each species depresses its own growth more than it depresses that of its competitor — because in this case, the two species will coexist. In all other cases, either species 1 wins (i.e., $N_1 \rightarrow K_1$, $N_2 \rightarrow 0$), or species 2 wins.

a. In the case of coexistence, species 1 and 2 eventually reach numbers $\hat{N}_1$ and $\hat{N}_2$, which remain constant thereafter; that is $dN_1/dt = dN_2/dt = 0$ when $N_1 = \hat{N}_1$ and $N_2 = \hat{N}_2$. Set $dN_1/dt = 0$ and $dN_2/dt = 0$ in the Lotka-Volterra equations and solve for N_1 and N_2 to show that $\hat{N}_1 = (K_1 - \alpha_{21}K_2)/(1 - \alpha_{21}\alpha_{12})$ and $\hat{N}_2 = (K_2 - \alpha_{12}K_1)/(1 - \alpha_{21}\alpha_{12})$.

b. Find $\hat{N}_1$ and $\hat{N}_2$ for $K_1 = 1500$, $K_2 = 2500$, $\alpha_{21} = .25$, and $\alpha_{12} = .50$. Since r_1 and r_2 do not affect the values of $\hat{N}_1$ and $\hat{N}_2$, what do they affect?

POPULATIONS

At this point it becomes essential to be more precise about the word "population." In ordinary, everyday usage, "population" means a group of individuals. Usage is sometimes the same in population genetics, though the group in question usually consists of members of the same species. Population genetics also requires a more precise term, however: namely, a term to refer to an actual, evolving population. Precise definition of the evolving unit in natural populations is complicated because of the almost universal presence of some sort of GEOGRAPHICAL STRUCTURE, some typically nonrandom pattern in the spatial distribution of organisms. Members of a species are rarely distributed homogeneously in space; there is almost always some sort of clumping or aggregation, some schooling, flocking, herding, or colony formation. Population subdivision is often caused by environmental patchiness, areas of favorable habitat intermixed with unfavorable areas. Such environmental patchiness is obvious in the case of, for example, terrestrial organisms on islands in an archipelago, but patchiness is a common feature of most habitats — freshwater lakes have shallow and deep areas, meadows have marshy and dry areas, forests have sunny and shady areas. Population subdivision can also be caused by social behavior, as when wolves form packs. Even the human population is clumped or aggregated — into towns and cities, away from deserts and mountains.

Population subdivision creates a problem in defining the basic unit of population genetics — the evolving population — because geographical structure usually influences mating patterns; individuals typically tend to mate with members of their own subpopulation. Subpopulations of a larger population that form locally interbreeding groups of organisms are often called LOCAL POPULATIONS, MENDELIAN POPULATIONS, or DEMES. We will use the word "population" in the sense of "local population," except where a broader meaning is clear from context.

NATURAL SELECTION

Population genetics is mainly concerned with how Mendel's laws and other genetic phenomena influence the evolutionary process. One of the ironies of the history of population genetics is that the theory of evolution by means of natural selection proposed by Charles Darwin (1809–1882) in his monumental book, *Origin of Species,* first published in 1859, immediately generated so much excitement that it completely overshadowed Mendel's discoveries. [Alfred Wallace (1832–1913) published ideas similar to Darwin's at about the same time.] Indeed, according to Iltis (1932), on the cold, clear winter evening of February 8, 1865, when Mendel read his paper to the 40 members of the Brünn Natural History Society, the minutes record not a single question or discussion of Mendel's paper, but they do record that the *Origin* was discussed! The irony is that the prevailing beliefs about heredity were subtly incompatible with Darwin's theory (Box F), though this was not realized at the time and was not corrected until much later when Mendelism was incorporated into the theory.

The theory of evolution by means of natural selection is based on a number of premises and the conclusions that follow from them. The first premise is that all species have more offspring than can possibly survive and reproduce;

F The Flaw in Darwin's Views of Heredity

In Darwin's time it was generally believed that blending inheritance was the rule; translated into modern terms, the theory of blending inheritance postulates that the alleles in an *Aa* heterozygote become altered in such a manner that all gametes produced by the heterozygote will carry an entirely new allele whose effect on phenotype is the average of the phenotypic effects of *A* and *a*. That blending inheritance is incompatible with evolution by means of natural selection was first pointed out in 1867 by H. C. Fleeming Jenkin, Professor of Engineering at Edinburgh (see Dunn, 1965). The occurrence of evolution by natural selection requires the continued availability of genetic variation on which selection can act. Jenkin realized that blending inheritance causes genetic variation to disappear extremely rapidly; he argued that any new mutation "will be swamped by numbers and after a few generations its peculiarity will be obliterated."

Jenkin's argument can be made quantitative by considering the fate of a newly arising mutation that affects a trait such as height. Since with blending inheritance the mutant allele will become altered as time goes on, it is convenient to denote the original mutant allele as A_0 (for the 0th, or initial, generation). Moreover, since the allele is a new mutation, it must occur in a heterozygote, A_0a, where *a* represents the normal allele at the locus. Suppose that the effect of the A_0 allele is to increase the height of A_0a individuals by an amount of x_0 units as compared to the height of *aa* individuals. From a conceptual point of view, the phenotype of an A_0a individual is the same as if the individual were homozygous for some novel allele, say A_1, in which each A_1 allele added $x_0/2$ units to height. The blending theory of inheritance claims that the above conceptual truth is true in fact — that the A_0a individuals produce gametes carrying a "blended" allele, A_1, the effect of which is to add $x_0/2$ units to phenotype. Thus, in the first generation after the occurrence of a new allele A_0 of effect x_0, the allele becomes A_1 and has effect $x_0/2$; the same reasoning as above implies that, in the second generation, the A_1 allele is altered to A_2, and the effect of A_2 is $x_0/4$. Indeed, assuming the *A* alleles remain so rare as to be found exclusively in heterozygotes, the alleles in generation *n* may be denoted A_n, and each A_n allele will have a phenotypic effect of $x_0/2^n$. In short, with blending inheritance, the effect of an individual, rare mutant allele is reduced by half in each generation; indeed, "after a few generations its peculiarity will be obliterated." F

With blending inheritance, in a large but nongrowing population, the numbers A_0, A_1, A_2, A_3, A_4, . . ., A_n alleles increase according to the sequence 1, 2, 4, 8, 16, . . ., 2^n as long as the mutant alleles are rare. Why? With nonblending inheritance and Mendelian segregation, on the other hand, the effect of a mutant allele, A_0, remains the same in all generations and its number remains constant. Why? (Note: this constancy of allele numbers or frequency in the absence of evolutionary forces is the single most important feature of Mendelian segregation as it pertains to population genetics; many of the implications are spelled out in Chapter 2.)

---

this premise follows from the observation that, under optimal environmental conditions, population growth tends to be exponential (see Figure 23A, for example). The second premise is that organisms vary in their ability to survive and reproduce, a postulate whose validity is obvious for the simple reason that some organisms do survive and reproduce and others do not. The third premise is that part of the variation in ability to survive and reproduce is hereditary; this premise is a key one, and Darwin took great pains in the *Origin* to document the occurrence of genetic variation affecting virtually all traits in natural populations and in populations of domesticated species.

The premise about genetic variation can be stated in modern terminology by saying that certain genotypes are more adapted to their environment than are other genotypes, as evidenced by their greater FITNESS — their greater ability to survive and reproduce in that environment. The conclusion from the three premises is the occurrence of NATURAL SELECTION, the process by which genotypes with greater fitness leave, on the average, more offspring than do less fit genotypes. Because of natural selection, those favorable alleles that promote higher fitness will be represented disproportionally in the next and succeeding generations. As these favorable alleles increase in frequency in the population, more of the fitter genotypes will be formed, so the types and frequencies of alleles in the population will gradually change so as to promote greater adaptation to the environment.

The process of natural selection, which leads to increased adaptation of organisms to their environment, is often called MICROEVOLUTION. Although population genetics is primarily interested in microevolution, especially in the effects of such factors as mutation, migration, mating systems, and random effects due to small population size, the study of microevolution overlaps with that of MACROEVOLUTION, a term referring to the processes that lead to the formation of new species, genera, families, orders, and higher taxonomic categories. That macroevolution can involve phenomena quite different from those involved in microevolution is evidenced by the frequent origin of new species of plants by hybridization and polyploidy, and it is also suggested by the frequency with which major chromosomal changes occur in both plant and animal macroevolution (White, 1978; De Grouchy et al., 1978). Nevertheless, it is not yet clear to what extent macroevolutionary events are the result of microevolutionary processes occurring gradually over thousands of generations and to what extent macroevolutionary events involve wholly new principles (Gould, 1977). In any case, it is clear that population genetics underpins our understanding of the process of microevolution, and it contributes to an understanding of macroevolution as well.

GENETIC VARIATION

Natural selection can alter the genetic composition of a population only insofar as there is preexisting genetic variation in the population. As will be emphasized repeatedly in this book, natural populations are a rich storehouse of genetic variation. Barring identical twins, two members of almost any species are likely to have different genotypes at many loci. The genetic variation found in a population reflects, in part, the population's evolutionary history. This variation also determines how rapidly the population will respond either to natural selection or to artificial selection for improvement of agriculturally important characteristics. It is therefore of some importance to discover how this variation is organized into genotypes, and to use this information to predict how a population will respond to selection.

Genetic variation, its origin, maintenance, and disposition, is the overall theme of this book. Chapter 2 focuses on genetic variation itself and how this is organized in populations, including the effects of particular patterns of mating, such as inbreeding. Chapter 3 deals with the evolutionary forces of migration, mutation, and natural selection, and it also includes the effects of small population size. Chapter 4 is concerned with the inheritance of complex traits influenced by many loci and with the consequences of artificial selection as practiced by plant and animal breeders. Chapter 5 is a kind of synthesis focusing on actual examples of microevolution; on modes of selection that do not act directly on the individual but rather on the individual's relatives (kin selection) or on the entire population (group selection); and finally on the important macroevolutionary process of speciation and its relationship to ecological opportunity.

Problems

1. In the mating *Aa* × *Aa* (where *A* represents a dominant allele), what proportion of offspring will be *Aa*? What proportion will be *Aa* among those that have the phenotype associated with the dominant allele?
2. Many rare dominant alleles that cause major phenotypic abnormalities when they are heterozygous are lethal when they are homozygous. Such is the case with the allele *T* (for short-tailedness) in the house mouse, for which *T*/*T* homozygotes die as embryos. What phenotypic ratio of tailless to normal would be expected in the mating *T*/+ × *T*/+? (The symbol + represents the normal allele at the locus.)
3. A mating of *Aa* × *Aa* produces six offspring. What is the probability that two or fewer are genotypically *aa*?
4. For the deck of cards in Figure 21, what is the probability that a randomly drawn face card will be a Jack? What is the probability that a randomly drawn Jack is the Jack of Hearts?

What is the probability that a randomly drawn face card is the Jack of Hearts? The probability of the last event is equal to the product of the probabilities of the other two. Why?

5. What genotypes of offspring and in what proportions would be expected from a mating of *Ab*/*aB* with *ab*/*ab*, when the recombination fraction between the *A* and *B* loci is $r = .12$?

6. A cross of *Ab*/*aB* with *ab*/*ab* produced 60 nonrecombinant and 40 recombinant offspring. Is there evidence of linkage? (Perform a χ^2 test.) If so, estimate the recombination fraction and its standard error (Box A).

7. Among families with two children in which one is known to be a girl, what is the probability that the other one is a boy? (Assume a sex ratio of 1/2.)

8. A woman whose brother is affected with a condition due to a rare, sex-linked recessive seeks genetic counseling. What is the probability that the woman is a carrier of the recessive allele? Assuming she is a carrier, what is the probability that she will have an affected son? What is the overall probability that she will have an affected son?

9. A recessive allele on a chromosome will usually be expressed phenotypically if the homologous chromosome carries a deficiency that includes the locus. Why? (This phenomenon is known as "pseudodominance.")

10. What is the maximum number of copies of a gene that can be generated by mispairing and crossing over (Figure 11) if one chromosome carries seven tandem repeats of the gene and the other carries five tandem repeats?

11. What kinds of chromosomes would result from a single crossover within the "loop" of an individual heterozygous for a pericentric inversion? (See Figure 13.) Does the answer help explain why pericentric inversions are rare in *Drosophila* populations while paracentric inversions are common?

12. What amino acid sequence would be formed from a DNA template with the nucleotide sequence TACTGTGAAA GATTTCCGACT? What would happen to the amino acid sequence following a base-substitution mutation that substituted an A for the first G in the template? What would happen following an insertion mutation in which an A was inserted after the first C in the template? (Note: the latter type of mutation is called a frameshift mutation.)

13. In the exponential model of population growth, what happens to population size if r becomes and remains negative? In the logistic model of population growth, what happens to population size if $N_0 > K$ (assuming that r is positive)?

14. Charles Darwin could have discovered segregation had he known what to look for, as Mendelian segregation occurred in at least one of his own experiments. Darwin (cited in Iltis, 1932) studied flower shape in the snapdragon (*Antirrhinum*). In a cross between a true-breeding strain with regular (peloric) flowers and a true-breeding strain with irregular (normal) flowers, all of the F_1's were normal. Crosses of $F_1 \times F_1$ yielded 88 normal and 37 peloric plants. Perform a χ^2 test assuming a 3:1 ratio in the F_2. Which allele is dominant, the one determining peloria or the normal one?

15. For a mating between triple heterozygotes at three unlinked loci, there are eight phenotypic classes among the offspring. What are the expected phenotypic ratios? Mendel carried out such an experiment (involving the loci *a, i,* and *r* in Figure 7 — he said that, of all experiments, this one required the most time and effort), and he obtained a phenotypic ratio of 269:98:86:88:30:34:27:7 (in a total of 639 progeny). Calculate the χ^2 and its associated probability value.

Further Readings

Cavalli-Sforza, L. L. and W. F. Bodmer. 1971. *The Genetics of Human Populations.* W. H. Freeman, San Francisco.

Crow, J. F. 1976. *Genetics Notes,* 7th ed. Burgess, Minneapolis.

Dunn, L. C. 1965. *A Short History of Genetics.* McGraw-Hill, New York.

Goodenough, U. 1978. *Genetics,* 2nd ed. Holt, Rinehart and Winston, New York.

Harris, H. 1970. *The Principles of Human Biochemical Genetics.* North Holland, Amsterdam and London.

Hartl, D. L. 1977. *Our Uncertain Heritage: Genetics and Human Diversity.* J. B. Lippincott, Philadelphia.

Iltis, H. 1932. *Life of Mendel,* trans. by E. and C. Paul. Norton, New York.

Jacob, F. 1977. Evolution and tinkering. *Science* 196:1161–1166.

McKusick, V. A. 1978. *Mendelian Inheritance in Man,* 5th ed. Johns Hopkins Univ. Press, Baltimore.

Snedecor, G. W. and W. G. Cochran. 1967. *Statistical Methods,* 6th ed. Iowa State Univ. Press, Ames, Iowa.

Sokal, R. R. and F. J. Rohlf. 1969. *Biometry.* W. H. Freeman, San Francisco.

Stent, G. S. and R. Calendar. 1978. *Molecular Genetics, An Introductory Narrative,* 2nd ed. W. H. Freeman, San Francisco.

Strickberger, M. W. 1976. *Genetics,* 2nd ed. Macmillan, New York.

Sturtevant, A. H. 1965. *A History of Genetics.* Harper & Row, New York.

Sutton, H. E. 1980. *An Introduction to Human Genetics,* 3rd ed. Saunders, Philadelphia.

Watson, J. D. 1976. *Molecular Biology of the Gene,* 3rd ed. W. A. Benjamin, Menlo Park, California.

Whitehouse, H. L. K. 1969. *Towards an Understanding of the Mechanism of Heredity.* St. Martin's Press, New York.

Wilson, E. O. and W. H. Bossert. 1971. *A Primer of Population Biology.* Sinauer, Sunderland, Massachusetts.

Wright, S. 1968. *Evolution and the Genetics of Populations,* Vol. 1: *Genetic and Biometric Foundations.* Univ. of Chicago Press, Chicago.

· 2 ·

Mendelian Populations

The field of population genetics has set for itself the tasks of determining how much genetic variation exists in natural populations and of explaining this variation in terms of its origin, maintenance, and evolutionary importance. Genetic variation, in the form of multiple alleles for many individual loci, exists in most natural populations; this conclusion can be documented by any number of methods, the most important of which are discussed in this chapter. On the other hand, all methods are limited to the study of a certain number of loci, usually a small number of loci in comparison to the total number in the organism. Extrapolation of the results for a small number of loci to the entire genome is therefore questionable, and because of the uncertainty as to whether the loci studied are truly representative of the genome, we still do not know, quantitatively, how much genetic variation is present in natural populations.

Whatever total amount of genetic variation exists, sufficient variation has already been found at the loci amenable to study that, for most populations, no two individuals (barring identical twins) could be expected to have the same genotype at all loci. Thus, it becomes important to describe how genetic variation in natural populations is organized into genotypes — to determine, for example, whether alleles at different loci are associated at random. As will be seen later in this chapter, the combination of particular genes into genotypes depends on the types and frequencies of mating that can occur, and on the amount of recombination between loci.

Genetic Diversity

ELECTROPHORESIS

The most useful procedure yet devised for revealing genetic variation is ELECTROPHORESIS, a technique that came into widespread use in the late 1960s (Lewontin and Hubby, 1966; Harris, 1966, 1969; Hubby and Lewontin, 1966). One laboratory setup for electrophoresis is illustrated schematically in Figure 1. To carry out a typical electrophoresis experiment, a small sample of tissue is ground up or a small amount of blood serum is drawn from each of a number of individuals in the population to be surveyed. Each sample is placed in a small slot near the edge of a rectangular, fairly thin slab of a jelly-like material, usually starch, polyacrylamide, or agarose. Electrodes are attached to sponge wicks soaked in buffer solution at each end of the gel, and a current is applied across the gel for several hours. Protein molecules in the sample move through the gel in response to the electric field. After electrophoresis, the gel can be stained for a particular enzyme by soaking it in a solution containing a substrate for the enzyme along with a dye that precipitates where the enzyme-catalyzed reaction occurs. A dark band thus appears in the gel, revealing the position of the enzyme (see Harris and Hopkinson, 1976, for detailed procedures).

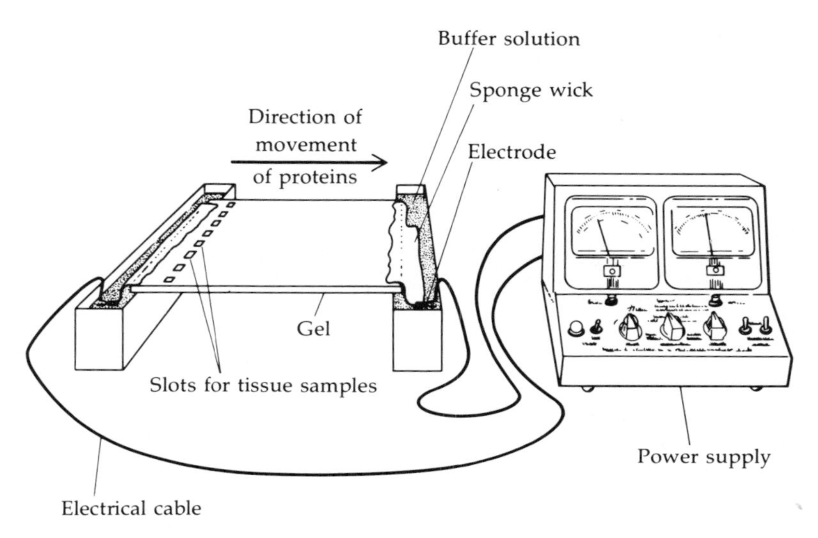

FIGURE 1. One type of laboratory apparatus for slab gel electrophoresis.

If an enzyme in an individual has an amino acid substitution that leads to a difference in the enzyme's overall ionic charge, then this enzyme will have a somewhat altered ELECTROPHORETIC MOBILITY — that is to say, it will move at a somewhat different rate. This change in mobility occurs because enzymes of the same size and shape move at a rate determined largely by the ratio of the number of positively charged amino acids (primarily lysine, arginine, and histidine) to the number of negatively charged ones (principally aspartic acid and glutamic acid). Electrophoresis can therefore be used to detect mutations that result in differences in electrophoretic mobility of the corresponding enzymes.

One possible result of an electrophoresis experiment is shown in the gel photograph in Figure 2; all individuals in this sample have an enzyme with the same electrophoretic mobility. Another kind of result is shown in Figure 3; here some individuals are homozygous for an allele, *F*, associated with a rapidly migrating enzyme (*fast*), some are homozygous for an allele, *S*, associated with an enzyme that migrates more slowly (*slow*), and some are heterozygous, *F*/*S*, for the alleles. Two enzyme bands appear in heterozygotes because the enzyme itself is MONOMERIC — it consists of a single polypeptide chain — and heterozygous individuals therefore produce two kinds of polypeptide chains, one corresponding to each allele. When the enzyme in question is

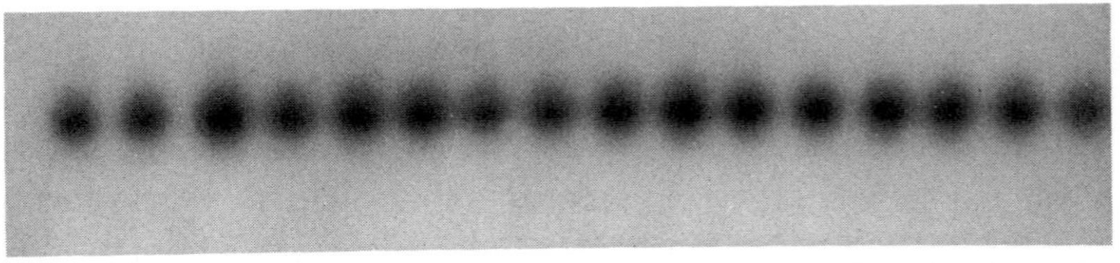

FIGURE 2. Results of electrophoresis of the enzyme phosphoglucomutase-1 from 16 cultured cell lines originating from individuals of the mouse, *Mus musculus*. The gene that codes for the enzyme is *Pgm-1*, and the position of the enzyme in the gel is indicated by the dark bands. In this population, all individuals have an enzyme with the same electrophoretic mobility. (Courtesy of S. E. Lewis and F. M. Johnson.)

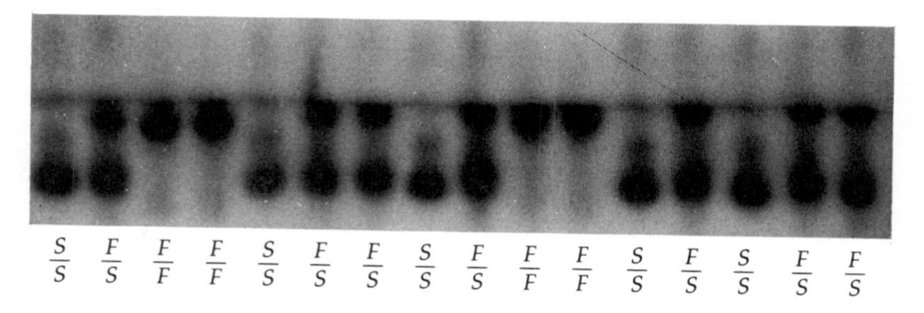

FIGURE 3. Results of electrophoresis of the enzyme glucose phosphate isomerase-1 from 16 cultured cell lines originating from individuals of the mouse, *Mus musculus*. The gene that codes for the enzyme is *Gpi-1*. In this population, some individuals are homozygous for an allele (*S*) corresponding to a slow-migrating enzyme, some are homozygous for an allele (*F*) corresponding to a fast-migrating enzyme, and the rest are heterozygous *F*/*S*. The genotypes of the cell lines are indicated beneath the enzyme bands. This enzyme is a monomer, so the heterozygotes exhibit two enzyme bands of differing mobility. (Courtesy of S. E. Lewis and F. M. Johnson.)

MULTIMERIC, say a dimer (consisting of two polypeptide chains), then the heterozygote will produce three types of dimers: rapidly migrating (*fast* + *fast*) and slowly migrating (*slow* + *slow*) HOMODIMERS, like those produced in the corresponding homozygotes, and a *fast* + *slow* HETERODIMER, which typically is found to have an intermediate electrophoretic mobility. Results obtained by electrophoresis of a dimeric enzyme are shown diagrammatically in Figure 4, where the homodimers are represented by light gray bands, the heterodimer by a dark gray band.

ALLELE FREQUENCY AND POLYMORPHISM

Enzymes differing in electrophoretic mobility as a result of allelic differences at a single locus are called ALLOZYMES. Thus, allozyme variation in a population is an indication of genetic variation, and it turns out that such genetic variation is very common. Extensive allozyme variation has been found in virtually all natural populations studied by electrophoresis, including organisms such as the bacterium *Esche-*

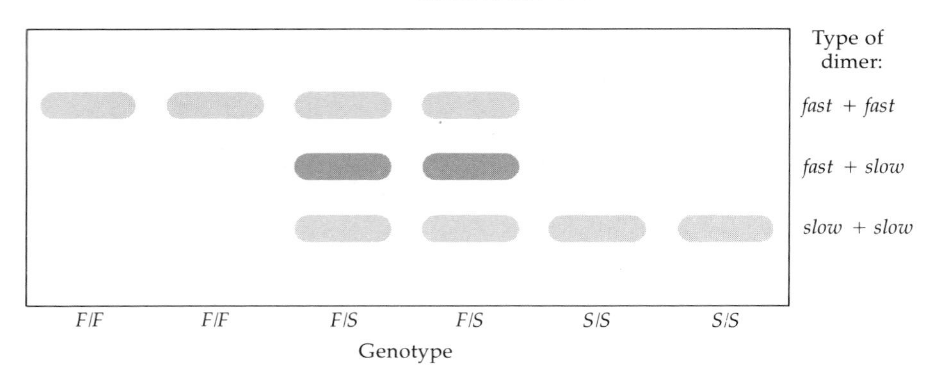

FIGURE 4. Typical gel pattern observed for dimeric enzymes.

richia coli (Milkman, 1973), plants (see Clegg and Allard, 1972; Hamrick and Allard, 1972; Levin, 1978, for representative examples), *Drosophila* (see the classic paper by Lewontin and Hubby, 1966, and those by Richmond, 1972; Prakash, 1977b; Marinković et al., 1978, for examples), the mouse (e.g., Selander et al., 1969), and humans (the classic study is by Harris, 1966, but see also Ruddle et al., 1969; Harris and Hopkinson, 1972).

In order to compare different loci and different populations, it is necessary to have some convenient quantitative measure of genetic variation. Genetic variation can be quantified using the concept of allele frequency. The ALLELE FREQUENCY of a prescribed allele among a group of individuals is simply the proportion of all alleles at the locus that are of the prescribed type. The frequency of any prescribed allele in a sample is therefore equal to twice the number of homozygotes for that allele (because each homozygote carries two copies of the allele) plus the number of heterozygotes for that allele (because each heterozygote carries one copy) divided by two times the number of individuals in the sample (because each individual carries two alleles at the locus). Among the small sample of 16 individuals in the gel shown in Figure 3, for example, there are 4 *F*/*F* homozygotes, 7 *F*/*S* heterozygotes, and 5 *S*/*S* homozygotes: the allele frequency of *F* is therefore $(2 \times 4 + 7)/(2 \times 16) = .469$, whereas the allele frequency of S is $(2 \times 5 + 7)/(2 \times 16) = .531$. (Detailed calculations are shown in Table I.) Note that the sum of all allele frequencies, in this case those for F and S, must be 1. The allele frequency in a sample of individuals from a population is only an estimate of the true allele frequency in the whole population, of course, but the estimate will usually be close to the true frequency if the sample is sufficiently large; for this reason, allele frequency estimates should be based on samples of 100 or more individuals. (Calculation of the precision of allele-frequency estimates involves the concept of standard error, which is discussed in Box A of Chapter 1.) One further comment about allele frequency: an allele frequency is often imprecisely called a GENE FREQUENCY if the allele in question is clear from context. (See Box A of this chapter for allele-frequency calculations involving actual data.)

The idea of allele frequency gives rise to the notion of polymorphism. A POLYMORPHIC LOCUS is a locus at which the most common allele has a frequency of less than .99. Conversely, a MONOMORPHIC LOCUS is one that is not polymorphic. The cutoff at .99 in the definition of polymorphism is arbitrary, but it serves to focus attention on those loci with common allelic variation. If you look long enough and hard enough in any large population, you are bound to find rare alleles at virtually every locus. RARE ALLELES are alleles with frequencies of less than .005, and between one and two

TABLE I. Calculation of allele frequencies.

	GENOTYPE			
	F/F	F/S	S/S	Total
No. of individuals	4	7	5	16
No. of F alleles	8	7	0	15
No. of S alleles	0	7	10	17
No. of $F + S$ alleles	8	14	10	32

Allele frequency of F = 15/32 = .469
Allele frequency of S = 17/32 = .531

A Allele Frequency Calculations

Levy and Levin (1975) used electrophoresis to study the phosphoglucose isomerase-2 locus in the evening primrose *Oenothera biennis,* a complex permanent translocation heterozygote of the type discussed in Figure 14 of Chapter 1. They observed two alleles affecting electrophoretic mobility of the enzyme, *PGI-2a* and *PGI-2b*. In 57 strains, they observed 35 *PGI-2a/PGI-2a,* 19 *PGI-2a/PGI-2b,* and 3 *PGI-2b/PGI-2b.*

a. Calculate the allele frequencies of *PGI-2a* and *PGI-2b.*

Mukai, Watanabe, and Yamaguchi (1974) studied 1158 second chromosomes from a population of *Drosophila melanogaster* in Raleigh, North Carolina. Two alleles affecting electrophoretic mobility were found at the α-glycerol-3-phosphate dehydrogenase-1 locus, denoted $\alpha Gpdh\text{-}1^F$ (for the allele determining the fast migrating enzyme) and $\alpha Gpdh\text{-}1^S$ (for the slow allele). Two electrophoretic alleles were also found at the α-amylase locus, denoted Amy^F and Amy^S. The chromosomes were also classified according to whether they carried the polymorphic paracentric inversion called the *NS* (Nova Scotia) inversion on the right arm of the chromosome. Results were as follows:

$\alpha Gpdh\text{-}1^F$	Amy^F	non-*NS*	726
$\alpha Gpdh\text{-}1^F$	Amy^F	*NS*	90
$\alpha Gpdh\text{-}1^F$	Amy^S	non-*NS*	111
$\alpha Gpdh\text{-}1^F$	Amy^S	*NS*	1
$\alpha Gpdh\text{-}1^S$	Amy^F	non-*NS*	172
$\alpha Gpdh\text{-}1^S$	Amy^F	*NS*	32
$\alpha Gpdh\text{-}1^S$	Amy^S	non-*NS*	26
$\alpha Gpdh\text{-}1^S$	Amy^S	*NS*	0

From these data calculate:

b. the frequency of the $\alpha Gpdh\text{-}1^F$ and $\alpha Gpdh\text{-}1^S$ alleles,

c. the frequency of the Amy^F and Amy^S alleles,

d. the frequency of *NS*-bearing and non-*NS*-bearing chromosomes.

individuals per thousand are heterozygous for rare alleles at any locus, as judged by an electrophoretic survey of 43 enzyme loci in 250,000 Europeans (Harris and Hopkinson,

1972). Many rare alleles are deleterious and are presumably maintained in populations by recurrent mutation, as will be discussed in Chapter 3. The definition of polymorphism is an attempt to focus on loci having alleles with frequencies too high to be explained solely by recurrent mutation. With the definition of polymorphism given above, at least two percent of the population will be heterozygous for the most common allele, for reasons that will become clear in a few pages.

ALLOZYME POLYMORPHISMS

Polymorphism for alleles that determine allozymes is extremely widespread. This is shown in Figure 5, which summarizes the results of electrophoretic surveys of up to 40 loci in local populations of 133 species. The data in Figure 5 are from Selander (1976), but see Nevo (1978) for another review and Brown (1979) for one focusing on plants. Each graph in Figure 5 gives the number of loci examined and the number of species, and the open bar (*P*) is the average percentage of loci found to be polymorphic in any one local population. The gray bar (*H*) is the percentage of loci that are heterozygous in an average individual, which is calculated by totaling the number of heterozygotes for each locus, dividing this by the total number of individuals in the sample, and averaging over all loci. The heterozygosity for the one locus in the small sample of Figure 3 is $7/16 = .438$. To further illustrate calculation of P and H, we may use data of Harris (1966), who studied ten enzyme loci in the English population. Of these ten loci, three were found to be polymorphic, so in this sample $P = 3/10 = .33$. The proportion of heterozygous individuals for each of the seven monomorphic loci was, of course, 0, and the proportion of heterozygous individuals for the three polymorphic loci was .509 (for red cell acid phosphatase), .385 (for phosphoglucomutase), and .095 (for adenylate kinase); thus, in this sample, $H = (.509 + .385 + .095 + 7 \times 0)/10 = .099$.

Figure 5 can be roughly summarized as follows: for continental species having a wide geographic range and large population size, about 25 to 50 percent of loci have allozyme

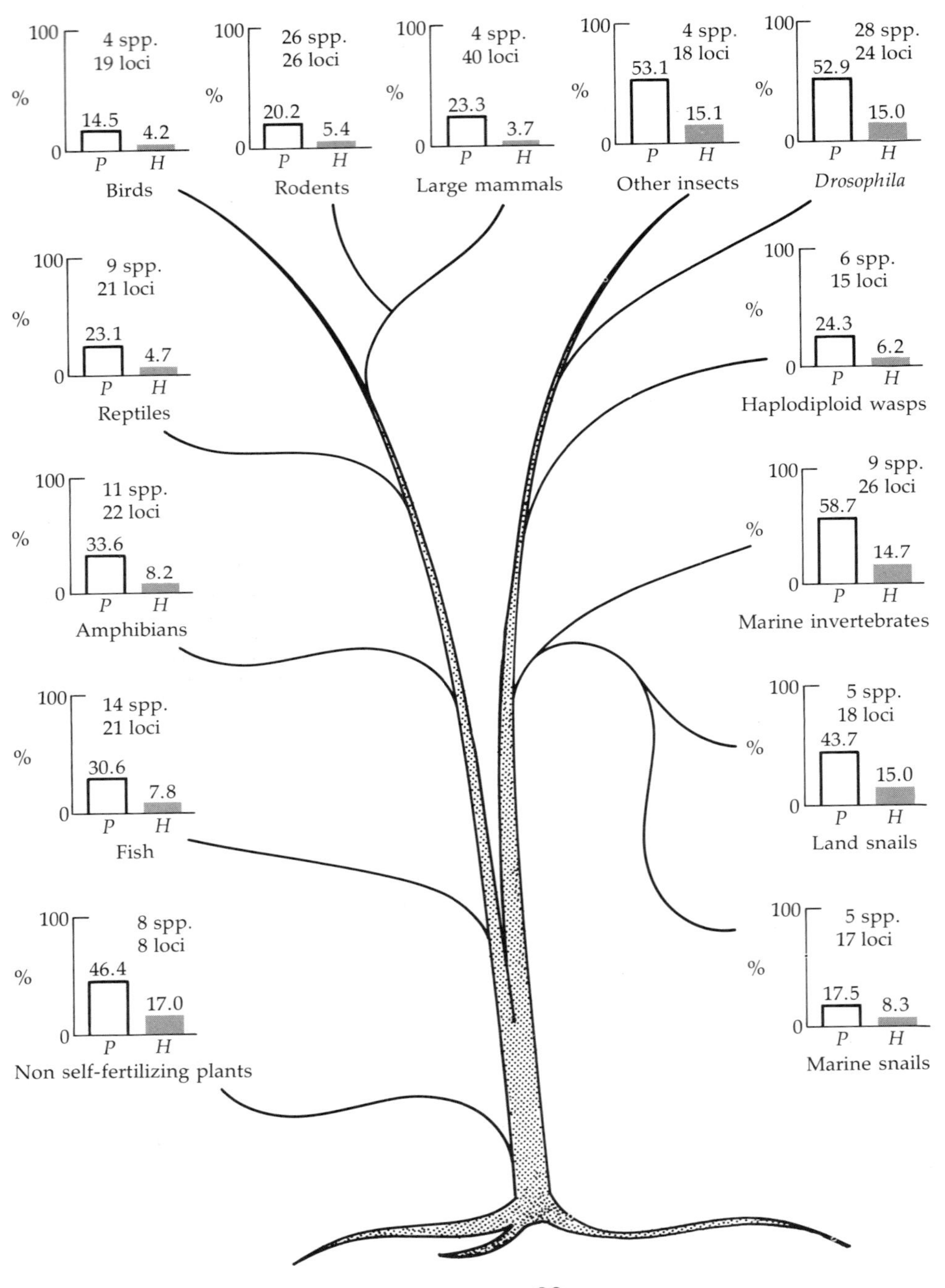
100
4 spp.
19 loci
%
14.5
4.2
0
P
H
Birds
100
26 spp.
26 loci
%
20.2
5.4
0
P
H
Rodents
100
4 spp.
40 loci
%
23.3
3.7
0
P
H
Large mammals
100
4 spp.
18 loci
%
53.1
15.1
0
P
H
Other insects
100
28 spp.
24 loci
%
52.9
15.0
0
P
H
Drosophila
100
9 spp.
21 loci
%
23.1
4.7
0
P
H
Reptiles
100
6 spp.
15 loci
%
24.3
6.2
0
P
H
Haplodiploid wasps
100
11 spp.
22 loci
%
33.6
8.2
0
P
H
Amphibians
100
9 spp.
26 loci
%
58.7
14.7
0
P
H
Marine invertebrates
100
14 spp.
21 loci
%
30.6
7.8
0
P
H
Fish
100
5 spp.
18 loci
%
43.7
15.0
0
P
H
Land snails
100
8 spp.
8 loci
%
46.4
17.0
0
P
H
Non self-fertilizing plants
100
5 spp.
17 loci
%
17.5
8.3
0
P
H
Marine snails

FIGURE 5. Proportion of polymorphic loci (P) and average heterozygosity (H), from Selander's (1976) summary of a large number of electrophoretic surveys. The lengths of the branches on the evolutionary "tree" are arbitrary.

polymorphisms (P = .25 to .50), and an average individual is heterozygous at 5 to 15 percent of its loci (H = .05 to .15). There is, however, significant variation from group to group. (Compare, for example, land snails and marine snails.) In general, although there are many individual exceptions, plants that regularly avoid self-fertilization and invertebrates have amounts of polymorphism and heterozygosity near the high end of the range ($P \approx .50, H \approx .15$), whereas vertebrates tend to be near the low end of the range ($P \approx .25, H \approx .05$). Humans are fairly typical of large mammals; an extensive survey of 87 human loci gave estimates of P = .38 and H = .07 (Harris et al., 1977). Allozyme polymorphisms are obviously widespread among both plants and animals. Indeed, even natural, gut-dwelling populations of the haploid bacterium *Escherichia coli* seem to have levels of polymorphism comparable to those in Figure 5 (Milkman, 1973). (Box B is an example of "raw data" collected in an allozyme survey of a hypothetical population.)

HOW REPRESENTATIVE ARE ALLOZYMES?

The validity of estimates of polymorphism based on electrophoresis is open to question. The amount of polymorphism may be underestimated because, at best, routine electrophoresis detects only those amino acid substitutions that result in charge differences in the protein, and the procedure may miss even some of these. The resolving power of electrophoresis can be enhanced by using softer, less concentrated gels or varying pH conditions (Ramshaw et al., 1980), by using other electrophoretic techniques (Johnson, 1977; Finnerty and Johnson, 1979; Leigh Brown and Langley, 1979), or by combining electrophoresis with other procedures. Some amino acid substitutions that do not alter electrophoretic mobility do render the enzyme sensitive to high tem-

B Electrophoretic Data: A Simple Example

The gel patterns shown are the results of electrophoresis of a perfectly representative sample of 50 individuals from a hypothetical local population. Tissue samples from each of the individuals were electrophoresed in five gels, and these were stained for enzymes corresponding to loci *A* through *E*. Breeding tests show that differences in enzyme mobility are in each case due to alleles at a single locus. The population is diploid and undergoing random mating with respect to these loci. All five enzymes are either monomers or dimers.

a. Which enzymes are monomeric, which dimeric, and which uncertain from the gels?

b. How many electrophoretically distinct alleles are there for each of the loci *A* through *E*?

c. What are the allele frequencies for each locus?

d. Which of the loci are polymorphic in this sample? (For present purposes, consider a locus as monomorphic if the frequency of the most common allele is equal to or greater than .99.) What is the proportion of polymorphic loci?

e. What is the average heterozygosity for each locus? What is the average heterozygosity for all five loci?

---

perature, for example (Bernstein et al., 1973; Trippa et al., 1976). Incubating the sample at high temperature therefore destroys enzyme activity. The diagrammatic example in Figure 6 shows gels stained for enzyme activity with or without heat treatment. One enzyme is seen to be temperature sensitive, indicating that it contains one or more amino acid substitutions. Use of techniques of this sort has increased the number of identified alleles at the *xanthine dehydrogenase* locus in *Drosophila pseudoobscura* from 6 to 37 and increased estimates of average heterozygosity from .44 to .73 (Singh et al., 1976). Bonhomme and Selander (1978) were able to identify about twice as many alleles in the house mouse as had been previously identified using routine electrophoresis. On the other hand, while the more precise techniques reveal additional alleles at loci known to be polymorphic,

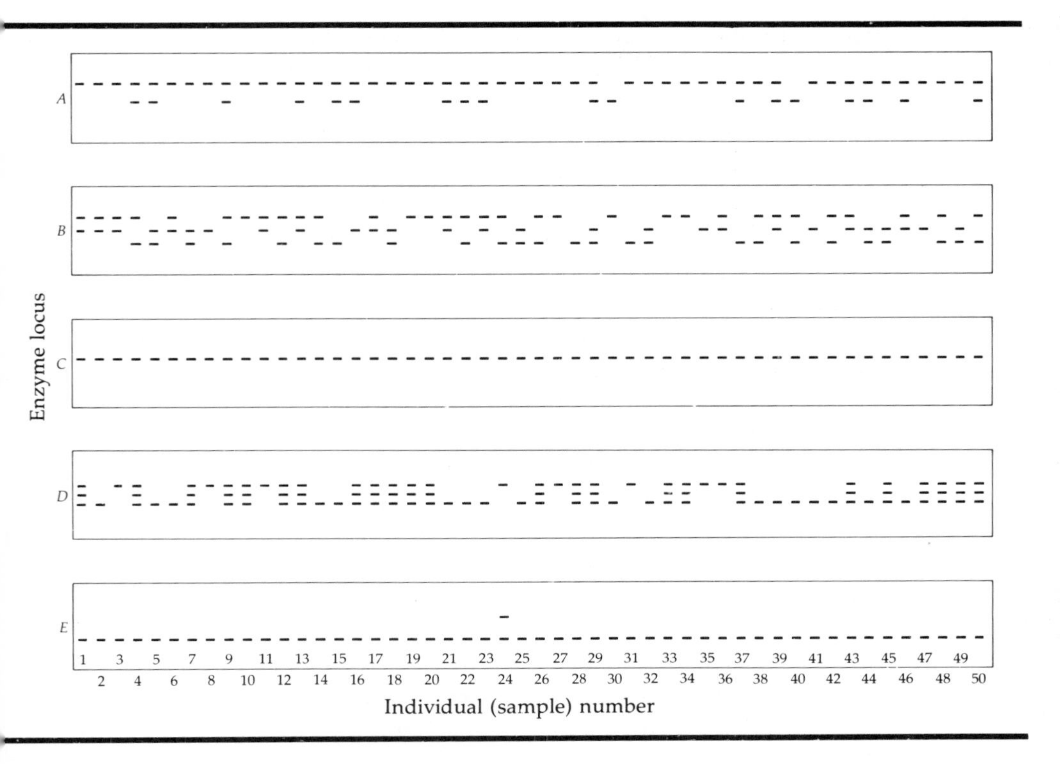

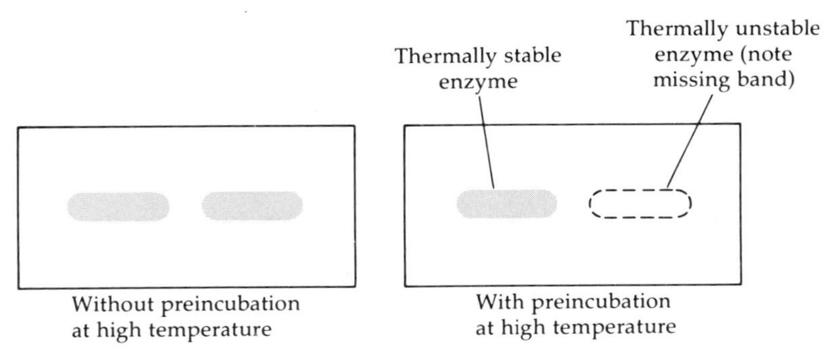

FIGURE 6. Expected gel patterns resulting from heat treatment of a thermally unstable (thermolabile) enzyme. The gel on the right has been incubated at high temperature before staining and the thermolabile enzyme has been rendered inactive.

thus increasing estimates of H, loci classified as monomorphic by means of routine electrophoresis tend to remain monomorphic even with more critical techniques (Coyne et al., 1978), so estimates of P remain much the same as before.

While routine electrophoretic surveys may, for the reasons given above, miss a significant amount of genetic variation, they could also overestimate the amount of polymorphism because the enzymes typically surveyed are those found in relatively high concentration in tissues or body fluids; such enzymes are often called GROUP II ENZYMES to distinguish them from more substrate-specific GROUP I ENZYMES involved in processes such as energy transformation (Gillespie and Kojima, 1968). Group II enzymes are hardly a random sample of loci, because the sample excludes not only group I enzymes but also regulatory loci and such loci as code for the several dozen ribosomal proteins and transfer RNAs. For example, Leigh Brown and Langley (1979) have used a powerful two-dimensional electrophoretic technique to study the 54 proteins found in highest concentration in *Drosophila*; these 54 loci included only six that were polymorphic ($P = 6/54 = .11$), and the average heterozygosity was $H = .04$ — values much smaller than those usually quoted for *Drosophila*. By contrast, among 11 group I loci studied by routine electrophoresis, the corresponding values were $P = .27$ and $H = .04$, but among ten group II enzymes, the values were $P = .70$ and $H = .24$ (Gillespie and Langley, 1974).

In short, while natural populations do contain substantial genetic variation — detectable as allozymes by electrophoresis — we still do not know what level of polymorphism and heterozygosity is characteristic of entire genomes. For certain loci the absolute amount of genetic variation may soon be amenable to study by the most powerful method of all — direct determination of the nucleotide sequences in the DNA.

VISIBLE GENETIC VARIATION IN NATURAL POPULATIONS

That genetic variation is widespread in natural populations is supported by the results of other kinds of studies, particularly by the generally deleterious effects of close inbreeding (discussed later in this chapter) and by the success of artificial selection in altering almost any morphological or physiological characteristic (discussed in Chapter 4). Special procedures are usually required to detect genetic variation because most of it is HIDDEN VARIATION not apparent at the phenotypic level. Among humans, for example, the obvious differences between normal people in hair color, eye color, skin color, stature, and other such traits are not usually traceable to individual loci; most of these traits are POLYGENIC (or MULTIFACTORIAL), influenced by several or many loci acting in concert, and usually also influenced by environment. Even abnormal phenotypes are difficult and sometimes impossible to trace to single loci. In humans, about five percent of liveborn babies have some serious physical or mental disability, yet many of these conditions have no known hereditary basis (lack of oxygen at the time of birth can be an environmental cause of mental retardation, for instance), or their mode of inheritance is polygenic and therefore complex. Similarly, about five percent of the individuals in natural populations of *Drosophila melanogaster* have phenotypic abnormalities (reviews in Spencer, 1947; Lewontin, 1974a). Many of the abnormalities resemble those found in special laboratory strains carrying particular mutant genes (for examples, see Figure 7). Yet breeding tests indicate that no more than a third of phenotypically abnormal flies owe their abnormality to individual mutant genes; the rest owe their phenotypic abnormality to more complex genetic causes or to environmental accidents. Because most genetic variation is hidden, it becomes important to determine how genetic variation is organized into genotypes. This is the subject of the next section.

Organization of Genetic Diversity

RANDOM MATING

Predicting genotype frequencies at a locus from knowledge of allele frequencies is quite straightforward, but there are a few minor wrinkles to be ironed out. Genotype frequencies are determined in part by mating patterns. For example,

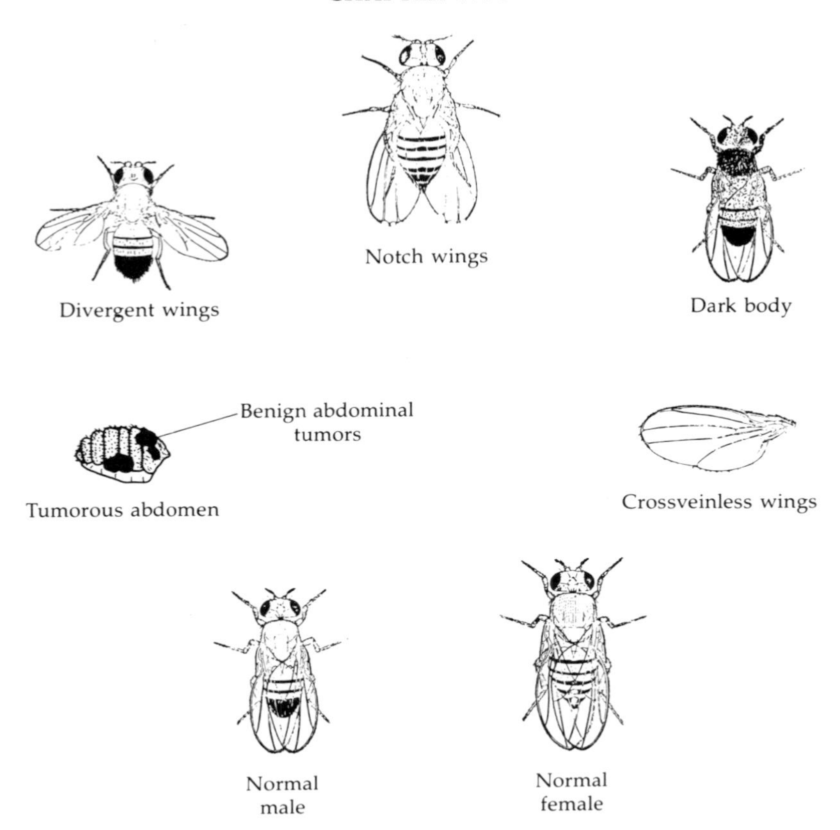

FIGURE 7. Examples of some phenotypic abnormalities found in natural *Drosophila melanogaster* populations; sketches of normal flies are presented for comparison. A relatively small proportion of such abnormal flies owe their abnormality to single mutant genes.

plant species that regularly undergo self-fertilization are expected to have very low levels of heterozygosity, for reasons that will become clear when self-fertilization is discussed later in this chapter. One of the simplest and most important mating patterns is RANDOM MATING, in which mating takes place at random with respect to the locus under consideration. With random mating, the chance that an individual mates with another having a prescribed genotype is equal to the frequency of that genotype in the population. For example, suppose that in some population a locus has genotypes *AA*, *Aa*, and *aa* in the proportions .16, .48, and .36, respectively; if mating is random, *AA* males will mate with *AA*, *Aa*, and *aa* females in the proportions .16, .48, and .36, respectively, and these same proportions will apply to the mates of *Aa* and *aa* males. In this section, we focus on random mating; effects of departures from random mating are taken up in later sections of this chapter.

It is important to keep in mind that mating can be random with respect to some traits, but at the same time, nonrandom with respect to others. In humans, for example, mating seems to be random with respect to blood groups, allozyme phenotypes, and many other such characteristics, but mating is nonrandom with respect to some other traits such as skin color and height. Genotype frequencies are also influenced by various evolutionary forces: among them mutation, migration, and natural selection. For the moment, these evolutionary forces will be assumed to be absent or at least negligibly small in magnitude; their effects will be discussed in Chapter 3. Additionally, genotype frequencies are affected by chance statistical fluctuations that occur in all small populations, also a subject of Chapter 3. For now, imagine that each local population is sufficiently large that small-population effects can be neglected. Although small-population effects do occur unless the population is infinite in size, the magnitude of the effects is sufficiently small that they can usually be neglected if population size is 500 or more.

AGE STRUCTURE OF POPULATIONS

There is one other complication in predicting genotype frequencies: virtually all populations have some sort of age structure, which refers to the age of individuals in the population. The AGE STRUCTURE of a population is a set of numbers that shows, for each age class in the population, what proportion (or number) of individuals are in that age class. Age in humans is commonly measured in years, but for convenience the one-year age groups are often pooled into

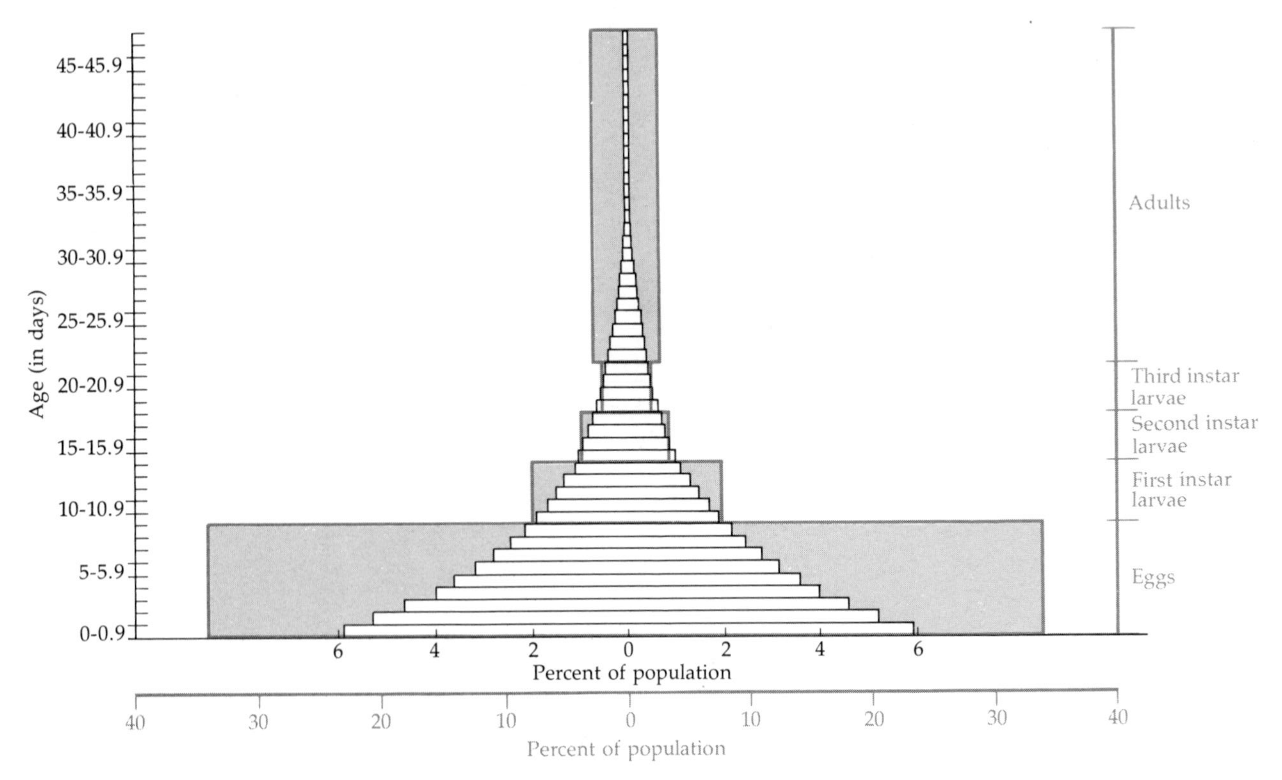
Age (in days)
45-45.9
40-40.9
35-35.9
30-30.9
25-25.9
20-20.9
15-15.9
10-10.9
5-5.9
0-0.9
6 4 2 0 2 4 6
Percent of population
40 30 20 10 0 10 20 30 40
Percent of population
Life-History stage
Adults
Third instar larvae
Second instar larvae
First instar larvae
Eggs

FIGURE 8. Theoretical stable age structure for the human louse, with age reckoned in days (left ordinate) or life-history stage (right ordinate). (Data from Evans and Smith, 1952.)

five-year classes, individuals of age 0–4 years, 5–9, 10–14, and so on. In some organisms, it is inconvenient to record age in terms of absolute time, so various life-history stages are used instead; for example, age classes in the human louse (Figure 8) may be taken to correspond to five life-history stages: fertilized eggs; first, second, and third instar larvae; and adults. The age structure of a population may change over time, and it is typically somewhat different for males and females, though here the sexes will be treated together. Age structures are frequently represented pictorially as a series of bars or rectangles stacked one on top of another, each bar denoting one age class and having its

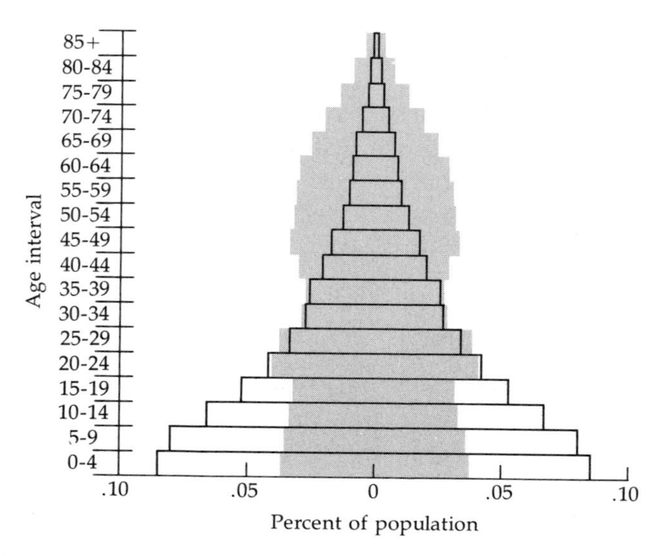

FIGURE 9. Age structures of Mexico (open bars) and Sweden (gray bars) in 1970. Sweden has a relatively stable population, as seen from its almost rectangular age structure, whereas Mexico's population is growing rather rapidly, as seen in its nearly triangular age structure. (Data from United Nations Demographic Yearbook, 1973.)

width proportional to the percentage (or number) of individuals in that age class. Examples of this kind of representation are shown in Figures 8 and 9. Note that the percentages are graphed symmetrically around zero, so, for example, the percentage of lice in the 0 to 0.9-day age class in Figure 8 is 5.9 + 5.9 = 11.8.

The age structure of a population at any time is determined by several factors: the age structure at some previous time, which must be short relative to the life span of the organism (in humans, this would correspond to the age structure in the previous five-year period); the AGE-SPECIFIC DEATH RATE for each age class, which is the probability that an individual who is alive at the beginning of a specified age class will die before reaching the next age class; the AGE-SPECIFIC BIRTH RATE for each age class, which is the average number of offspring born to an individual in a specified age class before dying or reaching the next age class; and the AGE-SPECIFIC EMIGRATION and AGE-SPECIFIC IMMIGRATION rates for each age class, which are the numbers of individuals in a specified age class who leave the population or enter it, respectively, due to migration. Charting the course of age-structure changes in a population is therefore quite difficult, especially because death rates, birth rates, and immigration rates can change very rapidly (examples from recent U.S. history include a huge influx of immigrants around 1900 and a dramatic increase in the birth rate following World War II).

STABLE AGE STRUCTURE

Despite the complexities in projecting a population's age structure, a few general comments can be made. If age-specific birth and death rates remain constant in a population with no emigration or immigration, then the population may eventually reach a STABLE AGE STRUCTURE — that is, an age structure that does not change through time. Population size may be increasing or decreasing or staying the same in a population with a stable age structure; stability of age structure means only that the relative proportions of individuals in the various age classes remain the same through time. (See Box C for stable age-structure calculations.) The stable age structure eventually reached with constant birth and death rates does not depend on the age structure in the initial population; all initial populations, whatever their makeup, will eventually converge to the same stable age structure, provided they have the same birth and death rates. The louse age structure in Figure 8, calculated from measured age-specific birth and death rates, is the stable age structure that the human louse population would eventually reach given constant environmental conditions as favorable as those in the laboratory in which the birth and death rates were measured; under these favorable conditions the louse population would have a DOUBLING TIME (the time required to double in size) of 6.24 days! In human populations, generally speaking, a triangular or flared-base age structure indicates rapid population growth; a more rectangular age structure is usually indicative of slow or no population growth. In Figure 9 the age structure of Mexico in 1970 is shown (black outline), a population with a doubling time of 19.8 years; for comparison, the 1970 age structure of Sweden, a more stable population with a doubling time of 173 years, is shown in gray.

Method for Calculating Stable Age Structures C

When age-specific birth and death rates in a population are constant and no migration occurs, the population will often attain a stable (i.e., unchanging) age structure. There is an elegant theory (Lotka, 1922) for predicting what this ultimate stable age structure will be. Let the number i ($i = 0, 1, \ldots, m$) refer to the age class $(i, i + 1)$, that is, i refers to individuals whose age is between i and $i + 1$. Since the limit of i is m, the maximum age that any individual can attain is $m + 1$. We are interested in the ultimate stable age structure, which is the relative number of individuals in each age class at a point when the age structure is no longer changing. To this end let s_i ($i = 0, 1, \ldots, m - 1$) be the probability that an individual in age class i survives to age class $i + 1$; setting $s_m = 0$ insures that the maximum age will be $m + 1$. Furthermore,

C let b_i ($i = 0, 1, \ldots, m$) be the average number of offspring born to an individual in age class i.

What follows now may seem like hocus-pocus because its theoretical foundation is beyond the scope of this book. (An excellent elementary treatment of the theory can be found in Crow and Kimura, 1970.) Nevertheless, to calculate the stable age structure, we first form the polynomial

$$\lambda^{m+1} - b_0\lambda^m - s_0b_1\lambda^{m-1} - s_0s_1b_2\lambda^{m-2} - s_0s_1s_2b_3\lambda^{m-3} - \ldots - s_0s_1s_2 \ldots s_{m-2}b_{m-1}\lambda - s_0s_1s_2 \ldots s_{m-1}b_m = 0$$

This equation is known as the CHARACTERISTIC EQUATION, and it can be shown to have a single positive real root, called the DOMINANT EIGENVALUE or the CHARACTERISTIC ROOT. (In practice, the characteristic root is usually found using a computer.)

When the characteristic root is found (call it λ), the stable age structure can be calculated. In the stable age structure, the population has the age classes $0, 1, 2, \ldots, m$ in the ratio $1{:}s_0\lambda^{-1}{:}s_0s_1\lambda^{-2}{:}s_0s_1s_2\lambda^{-3}{:} \ldots {:}s_0s_1s_2 \ldots s_{m-1}\lambda^{-m}$. Furthermore, λ is the rate of growth of the population after it has reached a stable age structure; specifically, if N_t represents the total size of a population with a stable age structure, then the size of the population after one additional time unit will be $N_{t+1} = N_t\lambda$.

a. Assume a population with five age classes, 0, 1, 2, 3, and 4. Let $s_0 = s_1 = s_2 = s_3 = .5$ and $s_4 = 0$; let $b_0 = b_4 = 0$, $b_1 = b_3 = 1$, and $b_2 = 1.5$. What will be the population's stable age structure? (Hint: try $\lambda = 1$ as a solution of the characteristic equation.)

b. From the equation $N_{t+1} = N_t\lambda$, show that $N_t = N_0\lambda^t$. Denote the population's doubling time (i.e. the time it takes to double in size) as t^*, and find an expression for λ in terms of t^*. (Hint: set $N_t = 2N_0$ and $t = t^*$.) Then calculate the λ for Sweden ($t^* = 173$ years) and for Mexico ($t^* = 19.8$ years), and note $\lambda - 1$ is the rate of increase in population size in a single year.

POPULATION MODELS

Population models that separately keep track of each of a large number of age classes in the population tend to be rather complex, as might well be imagined (see, for example, Charlesworth, 1970). In much of population genetics, the problem of age structure is sidestepped by focusing on pop-

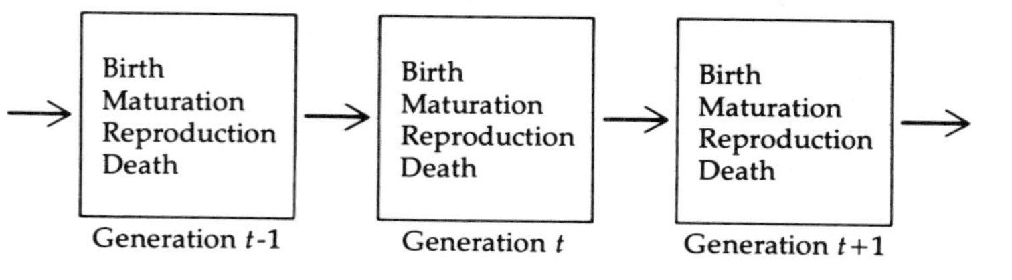

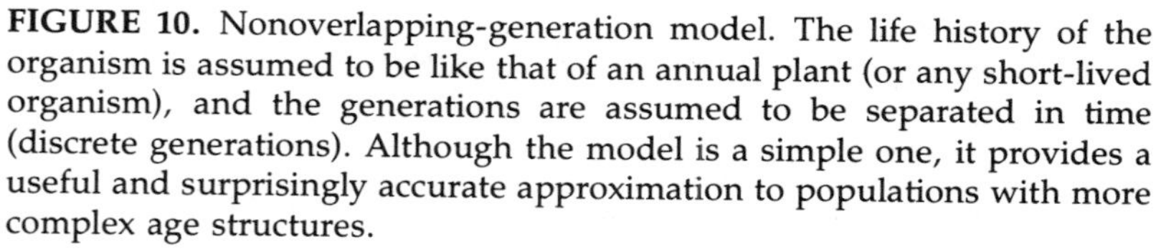

FIGURE 10. Nonoverlapping-generation model. The life history of the organism is assumed to be like that of an annual plant (or any short-lived organism), and the generations are assumed to be separated in time (discrete generations). Although the model is a simple one, it provides a useful and surprisingly accurate approximation to populations with more complex age structures.

ulations with a very special sort of life history. Imagine a population with a life history like that in annual plants with short growing seasons; all members of any generation germinate at about the same time, mature together, shed their pollen and are fertilized almost simultaneously, and die immediately after producing the new generation. Such a population has an exceedingly simple age structure: all individuals are of nearly the same age (Figure 10). This sort of hypothetical population, with its simple age structure, provides a model used in population genetics as a first approximation to populations with more complex age structures; the model is called a NONOVERLAPPING GENERATION MODEL. The nonoverlapping generation model turns out to be a surprisingly good approximation if the population in question is at or sufficiently near its stable age structure, or if the age structure is changing slowly. As will be seen shortly, calculations of expected genotype frequencies based on the nonoverlapping generation model are adequate for most purposes, and they usually provide satisfactory approximations even for the human population.

The Hardy–Weinberg Law

The assumptions we've made so far in developing the model for predicting genotype frequencies may be summarized as:

1. The organism in question is diploid.
2. Reproduction is sexual.
3. Generations are nonoverlapping.
4. Mating is random.
5. Population size is very large.
6. Migration is negligible.
7. Mutation can be ignored.
8. Natural selection does not affect the locus under consideration.

Collectively these assumptions constitute the Hardy–Weinberg model, the name given to genotype frequencies with random mating after their discovery in the early 1900s by G. H. Hardy (1908) and W. Weinberg (1908). (See Li, 1967a, for a discussion of the largely overlooked early contributions of W. E. Castle.) Once the model is carefully stated, as in the preceding list, the genotype frequencies for a locus with two alleles can be worked out very easily because random mating of individuals is equivalent to random union of gametes. (You can verify this for yourself in Box D.) Because of this equivalence, we can cross multiply the allele frequencies along the margins in the sort of Punnett square shown in Figure 11. The result is the so-called HARDY-WEINBERG LAW: with random mating, the frequencies of the genotypes (usually called GENOTYPE FREQUENCIES) of *AA*, *Aa*,

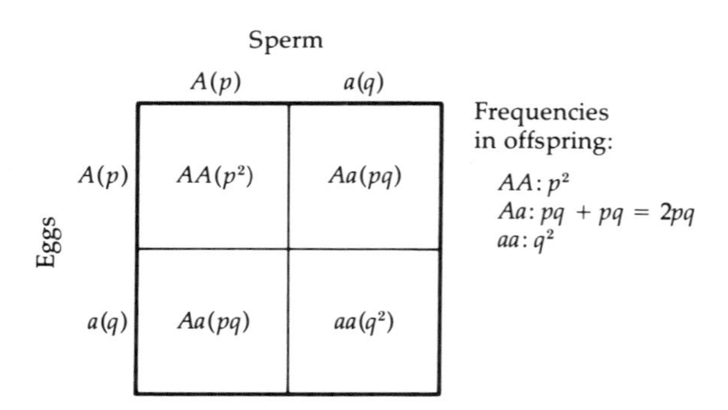

FIGURE 11. Punnett square showing Hardy–Weinberg frequencies generated by random mating with two alleles.

Further Implications of Random Mating

D

Random mating of individuals is equivalent to random union of gametes. To be precise, let P, Q, and R ($P + Q + R = 1$) be the frequencies of *AA*, *Aa*, and *aa* genotypes in a population. The frequency of the *A* allele is $p = P + Q/2$; that of the *a* allele is $q = Q/2 + R$. The table here shows the frequency of matings in the population and the offspring frequencies produced. In the next generation, the frequency of *AA* genotypes, call it P', will be $P' = P^2 \times 1 + 2PQ \times 1/2 + Q^2 \times 1/4 = (P + Q/2)^2 = p^2$, which is equivalent to random union of gametes with the frequency of the *A* allele equal to p. In the same way show that

a. $Q' = 2(P + Q/2)(Q/2 + R) = 2pq$, and
b. $R' = (Q/2 + R)^2 = q^2$

Mating	Frequency of mating	OFFSPRING FREQUENCIES		
		AA	*Aa*	*aa*
AA × *AA*	P^2	1	0	0
AA × *Aa*	$2PQ$	1/2	1/2	0
AA × *aa*	$2PR$	0	1	0
Aa × *Aa*	Q^2	1/4	1/2	1/4
Aa × *aa*	$2QR$	0	1/2	1/2
aa × *aa*	R^2	0	0	1

Assume random mating with respect to an autosomal locus having two alleles, *A* and *a*, with the allele frequency of *A* denoted as p and that of *a* as q.

c. Show that when *a* is rare (so that q^2 is much smaller than q), the frequency of heterozygotes in the population is approximately $2q$.

d. Show that the probability of a heterozygous offspring from a heterozygous parent is 1/2, irrespective of allele frequency.

e. Show that the square of the frequency of heterozygotes is equal to four times the product of the frequency of homozygotes.

and *aa* will be p^2, $2pq$, and q^2, respectively, where p represents the allele frequency of *A*, q the allele frequency of *a*, and, of course, $p + q = 1$ since there are only two alleles. Cross multiplication in a Punnett square is simply a systematic way of going through all the possibilities of gamete combination. The probability that a sperm or egg carries A is p; the probability that a sperm or egg carries a is q. With random combination of gametes, the chance that an A-bearing sperm fertilizes an A-bearing egg is $p \times p = p^2$; therefore, this is the frequency of AA genotypes. The probability that an A-bearing sperm fertilizes an a-bearing egg is $p \times q = pq$; and the probability that an a-bearing sperm fertilizes an A-bearing egg is $q \times p = qp$. Altogether the frequency of Aa heterozygotes is $pq + qp = 2pq$. Finally, the probability that an a-bearing sperm fertilizes an a-bearing egg is $q \times q = q^2$, which is the genotype frequency of aa. Note that $p^2 + 2pq + q^2 = (p + q)^2 = (1)^2 = 1$, thus accounting for all the offspring. Graphs of p^2, $2pq$, and q^2 for various values of p and q are shown in Figure 12.

It is perhaps a little hard to believe that so simple a result from so simple a model can hold so widely and be so important, but the Hardy–Weinberg law provides the foundation for many investigations in population genetics. Several implications should be spelled out immediately. (Others are discussed in Box D.) First, the allele frequencies in the model population remain constant through time. From the rule in Table I for calculating allele frequencies, the allele frequency of A among the offspring works out to be two times p^2 (because each AA homozygote carries two A alleles) plus $2pq$ (because each Aa heterozygote carries one A allele) divided by two (because each individual carries two alleles at the locus). In symbols, the allele frequency of A among the offspring is $(2p^2 + 2pq)/2 = p^2 + pq = p(p + q) = p \times 1 = p$. That is to say, the allele frequencies do not change from generation to generation; in any generation, therefore, the genotype frequencies will be p^2, $2pq$, and q^2 for AA, Aa, and aa, and these frequencies constitute what is often called the HARDY–WEINBERG EQUILIBRIUM for two alleles. The constancy of allele frequency (and therefore of the genotypic composition of the population) is the single most important implication of the Hardy–Weinberg law. This constancy of allele frequencies implies that, in the absence of specific evolutionary forces to change allele frequency (assumptions

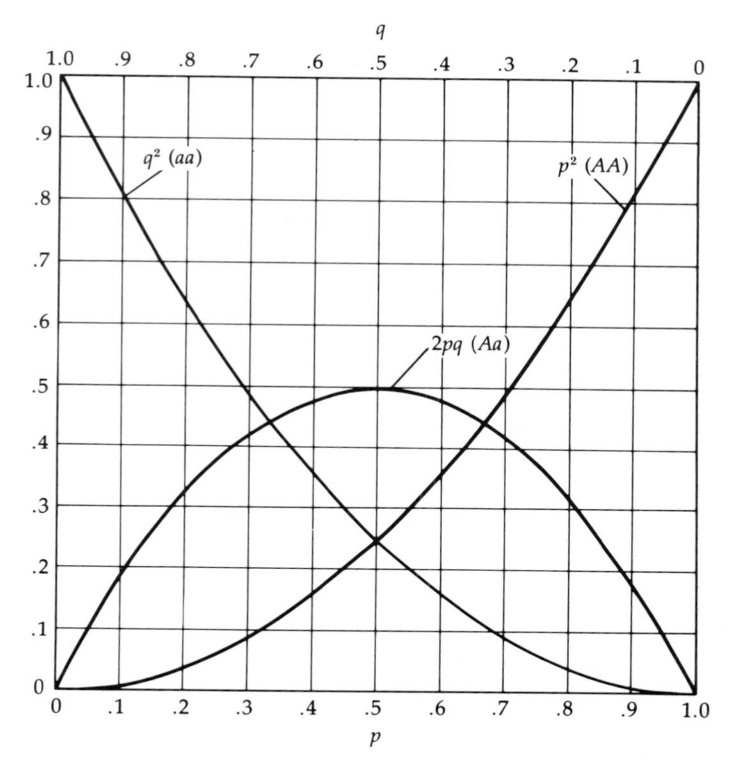

FIGURE 12. Graphs of p^2, $2pq$, and q^2. If the allele frequencies are between 1/3 and 2/3, heterozygotes will be the most common genotype in the population.

5, 6, 7, and 8 in the preceding list), the mechanism of Mendelian inheritance, by itself, will keep the allele frequencies constant and thus preserve genetic variation. A second item of interest is that the Hardy–Weinberg frequencies are attained in just one generation of random mating if the allele frequencies are the same in males and females. This, however, is true only with nonoverlapping generations; in age-structured populations, the Hardy–Weinberg frequencies are attained gradually over a period of several generations.

It is important to note here that the Hardy–Weinberg law is not very sensitive to certain kinds of departures from the assumptions numbered 3 to 8 in the earlier list, particularly numbers 5 to 8 (i.e., very large population size with no migration, mutation, or selection). That is to say, the mere fact that observed genotype frequencies may happen to fit the Hardy–Weinberg proportions cannot be taken as evidence that the assumptions in the list are valid. To be concrete, suppose that *AA*, *Aa*, and *aa* zygotes survive to adulthood in the proportions w_{11}, w_{12}, and w_{22}, respectively, but that the genotypes are otherwise completely equivalent and that assumptions 1 to 7 in the earlier list are satisfied. In such a case, *AA*, *Aa*, and *aa* adults will occur in Hardy–Weinberg proportions whenever $w_{12} = \sqrt{w_{11}w_{22}}$, whatever the actual values of w_{11}, w_{12}, and w_{22} may be; thus, for example, the Hardy–Weinberg proportions will hold even if $w_{11} = 1.0$, $w_{12} = .50$, and $w_{22} = .25$ — that is, even if three-fourths of the *aa* zygotes and one-half of the *Aa* zygotes die before reaching adulthood! If w_{12} does not equal $\sqrt{w_{11}w_{22}}$, there will be some deviation from Hardy–Weinberg proportions, but for plausible values of w_{11}, w_{12}, and w_{22} the deviations are expected to be small and detection therefore requires very large sample sizes (Lewontin and Cockerham, 1959).

HARDY-WEINBERG LAW IN OPERATION

Use of the Hardy–Weinberg law is exemplified in Table II, which gives the results of MN blood typing of 6129 American Caucasians: 1787 were genotypically *MM*, 3037 were *MN*, and 1305 *NN* (Mourant et al., 1976). The allele frequencies of *M* and *N* are calculated and the Hardy–Weinberg frequencies obtained as shown. The expected number of each genotype is obtained by multiplying the Hardy–Weinberg frequencies by the sample size (6129). The chi-squared value (χ^2) for goodness of fit is then calculated as the sum of (*observed number* − *expected number*)^2/*expected number* for each genotypic class, producing $\chi^2 = .04887$. (For an explanation of the chi-squared test for goodness of fit, see Chapter 1.)

Calculation of the degrees of freedom for the preceding χ^2 involves a subtlety not encountered in Chapter 1. Recall from the goodness-of-fit tests of Mendel's data in Chapter 1 that the expected number of individuals in each class of

TABLE II. Fit of Hardy–Weinberg frequencies to observed data.

	GENOTYPE[a]			
	MM	MN	NN	Total
No. of individuals	1787	3037	1305	6129
No. of M alleles	3574	3037	0	6611
No. of N alleles	0	3037	2610	5647
No. of $M + N$ alleles	3574	6074	2610	12258
	Allele frequency of M = 6611/12,258 = .53932 = p			
	Allele frequency of N = 5647/12,258 = .46068 = q			
Expected frequency	$p^2 = .29087$	$2pq = .49691$	$q^2 = .21222$	1.000
Expected no. (frequency × 6129)	1782.7	3045.6	1300.7	6129
Chi-squared value	$\frac{(1787 - 1782.7)^2}{1782.7}$ +	$\frac{(3037 - 3045.6)^2}{3045.6}$ +	$\frac{(1305 - 1300.7)^2}{1300.7}$ =	0.04887
	Degrees of freedom = 3 − 1 − 1 = 1			
	Probability associated with χ^2_1 of 0.04887 is about .90			

[a] The genotypes should more properly be denoted *M/M, M/N,* and *N/N,* but the symbols *MM, MN,* and *NN* are simpler and there is no ambiguity when only one locus is involved.

data was calculated from the total sample size and the appropriate Mendelian ratios; for such chi-squared tests, the number of degrees of freedom is simply equal to the number of classes of data minus 1, the 1 being deducted because the total sample size was used in calculating the expectations. For the type of chi-squared test in Table II, however, the data themselves are used to calculate the expectations; specifically, the data are first used to estimate the allele frequency of M ($p = .53932$; note that only p need be estimated from the data because q can be calculated from the relation $q = 1 - p$), and then this value of p is used in the Hardy–Weinberg formula along with the total sample size in order to obtain the expected number in each genotypic class. Since the value of p used in calculating the expectations is obtained from the data themselves, surely one should expect a better fit than would be the case were p obtained from some other source. Thus a smaller χ^2 should suffice to reject the hypothesis that the genotypes are in Hardy–Weinberg proportions, and the test should somehow take this into account. It turns out that the proper correction is an easy one — simply deduct one additional degree of freedom for estimating p from the data. For the chi-squared test in Table II, therefore, the appropriate number of degrees of freedom is 3 (the number of classes) minus 1 (for using the sample size in calculating expectations) minus 1 (for estimating p from the data), or $3 - 1 - 1 = 1$. [The general rule for chi-squared tests of the type in question is that the number of degrees of freedom equals the number of classes of data minus 1 (for using the sample size) minus the number of parameters estimated from the data.]

Returning now to Table II, the probability value associated with a χ^2 of .04887 with 1 degree of freedom is about .90 — a remarkably good fit. A χ^2 as large or larger would be obtained by chance 90 percent of the time in a population having its actual genotype frequencies as calculated from the Hardy–Weinberg law; thus there is no reason for thinking this population does not obey the Hardy–Weinberg law for the MN locus.

FREQUENCY OF HETEROZYGOTES

The MN blood groups are easy to work with in calculating allele frequencies because each genotype has a unique phenotype. However, in cases where one of two alleles is dominant, only two phenotypic classes can be recognized, the dominant type and the recessive type. One example is an antigen on the surface of red blood cells coded by the *D* allele in the Rh blood group system in humans. Genotypes *DD* and *Dd* are Rh^+ (positive), whereas *dd* is Rh^- (negative). Thus, *D* is dominant to *d*. In one test of 22,133 American Caucasians, 85.8 percent were found to be Rh^+ and 14.2 percent Rh^- (Mourant et al., 1976). Letting p stand for the allele frequency of *D* and q stand for the allele frequency of *d*, the allele frequencies can be calculated by noting that, with random mating, the frequency of homozygous recessives, in this case .142, must equal q^2. So $q^2 = .142$ or $q = \sqrt{.142} = .3768$. Then $p = 1 - q = 1 - .3768 = .6232$, and the frequencies of *DD*, *Dd*, and *dd* are expected to be $p^2 = (.6232)^2 = .3884$, $2pq = 2 \times .6232 \times .3768 = .4696$, and $q^2 = (.3768)^2 = .1420$, respectively. These values could have been approximated by reading off the curves in Figure 12. (Note from the discussion of degrees of freedom in the previous section that there is no chance for a goodness-of-fit test here because, with only two phenotypic classes, both degrees of freedom are lost, one because of fixed sample size and the other because q was estimated from the data.)

The above sort of estimation permits calculation of the frequency of heterozygous carriers of recessive alleles. If the observed frequency of recessive homozygotes is R, say, then with random mating $q = \sqrt{R}$ and the frequency of heterozygotes is estimated to be $2 \times (1 - \sqrt{R}) \times \sqrt{R}$. For cystic fibrosis in Caucasians, a severe disease involving abnormal glandular secretions due to an autosomal recessive, $R = .0006$ (approximately; Newcombe, 1964). The frequency of heterozygotes is therefore about $2 \times (1 - \sqrt{.0006}) \times \sqrt{.0006} = 2 \times .9755 \times .0245 = .0478$. Whereas only about one individual in 1667 (i.e., 1/.0006) is actually affected with cystic fibrosis, about one in 21 (i.e., 1/.0478) is a heterozygous carrier. This illustrates the important principle that the vast majority of rare recessive alleles are present in phenotypically normal heterozygotes and therefore concealed. (See Box D.)

Cases of Random Mating

RANDOM MATING: THREE OR MORE ALLELES

Genotypic frequencies under random mating for loci with three alleles are shown in Figure 13. Here it is convenient to label the alleles as A_1, A_2, and A_3 and the corresponding allele frequencies as p_1, p_2, and p_3. Since there are only three alleles, $p_1 + p_2 + p_3 = 1$. The genotype frequencies of A_1A_1, A_1A_2, A_2A_2, A_1A_3, A_2A_3, and A_3A_3 are p_1^2, $2p_1p_2$, p_2^2, $2p_1p_3$, $2p_2p_3$, and p_3^2, respectively, which can be obtained by expanding $(p_1A_1 + p_2A_2 + p_3A_3)^2$, a calculation that the Punnett square does automatically. In one test of 3977 Swiss from Zürich (Mourant et al., 1976) the allele frequencies of I^O, I^A, and I^B, which control the ABO blood groups, were estimated from the method in Box E to be $p_1 = .6708$ (for I^O), $p_2 = .2703$

Eggs \ Sperm	$A_1(p_1)$	$A_2(p_2)$	$A_3(p_3)$
$A_1(p_1)$	$A_1A_1(p_1^2)$	$A_1A_2(p_1p_2)$	$A_1A_3(p_1p_3)$
$A_2(p_2)$	$A_1A_2(p_1p_2)$	$A_2A_2(p_2^2)$	$A_2A_3(p_2p_3)$
$A_3(p_3)$	$A_1A_3(p_1p_3)$	$A_2A_3(p_2p_3)$	$A_3A_3(p_3^2)$

Frequencies in offspring:

A_1A_1: p_1^2
A_1A_2: $p_1p_2 + p_1p_2 = 2p_1p_2$
A_1A_3: $p_1p_3 + p_1p_3 = 2p_1p_3$
A_2A_2: p_2^2
A_2A_3: $p_2p_3 + p_2p_3 = 2p_2p_3$
A_3A_3: p_3^2

FIGURE 13. Punnett square showing Hardy–Weinberg frequencies for three autosomal alleles.

Allele Frequencies for ABO Blood-Group Locus

E

Estimation of allele frequencies at the human locus controlling the ABO blood group is troublesome because there are three principal alleles, I^A, I^B, and I^O, with both I^A and I^B dominant to I^O. Allele frequencies can be obtained with a computer by systematically searching for those allele frequencies that maximize the probability of obtaining the observed data. Such estimates are known as MAXIMUM LIKELIHOOD ESTIMATES, and they are statistically the best estimates. However, estimates of allele frequency that are very close to the maximum likelihood estimates can be calculated using a two-step procedure. Let p, q, and r denote the allele frequencies of I^A, I^B, and I^O, and let O, A, B, and AB denote the frequencies of the O, A, B, and AB blood-group phenotypes. Preliminary estimates are calculated as

$$r = \sqrt{\mathrm{O}}$$
$$p = 1 - \sqrt{\mathrm{B} + \mathrm{O}}$$
$$q = 1 - \sqrt{\mathrm{A} + \mathrm{O}}$$

Usually $p + q + r$ will not equal 1, but this can be corrected by calculating a correction factor $\Theta = 1 - p - q - r$ and forming the final estimates, $\hat{p}$, $\hat{q}$, $\hat{r}$, as

$$\hat{r} = \left(r + \frac{\Theta}{2}\right)\left(1 + \frac{\Theta}{2}\right)$$
$$\hat{p} = p\left(1 + \frac{\Theta}{2}\right)$$
$$\hat{q} = q\left(1 + \frac{\Theta}{2}\right)$$

Among 2060 Croatians, the following number of individuals of each phenotype were found (Mourant et al., 1976): 702 O, 862 A, 365 B, and 131 AB.

a. What are preliminary estimates of p, q, and r?

b. What are final estimates?

c. Assuming Hardy–Weinberg proportions, what are the expected numbers of O, A, B, and AB phenotypes?

d. Use a χ^2 test (there is one degree of freedom — why?) to determine whether the assumption of random mating seems justified.

TABLE III. Phenotype frequencies for ABO blood groups.

Genotype	Expected frequency	Blood type	Expected frequency	Observed frequency
I^O/I^O	$p_1^2 = (.6708)^2 = .4500$	O	.4500	.4493
I^O/I^A	$2p_1p_2 = 2\,(.6708)(.2703) = .3626$	A	.4357	.4365
I^A/I^A	$p_2^2 = (.2703)^2 = .0731$	A		
I^O/I^B	$2p_1p_3 = 2\,(.6708)(.0588) = .0789$	B	.0824	.0832
I^B/I^B	$p_3^2 = (.0588)^2 = .0035$	B		
I^A/I^B	$2p_2p_3 = 2\,(.2703)(.0588) = .0318$	AB	.0318	.0309

(for I^A), and $p_3 = .0588$ (for I^B). With random mating, the genotype frequencies can be calculated according to Figure 13, leading to the results shown in Table III. I^A and I^B are both dominant to I^O, so the expected frequency of blood type A is the sum of the genotype frequencies of I^O/I^A and I^A/I^A, and similarly for blood type B. As can be seen, the expected frequencies and the observed are in very close agreement.

In general, if there are n alleles — $A_1, A_2, \ldots, A_n$ — with frequencies $p_1, p_2, \ldots, p_n$ (with $p_1 + p_2 + \ldots + p_n = 1$, of course), then the genotype frequencies under random mating are p_i^2 for A_iA_i homozygotes and $2p_ip_j$ for A_iA_j heterozygotes. As an example of the application of this formula, we may use data from Taylor and Powell (1977), who carried out microscopic examination of the karyotypes (chromosomal constitutions) of larvae of *Drosophila persimilis* from Mather, California, and found four different gene arrangements of the third chromosome, the different gene arrangements being due to the presence of paracentric inversions. In one sample of 168 larvae, Taylor and Powell (1977) observed a frequency of .66 for the Whitney (*WH*) arrangement, .13 for Klamath (*KL*), .14 for Standard (*ST*), and .07 for Mendocino (*MD*). With four gene arrangements, there are four possible homokaryotypes (e.g., *WH*/*WH*) and six heterokaryotypes (e.g., *WH*/*KL*). In a random-mating population, the frequency of any homokaryotype is expected to be the square of the corresponding chromosomal frequency [i.e., *WH*/*WH* would have frequency $(.66)^2 = .4356$]; the frequency of any heterokaryotype is expected to be twice the product of the corresponding chromosomal frequencies (i.e., *WH*/*KL* would have frequency $2 \times .66 \times .13 = .1716$). Expected frequencies of all ten karyotypes can be obtained by multiplying out the expression $(.66WH + .13KL + .14ST + .07MD)^2$.

RANDOM MATING: SEX-LINKED LOCI

For loci on the human X chromosome, genotype frequencies differ in the sexes: in females, which have two X chromosomes, the genotype frequencies are as given by the Hardy–Weinberg law; in males, which have only one X chromosome, the genotype frequencies are simply equal to the allele frequencies. The rule for sex-linked loci is shown in Figure 14 for two alleles, and the rule applies to all organisms in which the females are XX and the males XY. This formulation holds only if allele frequencies for the sex-linked locus are the same in eggs and sperm, but it can be shown that such equality between the sexes is established quite rapidly, over a period of 10 or so generations even in age-structured populations (Box F). The important implication of Figure 14 is that conditions due to rare sex-linked recessives will be

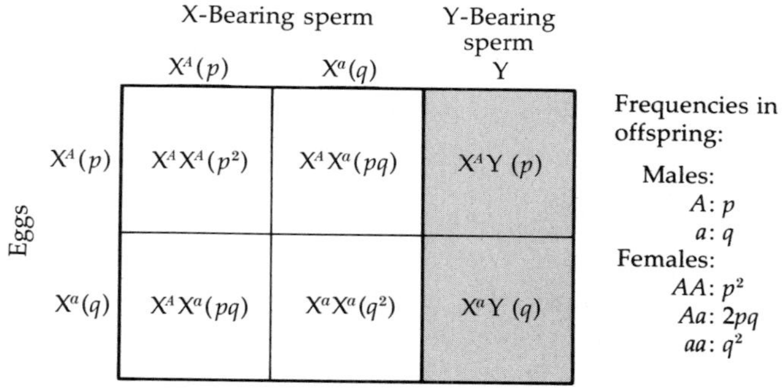

		X-Bearing sperm		Y-Bearing sperm
		$X^A (p)$	$X^a (q)$	Y
Eggs	$X^A (p)$	$X^A X^A (p^2)$	$X^A X^a (pq)$	$X^A Y (p)$
	$X^a (q)$	$X^A X^a (pq)$	$X^a X^a (q^2)$	$X^a Y (q)$

Frequencies in offspring:

Males:
- A: p
- a: q

Females:
- AA: p^2
- Aa: $2pq$
- aa: q^2

FIGURE 14. Consequences of random mating for sex-linked (i.e., X-linked) loci. Genotype frequencies in females will be the corresponding Hardy–Weinberg frequencies; genotype frequencies in males will be the corresponding allele frequencies.

F Hardy–Weinberg Equilibrium for Sex-Linked Loci

Attainment of Hardy–Weinberg equilibrium for sex-linked genes requires that allele frequencies be equal in the sexes. Establishment of allele-frequency equality occurs gradually over several generations, and it is governed by two principles resulting from the mechanism of sex-linked inheritance. If we consider a sex-linked locus with alleles A and a, and let m_n and f_n represent the frequency of the A allele in males and females in generation n, the principles are: (1) that $m_n = f_{n-1}$ — in other words, the frequency in males in any generation equals the frequency in females of the previous generation, because a male receives its X chromosome from its mother; and (2) that $f_n = (1/2)(m_{n-1} + f_{n-1})$ — in other words, the frequency in females in any generation equals the average allele frequency in the previous generation, because females receive one X chromosome from each parent.

a. Prove that the differences in allele frequency between the sexes becomes halved each generation, and that the differences alternate in sign, i.e., that $f_n - m_n = (-1/2)(f_{n-1} - m_{n-1})$.

b. Prove that the quantity $(2/3)(f_n) + (1/3)(m_n)$ remains constant; prove that if the allele frequency in both sexes equals this quantity, then the allele frequency will not change through time.

c. For initial frequencies $m_0 = .20$ and $f_0 = .80$, calculate m_i and f_i for $i = 1, 2, 3, 4, 5$, and 6. Calculate the ultimate allele frequency in both sexes (which is the expression in part b), and the ultimate frequencies of all possible male and female genotypes.

much more common in males than in females (because q^2 is much smaller than q itself when q is small). For example, about five percent of Western European males have the X-linked deuteranomaly type of color blindness, a defect in perception of green due to a reduced amount of a certain pigment in the retina of the eye. Consequently $q = .05$ for this condition, and the expected frequency of homozygous, and therefore color-blind, females is $(.05)^2 = .0025$. Affected males are 20 times more frequent than affected females. In the case of the X-linked protanomaly type of color blindness, which involves a different pigment, the frequency of affected males is about .01 in Western Europeans. The expected frequency of protanomalous females due to this X-linked allele is therefore $(.01)^2 = .0001$. In this case affected males are 100 times more frequent than affected females. These examples illustrate the principle that, for conditions due to rare sex-linked recessives, the rarer the condition, the more it will tend to be found exclusively in males.

RANDOM MATING: TWO LOCI

With random mating, the alleles at any locus will rapidly acquire random association into genotypes in the frequencies shown in Figures 11, 13, and 14. For two autosomal loci — loci on autosomes and not sex chromosomes — call them the A locus (with two alleles, A_1 and A_2, at frequencies denoted p_1 and p_2) and the B locus (also with two alleles, B_1 and B_2, at frequencies denoted q_1 and q_2), the genotype frequencies at the A locus will be p_1^2 for A_1A_1, $2p_1p_2$ for A_1A_2, and p_2^2 for A_2A_2; likewise the genotype frequencies at the B locus will be q_1^2 for B_1B_1, $2q_1q_2$ for B_1B_2, and q_2^2 for B_2B_2. However, the alleles at one locus may not be in random association in gametes with the alleles at the other locus. Given time, the alleles at the loci will reach random association in gametes, but the approach is gradual and the rate of approach is slower the smaller the recombination fraction (r) between the loci. The meaning of "random association in the gametes" is shown in Figure 15. Here the combinations of A's and B's in the squares refer to *gametes,* not to genotypes, as in earlier figures. A state of random gametic association between alleles at different loci is called LINKAGE EQUILIBRIUM. To be somewhat more precise about the effect that recombination has on the approach to linkage equilibrium, suppose that the actual frequencies of A_1B_1, A_1B_2, A_2B_1, and A_2B_2 gametes are P_{11}, P_{12}, P_{21}, and P_{22}, respectively, where $P_{11} + P_{12} + P_{21} + P_{22} = 1$. (Sorry about the double subscripts here, but this is really the simplest sort of notation for two loci.) In these symbols, linkage equilibrium is defined as the state in which $P_{11} = p_1q_1$, $P_{12} = p_1q_2$, $P_{21} = p_2q_1$, and $P_{22} = p_2q_2$.

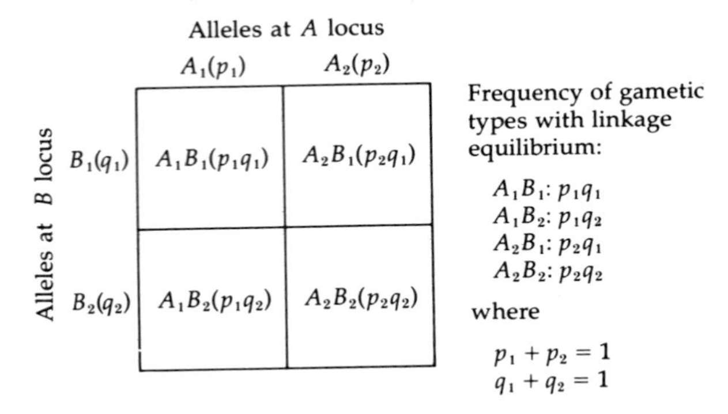

FIGURE 15. Random association between two alleles at each of two loci, showing expected gametic frequencies when the loci are in linkage equilibrium.

Suppose for the moment that the loci are *not* in linkage equilibrium. (Loci that are not in linkage equilibrium are said to be in LINKAGE DISEQUILIBRIUM.) We want to find out the gametic frequencies in the next generation. Consider first the A_1B_1 gamete. In any one generation, there are only two things that could have happened to a chromosome carrying A_1 and B_1: it could have undergone recombination between the loci (an event that would occur with probability r, where r denotes the recombination fraction), or the chromosome might have failed to undergo recombination between the loci (an event with probability $1 - r$). Among those chromosomes that did not undergo recombination, the frequency of A_1B_1 will have to be the same as it was in the previous generation; among the chromosomes that did undergo recombination, the frequency of A_1B_1 chromosomes will simply equal the frequency of $A_1\ ?/?\ B_1$ genotypes in the previous generation (the question mark denotes any allele at the locus), and this will be p_1q_1 because mating is random. Altogether, the frequency of A_1B_1 in generation n, call it $P_{11}^{(n)}$, will be $P_{11}^{(n)} = (1 - r)P_{11}^{(n-1)}$ [for the nonrecombinants] + rp_1q_1 [for the recombinants]. Here $P_{11}^{(n-1)}$ denotes the frequency of A_1B_1 in the previous generation. Subtraction of p_1q_1 from both sides leads to $P_{11}^{(n)} - p_1q_1 = (1 - r)(P_{11}^{(n-1)} - p_1q_1)$. The solution of this equation is straightforward:

$$P_{11}^{(n)} - p_1q_1 = (1 - r)(P_{11}^{(n-1)} - p_1q_1) = (1 - r)^2(P_{11}^{(n-2)} - p_1q_1)$$
$$= \ldots = (1 - r)^n(P_{11}^{(0)} - p_1q_1)$$

where $P_{11}^{(0)}$ is the frequency of A_1B_1 chromosomes in the initial generation in the founding population. Because $1 - r < 1$, $(1 - r)^n$ goes to zero as n gets large, but how fast $(1 - r)^n$ goes to zero depends on r; the closer r is to zero, the slower the rate. (Note here that $r = 1/2$ corresponds either to loci far apart on the same chromosome or to loci on different chromosomes.) In any case, since $(1 - r)^n$ goes to zero, $P_{11}^{(n)}$ goes to p_1q_1. Wholly analogous arguments hold for chromosomes carrying A_1B_2, A_2B_1, or A_2B_2, so $P_{12}^{(n)}$, $P_{21}^{(n)}$, and $P_{22}^{(n)}$ go to p_1q_2, p_2q_1, and p_2q_2, respectively. Thus, linkage equilibrium is attained at a rate determined by the value of r.

It is convenient to have a quantitative measure of the amount of linkage disequilibrium. One commonly used measure is the LINKAGE DISEQUILIBRIUM PARAMETER. Usually denoted by the symbol D, the linkage disequilibrium parameter is defined as $D = P_{11}P_{22} - P_{12}P_{21}$. From the expressions derived in the preceding paragraph, it can be shown (with a mass of algebra) that $D_n = (1 - r)D_{n-1}$ so that $D_n = (1 - r)^nD_0$, where D_n denotes the value of D in generation n. $D = 0$ therefore corresponds to linkage equilibrium. For prescribed allele frequencies p_1, p_2, q_1, q_2 the smallest possible value of D is $D_{min} = -p_1q_1$ or $-p_2q_2$, whichever is larger, and the largest possible value is $D_{max} = p_1q_2$ or p_2q_1, whichever is smaller. (See Box G.) Figure 16 shows how D goes to zero for various values of r.

So far — though the matter has not yet been studied extensively — it appears that values of D in natural populations are typically zero or very close to zero (indicating linkage equilibrium) unless the loci are very closely linked. (See Mukai et al., 1971; Langley et al., 1977, 1978. Cases where highly significant values of D are found in natural populations are provided by Clegg et al., 1972; Allard et al., 1972; Weir et al., 1972, for plants; Baker, 1975, for a species of *Drosophila*.)

An extreme, naturally occurring example of linkage dis-

G Linkage Disequilibrium of Two Loci

Suppose the four gametic types A_1B_1, A_1B_2, A_2B_1, and A_2B_2 have respective frequencies P_{11}, P_{12}, P_{21}, and P_{22}, with $P_{11} + P_{12} + P_{21} + P_{22} = 1$. The allele frequency of A_1 is therefore $p_1 = P_{11} + P_{12}$ and that of A_2 is $p_2 = P_{21} + P_{22}$; the allele frequency of B_1 is $q_1 = P_{11} + P_{21}$ and that of B_2 is $q_2 = P_{12} + P_{22}$. Define the disequilibrium parameter as $D = P_{11}P_{22} - P_{12}P_{21}$.

a. By direct substitution into the right-hand sides of the following expressions, show that

$$P_{11} = p_1q_1 + D$$
$$P_{12} = p_1q_2 - D$$
$$P_{21} = p_2q_1 - D$$
$$P_{22} = p_2q_2 + D$$

To calculate the maximum and minimum possible values of D (assuming the allele frequencies p_1 and q_1 are fixed), it is convenient to express the gametic frequencies in terms of P_{11} as follows: $P_{12} = p_1 - P_{11}$, $P_{21} = q_1 - P_{11}$, and $P_{22} = 1 - P_{11} - P_{12} - P_{21} = 1 - P_{11} - (p_1 - P_{11}) - (q_1 - P_{11}) = 1 - p_1 - q_1 + P_{11}$. Also, $D = P_{11}P_{22} - P_{12}P_{21} = P_{11}(1 - p_1 - q_1 + P_{11}) - (p_1 - P_{11})(q_1 - P_{11}) = P_{11} - p_1q_1$.

It is, of course, necessary that $P_{11} \geq 0$. Since $P_{12} = p_1 - P_{11} \geq 0$ also, $P_{11} \leq p_1$; because $P_{21} = q_1 - P_{11} \geq 0$, then $P_{11} \leq q_1$; and because $P_{22} = 1 - p_1 - q_1 + P_{11} \geq 0$, it follows that $P_{11} \geq p_1 + q_1 - 1$. Hence we can say that the minimum allowable value of P_{11} is the larger of 0 and $p_1 + q_1 - 1$, and the maximum allowable value of P_{11} is the smaller of p_1 and q_1.

b. Use the maximum and minimum values of P_{11} and the expression $D = P_{11} - p_1q_1$ to show that, for fixed p_1 and q_1, D_{max} equals p_1q_2 or p_2q_1, whichever is smaller; and that D_{min} equals $-p_1q_1$ or $-p_2q_2$, whichever is larger.

c. (This part is for students familiar with the definition of the correlation coefficient, either from a course in statistics or from looking ahead to Chapter 4.) Show that the correlation coefficient, ρ, between alleles on the same chromosome is $\rho = D/\sqrt{p_1p_2q_1q_2}$. (Hint: assign alleles A_1 and B_1 the arbitrary value 1 and alleles A_2 and B_2 the arbitrary value 0.)

d. For the data of Mukai et al. (1974) pertaining to the association between the *Amy* locus and the *NS* inversion in *Drosophila* (see the text), and for the data of Clegg et al. (1972) pertaining to two esterase loci in barley (see the text), verify numerically that $\chi^2 = \rho^2 N$, where N is the number of chromosomes in the sample. (There will be a small discrepancy between χ^2 and $\rho^2 N$, which is caused by rounding off in the calculations; the expression $\chi^2 = \rho^2 N$ is exact, so the discrepancy can be made as small as desired by carrying along sufficient decimal places.) G

e. For the data of Mukai et al. (1974) given in Box A, determine by a χ^2 test whether there is significant linkage disequilibrium between the locus of *αGpdh* and the *NS* inversion. (Note: the *αGpdh* locus is on the left arm of the second chromosome and recombines with a rate of 50 percent with the *NS* inversion.)

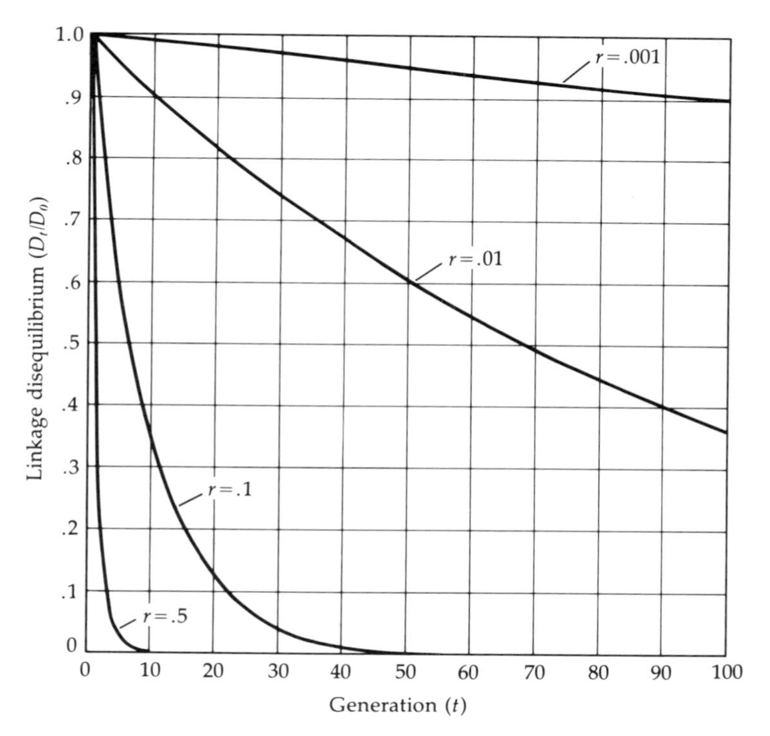

FIGURE 16. Decay of linkage disequilibrium parameter (D) with time, relative to its initial value (D_0), for various values of the recombination fraction between the loci (r). The graphs assume random mating. If there is substantial self-fertilization (as occurs in many plants), D decays much more slowly than shown here.

equilibrium involves the paracentric inversions mentioned in Chapter 1 that are polymorphic in populations of *D. pseudoobscura*. Because the inversions prevent recombination, each inversion represents a sort of "supergene," and natural selection accumulates beneficially interacting alleles (said to show GENETIC COADAPTATION) within each inversion (Prakash and Lewontin, 1968; Zouros, 1976; Prakash, 1977a; Voelker et al., 1978). (Wallace, 1968, provides an excellent general discussion of genetic coadaptation.)

A case of linkage disequilibrium between an allozyme locus (*Amy* = α-amylase) and a paracentric inversion (*NS* = Nova Scotia) in *D. melanogaster* is provided by the data of Mukai et al. (1974) in Box A. Among the 1158 chromosomes examined, the observed numbers of the four chromosomal types were

Amy^F	non-*NS*	898
Amy^F	*NS*	122
Amy^S	non-*NS*	137
Amy^S	*NS*	1

If we denote the allele frequency of Amy^F as p_1 and that of Amy^S as p_2, then $p_1 = (898 + 122)/1158 = .8808$ and $p_2 = 1 - p_1 = .1192$. Similarly, if the frequency of non-*NS*-bearing chromosomes is denoted q_1 and that of *NS*-bearing chromosomes as q_2, then $q_1 = (898 + 137)/1158 = .8938$ and $q_2 = 1 - q_1 = .1062$. With linkage equilibrium, the corresponding expected numbers of each of the chromosome types would be

Amy^F	non-*NS*	$1158 \times .8808 \times .8938 = 911.65$
Amy^F	*NS*	$1158 \times .8808 \times .1062 = 108.32$
Amy^S	non-*NS*	$1158 \times .1192 \times .8938 = 123.37$
Amy^S	*NS*	$1158 \times .1192 \times .1062 = 14.66$

The χ^2 value for the comparison of the observed data with these expectations is 16.17; the number of degrees of freedom for this χ^2 is 4 (the number of classes of data) minus 1 (for using the sample size in calculating expectations) minus 1 (for estimating p_1) minus 1 (for estimating q_1) or $4 - 1 - 1 - 1 = 1$. The probability level associated with a χ^2 of 16.17 with one degree of freedom is less than .0001, so we may conclude with considerable confidence that *Amy* and *NS* are not in linkage equilibrium.

As for the amount of linkage disequilibrium involved in the preceding example, let P_{11}, P_{12}, P_{21}, and P_{22} represent the frequencies of the four chromosomes types, given in the same order as the observed numbers. Then

$$\begin{aligned} P_{11} &= 898/1158 = .7755 \\ P_{12} &= 122/1158 = .1053 \\ P_{21} &= 137/1158 = .1183 \\ P_{22} &= 1/1158 = .0009 \end{aligned}$$

and $D = P_{11}P_{22} - P_{12}P_{21} = -.0118$. Since $-p_1q_1 = -.7873$ is smaller than $-p_2q_2 = -.0127$, $D_{min} = -.0127$, so the observed value of D is about 93 percent of its smallest possible value. (Incidentally, the *Amy* locus is not included within the *NS* inversion, but it lies very near the left-hand breakpoint; thus the amount of recombination between *Amy* and *NS* is very low.)

Two human examples showing linkage disequilibrium of very closely linked loci are shown in Table IV, where the data are drawn from the English population (Race and Sanger, 1975). One case involves the MN blood group locus and a closely linked locus (alleles *S* and *s*) that controls related blood cell antigens. The other example involves two loci

TABLE IV. Linkage disequilibrium for two human blood-group systems.

System	Gamete	Frequency	Allele	Frequency	D	D_{max}	$D/D_{max} \times 100$ (percent)
MNSs	*MS*	.247	*M*	.530	.0737	.1537	47.9
	Ms	.283	*N*	.470			
	NS	.080	*S*	.327			
	Ns	.390	*s*	.673			
Rh	*DE*	.144	*D*	.590	.0520	.0640	81.2
	De	.446	*d*	.410			
	dE	.012	*E*	.156			
	de	.398	*e*	.844			

involved in the Rh blood group system. In the former case, D is about 48 percent of its maximum possible value; in the Rh case, D is about 81 percent of its hypothetical maximum. (For examples of linkage disequilibrium in humans between the HL-A locus, which is involved in the immune response, and various loci causing increased susceptibility to such diseases as ankylosing spondylitis, see Bodmer and Bodmer, 1974, and Thomson and Bodmer, 1977.)

Such linkage disequilibrium as seen in the preceding examples can be caused by linkage disequilibrium in the founding population that has not yet had time to dissipate due to the small value of r. Another possible cause is admixture of populations with differing gametic frequencies. A third possibility is the occurrence of a sufficient intensity of natural selection in favor of certain heterozygous genotypes to overcome the natural tendency for D to go to zero. Here it should be noted that measures of linkage disequilibrium involving multiple alleles or involving three, four, or more loci have been developed (e.g., Bennett and Oertel, 1965; Smouse, 1974; Weir and Cockerham, 1978), but they are rather complex and need not detain us.

Assortative Mating

Although random mating is the most important mating system in many natural populations, there are certain departures from random mating that can also be important. Major types of departure are listed and defined in Table V. When choice of mates is based on phenotypes, mating is said to be "assortative." In POSITIVE ASSORTATIVE MATING (often called simply ASSORTATIVE MATING), individuals tend to choose mates that are phenotypically like themselves. In NEGATIVE ASSORTATIVE MATING (also called DISASSORTATIVE MATING), individuals tend to choose mates that are phenotypically unlike themselves. Of course, even with random mating, some mating pairs will be phenotypically similar or dissimilar, so assortative mating refers only to those situations in which mating partners are phenotypically more similar or dissimilar than would be expected by chance in a

TABLE V. Characteristics of several mating systems.

Mating system	Defining feature
Random mating	Choice of mates independent of genotype and phenotype
Positive assortative mating	Mates phenotypically more similar than would be expected by chance
Negative assortative mating (disassortative)	Mates phenotypically more dissimilar than would be expected by chance
Inbreeding	Mating between relatives

random-mating population. (For general discussions of various types of nonrandom mating see Lewontin et al., 1968, and Garrison et al., 1968.)

An example of negative assortative mating is found in the polymorphism known as HETEROSTYLY, which occurs in most species of primroses (*Primula*) and their relatives (Stebbins, 1971). In most primrose populations, there are approximately equal proportions of two types of flowers, one known as "pin," which has a tall style and short stamens, and the other known as "thrum," which has a short style and tall stamens. (In botanical terminology, recall, the style carries the stigma, which is the organ that receives pollen, whereas the stamens carry the anthers, which produce pollen.) In heterostyly, insect pollinators that work high on the flowers will pick up mostly thrum pollen and deposit it on pin stigmas, whereas pollinators that work low in the flowers will pick up mostly pin pollen and deposit it on thrum stigmas. Negative assortative mating therefore occurs because pins mate preferentially with thrums.

Pollination biology of flowering plants also provides examples of positive assortative mating. For example, if the length of time during which any individual plant flowers is short relative to the total duration of the flowering season, then plants that flower early in the season will preferentially be pollinated by other early flowering plants, and those that flower late will preferentially be pollinated by other late flowering ones. Thus, there will be positive assortative mating for flowering time.

In humans, positive assortative mating occurs for height, I.Q. score, and certain other traits, although the amount of assortative mating varies in different populations and does not occur in some. Negative assortative mating is apparently quite rare, so we shall ignore it, but one obvious example is sex; as regards sex, mating pairs tend to be phenotypically different.

In *Drosophila,* a curious type of nonrandom mating is found in a phenomenon called "minority male mating advantage," in which females mate preferentially with males having rare phenotypes (Petit and Ehrman, 1969; Ehrman and Parsons, 1976; Parsons, 1977; Ehrman and Probber, 1978). For example, in one study involving experimental populations of *D. pseudoobscura* containing flies homozygous for either the recessive *orange* eye-color mutation or the recessive *purple* eye-color mutation, Ehrman (1970) found that when 20 percent of the males were *orange,* the *orange*-eyed males participated in 30 percent of the observed matings; conversely, when 20 percent of the males were *purple,* the *purple*-eyed males participated in 40 percent of the observed matings. (There is a large literature on *Drosophila* mating behavior; for a sampler, see Spieth, 1968; Spiess, 1970; de Magalhães and Rodrigues Pereira, 1976; Markow, 1978; Molin, 1979.)

The consequences of positive assortative mating are complex and are dependent on the number of loci that influence the trait in question, on the number of different possible alleles at the loci, on the number of different phenotypes, on the sex performing the mate selection, and on the criteria for mate selection (for some models, see Karlin, 1969; Jacquard, 1974; O'Donald, 1978, 1979; Karlin and Raper, 1979). Rarely are traits with assortative mating determined by alleles at a single locus, however; most such traits are polygenic, so further discussion of positive assortative mating is best deferred to Chapter 4. Here we should note one obvious, qualitative consequence of positive assortative mating (see Chapter 4 for details and examples): since like phenotypes tend to mate, assortative mating will generally increase the frequency of homozygous genotypes in the population at the expense of heterozygous genotypes, and thus the phenotypic variance in the population will increase. (Negative assortative mating generally has the opposite effect.)

Inbreeding

THE INBREEDING COEFFICIENT

Mating between relatives is INBREEDING (Table V). Like positive assortative mating, inbreeding increases a population's homozygosity; unlike assortative mating, which affects only those loci on which mate selection is based (and loci in linkage disequilibrium with them), inbreeding affects all loci in the genome. In humans, the closest degree of inbreeding that commonly occurs in most societies is first cousin mating, but many plants regularly undergo self-fertilization.

The effect of inbreeding is usually measured by a quantity called the "inbreeding coefficient." To define the inbreeding coefficient, imagine the two alleles that occupy homologous loci in an inbred individual. These two alleles could be IDENTICAL BY DESCENT; that is, they could both have been derived by replication of a single allele in some ancestral population. (The concept of identity by descent is of fundamental importance in population genetics. See Cotterman, 1940, and Malécot, 1969, who developed the concept, and Gillois, 1966, for a discussion.) If the two alleles at the locus in question are identical by descent, then the individual is said to be AUTOZYGOUS at the locus. On the other hand, the alleles may not be replicas of a single ancestral allele, in which case the alleles are not identical by descent and the individual is said to be ALLOZYGOUS at the locus. Now, the INBREEDING COEFFICIENT, designated by the symbol F, is the probability that two alleles at a locus in an individual are identical by descent. The inbreeding coefficient is a relative concept; it measures the amount of autozygosity relative to that in some ancestral population. That is to say, we arbitrarily assume that all alleles in the ancestral population are unique (i.e., not identical by descent); the inbreeding coefficient of an individual in the present population is then the probability that the two alleles at a locus in the individual arose by replication of a single allele more recently than the time at which the ancestral population existed. The ancestral population need not be remote in time from the present one. Indeed, the ancestral population, presumed to be noninbred ($F = 0$), typically refers to the population existing just a few generations previous to the present one, and F in the present population then measures inbreeding that has occurred in the span of these few generations. Because the span of time involved is usually short, the possibility of mutation can safely be ignored. Autozygous individuals must therefore be homozygous for some allele at the locus under consideration. On the other hand, allozygous individuals can be either homozygous or heterozygous. Although, as shown in Figure 17, two distinct alleles of the same type (two A's or two a's, for example) may come together in an individual in the present generation and thereby make the individual homozygous, the alleles in the ancestral population are, by definition, not identical by descent, so the individual will be allozygous. Similarly, although a heterozygous individual must be allozygous (barring mutation), a homozygous individual may be either autozygous or allozygous. These concepts, illustrated in Figure 17, are fundamental to the understanding of inbreeding effects.

While the inbreeding coefficient, F, is the probability that the two alleles at a locus in an individual are identical by descent, the idea can readily be extended to alleles that are not necessarily in the same individual. That is to say, one could define another sort of "inbreeding coefficient" as the probability that *any* two randomly chosen alleles in a population (not necessarily in the same individual) are identical by descent. Furthermore, one could imagine defining a whole sequence of ancestral populations, each more remote than the one before. In such a framework, one could determine the probability of autozygosity of alleles in the present population relative to each of the ancestral populations in turn. These general comments are intended as a preliminary introduction to later chapters that use the probability of

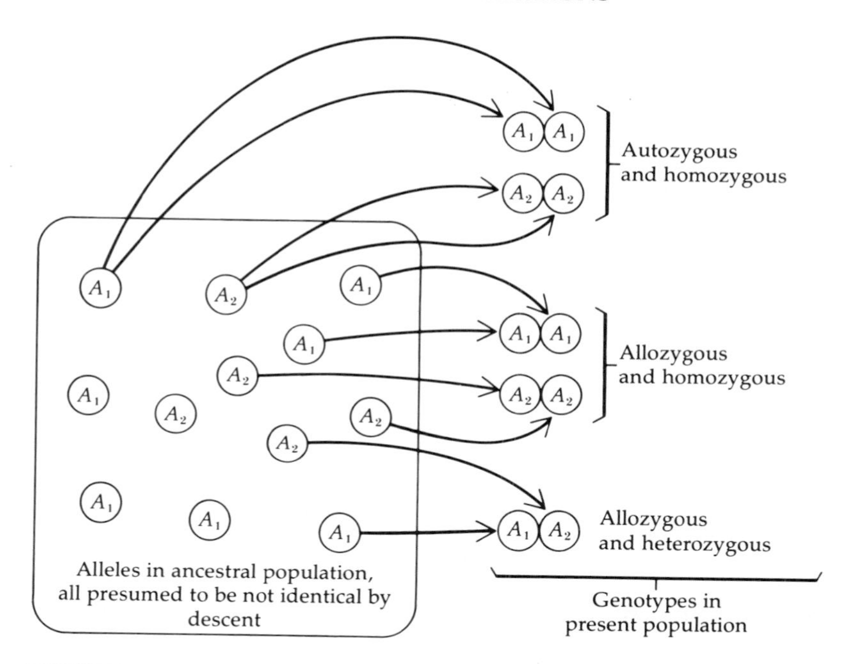

FIGURE 17. An autozygous individual is one whose alleles (at a locus under consideration) arose by DNA replication of a single ancestral allele. The alleles in an autozygous individual are said to be identical by descent. An allozygous individual, by contrast, is one whose alleles are not identical by descent. As shown here, allozygous individuals may be heterozygous or homozygous, but autozygous individuals must be homozygous (except in the unlikely event that one allele undergoes a mutation).

identity by descent in various contexts in order to measure genetic relationships within and between populations. For the moment we need the concept only in its simplest form: the inbreeding coefficient F is the probability that two alleles at a locus in an individual are identical by descent, relative to some one ancestral population.

GENOTYPE FREQUENCIES WITH INBREEDING

Imagine a population in which individuals have average inbreeding coefficient F. Focus on one individual and consider the alleles at any locus in that individual. Either of two things must be true: the alleles must be either allozygous (probability $1 - F$) or autozygous (probability F). If the locus is allozygous, then the probability that the individual has any particular genotype is simply the probability of that genotype in a random-mating population, because this particular set of loci was chosen to be allozygous and therefore noninbred ($F = 0$). On the other hand, if the locus is autozygous, then the individual must be homozygous, and the probability that the individual is homozygous for any particular allele is simply the frequency of that allele in the population as a whole. (Since the locus in question is presumed to be autozygous, knowing which allele is at the locus on one chromosome immediately tells you that an identical allele is on the homologous chromosome.) These considerations hold regardless of the number of alleles, but to simplify matters suppose there are only two alleles, *A* and *a*, at frequencies p and q (with $p + q = 1$). The probability that an individual is *AA* is therefore $p^2(1 - F)$ [for cases in which the locus is allozygous] $+ pF$ [for cases in which the locus is autozygous]. Similarly, the probability that the individual is *aa* is $q^2(1 - F) + qF$. Heterozygotes, *Aa*, then have frequency $2pq(1 - F)$, since heterozygous loci must be allozygous. These genotype frequencies with inbreeding are summarized in Table VI. When $F = 0$, the genotype frequencies become the familiar Hardy–Weinberg frequencies for random-mating populations. When $F = 1$, inbreeding is

TABLE VI. Genotype frequencies with inbreeding.

	FREQUENCY IN POPULATION				
Genotype	**With inbreeding coefficient *F***			**With $F = 0$ (random mating)**	**With $F = 1$ (complete inbreeding)**
AA	$p^2(1 - F)$	+	pF	p^2	p
Aa	$2pq(1 - F)$			$2pq$	0
aa	$q^2(1 - F)$	+	qF	q^2	q
	Allozygous loci		Autozygous loci		

complete; *AA* homozygotes will constitute a fraction p of the population and *aa* homozygotes will constitute a fraction q. One other item to note: the allele frequency of A in the population, calculated as in Table I, is $\{2[p^2(1 - F) + pF] + 2pq(1 - F)\}/2 = p^2(1 - F) + pF + pq(1 - F) = p(p + q)(1 - F) + pF = p(1)(1 - F) + pF = p - pF + pF = p$. The allele frequencies, therefore, do not change with inbreeding; inbreeding affects only the association of alleles into genotypes.

EFFECTS OF INBREEDING

In OUTCROSSING species — those that regularly avoid inbreeding — close inbreeding is generally harmful. The effects are seen most dramatically when inbreeding is complete or nearly complete. Although nearly complete autozygosity can be approached in most species by many generations of brother–sister mating, autozygosity of individual chromosomes can easily be accomplished in *Drosophila* by the sort of mating scheme shown in Figure 18. *Cy* (*Curly* wings) and *Pm* (*Plum*-colored eyes) are dominant mutations on chromosome 2 and are present on certain laboratory second chromosomes (denoted by gray and black) that carry several long inversions to prevent recombination. Single wild males are mated individually to *Cy*/*Pm* females, and a single *Curly*-winged son, carrying one wild-type second chromosome (hatched), is selected from each mating (Figure 18A). The *Curly*-winged male is backcrossed to *Cy*/*Pm* females, and *Curly*-winged heterozygous sons and daughters are selected (Figure 18B). These are mated with each other (Figure 18C1) and the offspring counted. Because *Cy*/*Cy* homozygotes are lethal, one-third of the surviving offspring are expected to be homozygous for the isolated wild-type chromosome, unless, of course, the wild-type chromosome carries recessive mutations that reduce viability. At the same time as the matings in Figure 18C1 are carried out, heterozygous males from one strain are mated with heterozygous females carrying a different wild-type chromosome (shown in stippling in Figure 18C2). In this case, one-third of the offspring are expected to be hetero-

(A) Mate and select single *Curly*-winged son.

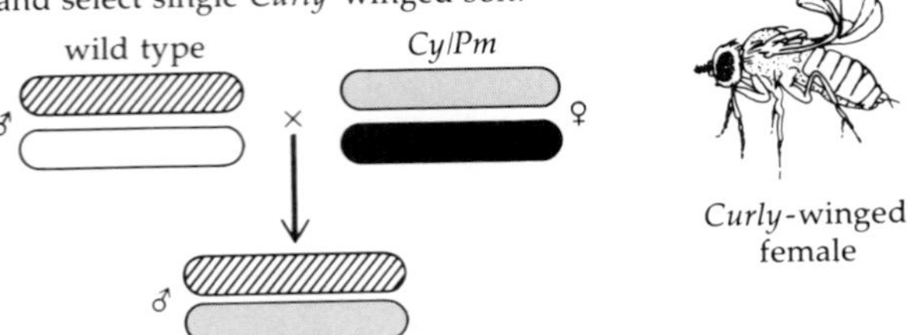

(B) Backcross the son from (A) and select *Curly* sons and daughters, which will be heterozygous.

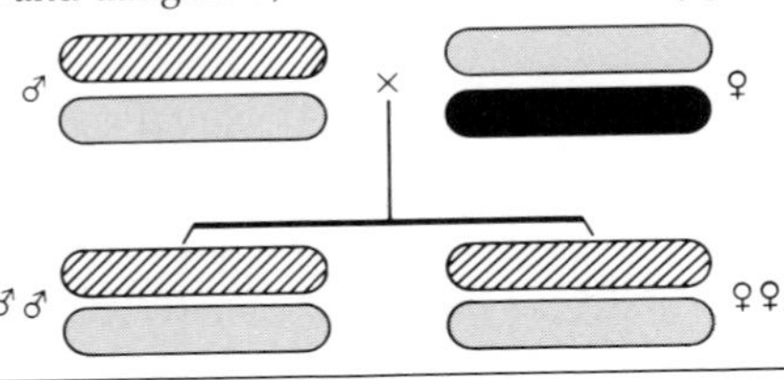

(C1) Mate heterozygotes from same strain and count proportion of *straight*-winged offspring.

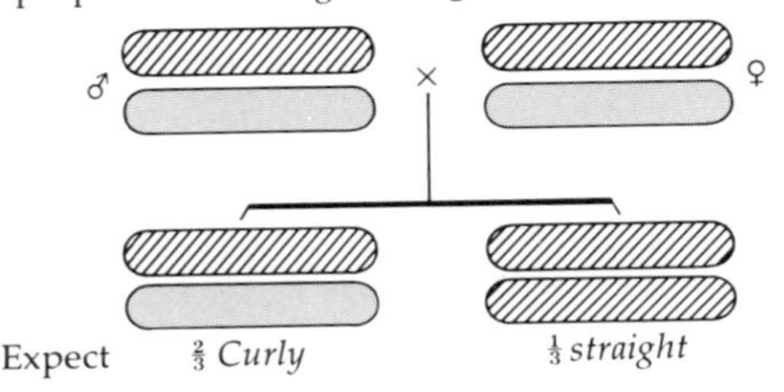

(C2) Mate heterozygotes from different strains and count proportion of *straight*-winged offspring.

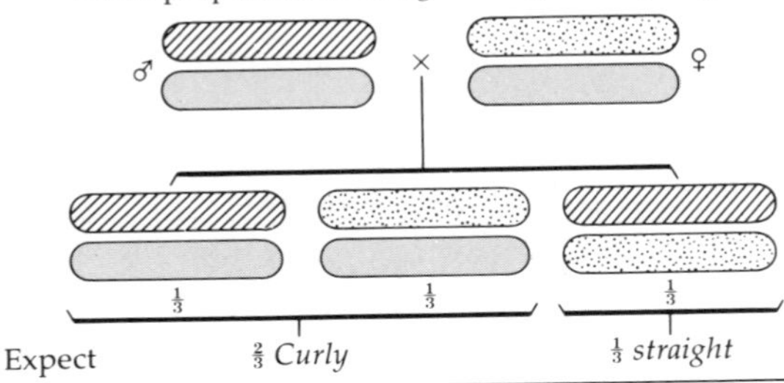

FIGURE 18. Mating scheme to extract single chromosomes (in this case, the second chromosome) from populations of *Drosophila melanogaster*. *Cy* (*Curly* wings) and *Pm* (*Plum* eyes) are dominant mutations carried on certain special laboratory chromosomes that have multiple inversions to prevent recombination. From each mating of the type in part A, a single

zygous — that is, to carry two *different* wild-type chromosomes. For matings both in C1 and C2 of Figure 18, the proportion of *straight*-winged flies divided by .333 therefore provides a measure of the viability of the homozygous or heterozygous genotype relative to the viability of *Cy*/+. (The + here denotes the wild-type second chromosome extracted from the natural population by this mating procedure.)

Data from an experiment using a procedure similar to that in Figure 18 are shown in Figure 19. As you can see, the homozygotes (gray outline) are relatively poor in viability. In fact, nearly 13 percent of the homozygotes are lethal. Moreover, among the homozygotes that have viabilities within the normal range of heterozygotes (black outline), virtually all can be shown to have reduced fertility (Sved and Ayala, 1970; Sved, 1971). Such close inbreeding is clearly very harmful indeed, and it provides a new dimension of genetic diversity. In the case of allozymes, genetic diversity is due to common alleles that do not perceptibly impair viability or fertility when homozygous. In the case of inbreeding, we are dealing with rare alleles that are severely detrimental when homozygous. (That the alleles are rare is shown by the small proportion of lethal or near-lethal heterozygotes.) Figure 19 shows that natural populations carry considerable hidden genetic variation in the form of rare deleterious recessive alleles. (See Dobzhansky et al., 1963; Crumpacker, 1967; Mukai and Yamaguchi, 1974, for more examples.)

son (carrying one of his father's second chromosomes, indicated by the hatching) is selected. This son is backcrossed (part B) in order to produce many replicas of that second chromosome. Brother–sister matings as in part C1 are expected to produce 1/4 *Cy*/*Cy*, 1/2 *Cy*/+, and 1/4 +/+ zygotes (the + denotes the wild-type second chromosome), but since *Cy*/*Cy* zygotes do not survive, the expected offspring are 2/3 *Cy*/+ (*Curly*-winged) and 1/3 +/+ (*straight*-winged). Matings as in part C2, between a male carrying one wild-type second chromosome (hatched) and a female carrying a different one (stippled), are also expected to produce 2/3 *Curly* and 1/3 *straight*-winged. However, in part C1, the *straight*-winged flies are homozygous for a single wild-type second chromosome; in part C2, the *straight*-winged flies are heterozygous for two different wild-type second chromosomes.

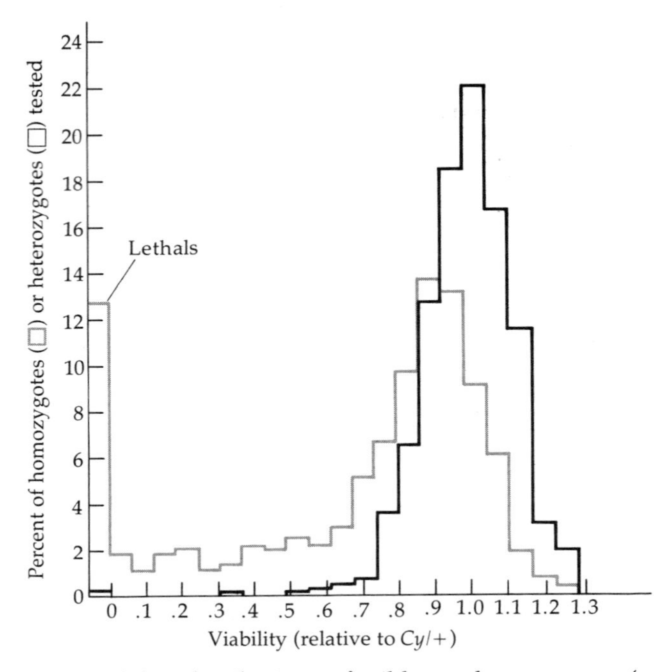

FIGURE 19. Viability distributions of wild-type homozygotes (gray outline) and wild-type heterozygotes (black outline) of second chromosomes extracted from *Drosophila pseudoobscura* according to a mating scheme similar to that in Figure 18. The histograms depict results of testing 1063 homozygous combinations and 1034 heterozygous combinations. Note that, in this sample, nearly 13 percent of the wild chromosomes are homozygous lethal, and many more have viabilities substantially below normal. (Data from Dobzhansky and Spassky, 1963.)

Detrimental effects of inbreeding (called INBREEDING DEPRESSION) are found in virtually all outcrossing species, and the more intense the inbreeding, the more harmful the effects (Figure 20). Inbreeding in humans is generally harmful, too, but the effect is difficult to measure because the degree of inbreeding is less than what is possible to achieve in experimental organisms, and the effects may also vary from population to population. Nevertheless, children of first cousin matings are on the average less capable than noninbred children in any number of ways (e.g., higher rate of mortality, lower I.Q. scores), though it should be emphasized that many such children are within a normal range of abilities (Morton et al., 1956; Neel and Schull, 1962; Schull and Neel, 1965; Morton, 1978). In most organisms, inbreeding depression is largely due to the increased homozygosity of rare recessive alleles, and inbreeding effects in humans are also seen most dramatically in the enhanced frequency of genetic abnormalities due to harmful recessive alleles among the children of first cousin matings. The enhanced frequency of such conditions results from the genotype frequencies given in Table VI; if *a* denotes a rare deleterious recessive, then among the children of first cousin matings,

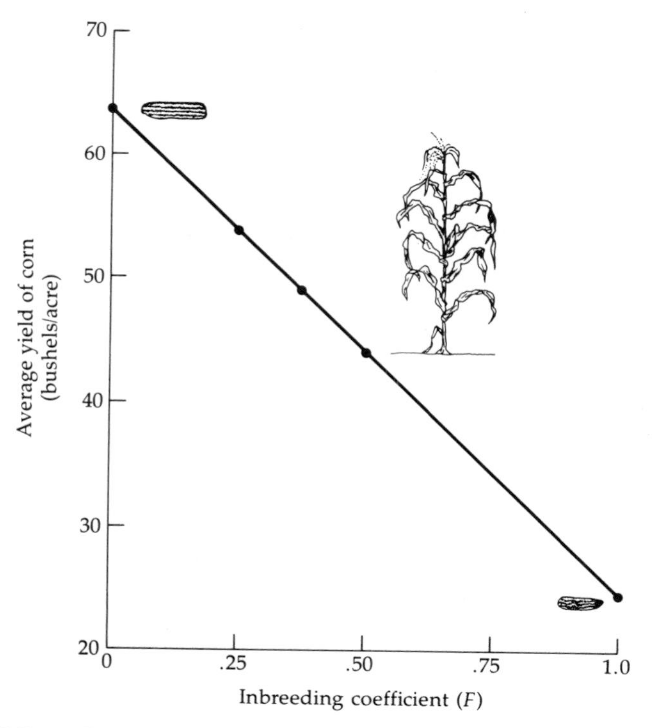

FIGURE 20. Decline of yield with inbreeding in corn (inbreeding depression). A linear inbreeding depression, as observed here, is expected to result from detrimental recessive alleles that are rendered homozygous by the inbreeding. (Data from Neal, 1935.)

the frequency of aa will be $q^2(1 - 1/16) + q(1/16)$, because for these children $F = 1/16$, as will be shown in the next section. If the proportion of first cousin matings in a population is c, then the total frequency of affected children will be $q^2(1 - c)$ (for the matings between nonrelatives) + $c[q^2(1 - 1/16) + q(1/16)]$ (for the first cousin matings). The proportion of all affected children that have first cousin parents will therefore equal the frequency of aa children from first cousin matings — $c[q^2(1 - 1/16) + q(1/16)]$ — divided by the total frequency of aa children given earlier. A little algebra then shows the proportion of aa children having first cousin parents to be $c(1 + 15q)/(c + 16q - cq)$, curves of which are shown in Figure 21 for $c = .01$ (the approximate value of c for Caucasians in the United States) and $c = .06$ (the approximate value in Japan). Also graphed are data on several rare autosomal-recessive conditions in Caucasians and Japanese. If there were no inbreeding effect, the proportion of affected children having CONSANGUINEOUS (related) parents would simply equal the proportion of consanguineous matings (.01 in the case of Caucasians, .06 for Japanese). There is clearly a dramatic inbreeding effect, and the rarer the frequency of the deleterious recessive allele, the greater the effect.

CALCULATION OF THE INBREEDING COEFFICIENT FROM PEDIGREES

Computation of F from a pedigree is simplified by drawing the pedigree in the form shown in Figure 22A, where the lines represent gametes contributed by parents to their offspring. The same pedigree is shown in conventional form in Figure 22B. The individuals in gray in Figure 22B are not represented in Figure 22A because they have no ancestors in common and therefore do not contribute to the inbreeding of individual I. The inbreeding coefficient of individual I, call it F_I, is the probability that I is autozygous at the autosomal locus under consideration. The first step in calculating F_I is to locate all the common ancestors in the pedigree, because an allele could become autozygous in I only if it were inherited through both of I's parents from a common

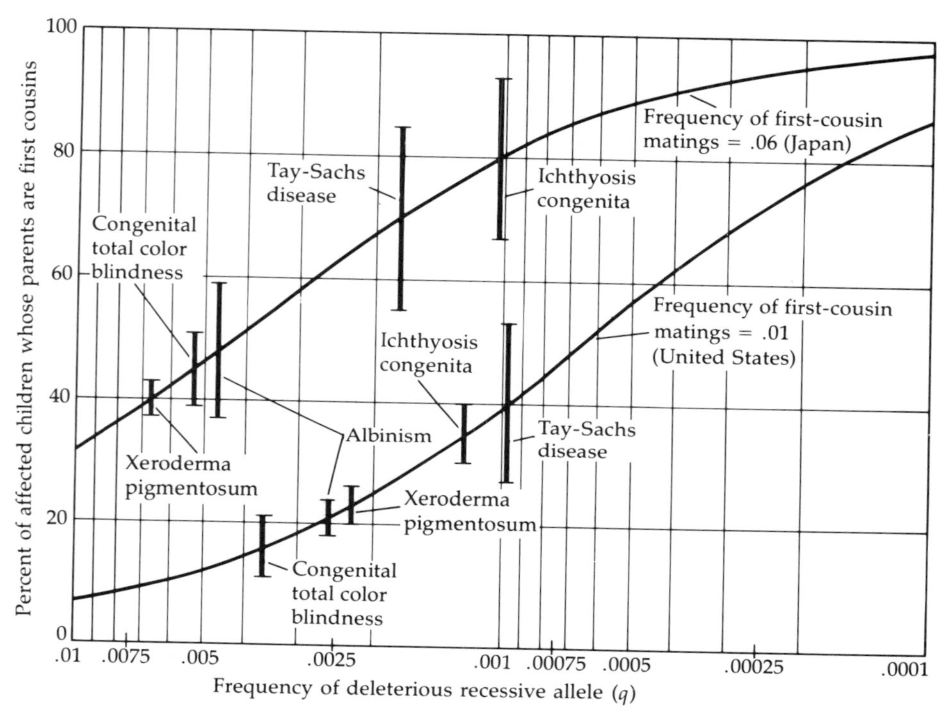

FIGURE 21. Curve of $c(1 + 15q)/(c + 16q - cq)$ for c (the proportion of first cousin matings) = .06 (Japan) and c = .01 (U.S.), against q (allele frequency of deleterious recessive). The ordinate is the proportion of affected children that have first cousin parents. Note that this proportion increases with the rarity of the harmful allele. Data pertaining to various diseases due to rare autosomal recessives are from Morton (1961).

ancestor; in this case, there is only one common ancestor, namely the individual labeled A. The next step in calculating F_I, carried out for each common ancestor, is to trace all the gametic paths that lead from one of I's parents back to the common ancestor and then down again to the other parent of I. These paths are the paths along which an allele in a common ancestor could become autozygous in I. In Figure 22A there is only one such path: DB$\underline{\text{A}}$CE (the common ancestor is underlined for bookkeeping purposes, an especially useful procedure in complex pedigrees).

The third step in calculating F_I is to calculate the proba-

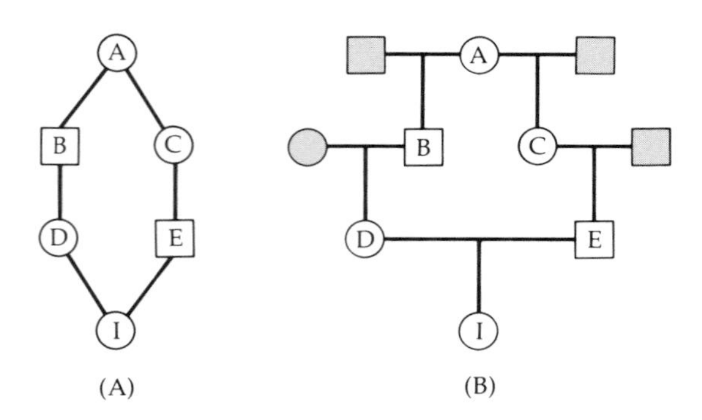

FIGURE 22. (A) Convenient way to represent pedigrees for calculation of the inbreeding coefficient. In this case, the pedigree shows a mating between half-first cousins. (B) Conventional representation of the same pedigree as in part A. Squares represent males, circles represent females, and the shaded individuals in part B are not depicted in part A because they do not contribute to the inbreeding of the individual designated I.

bility of autozygosity in I due to each of the paths in turn. For the path DB$\underline{A}$CE the reasoning involved is illustrated in Figure 23. Here the black dots represent alleles transmitted along the gametic paths, and the number associated with each loop is the probability of identity by descent of the alleles indicated. For all individuals except the common ancestor, the probability is one-half because of Mendelian segregation. (With Mendelian segregation, the probability that a particular allele present in a parent will be transmitted to a specified offspring is one-half.) To see why $(1/2)(1 + F_A)$ is the probability associated with the loop around the common ancestor, denote the alleles in the common ancestor as α_1 and α_2. (These symbols are used to avoid confusion with conventional allele symbols designating types of alleles, such as *A* and *a* for dominant and recessive, respectively.) The pair of alleles emanating from *A* could be $\alpha_1\alpha_1$, $\alpha_2\alpha_2$, $\alpha_1\alpha_2$, or $\alpha_2\alpha_1$, each with a probability of one-quarter because of Mendelian segregation. In the first two cases, the alleles are clearly identical by descent; in the second two cases, the alleles are identical by descent only if α_1 and α_2 are identical by descent, and α_1 and α_2 are identical by descent only if individual A is autozygous, which has probability F_A (the inbreeding coefficient of A). Altogether, the required probability for the loop around individual A is $1/4 + 1/4 + (1/4)F_A + (1/4)F_A = 1/2 + (1/2)F_A = (1/2)(1 + F_A)$. Now, since each of the loops in Figure 23 is independent of the others, the total probability of autozygosity in individual I due to this path is $1/2 \times 1/2 \times (1/2)(1 + F_A) \times 1/2 \times 1/2$, or $(1/2)^5(1 + F_A)$. Note that the exponent on the 1/2 is simply the number of individuals in the path. In general, if a path through a common ancestor A contains i individuals, the probability of autozygosity due to that path is

$$\left(\frac{1}{2}\right)^i (1 + F_A)$$

Thus the inbreeding coefficient of I in Figure 22A is $(1/2)^5(1 + F_A)$, or, if $F_A = 0$, simply $(1/2)^5 = 1/32$.

In more complex pedigrees, there will be more than a single path. The paths are mutually exclusive, however — if an individual is autozygous due to an allele inherited along one path, the individual cannot at the same time be autozygous due to an allele inherited along a different path.

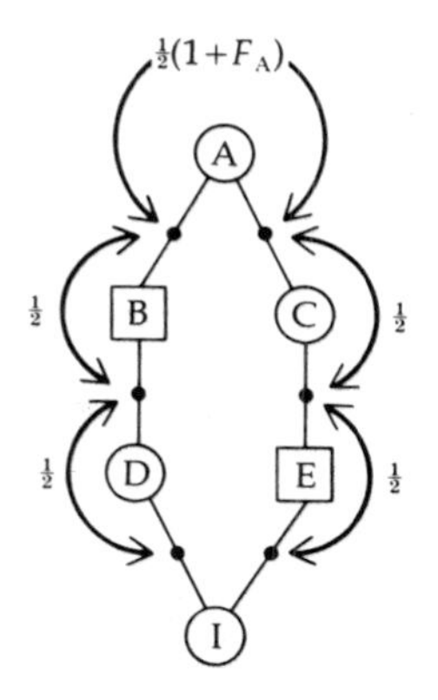

FIGURE 23. Loops for the pedigree in Figure 22A, showing probabilities that designated alleles (solid dots) are identical by descent. Each loop is independent of the others, so their probabilities multiply. Thus, the inbreeding coefficient of individual I is $F_I = (1/2)^5(1 + F_A)$, where F_A represents the inbreeding coefficient of the common ancestor.

Thus, the total inbreeding coefficient is the sum of the probability of autozygosity due to each separate path. The whole procedure for calculating F is summarized in the example in Figure 24. Here there are three common ancestors (A, B, and C) and four paths (one each with B or C and two with A as common ancestor). The total inbreeding coefficient of I is the sum of the four separate contributions shown in Figure 24. If A, B, and C are noninbred, then $F_A = F_B = F_C = 0$, and $F_I = (1/2)^3 + (1/2)^3 + (1/2)^5 + (1/2)^5$, which works out to $F_I = 5/16$. This is the probability that I is autozygous at a specified locus; alternatively, F_I can be interpreted as the average proportion of all loci in I that are autozygous (Franklin, 1977). (Rules for calculating F from irregular pedigrees or for sex-linked loci are summarized in Box H; calculation of F for populations with overlapping generations is discussed in Choy and Weir, 1978.)

REGULAR SYSTEMS OF MATING

In plant and animal breeding, it is often useful to know how fast the inbreeding coefficient increases when a strain is

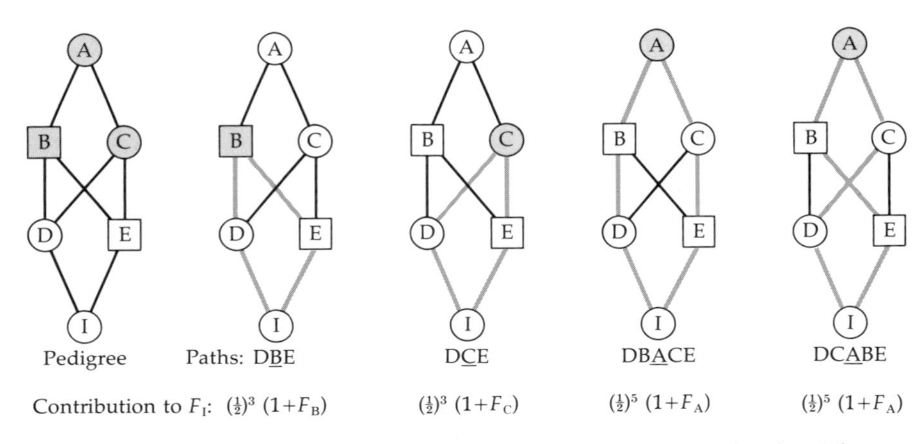

FIGURE 24. On the left is a pedigree of an individual I. On the right are the four paths through common ancestors (shaded lines) involved in calculating the inbreeding coefficient of I. Below each path is the contribution to F_I due to that path, calculated as in Figure 23. Each path is mutually exclusive of the others, so their probabilities add. Thus, the total inbreeding coefficient of I is the sum of the four separate contributions.

Calculation of F for Irregular Pedigrees

H

The usual formula for calculating the inbreeding coefficient F from a pedigree is $\Sigma(1/2)^n(1 + F_A)$, where the summation sign Σ means summation over all paths in the pedigree that trace back from the parents of the individual in question to any common ancestor, n is the number of individuals in each path (counting the parents of the individual in question, but not the individual itself), and F_A is the inbreeding coefficient of the common ancestor.

The same formula holds for X-linked genes, with the following modifications: (1) that $F = 1$ for any male (because males have only one X chromosome), (2) that paths connecting two consecutive males be excluded (because a male cannot transmit his X chromosome to his sons), and (3) that only females in a path be counted (because a male must transmit his X chromosome to his daughters).

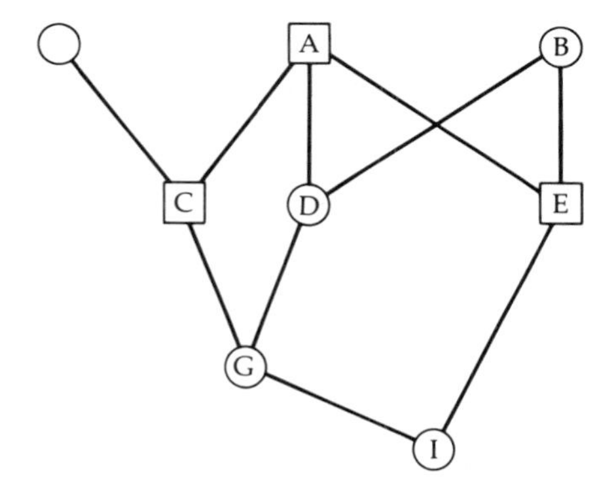

In the pedigree shown, assume $F_A = 1$ and $F_B = 1/4$, and calculate the inbreeding coefficient of individual I

a. for an autosomal gene,

b. for a sex-linked gene.

propagated by a regular system of mating, such as repeated self-fertilization, sib mating, or backcrossing to a standard strain (Fisher, 1949; Wright, 1958). The reasoning involved in calculating the inbreeding coefficient for any generation is illustrated in Figure 25 for repeated self-fertilization. (For

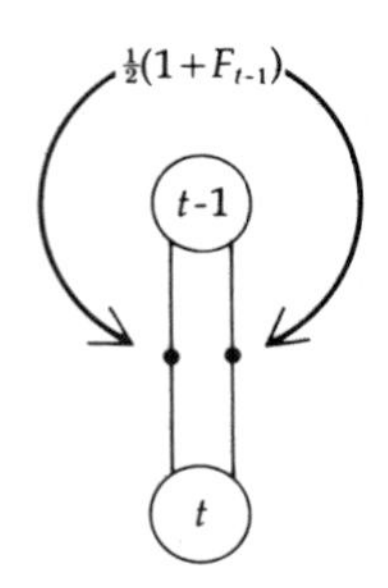

FIGURE 25. Increase of F due to continued self-fertilization. The individual in generation t is the offspring of self-fertilization of the individual in generation $t - 1$. The loop shows that $F_t = (1/2)(1 + F_{t-1})$.

sib mating, see Box I.) In Figure 25, the labels $t - 1$ and t refer to the individuals after $t - 1$ and t generations of self-fertilization. The loop around the individual in generation $t - 1$ designates the probability that the two indicated alleles are identical by descent. Here the standard formula $(1/2)^i(1 + F_A)$ applies with only one path and only one individual in that path, so $F_t = (1/2)^1(1 + F_{t-1})$, where F_t is the inbreeding coefficient in generation t. This equation is easy to solve in terms of the quantity $1 - F_t$, which is often called the PANMICTIC INDEX. (PANMIXIA is a synonym for random mating; with inbreeding, the heterozygosity is given in Table VI as $2pq(1 - F)$, so $1 - F$ measures the amount of heterozygosity relative to that in a panmictic population.) Multiplying both sides of the equation for F_t by -1 and then adding 1 to each side leads to $1 - F_t = 1 - (1/2)(1 + F_{t-1}) = 1 - (1/2) - (1/2)F_{t-1} = (1/2)(1 - F_{t-1})$, or

$$1 - F_t = \left(\frac{1}{2}\right)^t (1 - F_0)$$

where F_0 is the inbreeding coefficient in the initial generation. Self-fertilization therefore leads to an extremely rapid increase in the inbreeding coefficient. When $F_0 = 0$, then $F_1 = 1/2$, $F_2 = 3/4$, $F_3 = 7/8$, $F_4 = 15/16$, and so on. How F increases when $F_0 = 0$ under self-fertilization and several other regular systems of mating is shown in Figure 26.

Inbreeding with Continued Brother–Sister Mating

I

The pedigree shows three generations of recurrent sib mating. Letting F_t represent the inbreeding coefficient in generation t, it can be shown that $F_t = (1/2)F_{t-1} + (1/4)F_{t-2} + 1/4$ for this mating system, with $F_0 = F_1 = 0$. (Derivation of this equation entails methods beyond the scope of this book.) Let H_t represent the proportion of heterozygotes in generation t, and assume that the initial population (generation 0) is undergoing random mating. From Table VI in the text, $H_0 = 2pq$ and $H_t = 2pq(1 - F_t)$.

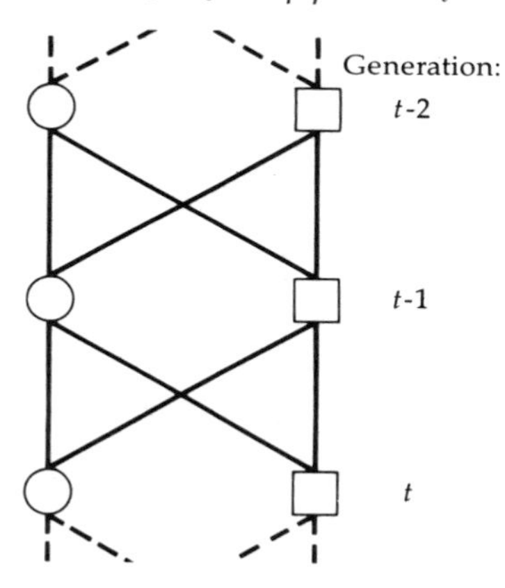

a. Show that $F_t = (H_0 - H_t)/H_0$ (i.e., the inbreeding coefficient measures the decrease in heterozygosity due to inbreeding).

b. By direct substitution, show that sib mating leads to the equation $H_t = (1/2)H_{t-1} + (1/4)H_{t-2}$. For $H_0 = H_1 = 1/2$, show that this expression leads to the sequence $H_2 = 3/8$, $H_3 = 5/16$, $H_4 = 8/32$, $H_5 = 13/64$. (Note for each value that the denominator is twice the preceding one and the numerator is the sum of the two preceding ones.)

c. Recurrent sib mating rather quickly reaches a state where the heterozygosity is reduced by a nearly constant fraction in each generation. Denote $H_t/H_{t-1} = H_{t-1}/H_{t-2}$ by the symbol λ, and use the expression in (b) to show that λ must satisfy the equation $4\lambda^2 - 2\lambda - 1 = 0$. Since λ is necessarily positive, show that the appropriate solution is $\lambda = (1 + \sqrt{5})/4 \approx .80902$. This means that after several generations of sib mating, the heterozygosity is reduced by $1 - \lambda \approx .19098$, or about 19 percent with each generation.

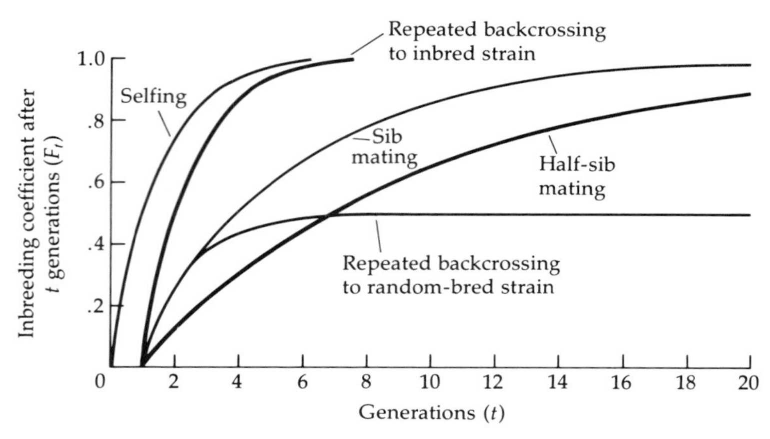

FIGURE 26. Increase of F for regular systems of mating: selfing, sib mating, half-sib mating, and repeated backcrossing.

Many plants reproduce predominantly by self-fertilization, including such crop plants as soybeans, sorghum, barley, and wheat. As expected of highly self-fertilizing species, each individual plant is highly homozygous for alleles such as those determining allozymes. Yet the proportion of polymorphic loci is comparable to that found in outcrossing species; self-fertilization simply reorganizes the genetic variation into homozygous genotypes (see Brown, 1979, and the following example from Clegg et al., 1972). On the other hand, self-fertilizing species carry fewer deleterious recessives than do outcrossing species, presumably because the increased homozygosity permits harmful recessives to be eliminated from the population by natural selection. One other important point about naturally self-fertilizing species: the high homozygosity for all loci implies that recombination will rarely lead to new gametic types not already present in the parent. Predominance of selfing in a species, therefore, has the effect of retarding the approach to linkage equilibrium, the reason being that the approach to linkage equilibrium is through recombination in double heterozygotes (AB/ab and Ab/aB in the case of two alleles at each locus), and with extreme inbreeding, such double heterozygotes will be rare. Indeed, the most extreme examples of linkage disequilibrium have been found in such predominantly self-fertilizing species as barley (*Hordeum vulgare*) and wild oats (*Avena barbata*).

Clegg et al. (1972) provide an example of extreme linkage disequilibrium between two unlinked ($r = 1/2$) esterase loci in *Hordeum vulgare,* a species that regularly undergoes more than 99 percent self-fertilization. The population had originated as a complex cross and had been maintained for 26 generations under normal agricultural conditions without conscious selection. At the esterase-A locus, they distinguished two alleles, A_1 and A_2; and at the esterase-D locus, they distinguished two alleles, D_1 and D_2. They observed the following numbers of each of the gametic types, although for practical purposes these numbers also refer to homozygous genotypes since there is such close inbreeding [the numbers in parentheses, calculated exactly as for the data of Mukai et al. (1974) involving the *Amy* locus and the *NS* inversion, are the expected numbers based on the assumption of linkage equilibrium]:

A_1D_1	1433	(1356.46)
A_1D_2	430	(506.54)
A_2D_1	787	(863.54)
A_2D_2	399	(322.46)

The χ^2 value in this case is 40.83, and (as with the data of Mukai et al., 1974), the χ^2 has 1 degree of freedom. This χ^2 has an associated probability of much less than .0001, so there certainly is significant linkage disequilibrium. For the above data, the linkage disequilibrium parameter is $D = .025$, which is about 15 percent of its hypothetical maximum. (For more about linkage disequilibrium in *Hordeum* and other plants, see Allard et al., 1972; Kahler et al., 1975; Allard, 1975; for some theoretical treatments of extreme linkage disequilibrium, see Franklin and Lewontin, 1970; Lewontin, 1970; Slatkin, 1972.)

One of the dramatic successes of plant breeding has come from the crossing of inbred lines to produce high-yielding hybrid corn, a procedure first suggested by G. H. Shull (1909). (See Hayes, 1963, for a history.) Yield of a genetically heterogeneous, outcrossing variety of corn can be improved by selecting the plants with the highest yields in each generation to be the progenitors of the next generation; such artificial selection results in only gradual improvement, however (see Chapter 4). If a large number of self-fertilized lines are established from a heterogeneous population, each line will decline in yield as inbreeding proceeds, due to the forced homozygosity of deleterious recessives (see Figure 20); many lines become so inferior that they have to be discontinued. Self-fertilized lines are not likely to become homozygous for exactly the same set of deleterious recessives, however, so when different lines are crossed to produce a hybrid, the hybrid will be heterozygous for these loci. Alleles favoring high yield in corn are generally dominant, and there may also be loci at which heterozygous genotypes have a more favorable effect on yield than do homozygous genotypes; in any case, the hybrid will have a

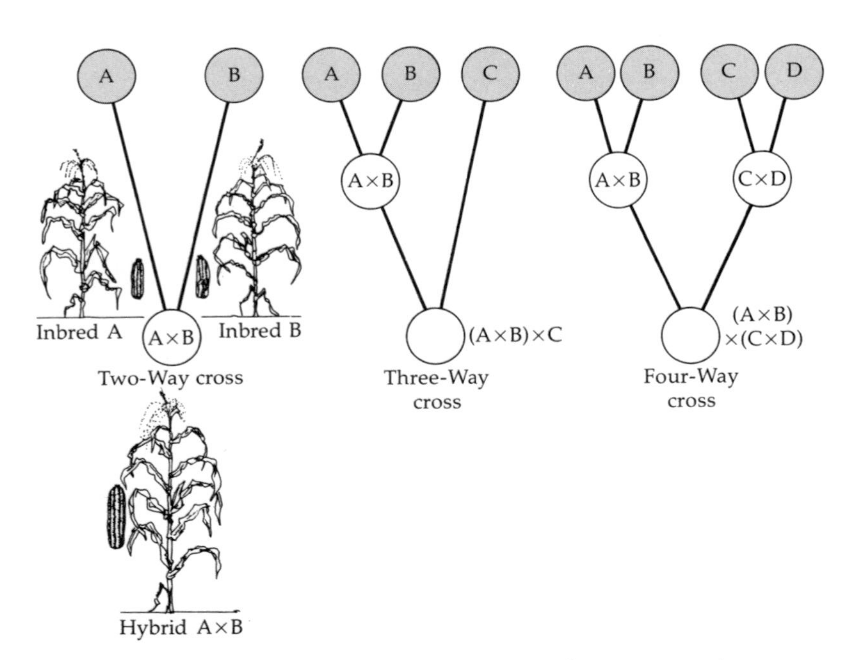

FIGURE 27. Mating schemes for inbred lines of corn to produce two-way, three-way, and four-way hybrids.

much higher yield than either inbred parent. This phenomenon of enhanced hybrid performance is called HYBRID VIGOR or HETEROSIS. In practice, inbred lines are crossed in many combinations to identify those that produce the best hybrids, and three-way or four-way crosses are sometimes used commercially (Figure 27). Yields of hybrid corn are typically 15 to 35 percent greater than yields of outcrossing varieties (Jenkins, 1978), and the successful introduction of hybrid corn has been remarkable. Virtually all corn acreage in the United States today is planted with hybrids, as compared to 0.4 percent of the acreage in 1933 (Sprague, 1978).

In extreme contrast to plants that regularly undergo self-fertilization are those, such as clover, that have evolved special loci with numerous alleles for the complete prevention of selfing. Such SELF-STERILITY ALLELES are one of the subjects of the next chapter.

Problems

1. If an enzyme is a tetramer composed of four polypeptide chains from the same locus, how many enzyme bands would you expect to see upon electrophoresis of tissue from a heterozygote?
2. For the data of Levy and Levin (1975) for *Oenothera biennis* in Box A, calculate the expected numbers of the three genotypes *PGI-2a/PGI-2a, PGI-2a/PGI-2b,* and *PGI-2b/PGI-2b,* assuming that the genotypes occur in Hardy–Weinberg proportions.
3. Kelus (cited in Mourant et al., 1976) reports a study of 3100 Poles, of which 1101 were *MM,* 1496 were *MN,* and 503 were *NN.* Calculate the allele frequencies of *M* and *N,* the expected numbers of the three genotypic classes (assuming random mating), and the χ^2 for goodness of fit. Interpret the χ^2.
4. Mourant et al. (1976) cite data on 400 Basques from Spain, of which 230 were Rh^+ and 170 were Rh^-. Calculate the allele frequencies of *D* and *d.* How many of the Rh^+ individuals would be expected to be heterozygous?
5. Phenylketonuria is a severe form of mental retardation due to a rare autosomal recessive. About one in 10,000 newborn Caucasians are affected with the disease. Calculate the frequency of carriers (i.e., heterozygotes).

6. The I^A "allele" for the ABO blood groups actually consists of two subtypes, I^{A_1} and I^{A_2}, either being considered "I^A". In Caucasians, about 3/4 of the I^A alleles are I^{A_1} and 1/4 are I^{A_2} (Cavalli-Sforza and Edwards, 1967). Among individuals of genotype I^AI^O, what fraction would be expected to be $I^{A_1}I^O$? What fraction $I^{A_2}I^O$? What would be the expected proportions of $I^{A_1}I^{A_1}$, $I^{A_1}I^{A_2}$, and $I^{A_2}I^{A_2}$ among I^AI^A individuals?

7. If the frequency of the "green" form of red–green color blindness (due to an X-linked locus) is 5 percent among males, what fraction of females would be affected? What fraction of females would be heterozygous?

8. Imagine an autosomal locus with four alleles, A_1, A_2, A_3, and A_4, at frequencies .1, .2, .3, and .4, respectively. Calculate the expected random-mating frequencies of all possible genotypes.

9. Consider a locus with two alleles (A_1 and A_2) and another locus with three alleles (B_1, B_2, B_3). Let $p_1 = .3$ be the allele frequency of A_1, $q_1 = .2$ be that of B_1, and $q_2 = .3$ be that of B_2. Calculate the frequencies of all possible *gametes*, assuming that the loci are in linkage equilibrium.

10. Given the following table of allele frequencies:

	Locus				
	1	2	3	4	5
allele 1	.63	.94	.995	1.0	.78
allele 2	.37	.06	.005	—	.12
allele 3	—	—	—	—	.06
allele 4	—	—	—	—	.04

What is the proportion (P) of polymorphic loci (using the definition in the text)? Assuming random mating and linkage equilibrium, what is the average heterozygosity (H) for the set of loci?

11. Calculate the inbreeding coefficient for the progeny of a parent–offspring mating; for a brother–sister mating.

12. Consider a harmful recessive allele at frequency .02 in a population in which 1.5 percent of the matings are between first cousins. Among homozygotes for the allele, what fraction will have parents who are first cousins?

13. The inbreeding coefficient of the offspring of second cousins

is $F = 1/64$. For an autosomal locus having alleles A and a at frequencies .995 and .005, respectively, what would be the frequencies of AA, Aa, and aa among the offspring of nonrelatives? Among the offspring of second cousins?

14. Suppose a maize two-way hybrid is allowed to undergo open pollination (i.e., random mating) for one generation. What is the inbreeding coefficient among the offspring? Answer the same question for the open-pollinated progeny of a three-way hybrid.

15. Calculate the analog of the Hardy–Weinberg equilibrium for a locus with two alleles in an autotetraploid (Bennett, 1954).

Further Readings

Ayala, F. J. (ed.). 1976. *Molecular Evolution.* Sinauer Associates, Sunderland, Massachusetts.

Cavalli-Sforza, L. L. and W. F. Bodmer. 1971. *The Genetics of Human Populations.* W. H. Freeman, San Francisco.

Cook, L. M. 1976. *Population Genetics.* Halsted Press, New York.

Crow, J. F. and M. Kimura. 1970. *An Introduction to Population Genetics Theory.* Harper and Row, New York.

Dobzhansky, Th. 1970. *Genetics of the Evolutionary Process.* Columbia Univ. Press, New York.

Dobzhansky, Th., F. J. Ayala, G. L. Stebbins and J. W. Valentine. 1977. *Evolution.* W. H. Freeman, San Francisco.

Elandt-Johnson, R. C. 1971. *Probability Models and Statistical Methods in Genetics.* John Wiley and Sons, New York.

Felsenstein, J. and B. Taylor (eds.). 1974. *A Bibliography of Theoretical Population Genetics.* A. E. C. Rep. No. RLO-2225-5-18, National Technical Information Service, Springfield, Virginia.

Hayes, H. K. 1963. *A Professor's Story of Hybrid Corn.* Burgess, Minneapolis.

Lewontin, R. C. 1974. *The Genetic Basis of Evolutionary Change.* Columbia Univ. Press, New York.

Li, C. C. 1976. *First Course in Population Genetics.* Boxwood, Pacific Grove, California.

Mettler, L. E. and T. G. Gregg. 1969. *Population Genetics and Evolution.* Prentice-Hall, Englewood Cliffs, New Jersey.

Morris, L. N. (ed.). 1971. *Human Populations, Genetic Variation, and Evolution.* Chandler, New York.

Mourant, A. E., A. C. Kopeć and K. Domaniewska-Sobczak. 1976. *The Distribution of Human Blood Groups and Other Polymorphisms,* 2nd ed. Oxford Univ. Press, New York.

Murphy, E. A. and G. A. Chase. 1975. *Principles of Genetic Counseling.* Year Book Medical Publishers, Chicago.

Provine, W. B. 1971. *The Origins of Theoretical Population Genetics.* Univ. of Chicago Press, Chicago.

Race, R. R. and R. Sanger. 1975. *Blood Groups in Man,* 6th ed. J. B. Lippincott, Philadelphia.

Schull, W. J. and J. V. Neel. 1965. *The Effects of Inbreeding on Japanese Children.* Harper and Row, New York.

Spiess, E. B. 1977. *Genes in Populations.* John Wiley and Sons, New York.

Wright, S. 1969. *Evolution and the Genetics of Populations,* Vol. 2: *The Theory of Gene Frequencies.* Univ. of Chicago Press, Chicago.

· 3 ·

The Causes of Evolution

Populations are extremely complex entities. The Hardy-Weinberg model discussed in Chapter 2, useful as it is as a first approximation, ignores most of the complexities of actual populations. Populations are not infinitely large, and their sizes are rarely constant, so fluctuations in allele frequency can occur by chance. Populations are also subject to the systematic evolutionary forces of migration, mutation, and natural selection, all of which cause nonrandom or directional changes in allele frequency. Changes in allele frequencies are changes in the genetic makeup of a population. Since EVOLUTION is often defined as cumulative change in the genetic makeup of a population, the fundamental process in evolution is change in allele frequencies. In this chapter we focus on the various forces that change allele frequency. These forces hold the key to understanding the origin and maintenance of genetic variation.

Random Genetic Drift

RANDOM GENETIC DRIFT refers to chance fluctuations in allele frequency; such fluctuations occur particularly in small populations as a result of random sampling among gametes. To be specific, a population of nine diploid organisms arises from a sample of just 18 gametes out of an essentially infinite pool of gametes. Because small samples are frequently not representative, an allele frequency in the sample may differ from that in the entire pool of gametes. In fact, if the number of gametes in a sample is represented as $2N$ (in the above example, $2N = 18$), the probability that the sample contains exactly i alleles of type A is the binomial probability

$$\binom{2N}{i} p^i q^{2N-i}$$

where $\binom{2N}{i}$ means $(2N)!/i!(2N - i)!$ (refer to Chapter 1 for a discussion of binomial probabilities); p and q are, respectively, the allele frequencies of A and a in the entire pool of gametes ($p + q = 1$); and i takes on any value between 0 and $2N$. The new allele frequency in the population (call it p') is therefore $i/2N$ because, by definition, the allele frequency of A equals the number of A alleles (in this case, i) divided by the total (in this case, $2N$). In the next generation, the sampling process occurs anew, and the new probability of a prescribed number of A alleles occurring in the $2N$ gametes is given by the binomial probability above, with p now replaced by p' and q by $1 - p'$. Thus, the allele frequency may change at random from generation to generation. A computer-generated example based on random numbers is shown in Figure 1. Each small graph gives the number of A alleles in 19 successive generations of random genetic drift in a population of size $N = 9$ (so $2N = 18$). As you can see, individual populations behave very erratically. In seven populations, the A allele became FIXED (that is, $p = 1$); in six populations, A became LOST (that is, $p = 0$). The other 11 populations remained UNFIXED (A was neither fixed nor lost), but the final allele frequency among the unfixed populations was as likely to be one value as another. The principal conclusion from Figure 1 is that allele frequencies

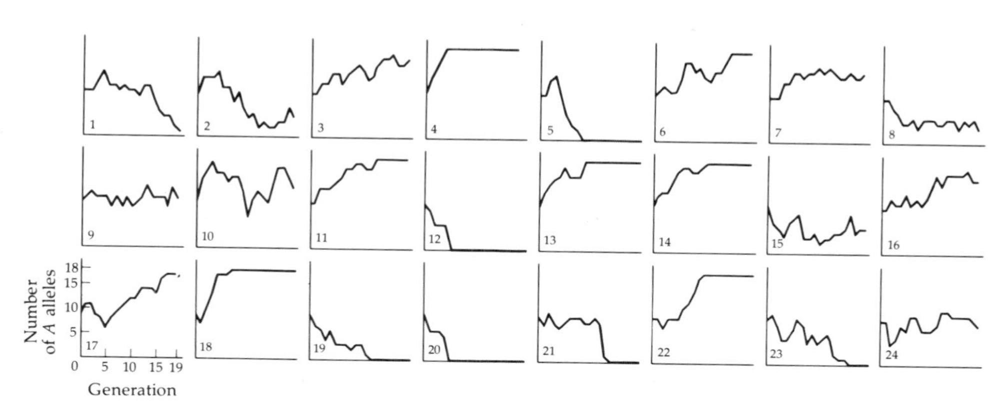

FIGURE 1. Change of allele frequency by random genetic drift over 19 generations in 24 hypothetical populations of size $N = 9$.

behave so erratically in any one population that prediction is virtually impossible.

Although changes in allele frequency due to random genetic drift in any individual population may defy prediction, the average behavior of allele frequencies in a large number of populations can be predicted. A framework for thinking about the problem is illustrated in the model in Figure 2.

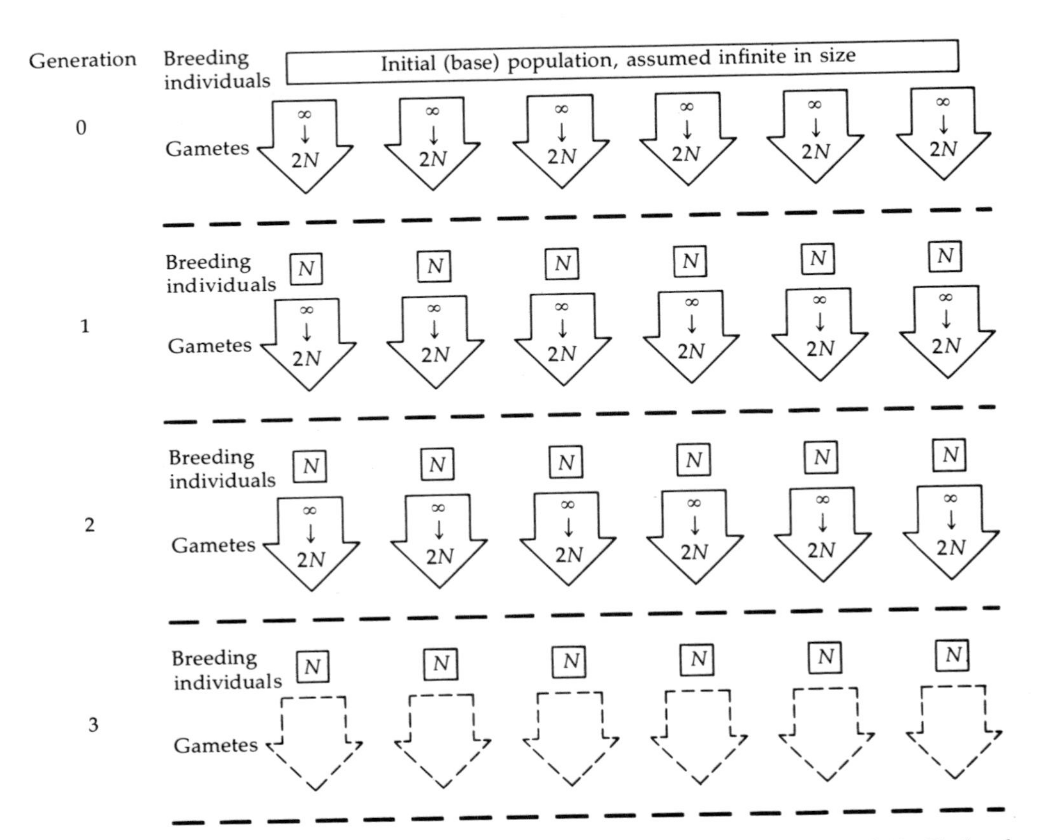

FIGURE 2. Model for analyzing effects of random genetic drift. Each of the subpopulations founded from the initial large population (shown by vertical columns of boxes and arrows) is assumed to be genetically isolated from the others. Each subpopulation produces an infinite number of gametes, of which $2N$ are "chosen" by random sampling to form the next generation's breeding population; random genetic drift results from sampling error in this process.

Here an infinitely large initial population is imagined to be split up into a large number of local populations or subpopulations, each of size N, which correspond to the populations in Figure 1. (As will be explained shortly, N is technically the "effective population number.") Except for their finite size, the subpopulations are assumed to satisfy all the assumptions of the Hardy–Weinberg model, as summarized in Table I, with the additional stipulations that the number of males and females is equal and that each individual has an equal chance of contributing successful gametes to the next generation. In short, the subpopulations in Figure 2 are theoretically uncomplicated or "ideal," and reference will be made to such ideal populations throughout this chapter. The importance of the concept of ideal populations is that ideal populations provide a standard of comparison for other populations that violate the ideal-population assumptions.

Results of the kind of population structure in Figure 2 are shown for an actual case in Figure 3, which recounts the history of 19 generations of random genetic drift in 107 subpopulations of *Drosophila melanogaster*, each population initiated with 16 bw^{75}/bw (bw = *brown* eyes) heterozygotes and maintained at a constant size of 16 by randomly choosing 8 males and 8 females as the breeding population. Each histogram in Figure 3 gives the number of populations having 0, 1, 2, . . ., 32 bw^{75} alleles. The overall pattern of change in allele frequency in Figure 3 is obvious: the initially humped distribution of allele frequency gradually becomes

TABLE I. Assumptions of model of random genetic drift.

(1) Diploid organism
(2) Sexual reproduction
(3) Nonoverlapping generations
(4) Many independent subpopulations, each of constant size N
(5) Random mating within each subpopulation
(6) No migration between subpopulations
(7) No mutation
(8) No selection

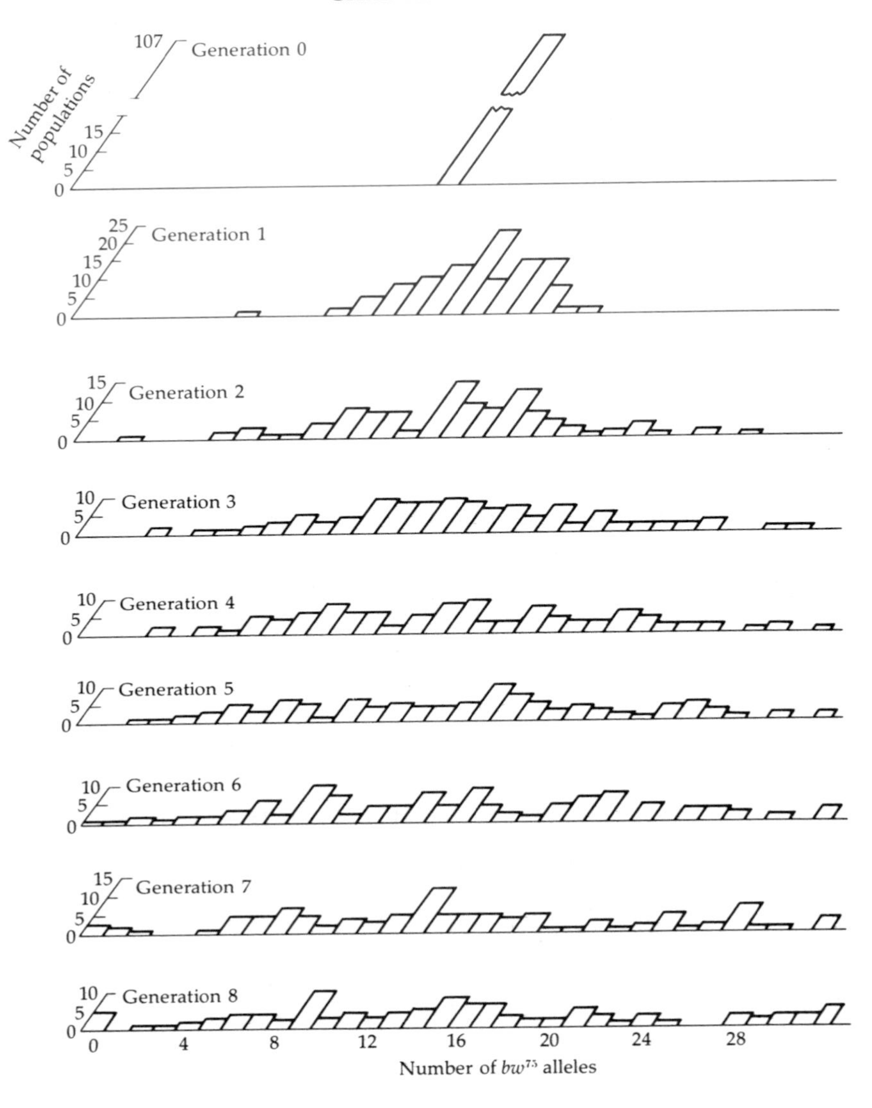

FIGURE 3. Random genetic drift in 107 actual populations of *Drosophila melanogaster*. Each of the initial 107 populations consisted of 16 bw^{75}/bw heterozygotes (N = 16; *bw* = *brown* eyes). From among the progeny in each generation, 8 males and 8 females were chosen at random to be the parents of the next generation. The abscissa of each histogram gives the number of bw^{75} alleles in the population, and the ordinate gives the corresponding number of populations. (Data from Buri, 1956.)

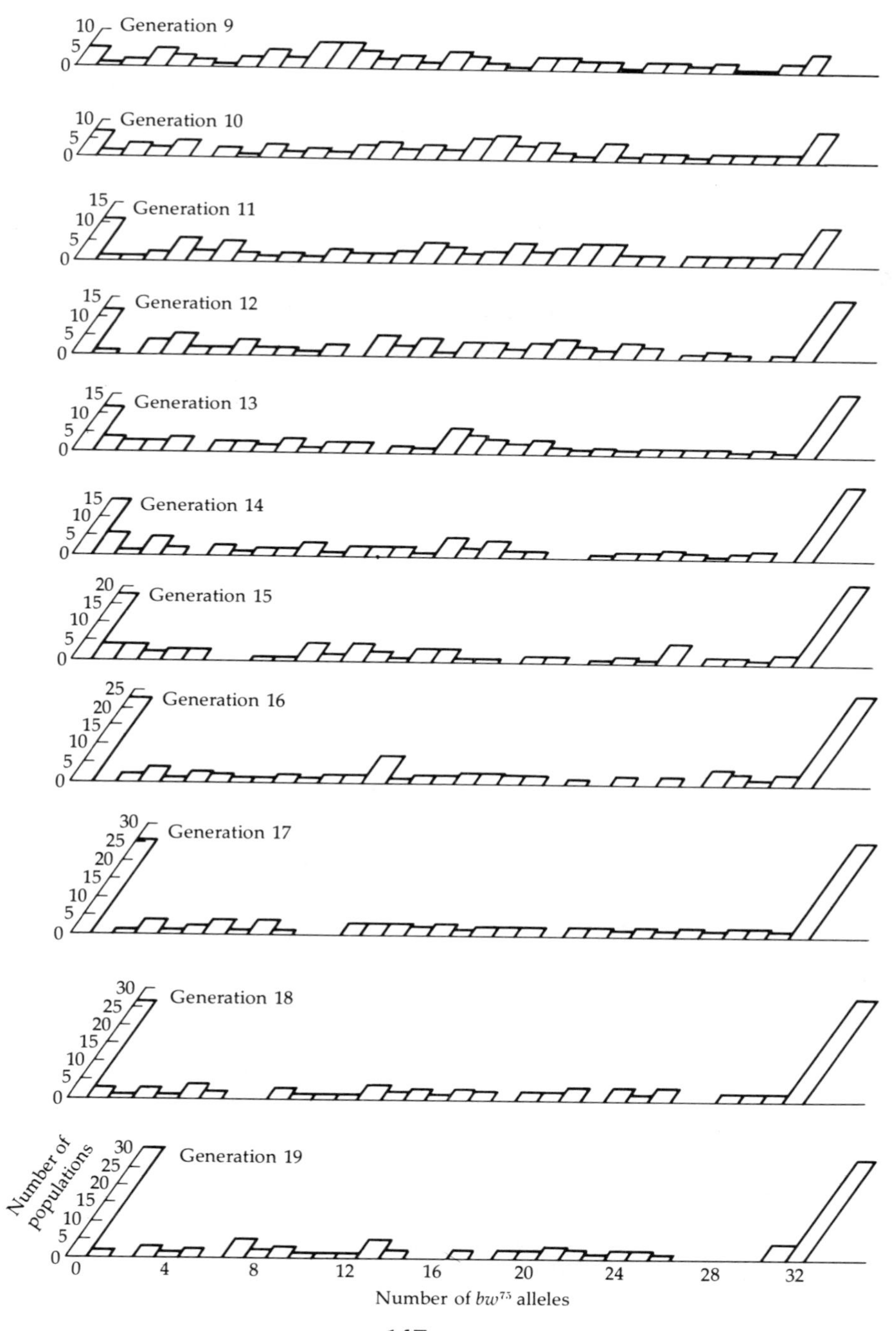
Generation 9
Generation 10
Generation 11
Generation 12
Generation 13
Generation 14
Generation 15
Generation 16
Generation 17
Generation 18
Generation 19
Number of populations
0
5
10
15
20
25
30
0
4
8
12
16
20
24
28
32
Number of bw^{75} alleles

flat as populations fixed for bw^{75} or bw begin to pile up at the ends. (Once an allele has been fixed or lost in this type of example, it remains fixed or lost because mutation is negligible over such a small number of generations in small populations.) By about 18 generations, half the populations are fixed for one allele or the other, and among the unfixed populations, the distribution of allele frequencies is essentially flat.

CONSEQUENCES OF RANDOM GENETIC DRIFT

The pattern of change in allele frequency shown in Figure 3 is very nearly that expected theoretically for an ideal pop-

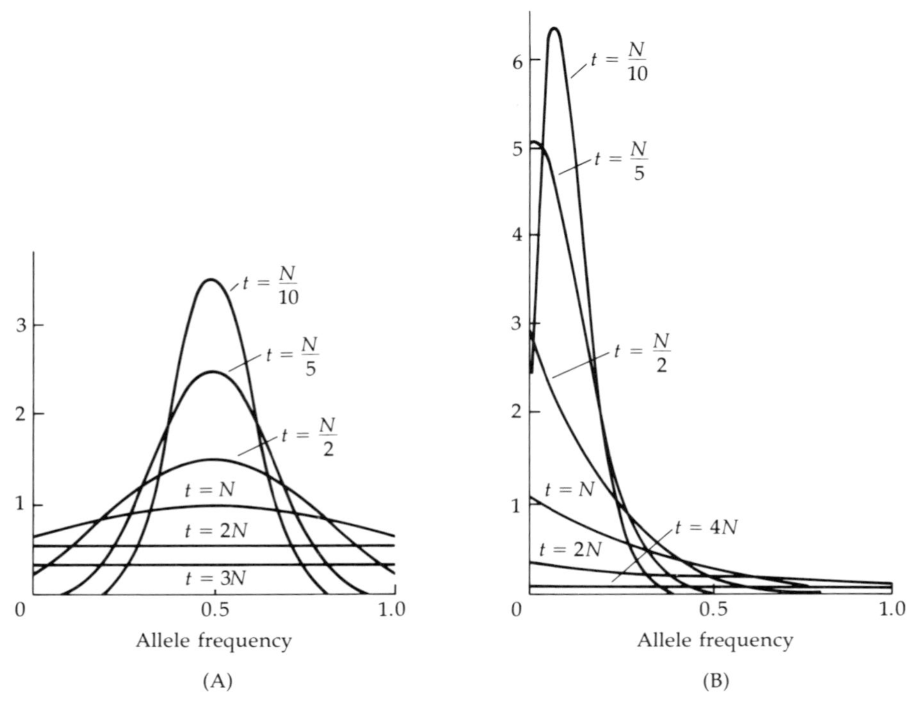

FIGURE 4. Theoretical results of random genetic drift. (A) Initial allele frequency = .5; (B) Initial allele frequency = .1. The area under each curve is equal to the proportion of populations in which fixation or loss has not yet occurred. The curves are the distributions of allele frequencies in those unfixed populations. (From Kimura, 1955.)

ulation, although the full-blown theory of random genetic drift requires mathematics beyond the scope of this book (see Kimura, 1964, 1976; Crow and Kimura, 1970; Kimura and Ohta, 1971). The two families of curves in Figure 4 are the theoretical distributions of allele frequency among unfixed populations after various times (t) measured in units of N generations. In Figure 4A, all populations have an initial allele frequency of .5, as in the actual populations in Figure 3; after about $t = 2N$ generations, the distribution of allele frequency is essentially flat, and by this time about half the populations are still unfixed. The distributions in Figure 4 refer only to those populations that are unfixed; as time goes on, of course, more and more of the populations do become fixed, thus the distributions progressively pile up at 0 and 1, as in the histograms in Figure 3. Indeed, in Figure 4, the area under each curve is equal to the proportion of unfixed populations, and this area becomes progressively smaller as time goes on. Figure 4B shows what happens when the initial allele frequency is 0.1; here the distributions are highly asymmetrical, and the distribution of allele frequency does not become flat until about $t = 4N$ generations, by which time only about 10 percent of the populations remain unfixed. Once a flat distribution of allele frequency is reached, the distribution remains flat, but random drift continues until fixation or loss has occurred in all populations. The average time required for fixation or loss is shown in Figure 5, which gives the average time in units of N generations that an allele, initially at frequency p_0, remains in the population before it is fixed or lost. When $p_0 = .5$, the average time that a population remains unfixed is about $2.8N$ generations.

Many of the important consequences of random genetic drift follow from the fact that the population structure in Figure 2 entails a peculiar sort of "inbreeding," but the nature of the inbreeding is very subtle because the population structure in Figure 2 can be considered on two distinct levels — the level of the individual subpopulation and the level of the total population, which is the aggregate of all subpopulations.

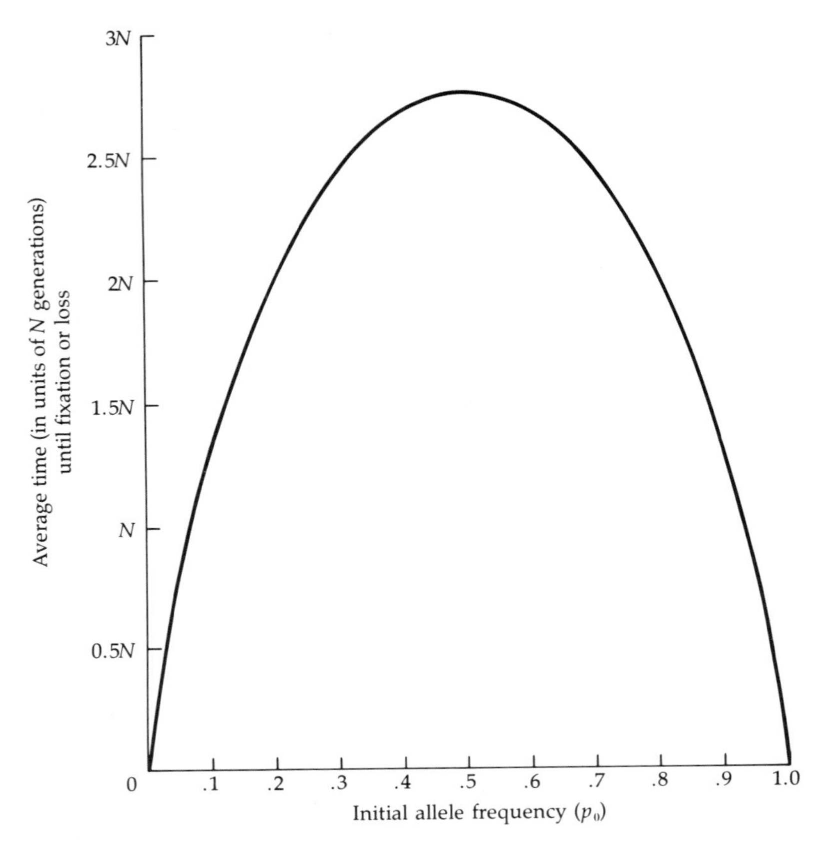

FIGURE 5. Average persistence of a neutral allele in an ideal population of size N, against initial allele frequency.

Consider first any one of the subpopulations in Figure 2. Within this subpopulation (call it subpopulation number i), mating is random because of the assumptions in Table I; if the allele frequencies of A and a in the ith subpopulation are denoted p_i and q_i, then the genotype frequencies of AA, Aa, and aa will be given by the familiar Hardy-Weinberg law as p_i^2, $2p_iq_i$, and q_i^2. Furthermore, picture the situation in Figure 2 at a time so advanced that all subpopulations will be fixed for one allele or the other. Within the ith subpopulation, therefore, either p_i equals 0 or p_i equals 1. The genotype frequencies of *AA*, *Aa*, and *aa* in that subpopulation will then be either 0, 0, and 1 (if $p_i = 0$), or they will be 1, 0, and 0 (if $p_i = 1$). These genotype frequencies, though extreme, still satisfy the Hardy-Weinberg law. Thus, within any one subpopulation in Figure 2, the frequency of heterozygotes is that expected with random mating.

The situation regarding the total population in Figure 2 is very different, however, as there is an overall deficiency of heterozygotes. The meaning of "total population" in this context can be made clear by considering a simple example. Suppose the subpopulations in question are colonies of mice in a large barn. (Were the mice real, we would have to worry about migration between subpopulations, but these are hypothetical mice.) Suppose further that we are unaware of the existence of such subpopulations but instead think that the barn contains a single randomly mating population. To study the "total population" of the barn, we trap mice at random. If time is so advanced that a fraction p of the subpopulations are fixed for A and a fraction q are fixed for a (remember, $q = 1 - p$), then a fraction p of the time, we will trap an *AA* mouse and a fraction q of the time, an *aa* mouse. The overall allele frequency of A among the trapped animals would then be p, and we would naively expect a fraction $2pq$ of the animals to be heterozygous. In fact, we would have caught no heterozygotes at all!

This rather paradoxical result — that there is a deficiency of heterozygotes in the total population even though random mating occurs within each subpopulation — is a consequence of the random genetic drift of allele frequencies among subpopulations due to their finite size. Were the subpopulations so large that random drift could be ignored, each subpopulation would have the same allele frequencies and Hardy-Weinberg genotype frequencies as any other. In such a case, to return to the mouse example, it would not matter to the proportion of heterozygotes if mice were sampled from the total population or from any one of the subpopulations. For the model in Figure 2, however, random genetic drift is important. The total population has a deficiency of heterozygotes, much as if there were inbreeding.

It is this inbreeding-like effect of population subdivision that we now set out to quantify.

Because Figure 2 entails an inbreeding-like effect, it is convenient to measure the effect in terms of the probability of autozygosity (i.e., identity of alleles by descent). Accordingly, let F_t represent the probability that two alleles chosen at random from within the same subpopulation in generation t are identical by descent. In the present context, the symbol F_{ST} is often used in place of F, but we will simply use F to avoid a proliferation of subscripts. It is worth noting that F_{ST} is one of the important extensions of the concept of probability of identity by descent foreshadowed in Chapter 2. F_{ST} is therefore a kind of inbreeding coefficient, and the subscript S stands for "subpopulation," while the T stands for "total population"; F_{ST} is an appropriate symbol because it measures the probability of identity by descent of alleles within a subpopulation relative to the total population, which is taken to be equivalent to the ancestral or base population in Figure 2.

Even though random mating occurs within each subpopulation of Figure 2 because gametes do combine at random, any two alleles in a subpopulation may be identical by descent due to the limited population size. Thus F_t does not equal zero. The actual value of F_t can be calculated as in Figure 6. This figure shows the $2N$ alleles in a breeding population of generation $t - 1$ (labeled $\alpha_1, \alpha_2, \alpha_3, \ldots, \alpha_i, \ldots, \alpha_j, \ldots, \alpha_{2N}$ to avoid confusion with conventional allele symbols), the gametes derived from generation $t - 1$ (each gametic type representing a fraction $1/2N$ of the entire gametic pool), and two randomly chosen pairs of alleles in the breeding population of generation t. The probability that the second chosen allele is of the same type as the first is $1/2N$, because this is the frequency of each allelic type in the gametic pool; the probability that the second chosen allele is of a different type from the first is accordingly $1 - 1/2N$. In the first case ($\alpha_i\alpha_i$), the probability of identity by descent is 1; in the second case ($\alpha_i\alpha_j$), it is F_{t-1}. Altogether $F_t = 1/2N + (1 - 1/2N)F_{t-1}$. Multiplying both sides by -1 and adding 1 leads to $1 - F_t = 1 - 1/2N - (1 - 1/2N)F_{t-1} = (1 - 1/2N)$

$(1 - F_{t-1})$, so $1 - F_t = (1 - 1/2N)^t(1 - F_0)$, or simply

$$F_t = 1 - \left(1 - \frac{1}{2N}\right)^t$$

when $F_0 = 0$. Figure 7 shows the rapid increase of F_t in small populations.

Strictly speaking, the equation $1 - F_t = (1 - 1/2N)^t(1 - F_0)$ applies only to populations (many plants, for example) that can undergo self-fertilization, as the theoretical considerations in Figure 6 make no stipulation that half the successful gametes in any generation must come from males

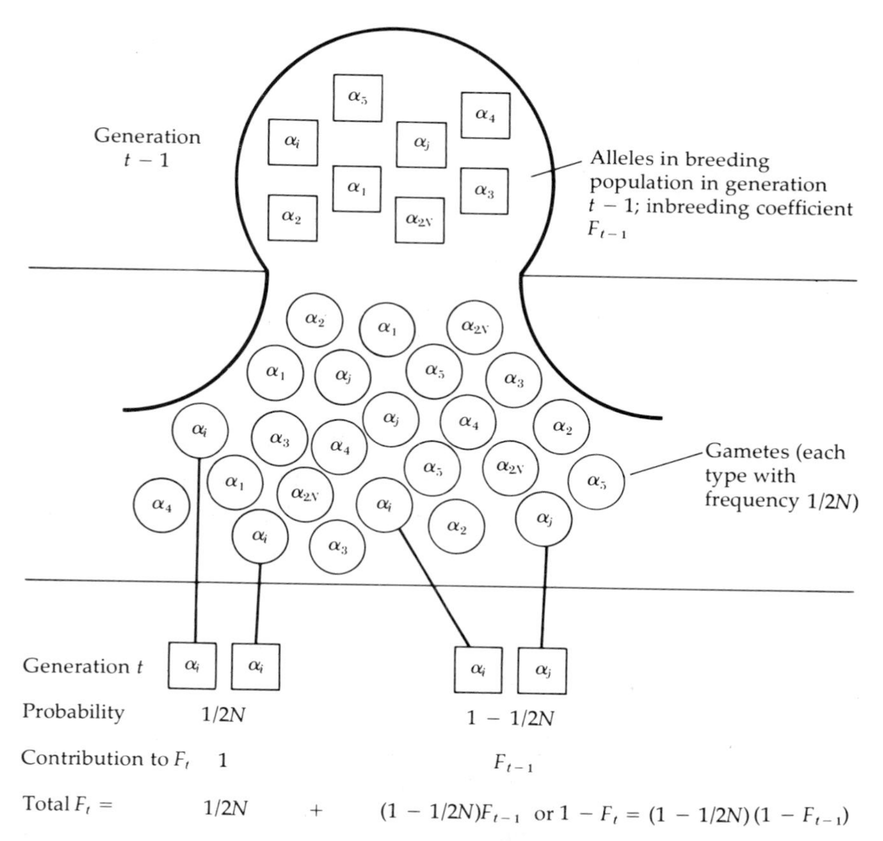

FIGURE 6. Random sampling in each of the subpopulations in Figure 2 leads to the relation $F_t = (1/2N) + [1 - (1/2N)]F_{t-1}$.

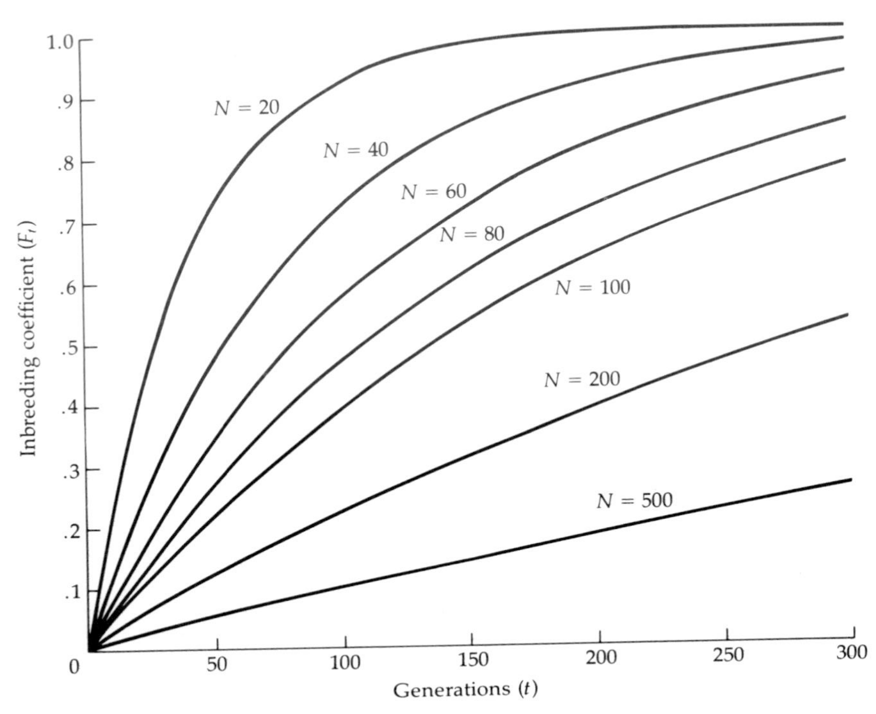

FIGURE 7. Increase of F_t in ideal populations as a function of time and effective population size N.

and the other half from females. The correction for organisms such as those with separate sexes that cannot undergo self-fertilization is minor, however: if there are equal numbers of males and females, simply replace N in the above equation by $N + 1/2$ (see Crow and Kimura, 1970). Since $1/2(N + 1/2)$ is very nearly equal to $1/2N$ for realistic values of N, the equation for F_t above (without the correction) may be taken as an excellent approximation for organisms with separate sexes.

As an example of how the formula for F_t can be used, consider the following question: What is the inbreeding coefficient (i.e., the probability of autozygosity) of a mouse in a colony that was established from a much larger colony 30 generations ago and since that time has maintained a constant size of 20 individuals each generation? For the sake of simplicity, assume that the colony is "ideal" in the sense discussed earlier. The inbreeding coefficient of a mouse is the probability that the alleles at a locus in the mouse are identical by descent; because mating within the colony is random, this probability is equal to the probability that two randomly chosen gametes carry alleles that are identical by descent. The latter probability is, by definition, F_t, so we may use $1 - F_t = (1 - 1/2N)^t(1 - F_0)$ with $N = 20$ (as noted above, use of 20.5 would be a bit more accurate), $t = 30$, and $F_0 = 0$ (justified because the base population was said to be large). Thus $1 - F_t = (1 - 1/40)^{30} = .468$, or $F_t = 1 - .468 = .532$. Note that there is substantial inbreeding even though mating is random and genotypic frequencies within the colony are given by the Hardy–Weinberg law. The inbreeding does not arise from actual nonrandom mating; it arises because the population is small in size.

Several important consequences of the population structure in Figure 2 are summarized in Table II. First, although each subpopulation is of finite size, we can imagine so many

TABLE II. Consequences of population subdivision and random genetic drift on genotype frequency[a].

		Generation		
		0	**t**	∞
Inbreeding coefficient (F_t) (Average over all populations)		0	$1 - \left(1 - \frac{1}{2N}\right)^t$	1
Genotype frequency (Average over all populations)	*AA*:	p_0^2	$p_0^2(1 - F_t) + p_0F_t$	p_0
	Aa:	$2p_0q_0$	$2p_0q_0(1 - F_t)$	0
	aa:	q_0^2	$q_0^2(1 - F_t) + q_0F_t$	q_0
Allele frequency (Average over all populations)	*A*:	p_0	p_0	p_0
	a:	q_0	q_0	q_0

[a] The table refers to the effects of population subdivision and random genetic drift on genotype frequencies at a single locus. For examples of the effects of random genetic drift on linkage disequilibrium of two loci, see Sved, 1968; Feldman and Christiansen, 1975; Hill, 1976; and Avery and Hill, 1979.

of them that the size of the total population is effectively infinite. For an infinite population that obeys the assumptions in Table I, the allele frequencies must remain constant. That is to say, even though the allele frequency in any individual subpopulation may change willy-nilly due to random genetic drift, the overall average allele frequency of A among subpopulations will always be p_0, where p_0 represents the allele frequency of A in the base population. Figure 8B gives an experimental demonstration of the constancy of average allele frequency.

Secondly, we have shown that F_t is the probability of autozygosity at a locus in an individual in generation t. For an individual chosen at random from the total population in generation t, therefore, the probability of allozygosity is $1 - F_t$. Because p_0 is the overall allele frequency of A, the probability that a randomly chosen individual will be genotypically AA is $p_0^2(1 - F_t)$ [for the case of allozygosity] + p_0F_t [for the case of autozygosity]. Similarly, the probability that the individual will be Aa equals $2p_0q_0(1 - F_t)$; and the probability that the individual will be aa equals $q_0^2(1 - F_t) + q_0F_t$. Thus the genotypic frequencies in the total population are given by the usual formula for inbreeding (see Table VI in Chapter 2). However, within any one subpopulation, the genotypic frequencies obey the Hardy–Weinberg law because of random mating. Substituting for F_t in the expression for the frequency of heterozygotes, we see that the average heterozygosity among subpopulations at time t equals $2p_0q_0(1 - F_t) = 2p_0q_0(1 - 1/2N)^t$; this is the theoretical curve plotted in Figure 8A (with $p_0 = q_0 = .5$).

Third and finally, since F_t eventually goes to 1, all subpopulations eventually become fixed for one allele or the other. Because the average allele frequency of A must remain p_0 even when all subpopulations have become fixed, the proportion of subpopulations that eventually become fixed for A must be p_0 (and the proportion that eventually become fixed for a must be q_0). Stated another way, the probability of ultimate fixation of an allele in any ideal subpopulation is equal to the frequency of that allele in the initial population. This point is illustrated for an actual case in Figure 3,

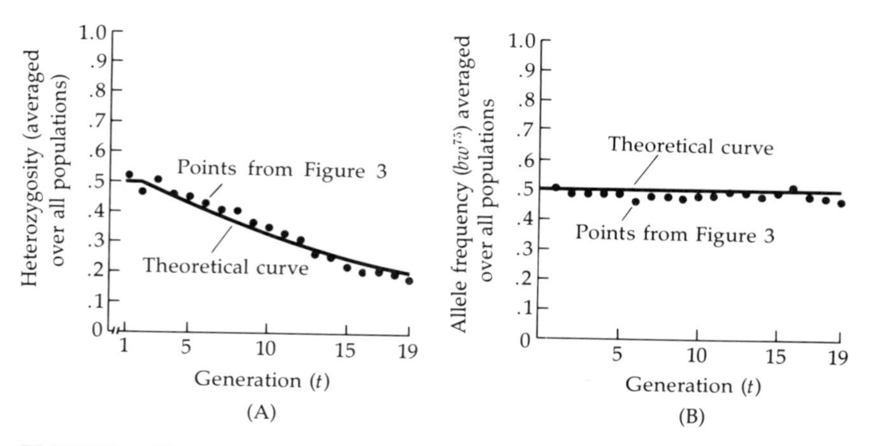

FIGURE 8. Theoretical curves for average heterozygosity (A) and average allele frequency (B) along with actual values from the experiment in Figure 3. (Data from Buri, 1956.)

where $p_0 = .5$; by generation 19, 58 populations have become fixed, 30 for the *bw* allele and 28 for bw^{75}.

EFFECTIVE POPULATION NUMBER

The effective number of a population usually differs from the actual number (the effective number is almost always smaller) because no real population obeys all the assumptions in Table I exactly. In any actual case, there must be corrections for such complications as age structure (Felsenstein, 1971; Hill, 1972), unequal numbers of males and females, and unequal family size (see Crow and Kimura, 1970, for a good review). To illustrate calculation of effective population size in one case, we will determine the effective size of a population whose size varies from generation to generation. This is an important example because natural populations do fluctuate in size, sometimes by a factor of 10 or more in a single generation (see Figure 9 for two spectacular examples, and also Figure 23B in Chapter 1). For the sake of simplicity, assume that the population is ideal in all respects except that its size is not constant. We will consider the situation over just two generations, reserving the more

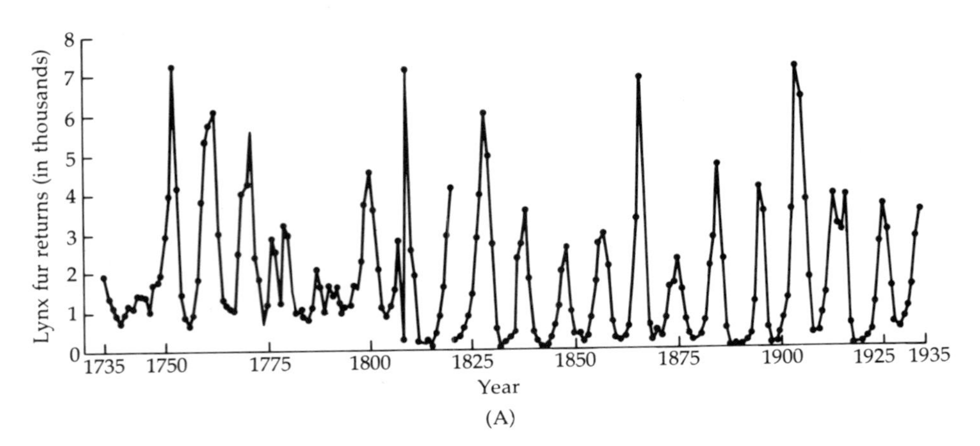

FIGURE 9. Fluctuations in population size. (A) Famous cycles of population density in the Canadian lynx (*Lynx canadensis*) as reflected by the number of lynx furs brought to trading posts of the Hudson's Bay Company by trappers. The record covers 200 years and is one of the longest population records known. Values for 1735–1820 are from Hudson's Bay Company records for an area that is now Manitoba, northern Ontario, western Quebec, and part of Saskatchewan; points for 1821–1934 are from the MacKenzie River District, which included the western part of the District of MacKenzie in the modern Northwest Territories and small parts of Alaska, the Yukon, British Columbia, and Alberta. Juxtaposing the graphs is justified because lynx cycles are synchronized throughout continental Canada, though the "troughs" prior to 1821 are not as low as later ones because the trapping areas in the first and last parts of the graph are different. Note the regular fluctuations in population density (by a factor of 50 or more) with a period of 8 to 10 years (average 9.6). Why lynx populations are cyclic is not known for certain. It is interesting that the lynx eats mainly rabbits — the snowshoe hare, *Lepus americanus*

general theory for Box A. Accordingly, suppose that the population's size in two successive generations is N_1 and N_2. The arguments laid out in Figure 6 imply that $1 - F_2 = (1 - 1/2N_2)(1 - F_1)$ and $1 - F_1 = (1 - 1/2N_1)(1 - F_0)$; substituting from the second equation into the first leads to $1 - F_2 = (1 - 1/2N_2)(1 - 1/2N_1)(1 - F_0)$. We want to express this in the general form $1 - F_t = (1 - 1/2N)^t(1 - F_0)$, where N is the effective population number. In our example $t = 2$, so $1 - F_2 = (1 - 1/2N)^2(1 - F_0)$. Setting the two expressions

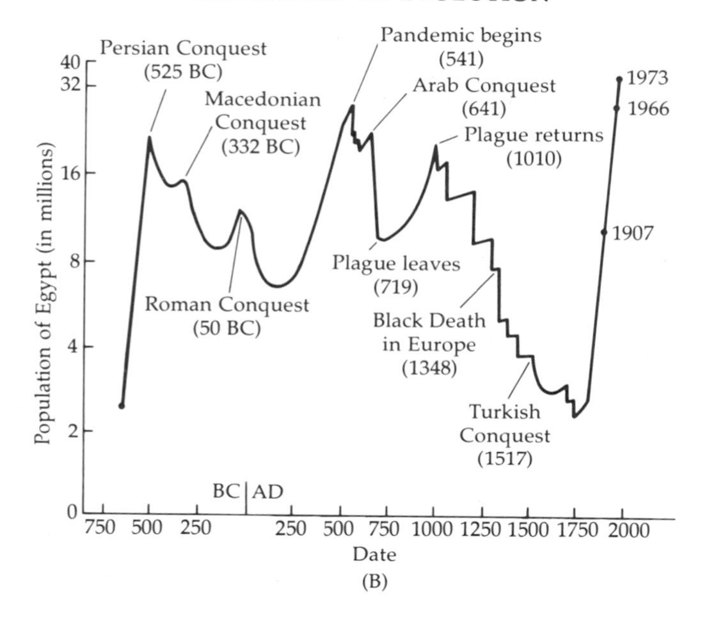

("It lives on Rabbits, follows the Rabbits, thinks Rabbits, tastes like Rabbits, increases with them, and on their failure dies of starvation in the unrabbited woods," says hare observer E. T. Seton) — and the rabbits cycle too, with about the same period. (Data and quote from Elton and Nicholson, 1942.) (B) Historical demographer's reconstruction of the population history of Egypt. Note large fluctuations due to plagues and conquests, and the recent upsurge. (From Hollingsworth, 1969.) Such extreme fluctuations in population size as shown here are not atypical, and their relevance to population genetics is that the effective size of a fluctuating population is markedly reduced by recurrent episodes of small population number.

for $1 - F_2$ equal to each other we obtain $(1 - 1/2N)^2 = (1 - 1/2N_1)(1 - 1/2N_2)$, for which $1/N = (1/2)(1/N_1 + 1/N_2)$ turns out to be an excellent approximation (see Box A for justification). In general,

$$\frac{1}{N} = \left(\frac{1}{t}\right)\left(\frac{1}{N_1} + \frac{1}{N_2} + \cdots + \frac{1}{N_t}\right)$$

so the effective number N is the so-called HARMONIC MEAN of the actual numbers — the reciprocal of the average of

A Effective Population Number

The effective number of a nonideal population is the size of an ideal population having the same rate of increase in F per generation as the nonideal population. In this box, we derive expressions for effective population number in two important cases: (a) when the actual size varies from generation to generation, and (b) when there are unequal numbers of males and females. The actual populations are assumed to be ideal in all respects except the ones specified.

a. Because $e^{-x} = 1 - x + x^2/2! - x^3/3! + \ldots$, we may write $e^{-x} = 1 - x$ if x^2 is small compared to x. Motivated by the expression $(1 - 1/2N)^2 = (1 - 1/2N_1)(1 - 1/2N_2)$ for the average effective size of a population over two generations (see the text), put $x = 1/2N$, $x_1 = 1/2N_1$, and $x_2 = 1/2N_2$, and assume that the N's are large enough that x^2, x_1^2, and x_2^2 are indeed much smaller than x, x_1, and x_2. Then the above expression becomes approximately $(e^{-x})^2 = (e^{-x_1})(e^{-x_2})$. Show from this that $2x = x_1 + x_2$, or $1/N = (1/2)(1/N_1 + 1/N_2)$. [In general, $1/N = (1/t)(1/N_1 + 1/N_2 + \ldots + 1/N_t)$, where t is the number of generations.] Suppose a population of insects in four consecutive generations has effective sizes of 10, 10^2, 10^3, 10^4. What is the average effective size over the whole period?

b. In Figure 6, it may be noted that the effective population number may be defined as the reciprocal of the probability that two uniting gametes come from the same parent. When there are separate sexes, of course, uniting gametes must come from different parents; in such a case, the effective number may be defined as the reciprocal of the probability that two uniting gametes come from the same grandparent. Because both sexes must contribute equally to offspring, the probability that uniting gametes come from a male grandparent is 1/4. Given that the gametes both come from a male, the probability that they come from the *same* male is $1/N_m$, where N_m is the number of males in the grandparental generation; $1/N_f$ is the corresponding probability for a female grandparent. Altogether, then, $1/N = 1/4N_m + 1/4N_f$, or $N = 4N_mN_f/(N_m + N_f)$ among the grandparents. What is the effective size of a population of 100 cows fertilized by artificial insemination with the sperm of four bulls? What would it be if there were 200 cows?

reciprocals. The harmonic mean tends to be dominated by the smallest terms. Suppose, for example, that $N_1 = 1000$, $N_2 = 10$, and $N_3 = 1000$ in a population that underwent a severe temporary reduction in size (a BOTTLENECK) in generation 2. Then $1/N = (1/3)(1/1000 + 1/10 + 1/1000) = .034$, or $N = 1/.034 = 29.4$. The average effective number over the three-generation period is only 29.4, whereas the average actual number is $(1/3)(1000 + 10 + 1000) = 670$. A severe population bottleneck often occurs in nature when a small group of emigrants from an established subpopulation founds a new subpopulation; the accompanying random genetic drift is then known as a FOUNDER EFFECT (see Holgate, 1966; Nei et al., 1975; Chakraborty and Nei, 1977; Neel and Thompson, 1978).

A second important case in which the effective number of a nonideal population can readily be calculated concerns sexual populations in which the number of males and females is unequal. This inequality creates a peculiar sort of "bottleneck"; because half of the alleles in any generation must come from each sex, any departure of the sex ratio from equality will enhance the opportunity for random genetic drift. This situation is important in wildlife management, where, for many game animals (pheasants and deer come immediately to mind) the legal bag limit for males is much larger than for females. Although some management desiderata are served by such hunting regulations (for example, the species involved are usually polygamous, so one male can fertilize many females and overall actual population size can be maintained), it must be remembered that the resultant inequality in sex ratio reduces the effective population number. Specifically, if a sexual population consists of N_m males and N_f females, the actual size is $N_a = N_m + N_f$. However, as shown in Box A, the effective population number is $N = 4N_mN_f/(N_m + N_f)$. To take a realistic example, if hunting is permitted to a level at which the number of surviving males is one-tenth the number of females, then the effective population number will be a mere one-third of the actual population number. (To see this for yourself, set $N_m = .1N_f$ in the formulas for N_a and N.)

GENETIC DIFFERENTIATION OF POPULATIONS

The principal effect of random genetic drift is divergence in allele frequency between subpopulations. In nature, genetic divergence involves such complications as mutation, migration, and natural selection, but the most useful measures of genetic divergence are motivated by the theory of random genetic drift. One very useful measure is the quantity called F in Table II, which as a measure of genetic differentiation is usually called the FIXATION INDEX and symbolized F_{ST}. Amounts of genetic divergence between human subpopulations and between subpopulations of several other species are presented in Table III. F_{ST} is the same as F_t in Table II, and the symbols H_T and H_S can also be explained with reference to Table II. (In Table III, the quantities H_T and H_S have been averaged over all loci involved, mostly allozyme loci.) H_T is called the TOTAL HETEROZYGOSITY and is the probability that two gametes chosen at random from the total population will carry different alleles. The average allele frequencies of A and a in the total population are p_0 and q_0 (see Table II), so $H_T = 2p_0q_0$. H_S is the SUBPOPULATION HETEROZYGOSITY, defined as the average heterozygosity among subpopulations; from Table II, $H_S = 2p_0q_0(1 - F_{ST})$. Now,

$$\frac{H_T - H_S}{H_T} = \frac{2p_0q_0 - 2p_0q_0(1 - F_{ST})}{2p_0q_0} = 1 - (1 - F_{ST}) = F_{ST}$$

In the above equation, the appropriateness of the symbol F_{ST} is manifest: population subdivision and its attendant random genetic drift causes the average subpopulation heterozygosity (H_S) to be smaller than what the heterozygosity would be were the subpopulations combined into a single, large randomly mating unit (H_T), and F_{ST} measures the extent of this reduction in heterozygosity. To say the same thing in a slightly different way, the quantity F_{ST} measures the amount of genetic variation in the whole population that is attributable to genetic differentiation among subpopulations; when $F_{ST} = 0$, for example, there is no genetic differentiation among subpopulations. (See Box B for more detail on the calculation and interpretation of F_{ST}; also Spielman

TABLE III. Total heterozygosity, average heterozygosity among subpopulations, and fixation index for various organisms.

Organism	Number of populations	Number of loci	Total heterozygosity (H_T)	Subpopulation heterozygosity (H_S)	Fixation index[a] (F_{ST})
Human	3 (major races)	35	.130	.121	.069
Human, Yanomama Indians	37 (villages)	15	.039	.036	.077
House mouse (*Mus musculus*)	4	40	.097	.086	.113
Jumping rodent (*Dipodomys ordii*)	9	18	.037	.012	.676
Drosophila equinoxialis	5	27	.201	.179	.109
Horseshoe crab (*Limulus*)	4	25	.066	.061	.076
Lycopod plant (*Lycopodium lucidulum*)	4	13	.071	.051	.282

[a] F_{ST} is calculated as $(H_T - H_S)/H_T$. (Data from Nei, 1975.)

et al., 1977; Yamazaki, 1977; and Chakraborty et al., 1978.)

Values of F_{ST} in Table III imply that genetic divergence between human subpopulations is quite small. Of the total genetic variation found in three major races (Caucasoid, Negroid, and Mongoloid), only .07 = 7 percent is ascribable to genetic differences *among* races. That is to say, 93 percent of the total genetic variation is found *within* races. Again, of the total genetic variation found in the native Yanomama Indians of Venezuela and Brazil, only .077 = 7.7 percent is due to differences in allele frequency among villages; 92.3 percent of the total genetic variation is found within any single village (Neel, 1978). Values of F_{ST} for other organisms are quite variable. In general, F_{ST} is influenced by the effective size of the subpopulations, by the amount and pattern of migration between subpopulations, and by other factors, including natural selection.

B Calculation and Interpretation of F_{ST}

The data in part (b) of this box are the allele frequencies at the *Esterase*-4 locus in *Drosophila willistoni* (subpopulation 1, studied by Ayala et al., 1971) and *Drosophila equinoxialis* (subpopulation 2, studied by Ayala et al., 1972). The purpose of this box is to show how F_{ST} can be used to measure the amount of genetic differentiation at this locus. Since the data in part (b) involve multiple alleles, we begin with a simpler hypothetical situation involving only two alleles.

The fixation index F_{ST} measures the amount of genetic differentiation among subpopulations relative to a hypothetical group of subpopulations, each homozygous, but having the same overall average allele frequency as the real subpopulations. Qualitatively speaking, the range .05 to .15 for F_{ST} may be considered to indicate moderate differentiation, .15 to .25 to indicate great differentiation, and above .25 to indicate very great differentiation. However, to quote Wright (1978), "Differentiation is by no means negligible if F_{ST} is as small as .05 or even less."

a. Calculate F_{ST} for each of the groups of populations (group 5 is the hypothetical group mentioned earlier), where the numbers in the body of the table are allele frequencies in the corresponding subpopulations. Assume two alleles and random mating within subpopulations; thus, the heterozygosity within a subpopulation having allele frequency p is $H_S = 2p(1-p)$. Use the formula $F_{ST} = (H_T - \overline{H}_S)/H_T$, where $H_T = 2\overline{p}(1-\overline{p})$ is the total heterozygosity in the group and $\overline{H}_S$ is the average of the within-population heterozygosities for that group. Then calculate F_{ST} for each group using the formulas $F_{ST} = \sigma^2/\overline{p}(1-\overline{p})$ and $\sigma^2 = \overline{p^2} - \overline{p}^2$ (i.e., the average of the squares minus the square of the average). B

	Group				
Subpopulation	**1**	**2**	**3**	**4**	**5**
1	.15	.10	.05	.01	1.0
2	.20	.15	.10	.05	0
3	.30	.35	.40	.45	0
4	.35	.40	.45	.49	0
Average ($\overline{p}$)	.25	.25	.25	.25	.25

b. For multiple alleles and random mating within subpopulations, let $H_S^{(i)} = 2p_i(1-p_i)$ represent the heterozygosity for allele A_i in a subpopulation with allele frequency p_i, and let $\overline{H}_S^{(i)}$ denote the average of $H_S^{(i)}$ over all subpopulations. Similarly, let $H_T^{(i)} = 2\overline{p}_i(1-\overline{p}_i)$ denote the total heterozygosity for the corresponding allele. Then $F_{ST}^{(i)} = [H_T^{(i)} - \overline{H}_S^{(i)}]/H_T^{(i)}$ may be called the fixation index of allele i. The total heterozygosity in a subpopulation is $H_S = 1 - \Sigma p_i^2$ (summation over all alleles at the locus), and the average of H_S over all subpopulations is denoted $\overline{H}_S$; the total heterozygosity is $H_T = 1 - \Sigma\overline{p}_i^2$. Overall, $F_{ST} = [H_T - \overline{H}_S]/H_T$. For the data below, calculate $F_{ST}^{(i)}$ for $i = 1, 2, 3$, and 4, and F_{ST}. Verify that F_{ST} is the weighted average: $F_{ST} = \Sigma\overline{p}_i(1-\overline{p}_i)F_{ST}^{(i)}/\Sigma\overline{p}_i(1-\overline{p}_i)$. Also calculate $F_{ST}^{(i)}$ as $\sigma_{(i)}^2/\overline{p}_i(1-\overline{p}_i)$, where $\sigma_{(i)}^2 = \overline{p_i^2} - \overline{p}_i^2$, the averages being over all subpopulations, and calculate F_{ST} as $\Sigma\sigma_{(i)}^2/\Sigma\overline{p}_i(1-\overline{p}_i)$.

Allele	Frequency	Subpopulation 1	Subpopulation 2
Est-4[1]	$p_1 =$	.011	.150
Est-4[2]	$p_2 =$	.169	.769
Est-4[3]	$p_3 =$	.801	.081
Est-4[4]	$p_4 =$	.019	0

RACE

A RACE is a group of individuals in a species who are genetically more similar to each other than they are to members of other such groups. Populations that have undergone some degree of genetic divergence as measured by, for example, F_{ST}, therefore qualify as races. Using this definition, the human population contains many races. Each Yanomama village represents, in a certain sense, a separate "race," and the Yanomama as a whole also form a distinct "race." Such fine distinctions are rarely useful, however; it is usually more convenient to group populations into larger units that still qualify as races in the definition above. These larger units often coincide with races based on physical characteristics such as skin color, hair color and texture, facial features, and body conformation, as defined by anthropologists (modern anthropologists also take cultural and linguistic similarities into account).

Here it must be pointed out that the data in Table III, which indicate much more genetic variation within than among human races, may be misleading. The conclusion was based primarily on genes determining allozymes, recall; it certainly is not true for genes influencing skin color, hair color, hair texture, and other traits that most people think of in connection with the word "race." However, skin color and other prominent racial characteristics are used to delineate races precisely because racial differences for these traits are rather large, so the genes involved cannot be representative of the whole genome. On the other hand, allozyme loci may not be very representative either.

In any case, grouping of populations can be accomplished using any good measure of the amount of genetic divergence (often called GENETIC DISTANCE) between populations. (Most measures of genetic distance are related to F_{ST} in one way or another; some of these are discussed in Box C.) Once a suitable measure of genetic distance is chosen, the genetic distance is calculated between all possible pairs of populations. (For n populations, there are $n(n - 1)/2$ such pairs.) Then the two populations with the smallest distance are

Some Measures of Genetic Distance

C

Many measures of genetic distance have been proposed (see Edwards, 1971; Crow and Denniston, 1974; Hedrick, 1975; Nei, 1977; and Smith, 1977, for examples and discussion), but most of them can be related to F_{ST} in one way or another. In part (a), we investigate the relationship between F_{ST} and d (defined below), a measure of genetic distance based on geometrical considerations. Part (b) shows how the measure d can be used to construct "trees" such as that in Figure 10. Part (c) introduces a useful measure of the absolute amount of genetic divergence between populations.

a. As noted above, d is a measure of genetic divergence between populations based on geometrical reasoning; the reasoning can be illustrated by considering a single locus with two alleles in each of two populations. Let the allele frequencies be denoted p_1 and q_1 in one population, and let the corresponding frequencies in the other be p_2 and q_2. Geometrically, any two populations will fall on the circumference of a quarter circle of radius 1 with axes $\sqrt{p}$ and $\sqrt{q}$, as shown in the figure. The length of the chord connecting two populations is $\sqrt{2}d$, where $d = \sqrt{1 - \cos \theta}$, θ being the angle between the corresponding radii. The quantity d serves as a satisfactory measure of genetic distance, and it can be shown that $d^2 = 1 - \sqrt{p_1p_2} - \sqrt{q_1q_2}$.

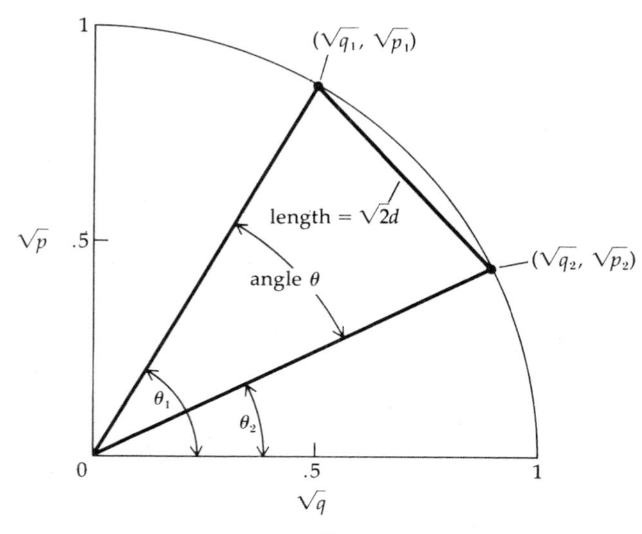

C If the allele frequencies in the two populations are not too different, $2d^2 = F_{ST}$, approximately. To illustrate this point, let $p_1 = .5 + x$, $q_1 = .5 - x$, $p_2 = .5 - x$, $q_2 = .5 + x$. Use $F_{ST} = \sigma^2/\bar{p}\bar{q}$ to show that $F_{ST} = 4x^2$, and use the approximation $\sqrt{.25 - x^2} = .5 - x^2$ (valid for small x) to show $2d^2 = 4x^2$ also. Thus, for small x, $2d^2 = F_{ST}$. Then calculate F_{ST} and the exact value of $2d^2$ for $x = .1$, .2, .3, .35, .4, and .49.

(Optional for students who like trigonometry: use the elementary definitions of sin and cos, the identity $\cos(\theta_1 - \theta_2) = \cos\theta_1\cos\theta_2 + \sin\theta_1\sin\theta_2$, and the formula for length, ℓ, of a chord subtending an angle θ on a unit circle $\ell = 2\sin\theta/2 = 2\sqrt{(1 - \cos\theta)/2}$, to derive $d^2 = 1 - \sqrt{p_1p_2} - \sqrt{q_1q_2} = 1 - \cos\theta$ and $\ell = d\sqrt{2}$.)

b. The measure d is frequently used for grouping of populations based on genetic similarity (so-called "cluster analysis"). Such analyses must involve many loci and many populations, but for illustrative purposes, one locus with two alleles will suffice. To construct a tree like that in Figure 10 for the four populations below, first calculate the d's for all six pairwise comparisons; the pair with the smallest d represents a "cluster," and the populations involved should be pooled together by averaging their allele frequencies. Then repeat the procedure using the two remaining populations along with the pooled one. Finally, draw the tree so that the populations with the smallest d's branch off last. (You may make the branches of any convenient length, but for ease in comparing your answer with that in the back of the book, draw the tree rooted at the left and growing to the right, and orient the branches so that the sequence of populations from top to bottom will be 1, 2, 3, 4.)

Allele frequency	Population			
	1	2	3	4
p	.15	.25	.30	.40
q	.85	.75	.70	.60

c. Another measure of genetic divergence is $\bar{D}_m = n(H_T - H_S)/(n - 1)$, where n is the number of subpopulations. $\bar{D}_m$ is a measure of the absolute amount of genetic divergence among populations, whereas $F_{ST} = (H_T - H_S)/H_T$ is a measure of the relative amount of divergence. (Nei, 1975, has argued that $\bar{D}_m$ is perhaps the best measure of divergence between species.) Calculate $\bar{D}_m$ for each of the groups in Table III.

grouped together and treated as a single "population," and the genetic distance between this new "population" and each of the remaining ones is calculated. The process is repeated until only two such groups remain. (The procedures are spelled out for a numerical example in Box C.)

Results of such an analysis are presented in Figure 10, where the length of each line segment is proportional to genetic distance. Among the three major human groups, Caucasoid, Negroid, and Mongoloid, genetic divergence is consistently found to be greatest between Negroids and Mongoloids. A point raised earlier should be emphasized again, however: no group of loci that can be studied is necessarily representative of the entire genome. (For further discussion, see Lewontin, 1972; and Nei, 1978.)

Mutation

ORIGIN OF GENETIC VARIATION

Mutation is the ultimate source of genetic variation. This is an important statement, but hardly a profound one. Genetic

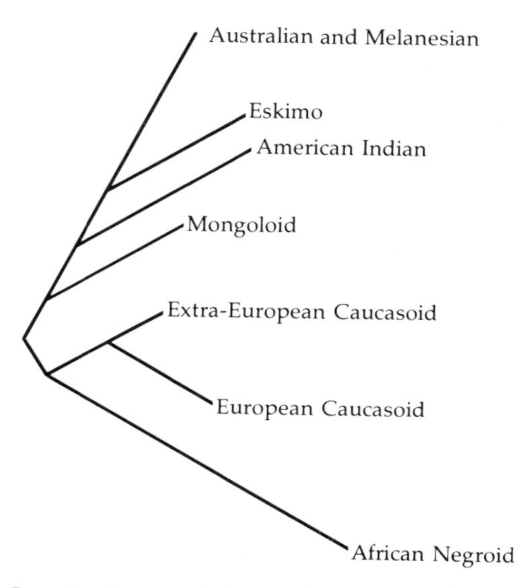

FIGURE 10. Genetic distance between human populations based on allele frequencies. Each line segment is proportional to genetic distance. (After Cavalli-Sforza, 1966.)

variation can arise only if there is some change in the genetic material, and any heritable change in the genetic material is, by definition, MUTATION. The word "mutation" is here used in its widest sense to mean all genetic changes, including visible chromosome abnormalities, aneuploidy, and polyploidy. However, the vast bulk of genetic variation within virtually all natural populations consists of subtle, invisible changes in DNA — particularly mutations involving nucleotide substitutions, such as those responsible for allozymes, and perhaps small deletions or duplications.

Because spontaneous mutation rates are typically quite small, on the order of 10^{-4} to 10^{-6} mutations per locus per generation, the tendency for allele frequencies to change as a result of recurrent mutation (mutation pressure) is very small over the course of a few generations. On the other hand, the cumulative effects of mutation over long periods of time can become appreciable. A useful model for thinking about the problem is summarized in Table IV, essentially the Hardy-Weinberg model, but with mutation allowed. For the moment we focus on mutations that are SELECTIVELY NEUTRAL, which is to say mutations that have so little effect on the ability of the organism to survive and reproduce that natural selection does not bring about any change in their frequency.

Consider a locus with two alleles, *A* and *a*, and suppose the mutation rate per generation from *A* to *a* is μ and that the rate from *a* to *A* is ν. Let p_t and q_t (with $q_t = 1 - p_t$) denote the allele frequencies of *A* and *a* in generation *t*. An *A* allele in generation *t* can originate in only two ways: it could have been an *A* allele in generation $t - 1$ that escaped

TABLE IV. Assumptions of the mutation model.

(1) Nonoverlapping generations
(2) Infinite population size
(3) Mutation rates constant
(4) No migration
(5) No selection

mutation to a (probability $1 - \mu$), or it could have been an a allele that mutated to A (probability ν). Symbolically, $p_t = p_{t-1}(1 - \mu) + (1 - p_{t-1})\nu = p_{t-1} - p_{t-1}\mu + (1 - p_{t-1})\nu$. Now, if p_{t-1} is close to 1, $(1 - p_{t-1})\nu$ will be the product of two small numbers, so it will be very small indeed, small enough to ignore. At the same time, $p_{t-1}\mu$ will be very close to μ. Altogether, when p_{t-1} is close to 1, $p_t = p_{t-1} - \mu$ to a close approximation. Therefore, $p_1 = p_0 - \mu$ and $p_2 = p_1 - \mu = (p_0 - \mu) - \mu = p_0 - 2\mu$, and so on. In general, $p_t = p_0 - t\mu$ for the first hundred generations or so. Because $q_t = 1 - p_t$, we can just as well write $q_t = q_0 + t\mu$. Thus q_t should increase linearly with time, and the slope of the line should be μ. Because μ is small, this linear increase in q_t is difficult to detect unless population size is huge. Such population sizes can be attained in a bacterial CHEMOSTAT, a device for maintaining populations of bacteria in a continuous state of growth and cell division (Figure 11). The linear increase in

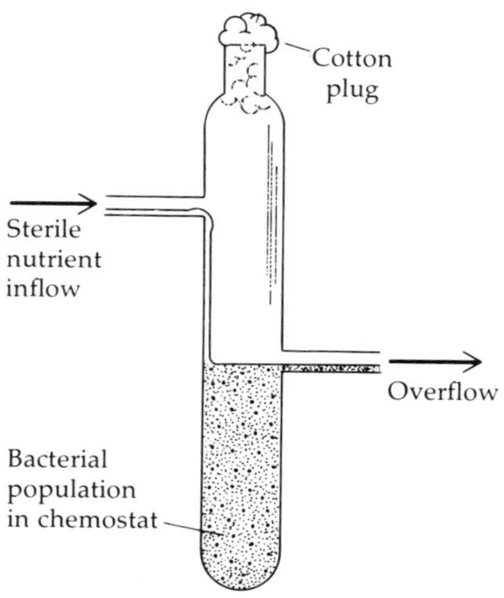

FIGURE 11. Simplified diagram of a bacterial chemostat. The actual apparatus is more complicated and has facilities for aeration and precautions against contamination. At the steady state, the rate of inflow of nutrient equals the rate of outflow. Cells within the chemostat are in a continuous state of division, but the population does not increase in size because, in any interval of time, the number of new cells produced by division is balanced by the number washing out through the outflow.

q_t that occurs in chemostats due to mutation pressure is shown in Figure 12. Note the abrupt increase in mutation rate (indicated by the increase in slope) shortly after the addition of caffeine, a bacterial mutagen. (For more on chemostats and their use in population genetics, see Cox and Gibson, 1974; Dykhuizen, 1978.)

To see what happens to allele frequency in the long run, return again to the expression $p_t = p_{t-1}(1 - \mu) + (1 - p_{t-1})\nu$. Suppose one waits until p_t no longer changes, that is, until $p_t = p_{t-1}$. (Any value of p_t with the property that $p_t = p_{t-1}$ is called an EQUILIBRIUM.) Denote the equilibrium by $\hat{p}$. The value of $\hat{p}$ can be found by substituting it into the above expression: $\hat{p} = \hat{p}(1 - \mu) + (1 - \hat{p})\nu = \hat{p} - \hat{p}\mu + \nu - \hat{p}\nu = \hat{p} + \nu - \hat{p}(\mu + \nu)$, or, with a little cancellation and rearrangement, $\hat{p} = \nu/(\mu + \nu)$. Curves of p_t with $\mu = 10^{-4}$ and $\nu =$

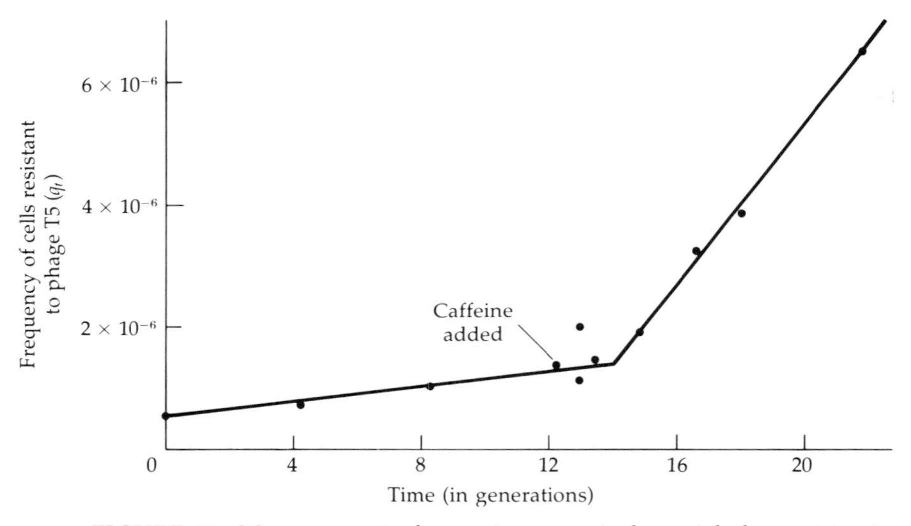

FIGURE 12. Measurement of mutation rates in bacterial chemostats, in this case the rate of mutation to resistance to bacteriophage T5. The mutation rate is estimated as the slope of the straight-line segments. Prior to addition of caffeine, the slope is 1.3×10^{-8} per hour. After addition of caffeine (concentration = 150 mg/1), $\mu = 12 \times 10^{-8}$ per hour. Mutation rates per hour in bacterial chemostats are constant for generation times between 2 and 12 hours; the present experiment had a generation time of 5.5 hours. (From Novick, 1955.)

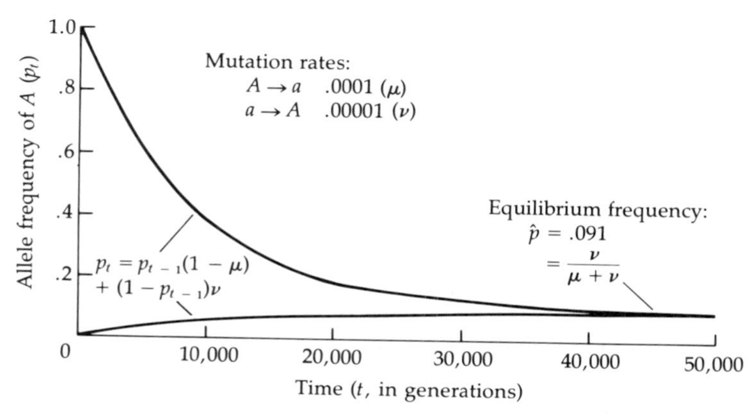

FIGURE 13. Theoretical change in allele frequency under pressure of reversible mutation. Note that attainment of near-equilibrium values requires tens of thousands of generations for realistic mutation rates.

10^{-5} are shown in Figure 13. Note that whatever the initial allele frequency of A, the allele frequency of A eventually goes to $\hat{p} = \nu/(\mu + \nu)$, which in the case in the figure is $\hat{p} = .00001/(.0001 + .00001) = 1/11 = .091$. Figure 13 shows that mutation pressure is a very weak force inasmuch as the population requires thousands or tens of thousands of generations to reach equilibrium. (An explicit expression for p_t in this "reversible mutation" model is derived in Box H later in this chapter.)

NUMBER OF ALLELES MAINTAINED IN POPULATIONS

Recall from Chapter 2 that loci controlling allozymes often have more than two alleles. It is therefore of some importance to determine how many alleles can be maintained by mutation pressure. If allozyme loci tend to have more alleles than would be expected from mutation pressure alone, then such other forces as are operative in nature must tend to accumulate alleles; on the other hand, if there are fewer alleles than expected, then the forces must tend to eliminate alleles. There is a technical problem in calculating the number of alleles that can be maintained by mutation. If the procedure for distinguishing among alleles has low resolving power, few alleles will be detected no matter how many may actually be present in the population. Electrophoresis augmented with other procedures may have satisfactory resolving power for some types of studies, but many alleles with synonymous base substitutions (involving primarily the third nucleotide of a codon) will remain undetected. Nevertheless, in principle, we can imagine for the sake of argument that all alleles are distinguishable, as would be the case were their DNA nucleotide sequences to be determined. Since an average protein contains roughly 300 amino acids, the average length of the amino acid coding part of an average gene must be roughly 900 nucleotides. The number of possible alleles is therefore staggering (4^{900} to be exact, which equals about 10^{542}). Thus we can suppose that every time a mutation occurs it creates a new allele, one that does not already exist in the population. This is called the INFINITE-ALLELES MODEL (Kimura and Crow, 1964). The infinite-alleles model is but one way to specify the characteristics of new mutations. Another model is called the STEPWISE-MUTATION MODEL; stemming from electrophoretic studies, the stepwise-mutation model assumes that a new mutation so alters the corresponding protein that its rate of migration in a gel is increased or decreased by a single unit (Maruyama and Kimura, 1978; Kimura and Ohta, 1978; Ramshaw and Eanes, 1978). Although the infinite-alleles model represents a somewhat simplified view of mutation, it nevertheless provides a useful standard of comparison for other models or for observed allele frequencies.

In the infinite-alleles model, we must also assume that the population in question is finite because new alleles would continue to accumulate forever in an infinite population. So the model of interest is the one in Table I, but with mutation permitted. In order to learn whether actual populations have more or fewer alleles than would be expected from mutation pressure alone in the infinite-alleles model, we must determine how many alleles can be maintained in the infinite-alleles model. Thus, the question is, for what number of alleles will the creation of new alleles by mutation be exactly balanced by the loss of old alleles due to random genetic drift? For the infinite-alleles model, the answer is surprisingly simple. We can get it by expressing the proportion of homozygous genotypes in two ways — in terms of the allele frequencies and in terms of the fixation index (F_t in Figure 6). Setting these two proportions equal to each other will yield the desired result. Accordingly, let the effective population number be N and the mutation rate be μ. Because of the infinite-alleles assumption, each allele in the population arises only once, and homozygotes for any allele must therefore be autozygous. For n alleles with frequencies $p_1, p_2, p_3, \ldots, p_n$, the homozygosity in any subpopulation will be $\Sigma p_i^2 = p_1^2 + p_2^2 + \ldots + p_n^2$ because in each subpopulation mating is random; Σp_i^2 is thus the homozygosity in terms of allele frequency. On the other hand, since homozygotes are autozygous in the infinite-alleles model, the homozygosity also equals F, the fixation index. The calculation of F_t in Figure 6 is still correct for this model if neither allele (α_i or α_j) underwent mutation in the passage of one generation. Therefore $F_t = [1/2N + (1 - 1/2N)F_{t-1}](1 - \mu)^2$. The equilibrium value of F_t, call it $\hat{F}$, is found by solving $\hat{F} = [1/2N + (1 - 1/2N)\hat{F}](1 - \mu)^2$, from which

$$\hat{F} = \frac{1}{4N\mu + 1}$$

to an excellent approximation (the approximation is derived in Box D). Therefore the number of selectively neutral alleles increases under mutation pressure until $\hat{F} = 1/(4N\mu + 1)$; $\hat{F}$ is the expression for homozygosity in terms of the fixation index.

We thus have two measures of homozygosity: Σp_i^2 and $\hat{F}$. Since they measure the same thing they must be equal, so $\Sigma p_i^2 = \hat{F} = 1/(4N\mu + 1)$. However, any number of distributions of allele frequency can lead to the same homozygosity. For example, an equilibrium population with four alleles at frequencies $p_1 = .7$, $p_2 = .1$, $p_3 = .1$, $p_4 = .1$ has a homozygosity of $\hat{F} = p_1^2 + p_2^2 + p_3^2 + p_4^2 = (.7)^2 + (.1)^2 + (.1)^2 + (.1)^2 = .52$; likewise a population with two alleles at frequencies $p_1 = .6$, $p_2 = .4$ has the same homozygosity, namely

D Equilibrium Autozygosity for the Infinite-Alleles Model

Part (a) of this box justifies the approximation $\hat{F} = 1/(4N\mu + 1)$ for the equilibrium autozygosity in the infinite-alleles model of mutation. Parts (b) and (c) show that migration affects $\hat{F}$ in exactly the same manner as does μ. Migration is measured here by m, which is the probability that a randomly chosen allele in a subpopulation in any generation will come from a migrant. The precise model of migration is the "island model," which is discussed in detail in the section on migration in the text.

a. From the expression $\hat{F} = [1/2N + (1 - 1/2N)\hat{F}](1 - \mu)^2$ for the infinite-alleles model of mutation, show that

$$\hat{F} = \frac{1}{2N\left[\frac{1}{(1-\mu)^2} - 1\right] + 1}$$

exactly. Then, using the fact that $1/(1 - \mu)^2 = 1 + 2\mu + 3\mu^2 + 4\mu^3 + \ldots$ (for $\mu < 1$) and assuming that μ^2 and higher powers are negligible, show that $\hat{F} \approx 1/(4N\mu + 1)$.

b. The analogous expression for $\hat{F}$ with migration is $\hat{F} = [1/2N + (1 - 1/2N)\hat{F}](1 - m)^2$, because $(1 - m)^2$ is the probability that neither of the two randomly chosen alleles comes from a migrant. Why is it obvious from (a) above that $\hat{F} \approx 1/(4Nm + 1)$, provided m is not too large?

c. Assume an infinite-alleles model of mutation in a geographically structured population with migration as in (b). Show that $\hat{F} \approx 1/[4N(m + \mu) + 1]$ if m is not too large. (Hint: start by justifying $F_t = [1/2N + (1 - 1/2N)F_{t-1}](1 - m - \mu)^2$; then proceed as in (a) above.)

$\hat{F} = p_1^2 + p_2^2 = (.6)^2 + (.4)^2 = .52$. The problem that many distributions of allele frequency can lead to the same homozygosity can be sidestepped by imagining that all alleles are equally frequent. If the population has, say, n_e equally frequent alleles (the meaning of the subscript will be discussed in a moment), then $p_1 = p_2 = p_3 = \ldots = p_{ne} = 1/n_e$, and the homozygosity will be $\Sigma p_i^2 = n_e(1/n_e)^2 = 1/n_e$. At equilibrium, therefore, $1/n_e = \hat{F} = 1/(4N\mu + 1)$, or $n_e = 4N\mu + 1$. (The number n_e is called the EFFECTIVE NUMBER OF ALLELES, which explains the subscript *e*.) Diverse distributions of allele frequency can therefore be compared in terms of their effective number of alleles. The four-allele population and the two-allele population above that have identical homozygosities of .52 have the same effective number of alleles, namely $n_e = 1/.52 = 1.92$. Biologically speaking, n_e is the number of equally frequent alleles that would be required to produce the same homozygosity as in an actual population. At equilibrium, as noted above, $n_e = 4N\mu + 1$, and Figure 14 shows how n_e changes with $N\mu$.

THE NEUTRALITY HYPOTHESIS

Turning now to data from real populations, let us suppose that observed allozyme polymorphisms are due to selectively neutral alleles maintained in a balance between mutation and random genetic drift, a supposition called the

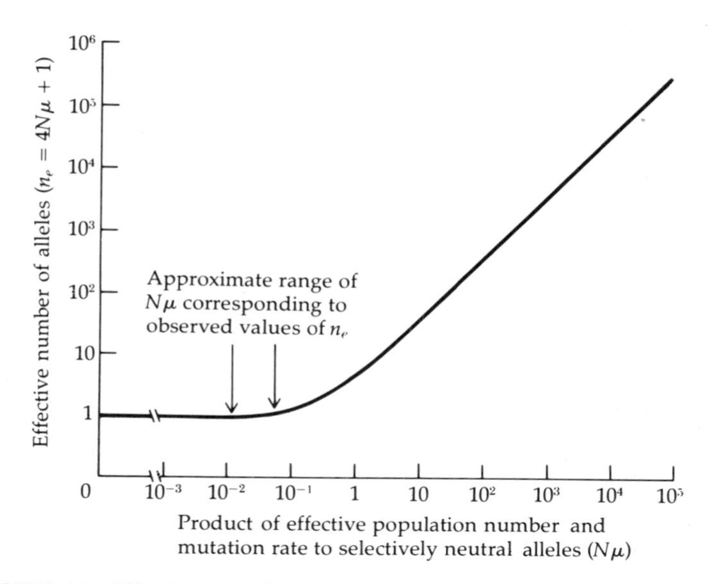

FIGURE 14. Effective number of alleles (n_e) against $N\mu$, showing that observed values of n_e correspond to a remarkably small range of $N\mu$. (After Lewontin, 1974a.)

NEUTRALITY HYPOTHESIS or the HYPOTHESIS OF SELECTIVE NEUTRALITY (Kimura, 1968; King and Jukes, 1969). In essence, the neutrality hypothesis states that many mutations have so little effect on the organism that they do not influence survival or reproduction. The frequencies of such alleles are not, therefore, determined by the forces of natural selection, but are instead determined by such other forces as migration (discussed later) and random genetic drift. Because the neutrality hypothesis is of fundamental importance in population genetics and evolution, it has been (and still is) a subject of considerable discussion (see the books by Lewontin, 1974a; Nei, 1975; and Ayala, 1976, for examples). If the hypothesis is true, or approximately true, then observed polymorphisms at, for example, allozyme loci have no particular significance in the adaptation of organisms to their environment; such polymorphisms are mere evolutionary "noise," and regardless of how much their study may reveal about population structure and random genetic drift, they tell us little or nothing about adaptive changes in evolution. If, on the other hand, the neutrality hypothesis is false, the next important task would be to study in detail the actual adaptive significance of polymorphisms.

To assess the plausibility of the neutrality hypothesis, many aspects of the hypothesis must be compared with the situation in actual populations. One aspect of the hypothesis developed in the preceding section concerns the homozygosity to be expected with the infinite-alleles model. Using observed homozygosities, we can calculate the effective number of alleles (n_e), and from the expression $n_e = 4N\mu + 1$ infer the corresponding values of $N\mu$. If the resulting values are grossly unreasonable, we can safely reject the infinite-alleles version of the neutrality hypothesis (or at least argue that actual populations cannot be at equilibrium).

Recall from Chapter 2 that observed values of heterozygosity for allozyme loci range from .05 to .15 in most organisms; observed homozygosities therefore range from $1 - .05 = .95$ to $1 - .15 = .85$, which corresponds to n_e in the range $n_e = 1/.95 = 1.052$ to $1/.85 = 1.176$. $N\mu$, which equals $(n_e - 1)/4$, therefore ranges from .013 to .044, as indicated in Figure 14. The maximum inferred value of N therefore differs from the minimum by a factor of roughly three, a surprisingly small range inasmuch as population size of different species varies by a factor of 10,000 or more. The distribution of homozygosities for allozyme loci thus appears to be too uniform among diverse organisms, and this finding has been interpreted as meaning that the neutrality hypothesis is wrong. Such a conclusion implies that natural selection is somehow involved in the maintenance of genetic polymorphisms. On the other hand, rejection of the neutrality hypothesis on grounds of the observed range of n_e is probably premature because routine electrophoresis does not distinguish all alleles (see Chapter 2 for discussion). Beyond that, estimates of effective population size (N) in nature are highly uncertain because the types of studies involved are very difficult, and estimates of μ (which in this case is the mutation rate to *neutral* alleles) are even more uncertain.

Figure 15 shows a second type of test of the adequacy of the neutrality hypothesis in explaining observed levels of

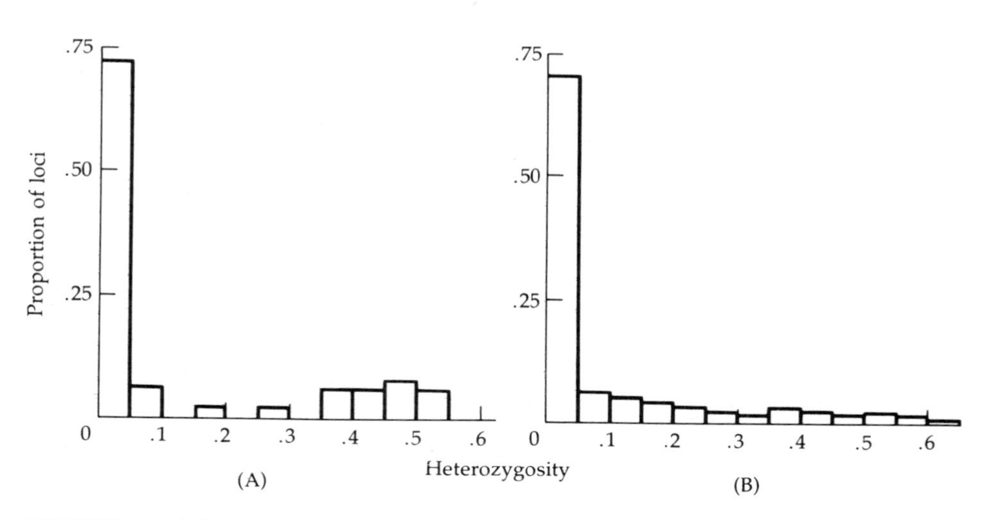

FIGURE 15. (A) Observed distribution of heterozygosity among loci in Caucasians. (B) Theoretical distribution for selectively neutral alleles. (From Nei et al., 1976.)

genetic variation at allozyme loci. The figure shows the observed distribution of heterozygosity at 74 loci in Caucasians (part A) along with the computer-generated theoretical distribution of the infinite-alleles model (part B); n_e is 1.11 in A and 1.10 in B, so the average heterozygosity is $1 - \hat{F} = 1 - 1/n_e = (n_e - 1)/n_e = (1.11 - 1)/1.11 = .099$ in A and .091 in B. The correspondence between the histograms is fairly good, but the observed distribution seems to have too many loci with heterozygosities in the range of .35 to .55.

A third type of test of the neutrality hypothesis is shown in Figure 16, which presents data on the mean and variance of heterozygosity in 77 vertebrate species (part A) and 46 invertebrate species (part B). The curves are those theoretically expected from the infinite-alleles model when the rate of selectively neutral mutation varies among loci (Nei et al., 1976). (The variance, as its name implies, measures how much variation there is among a set of numbers, the numbers in Figure 16 being heterozygosities. The variance can be calculated as the mean of the squares minus the square of the mean. To take a concrete example, suppose there are only three loci with heterozygosities of 0, .05, and .25. The mean heterozygosity is $(0 + .05 + .25)/3 = .100$, and the mean of the squares is $[(0)^2 + (.05)^2 + (.25)^2]/3 = .022$; the variance is therefore $.022 - (.100)^2 = .012$. Means and variances in Figure 16 have been calculated in this manner, but many more loci are involved.) At first glance, the fit in Figure 16 is impressive. On the other hand, the observed points are sufficiently scattered that any number of other curves could fit at least as well, and it is particularly troublesome that almost all vertebrate species with average heterozygosities less than .05 fall below the theoretical prediction. Here again, it is ambiguous as to whether actual data support or refute the neutrality hypothesis.

EVOLUTION OF AMINO ACID SEQUENCES

In the course of evolutionary time, proteins undergo changes in their amino acid sequence. For example, the amino acid sequence of the respiratory protein cytochrome *c* in humans differs from the cytochrome *c* sequence of rhe-

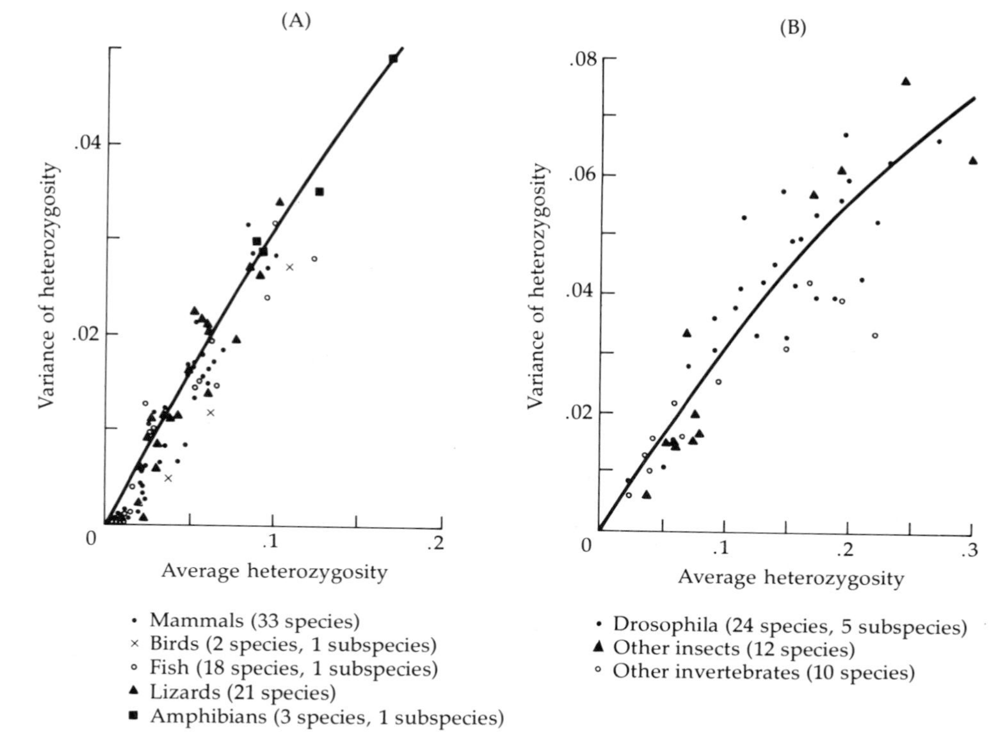

FIGURE 16. Mean and variance of heterozygosity among loci for (A) vertebrates and (B) invertebrates. The variance of a set of numbers is a convenient measure of how closely the individual numbers are clustered around the mean, small variances being associated with substantial clustering. The solid line in both (A) and (B) is the theoretical curve derived by Nei et al. (1976) for the infinite-alleles model when the mutation rate to neutral alleles varies among loci in such a manner that the variance in mutation rate equals the square of the mean mutation rate. Actual formulas for the theoretical curves are rather complex and are not presented. (After Nei et al., 1976.)

sus monkeys, kangaroos, ducks, tuna, moths, and *Neurospora* at 1, 12, 17, 31, 36, and 63 positions, respectively, out of the total 104 amino acids in the human molecule. (These numbers mean, for example, that human and kangaroo cytochrome *c* have identical amino acids at $104 - 12 = 92$ positions and different amino acids at 12 positions.) The relative "distances" between humans and the other species, as measured by the number of amino acid differences in cytochrome *c*, are in accord with the relative times since the evolutionary divergence between humans and the other species from their common ancestor, as estimated from the fossil record. A somewhat better measure of evolutionary divergence is based on the number of nucleotide differences (NUCLEOTIDE SUBSTITUTIONS) in DNA required to account for the observed amino acid differences (AMINO ACID SUBSTITUTIONS) because some amino acid substitutions require two nucleotide substitutions. [For example, from the genetic code (Table I in Chapter 1), it can be seen that at least two nucleotide substitutions are required to change a codon for methionine into one for tryptophan.]

The neutrality hypothesis provides a simple expression for the expected rate of nucleotide substitution. Suppose the frequency of selectively neutral mutations is μ in a population of effective size N. Any newly arisen mutation has an initial allele frequency of $1/2N$, because there is only one mutant allele among a total of $2N$ alleles at the locus in the population. Now, the probability that a neutral allele eventually becomes fixed is equal to its initial allele frequency, as shown in the section on random genetic drift earlier in this chapter, so the probability that a newly arisen neutral mutation will eventually become fixed is $1/2N$, a process that requires an average of $4N$ generations. (See Box E for more detail on the probability of fixation and loss in various sit-

E Random Genetic Drift for Neutral and Nonneutral Alleles

This box provides equations for calculating the probability that an allele will ultimately become fixed (or ultimately become lost) in a finite population and for calculating the average time required for fixation (or loss). (Methods for derivation of the equations can be found in Crow and Kimura, 1970.) Part (a) focuses on alleles that are neutral, (b) on beneficial alleles, and (c) on harmful alleles. Throughout we use the symbol N_a for actual population size and N_e for effective size.

a. Consider a newly arising mutation in a diploid population with an equal number of males and females; let the population have actual size N_a and effective size N_e. Since the mutation just arose, its initial frequency will be $p = 1/2N_a$. If the mutation is selectively neutral, the probability that it will eventually become fixed is $1/2N_a$, and the probability of its eventual loss is $1 - (1/2N_a)$. Moreover, newly arising neutral alleles that are destined to become fixed require a long time to become fixed, the average being $4N_e$ generations; however, those alleles destined to be lost are lost very quickly, on the average in $2(N_e/N_a)\ln(2N_a)$ generations. For $N_a = 5000$ and $N_e = 4000$, calculate the probability of and average time to fixation of a newly arising neutral mutation, and calculate the probability of and average time to loss. E

b. Assume a newly arising favorable allele with fitness $1 + s/2$ in heterozygotes and $1 + s$ in homozygotes, relative to a value of 1 for the nonmutant homozygotes. Thus the favorable allele is assumed to be additive in its effects on fitness. (For example, if A is the favored allele and $s = .02$, then AA, Aa, and aa zygotes survive to maturity in the ratio $1 + s{:}1 + s/2{:}1$, or 1.02:1.01:1.00.) The probability of eventual fixation of the favorable mutation is approximately $2s(N_e/N_a)$, the approximation being valid provided $2s(N_e/N_a)$ is small and $4N_es \gg 1$. (The symbol $\gg$ means "much greater than.") For an ideal population ($N_e = N_a$), therefore, the probability of eventual fixation is approximately twice the selection coefficient. For $N_a = 5000$ and $N_e = 4000$, calculate the probability of fixation of a newly arising additive allele with $s = .01$.

c. Assume a newly arising harmful allele with fitness $1 - s/2$ in heterozygotes and $1 - s$ in homozygotes. If $2N_es \gg 1$, then virtually all such alleles will be lost, and the average persistence in the population before final loss is $2(N_e/N_a)[\ln(2N_a/2N_es) + 1 - \gamma]$ generations, where γ is Euler's constant and equals .5772157. . . . For $N_a = 5000$ and $N_e = 4000$, what is the average time to extinction of a newly arising harmful allele with $s = .01$? (Compare answer with that for neutral alleles in (a).)

---

uations.) On the other hand, the number of newly arising neutral mutations in any generation equals $2N \times \mu$, so in any generation the average number of newly arising mutations that are destined to be fixed eventually is $1/2N \times 2N\mu = \mu$, the neutral mutation rate. At equilibrium, the number of mutations that are just beginning their random genetic drift to fixation must equal the number that are just completing their fortuitous drift; thus, at equilibrium the average number of new allele fixations per generation must also equal μ.

Observed rates of nucleotide substitution vary considerably in different protein molecules, in different regions of the same molecule, at different times in the evolution of the same protein, and at different sites in codons (Table V). Those who espouse the neutrality hypothesis interpret this variability in substitution rates as due to differences in rates of neutral mutation, that is, to differences in the probability that a mutation leading to a nucleotide substitution will cause a harmful (and therefore nonneutral) amino acid change (Kimura and Ohta, 1974; Kimura, 1977). For example, the amino acids that occupy the shaded positions in the portion of human apolipoprotein A shown in Figure 17 tend to be of the same chemical type (e.g., acidic, basic, hydrophilic, hydrophobic) in all 13 tandem repeats of a 33-nucleotide (11-amino acid) repetition within the molecule, a finding that could be interpreted as meaning that these particular amino acids are required to produce the proper molecular conformation. The rate of neutral mutation would therefore be much smaller in the codons for these critical amino acid sites than in the codons for other, even adjacent, sites in the same molecule, as most mutations in the critical codons would be harmful. On the other hand, those who deny the neutrality hypothesis interpret variation in rates of nucleotide substitutions that change amino acids as being due to differences in the rate of occurrence of favorable mutations that become fixed by natural selection (Clarke, 1970; Zuckerkandl, 1976). (The fate of alleles subject to natural selection is discussed later in this chapter; for the probability of fixation, see Box E.)

Despite variability in rates of nucleotide substitution, the rate turns out to be approximately constant when averaged over a sufficient time or number of proteins (Figure 18; note that the rate is approximately constant *per year,* not per generation). Thus, the average rate of nucleotide substitu-

TABLE V. Variation in evolutionary rates of nucleotide substitutions.

Comparison	Protein	Time involved (years × 10^6)	Amino acid changing nucleotide substitutions per nucleotide site per 10^9 years
Between proteins	Fibrinopeptides[a]	90	3.03
	Hemoglobin[a]	480	.57
	Cytochrome *c*[a]	580	.17
	Average of 7 proteins[b]	120	.45
Within proteins	α Hemoglobin, total molecule[c]	300	.37
	Heme contacts[c]	300	.07
	$\alpha_1 - \beta_2$ contacts[c]	300	.03
	$\alpha_1 - \beta_1$ contacts[c]	300	.43
Between times	Hemoglobin family (myoglobin, α, β, γ, ...)		
	From invertebrate–vertebrate divergence to bird–mammal divergence[c]	380	1.53
	From bird–mammal divergence to present[c]	300	.50
Between nucleotide sites	Amino acid changing substitutions (primarily codon positions 1 and 2), average of 7 proteins[b]	120	.45
	Synonymous substitutions at third codon position (portions of α and β hemoglobin and histone H2a)[b]	120	(5.82)

[a] From Dickerson, 1971. [b] From Fitch, 1976. [c] From Goodman, 1976.

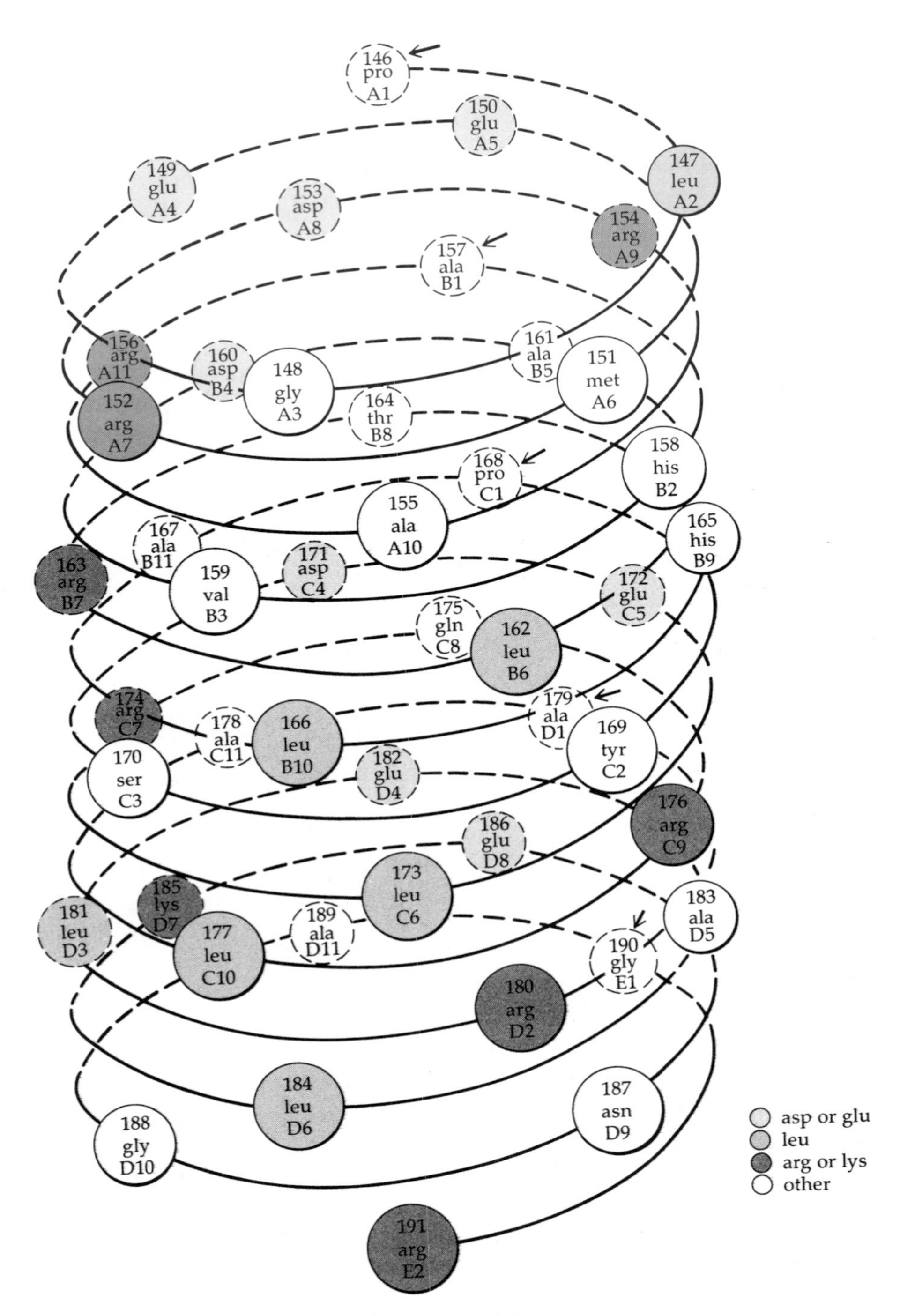
146 pro A1
147 leu A2
148 gly A3
149 glu A4
150 glu A5
151 met A6
152 arg A7
153 asp A8
154 arg A9
155 ala A10
156 arg A11
157 ala B1
158 his B2
159 val B3
160 asp B4
161 ala B5
162 leu B6
163 arg B7
164 thr B8
165 his B9
166 leu B10
167 ala B11
168 pro C1
169 tyr C2
170 ser C3
171 asp C4
172 glu C5
173 leu C6
174 arg C7
175 gln C8
176 arg C9
177 leu C10
178 ala C11
179 ala D1
180 arg D2
181 leu D3
182 glu D4
183 ala D5
184 leu D6
185 lys D7
186 glu D8
187 asn D9
188 gly D10
189 ala D11
190 gly E1
191 arg E2
asp or glu
leu
arg or lys
other

FIGURE 17. Helical structure of part of human apolipoprotein A-I showing four of the 13 tandem repeats of an 11 amino acid segment in the molecule. Arrows show the first amino acid in each repeat. Small numbers represent the numbers of the amino acid residues in the total protein; numbers prefixed by capital letters designate the numbers of amino acid residues in duplicated segments A, B, C, and so on. Note that the basic amino acids (arg, lys) are found mainly along the left- and right-hand sides of the molecule, the acidic amino acids (asp, glu) are found in the back, and the neutral amino acids (leu, gly, ala) are found mainly in front. This arrangement of amino acids presumably facilitates interaction of the hydrophobic front of this part of the molecule with the membrane and the hydrophilic rear of the molecule with the relatively water-rich cytoplasm. (Data from Fitch, 1977.)

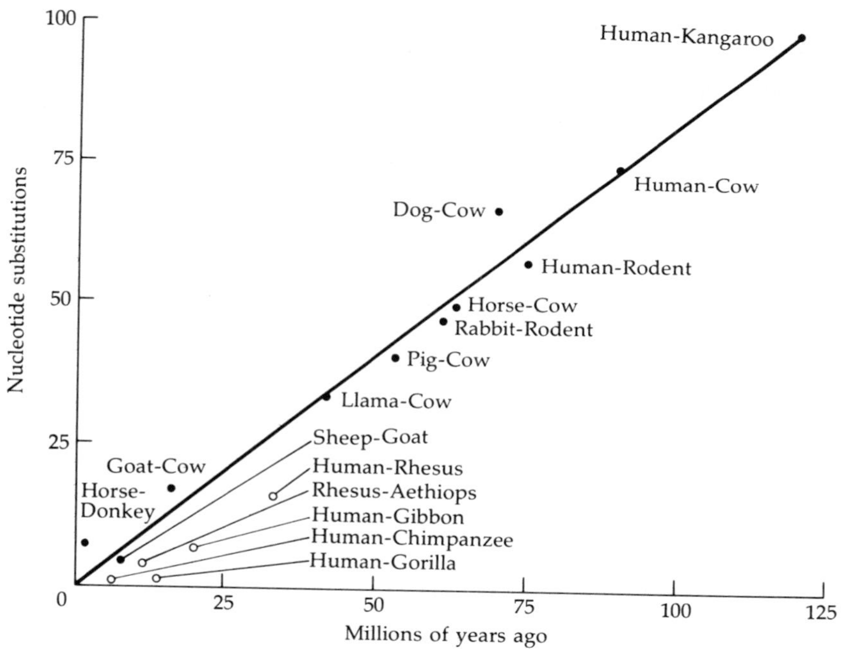

FIGURE 18. Estimated number of nucleotide substitutions in the gene for cytochrome *c* against estimated time since divergence of various pairs of different organisms. Note the linearity of the curve, indicating an approximately constant rate of substitution per year (or per million years). Open circles refer to primates, for which the rate of substitution is somewhat less than expected. (From Fitch and Langley, 1976.)

tion becomes a kind of MOLECULAR CLOCK; that is to say, the average rate of nucleotide substitution can be used as a measure of time since evolutionary divergence. (There is a large literature on molecular clocks. For a sampler, see Fitch, 1973; Fitch and Langley, 1976; Radinsky, 1978; Hartl and Dykhuizen, 1979.) Because of the molecular clock, differences in amino acid sequence (or inferred differences in nucleotide sequence) can be used as the basis for reconstructing evolutionary relationships (Fitch and Margoliash, 1967, 1970). Although such reconstructed evolutionary relationships are, by and large, in good agreement with those reconstructed from the fossil record (Figure 19), the actual amount of protein evolution is a very poor indicator of the amount of morphological evolution. Humans, for example, are morphologically very different from chimpanzees, yet the amount of protein divergence between humans and chimpanzees is extraordinarily small. This finding has led to the hypothesis that morphological evolution is primarily related to changes in regulatory genes, not to changes in structural genes of the sort studied for comparison of amino acid sequences (Wilson et al., 1974, 1977).

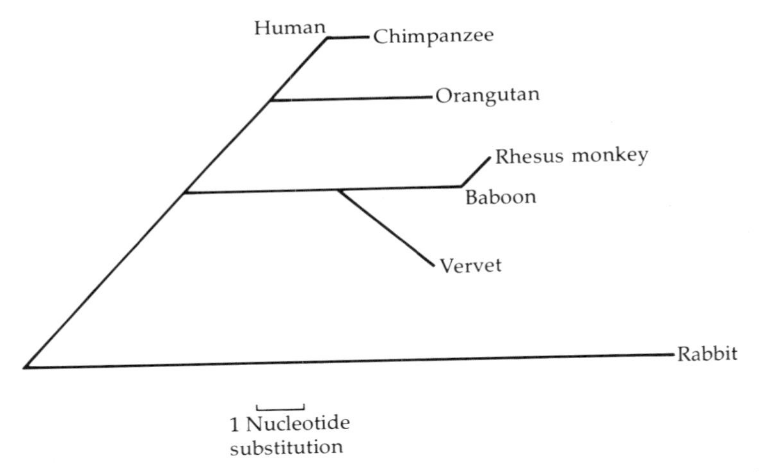

FIGURE 19. Estimated number of nucleotide substitutions in the gene for carbonic anhydrase between various organisms, scaled according to number of substitutions. (After Tashian et al., 1976.)

Comparisons of amino acid sequences can also be used to reconstruct evolutionary relationships among families of related proteins derived from common ancestral genes by duplication and subsequent evolutionary divergence. DUPLICATION AND DIVERGENCE, as illustrated for the hemoglobin family in Figure 20, is apparently a principal way that genes with new functions are acquired by eukaryotes in the course of evolution. A duplicate gene (i.e., one that is a duplicate of another gene in the same genome) can evolve more freely than a unique gene because the original function of the duplicate gene continues to be provided by its unmutated counterpart. Examples of duplication and divergence include the hemoglobin family in Figure 20, trypsin and chymotrypsin, the insulin polypeptides A and B, and other examples mentioned in Chapter 1.

Migration

In a subdivided population like the one in Figure 2, random genetic drift leads to genetic divergence of subpopulations. MIGRATION, which refers to the movement of individuals between subpopulations, is the "glue" that holds subpopulations together, that sets a limit to how much genetic divergence can occur. To see the homogenizing effects of migration, it is best to imagine first a group of subpopulations of infinite (not finite) size with differences in allele frequency among them. Suppose there are two alleles, A and a, with average allele frequencies denoted $\bar{p}$ and $\bar{q}$, respectively. Suppose that migration occurs and that the migrants are representative of the subpopulations; that is to say, among the migrants the allele frequencies of A and a are $\bar{p}$ and $\bar{q}$. The amount of migration can be measured by a number m representing the probability that a randomly chosen allele in a subpopulation comes from a migrant. Focus now on a randomly chosen allele in any subpopulation in generation t. This allele could have come from the same subpopulation in generation $t - 1$ (probability $1 - m$), in which case it will be an A allele with probability p_{t-1}, where p_{t-1} represents the allele frequency of A in the sub-

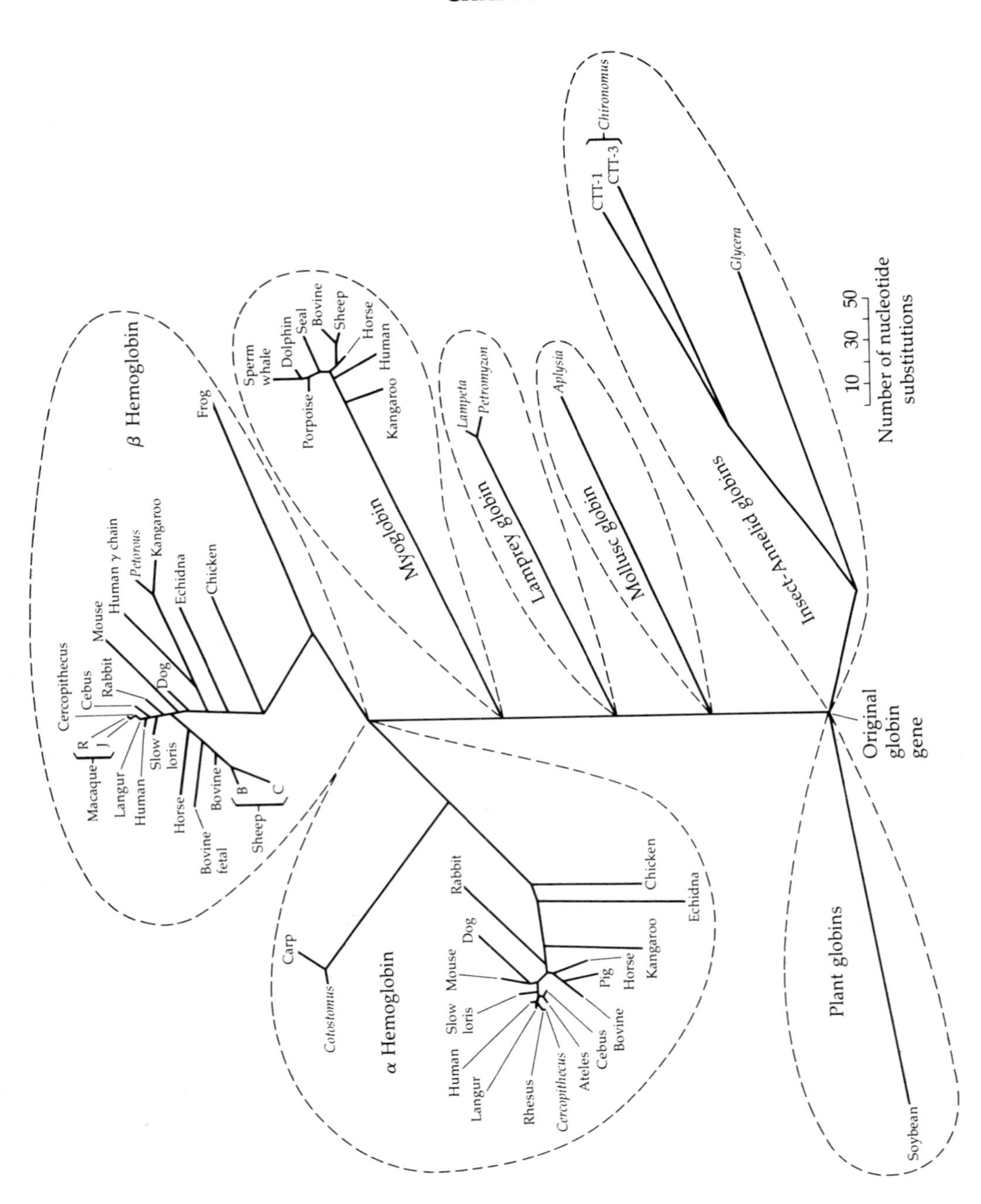

β Hemoglobin
Frog
Human γ chain
Petorous
Kangaroo
Echidna
Chicken
Mouse
Cercopithecus
Cebus
Rabbit
Dog
R
J
Macaque
Langur
Human
Slow
loris
Horse
Bovine
Bovine
fetal
B
C
Sheep
Myoglobin
Sperm
whale
Dolphin
Seal
Bovine
Sheep
Horse
Human
Kangaroo
Porpoise
Lamprey globin
Lampeta
Petromyzon
Mollusc globin
Aplysia
Insect-Annelid globins
CTT-1
CTT-3
Chironomus
Glycera
10
30
50
Number of nucleotide
substitutions
Original
globin
gene
α Hemoglobin
Carp
Cotostomus
Rabbit
Dog
Mouse
Slow
loris
Human
Langur
Rhesus
Cercopithecus
Ateles
Cebus
Bovine
Pig
Horse
Kangaroo
Chicken
Echidna
Plant globins
Soybean

FIGURE 20. Evolution of the globin family of proteins, scaled according to the estimated number of nucleotide substitutions required to account for the differences in amino acid sequence. (Data from Goodman, 1976.)

population in question in generation $t - 1$; the allele could also have come from a migrant in generation $t - 1$ (probability m), in which case it will be an A allele with probability $\bar{p}$. (Since random genetic drift, mutation, and natural selection are being ignored, $\bar{p}$ stays the same in all generations.) Altogether, $p_t = p_{t-1}(1 - m)$ [for the nonmigrants] $+ \bar{p}m$ [for the migrants].

We thus have an equation for p_t in terms of p_{t-1}, and a solution for this equation is one in which p_t is expressed in terms of p_0. A later box (H) outlines a method for finding the solution; it is $p_t = \bar{p} + (p_0 - \bar{p})(1 - m)^t$, where p_0 is the initial frequency of A in the subpopulation in question. As an example, suppose there are just two populations with initial allele frequencies of A of .2 and .8, respectively, and $m = .10$ [i.e., 10 percent of the individuals in either subpopulation in any generation are migrants having an allele frequency of A of $\bar{p} = (.2 + .8)/2 = .5$]. What would be the allele frequency of A in the two populations after 10 generations? For the population with initial allele frequency .2, we substitute $p_0 = .2$, $\bar{p} = .5$, and $m = .10$ into the above formula to obtain $p_{10} = .5 + (.2 - .5)(1 - .10)^{10}$, or $p_{10} = .395$; for the other population, we substitute $p_0 = .8$, $\bar{p} = .5$, and $m = .10$, so $p_{10} = .5 + (.8 - .5)(1 - .10)^{10} = .605$. Another example using the formula for p_t is shown in Figure 21, where there are five subpopulations (initial frequencies 1, .75, .50, .25, and 0) with $m = .10$; note how rapidly the allele frequencies converge to the same value (in this case, .5).

ISOLATE BREAKING AND WAHLUND'S PRINCIPLE

ISOLATE BREAKING refers to the fusion of formerly isolated subpopulations by migration. Fusion of populations reduces the frequency of homozygous genotypes, a phenomenon called WAHLUND'S PRINCIPLE (Wahlund, 1928). To reinforce

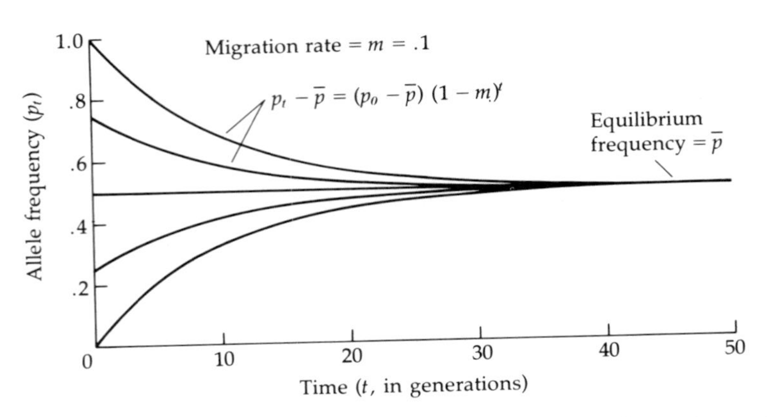

FIGURE 21. Change of allele frequency against time in five subpopulations exchanging migrants at the rate $m = .1$ per generation. Note rapid convergence to a common equilibrium frequency.

this idea, imagine two isolated subpopulations having allele frequencies of *a* of q_1 and q_2. The average frequency of *aa* homozygotes in the subpopulations is, from Table II, $\bar{q}^2(1 - F) + \bar{q}F$, where $\bar{q}$ represents the average allele frequency of *a* over all subpopulations (called q_0 in Table II) and F represents the fixation index (sometimes denoted F_{ST}). Were the subpopulations to fuse, the allele frequency of *a* in the fused population would be $\bar{q}$. After one generation of random mating, the frequency of *aa* homozygotes would be $\bar{q}^2$. Now, the frequency of *aa* before fusion is always greater than the frequency after fusion in the amount $\bar{q}^2(1 - F) + \bar{q}F$ [before fusion] $- \bar{q}^2$ [after fusion] $= -\bar{q}^2F + \bar{q}F = \bar{q}(1 - \bar{q})F = \bar{p}\bar{q}F$.

In human populations, the main effect of isolate breaking (fusion of populations) is to decrease the overall frequency of children born with genetic defects due to homozygous recessives, particularly for deleterious recessives having relatively high frequency in one of the populations. Examples of harmful recessives at high frequency in certain populations include the alleles for α_1-antitrypsin deficiency ($q = .024$) and cystic fibrosis ($q = .022$) in Caucasians, sickle-cell anemia in many Negro populations ($q = .05$ in American blacks, up to $q = .1$ in some African populations), albinism in the Hopi and some other southwest American Indians ($q = .07$), and Tay-Sachs disease among Ashkenazi Jews ($q = .02$). (Values of q in these examples are approximate.)

To illustrate the effect of isolate breaking, we may consider albinism among the Hopi. For this trait $q = .07$, so the frequency of affected children among the Hopi is $q^2 = (.07)^2 = .0049$, or about one in 200. For those Hopi who mate with members of an outside group, the situation is rather different. Suppose the outside group in this case is Caucasian; among Caucasians the frequency of albinism is about one in 20,000, so for Caucasians $q = \sqrt{1/20{,}000} = .007$. For Hopi that intermarry with Caucasians, the average allele frequency among mates is therefore $\bar{q} = (.07 + .007)/2 = .0385$, and the frequency of albino offspring would be $\bar{q}^2 = (.0385)^2 = .0015$, or about one in 675. The incidence of albinism is reduced by $.0015/.0049 = .31$, or roughly a third. (Of course, population fusion should not necessarily be advocated merely because it reduces the frequency of rare homozygous recessives, as social, cultural, and many other factors are often paramount in determining human mating patterns.)

As shown above, the reduction in frequency of recessive homozygotes due to population fusion can be expressed in terms of the fixation index as $\bar{p}\bar{q}F$. The reduction in frequency can also be expressed in terms of the variance in allele frequency among subpopulations. Recall that the VARIANCE among any set of numbers can be calculated as the mean of the squares (often called the MEAN SQUARE) minus the square of the mean, and it measures how closely the individual numbers are clustered around the mean. To take an extreme example, imagine two subpopulations with allele frequencies of 1.0 and 0, respectively; the mean square is $(1/2)[(1)^2 + (0)^2] = .5$ and the mean is $(1/2)[1 + 0] = .5$, so the variance is equal to $\sigma^2 = .5$ [the mean square] $- (.5)^2$ [the squared mean] $= .25$. (The symbol σ^2 is conventionally used for variance.) The allele frequencies in the two populations in this example are as different as they can possibly be, so the maximum possible variance in allele frequency for two populations is .25. As another example, suppose the two allele frequencies are .75 and .25; then $\sigma^2 = (1/2)[(.75)^2 + (.25)^2] - [(1/2)(.75 + .25)]^2 = .0625$. (Note in this example that the deviations of the allele frequencies from the mean are half as large as in the previous example — .25 instead of .50 — but the variance is only one-fourth as large; this disproportionate reduction in the variance occurs because the variance depends on the square of the deviation of each allele frequency from the mean.) As a final example, suppose the allele frequencies in the two populations are both .5. Then $\sigma^2 = (1/2)[(.5)^2 + (.5)^2] - [(1/2)(.5 + .5)]^2 = 0$, which is simply the mathematical manner of stating that the populations have the same allele frequency (i.e., there is no variation).

We are now in a position to express Wahlund's principle in terms of the variance in allele frequency. Consider the two populations in Figure 22 before fusion. The mean (or average) allele frequency, denoted $\bar{q}$, is $(1/2)(q_1 + q_2)$. The mean square, denoted $\overline{q^2}$, is the average of the squares, that is $(1/2)(q_1^2 + q_2^2)$. The variance in allele frequency, again denoted σ^2, is numerically equal to the mean square minus the squared mean, so, in symbols, $\sigma^2 = \overline{q^2} - \bar{q}^2$. As shown in Figure 22, the average frequency of *aa* homozygotes before population fusion is greater than the frequency after fusion

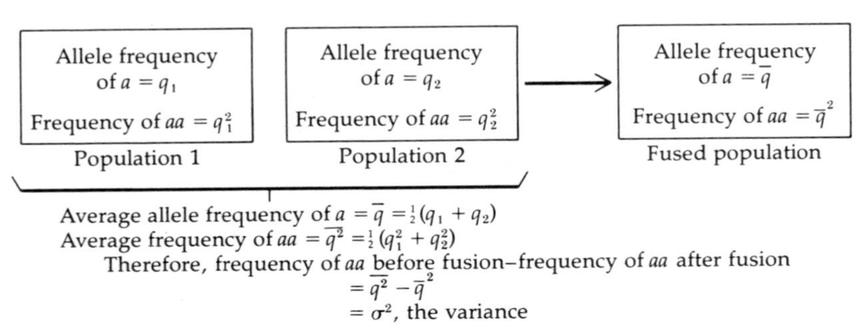

FIGURE 22. Illustration of Wahlund's principle (isolate breaking) — the frequency of homozygous recessives upon population fusion and random mating is less than the average frequency before fusion by an amount equal to the variance in allele frequency.

by the amount σ^2. Because the decrease in homozygosity with population fusion must be the same whether stated in terms of F or σ^2, it must be that $\overline{pq}F = \sigma^2$, or

$$F = \frac{\sigma^2}{\overline{pq}}$$

This expression provides a useful way of estimating F from allele frequency data. For example, in the three pairs of hypothetical populations in the paragraph above, $\bar{p} = \bar{q} = .5$; for the populations with allele frequencies of 1 and 0, $F = \sigma^2/\overline{pq} = .25/(.5)(.5) = 1$, which is simply a statement in terms of the fixation index F that the populations are as different as they can be. Similarly, for the populations with allele frequencies of .75 and .25, $F = .0625/(.5)(.5) = .25$; and for the populations with allele frequencies of .5 and .5, $F = 0/(.5)(.5) = 0$. Box B provides an opportunity to use the expression $F = \sigma^2/\overline{pq}$ in more complex situations.

HOW MIGRATION LIMITS GENETIC DIVERGENCE

It is remarkable how little migration is required to prevent significant genetic divergence among subpopulations due to random genetic drift. The effect can be seen quantitatively by considering the model of random drift summarized in Table I, but permitting migration at a rate m. (This model, which permits migration, is known as the ISLAND MODEL because it is easiest to visualize if each subpopulation is assumed to inhabit a little island of its own.) The expression F_t derived in Figure 6 is still valid, provided neither allele α_i or α_j has been replaced by a migrant allele. Therefore $F_t = [1/2N + (1 - 1/2N)F_{t-1}](1 - m)^2$. At equilibrium, $F_t = F_{t-1} = \hat{F}$, so $\hat{F} = [1/2N + (1 - 1/2N)\hat{F}](1 - m)^2$, which, for a sufficiently small m, works out to

$$\hat{F} = \frac{1}{4Nm + 1}$$

approximately. (Derivation of this approximation using simple algebra is provided as an exercise in Box D.) The actual *number* of migrants per generation is Nm, so $\hat{F}$ decreases as the number of migrants increases, as shown in Figure 23.

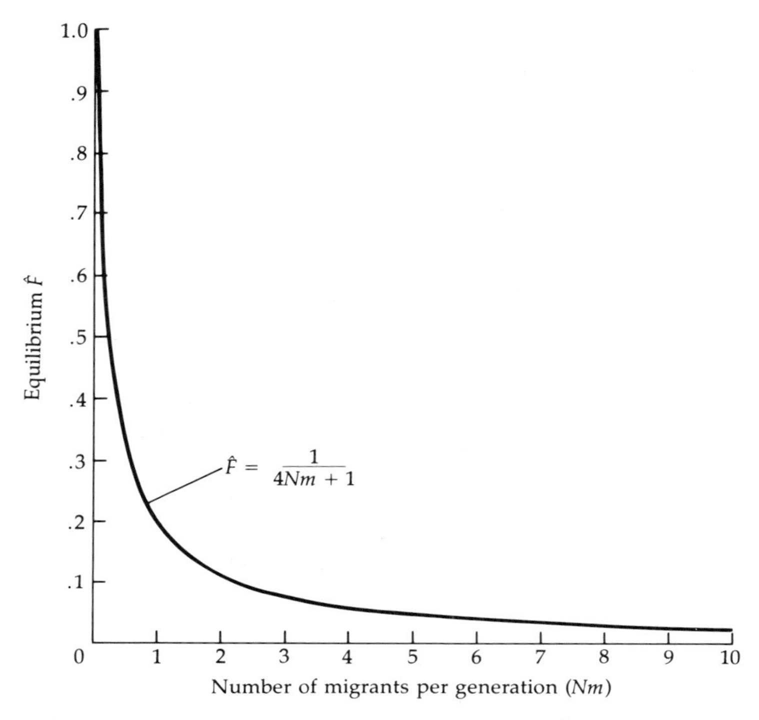

FIGURE 23. Decrease of equilibrium autozygosity ($\hat{F}$) against number of migrants per generation (Nm), showing how few migrants are necessary to reduce $\hat{F}$, and thus population differentiation, to very small levels.

(A number of migrants equal to, say, .5 in Figure 23 means one migrant individual every second generation.) When there is no migration, $Nm = 0$ and $\hat{F} = 1/(4 \times 0 + 1) = 1$; a single migrant into each subpopulation each generation ($Nm = 1$) reduces $\hat{F}$ from 1 to $\hat{F} = 1/(4 \times 1 + 1) = .2$, a decrease of 80 percent; one additional migrant each generation (so $Nm = 2$) reduces $\hat{F}$ to $1/(4 \times 2 + 1) = .11$, a further decrease of 45 percent. In short, migration is a potent force acting against genetic divergence due to genetic drift in subpopulations.

PATTERNS OF MIGRATION

Migration in actual populations is more complex than is assumed in the island model of migration. In nature, migrants come primarily from nearby populations; to the extent that nearby populations have similar allele frequencies, the effects of migration will be smaller, sometimes much smaller, than predicted by the island model. Populations in nature can be strung out along one dimension, such as a river bank; populations can also be distributed regularly in two dimensions; or there may be one large population with an internal genetic structure caused by the tendency for mating to occur between individuals born in the same region. Analysis of the effects of migration in such complex population structures is usually very difficult, but some of the important conclusions are summarized in Box F. Among humans, migration rates depend on age, sex, marital status, socioeconomic status, population density, and many other factors. Migration rates can change rapidly, moreover, so a full-blown theory of migration has to be extremely complex.

Models of Population Structure Involving Migration

F

Recall from Box D that $\hat{F} = 1/[4N(m + \mu) + 1]$ when mutation follows the infinite-alleles model and migration follows the island model, provided that $m + \mu$ is sufficiently small. Population structure in the island model is illustrated on the next page; the points represent individuals, which are found in discrete "colonies" representing populations. In this box, expressions for $\hat{F}$ are provided for more complex types of population structure shown in the figure. Part (a) considers the one-dimensional stepping-stone model, in which there are discrete colonies, but these are spread out along a line. Part (b) considers a uniform distribution in one dimension; in this case there are no discrete colonies, but individuals are spread out in one dimension. Part (c) deals with a uniform distribution in two dimensions; not only are there no discrete colonies, but the individuals are spread out uniformly in two dimensions.

a. One-dimensional stepping-stone model. Imagine a number of populations, each of effective size N, positioned at intervals along a linear habitat — a river bank, for example. If the rate of migration between adjacent colonies is $m/2$ and there is no migra-

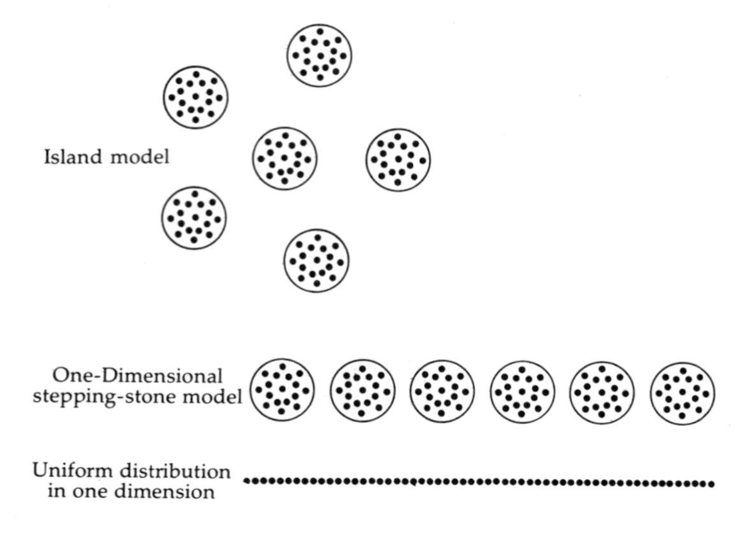

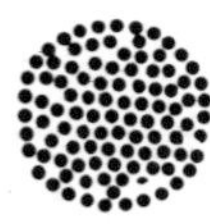

tion between nonadjacent colonies, then for an individual (or for two alleles chosen at random from within the same colony), $\hat{F} \approx (4N\sqrt{2m\mu} + 1)^{-1}$, where we assume that μ is much smaller than m and both are much smaller than 1 (Kimura and Weiss, 1964). For the island model, $\hat{F} \approx [4N(m + \mu) + 1]^{-1}$ (see Box D). Let $\mu = 10^{-6}$, $m = .01$, and $N = 50$; calculate $\hat{F}$ for the one-dimensional stepping-stone model and for the island model. By how much has restriction of migration to adjacent populations increased $\hat{F}$?

b. For a population distributed uniformly in one dimension (rather than in discrete colonies), let σ be the standard deviation of the distance between birthplaces of parents and their offspring, and let δ be the number of individuals per unit distance. (Migration is assumed to follow a "normal distribution" — see Chapter 4; "standard deviation" is defined as the square root of the variance, so the variance of the migration distance in this model is σ^2.) With these definitions, $\hat{F} \approx (1 + 4\delta\sigma\sqrt{2\mu})^{-1}$ (Malécot, 1967). Show that $\hat{F}$ is the same as in the one-dimensional stepping-stone model if $\delta = N$ and $\sigma = \sqrt{m}$. Verify for the values $\mu = 10^{-6}$, $m = .01$, and $N = 50$ as above, so that $\delta = 50$ and $\sigma = .10$. (Note for later use that the average number of individuals within a distance σ of any point is $2\sigma\delta = 10.0$.) F

c. For a population distributed uniformly in two dimensions, let σ^2 be the variance in the distance between the birthplaces of parents and their offspring in one dimension, as in (b) above; $2\sigma^2$ is thus the variance in two dimensions, so the two-dimensional standard deviation is $\sqrt{2}\sigma$. Migration is assumed to be "isotropic," i.e., to follow a normal distribution in both dimensions. Let δ be the number of individuals per unit area. For this model, $\hat{F} \approx [1 - 8\pi\delta\sigma^2/\ln(2\mu)]^{-1}$, where $\pi = 3.14159\ldots$ (Malécot, 1950, 1972). Solve for $\hat{F}$ in the case $\mu = 10^{-6}$, $\delta = 325$, and $\sigma = .07$. [In this case, the number of individuals within a circle of radius $r = \sqrt{2}\sigma$ around any point is $\pi r^2\delta = 2\pi\delta\sigma^2 = 10.0$, the same as in (b).] (For more on this model of population structure, see Morton et al., 1968, 1977; Morton, 1969, 1977; Lalouel, 1977.)

d. For a population of effective size N, the quantity $1/N$ is the probability that two gametes chosen at random come from the same individual. In populations that are distributed uniformly rather than in discrete colonies, calculation of the latter probability therefore yields an expression whose reciprocal is N, which in this context is called the "neighborhood size." In the one-dimensional case (b), $N = 2\sqrt{\pi}\delta\sigma$; in the two-dimensional case (c), $N = 4\pi\sigma^2\delta$ (Wright, 1969). Calculate the neighborhood size for the one-dimensional example in (b) and compare it with the neighborhood size for the two-dimensional example in (c).

Natural Selection

THE MEANING OF FITNESS

So far in this book the term NATURAL SELECTION has been used in its informal, intuitive sense, in the sense used by Darwin over a century ago in *The Origin of Species*: "Owing to this struggle for life, variations, however slight and from whatever cause proceeding, if they be in any degree profitable to the individuals of a species, in their infinitely complex relations to other organic beings and to their physical conditions of life, will tend to the preservation of such individuals, and will generally be inherited by the offspring.

The offspring, also, will thus have a better chance of surviving, for, of the many individuals of any species which are periodically born, but a small number can survive. I have called this principle, by which each slight variation, if useful, is preserved, by the term Natural Selection."

In this section, it is necessary to become more precise, more quantitative. First, it is important to note that natural selection is the driving force of evolution, the process that leads to greater adaptation of organisms to their environment. Subpopulations of a species are bound to live in somewhat different environments. If the environments are substantially different, natural selection, by increasing the adaptation of each subpopulation to its own environment, will promote genetic divergence of the subpopulations. If the environments are substantially similar, to consider the other extreme, natural selection, again by increasing adaptation, will tend to prevent genetic divergence of subpopulations. Random genetic drift enhances subpopulation divergence, whereas migration hinders it; so in any actual species, genetic variation produced by mutation will be organized, maintained, eliminated, or dispersed among subpopulations according to the precise and complex balance between natural selection, migration, and random genetic drift.

Natural selection acts on phenotypes, not on genotypes, and it acts on the whole phenotype as determined by many loci and countless environmental factors in one of three fundamentally different ways. These are illustrated in Figure 24. When selection favors phenotypes at one extreme of the range, selection is said to be DIRECTIONAL. (Directional selection will be discussed in detail in Chapter 4.) Selection favoring intermediate phenotypes is said to be STABILIZING (or NORMALIZING; for experimental examples see Prout, 1962; Kaufman et al., 1977). Selection simultaneously favoring phenotypes at *both* extremes of the range is said to be DISRUPTIVE (see Thoday and Boam, 1959; Thoday, 1972). Although selection occurs on phenotype, for purposes of exploring the consequences of selection, it is easiest to focus on how selection changes allele frequency at a single locus.

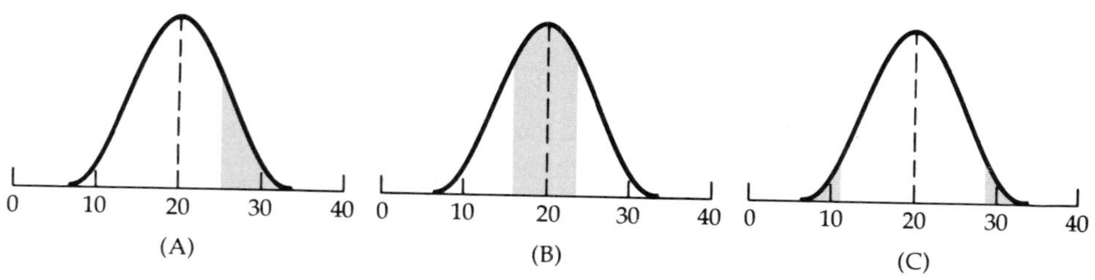

FIGURE 24. Three fundamentally different modes of selection. The curve represents the distribution of phenotypes in the population; and in each case, the shading represents those individuals favored by natural selection. (A) Directional selection. (B) Stabilizing selection. (C) Disruptive selection.

Quantitation of natural selection is accomplished by means of the concept of fitness. The FITNESS of a genotype is the average number of offspring produced by individuals of that genotype. This definition involves a number of subtleties that should be spelled out. First, the concept of fitness includes both VIABILITY (probability of survival to reproductive age) and FERTILITY (average number of offspring per survivor, allocating every offspring as belonging half to each parent). Table VI shows a hypothetical example involving four *AA*, six *Aa*, and four *aa* genotypes. (No actual case would involve such limited data.) Among the *AA* genotypes, 3/4 survived to reproductive age (viability) and the average number of offspring per survivor (fertility, often called FECUNDITY) is 6/3; the fitness of *AA* is therefore 3/4 [viability] × 6/3 [fertility] = 1.5. Among *Aa*'s the fitness is 5/6 [viability] × 9/5 [fertility] = 1.5. Finally, the fitness of *aa* is 3/4 × 2/3 = .5. Fitnesses calculated in the manner of Table VI are known as ABSOLUTE FITNESSES. In population genetics, it is usually convenient to deal in terms of the ratios of absolute fitnesses, which are called RELATIVE FITNESSES. The relative fitnesses of *AA, Aa,* and *aa* in Table VI are 1.5:1.5:.5. Because ratios remain unchanged when each value is divided by the same positive number, the relative fitnesses of *AA, Aa,* and *aa* can also be written as 1.5/.5:1.5/.5:.5/.5 = 3:3:1 or as 1.5/1.5:1.5/1.5:.5/1.5 = 1:1:.333. We will denote the relative fitnesses of *AA*:*Aa*:*aa* by use of the conventional symbols

TABLE VI. Estimation of fitness: a simple case.

Genotype	Survival	Number of offspring
AA	+	1
AA	+	2
AA	+	3
AA	−	−
	Viability = 3/4	Fertility = 6/3
	Fitness of *AA* = 3/4 × 6/3 = 1.5	
Aa	+	1
Aa	+	2
Aa	+	1
Aa	+	3
Aa	−	−
Aa	+	2
	Viability = 5/6	Fertility = 9/5
	Fitness of *Aa* = 5/6 × 9/5 = 1.5	
aa	−	−
aa	+	2
aa	+	0
aa	+	0
	Viability = 3/4	Fertility = 2/3
	Fitness of *aa* = 3/4 × 2/3 = .5	

$w_{11}:w_{12}:w_{22}$. Also, in interpreting theoretical models of selection, we will use the word "fitness" as a kind of synonym for viability, neglecting fertility for the sake of simplicity and brevity.

A second subtlety involving fitness is that fitness is exceedingly difficult to measure in practice. Fitness depends on the entire genotype, not merely on one locus. For the locus in question, therefore, one should study as many individuals of each genotype as possible in the hope that the effects of genetic background will average out to be equal in the genotypes. Then, too, genotypes should be counted at the same age in successive generations, otherwise possible sources of selection might be overlooked. Ideally, genotypes are identified in zygotes and the offspring also counted as zygotes, because zygotic genotype frequencies most accurately reflect the allele frequencies in gametes. Since identification of genotypes in zygotes is rarely possible, zygotic genotypes must be inferred from the observed genotypic frequencies in adults combined with the observed fertilities. Estimation of fitness is particularly difficult in age-structured populations. As noted in Chapter 2, each age class has its own age-specific death rate and birth rate, so overall viability and fertility of each genotype must be calculated as a sort of average viability and fertility over all age classes. Because of these complexities, actual estimates of fitness, when available, tend to be highly imprecise and uncertain unless fitness differences are very large. Another problem involving fitness is that fitness is rarely constant, at least not for appreciable lengths of time. The fitness of a genotype depends on the environment, so fitness can differ from subpopulation to subpopulation; or fitness can differ from generation to generation in the same subpopulation, changing as the environment changes. (There is a huge literature on the estimation of fitness in various situations or from various types of data. For examples, see Spiess and Schuellein, 1956; Prout, 1965, 1969, 1971a, 1971b; Bundgaard and Christiansen, 1972; Christiansen and Frydenberg, 1973; Christiansen et al., 1977a, 1977b; Cockerham and Mukai, 1978; Katz and Cardellino, 1978.)

EFFECTS OF SELECTION

Despite the many complexities of fitness mentioned above, the consequences of selection are most conveniently explored using the simple model outlined in Table VII. The model is essentially the Hardy–Weinberg model, but selection is permitted. Selection is assumed to occur on the diploid genotypes, not on the gametes, and segregation is assumed to be Mendelian. How allele frequencies change during one generation of selection is shown in Table VIII. The gametic frequencies of A and a in generation t are p_t and q_t, denoted p and q for brevity. Genotype frequencies in zygotes are given by the Hardy–Weinberg law (see Chapter

TABLE VII. Assumptions of constant selection model.

(1) Diploid organism
(2) Sexual reproduction
(3) Nonoverlapping generations
(4) Infinite population size
(5) Random mating
(6) No migration
(7) No mutation
(8) Selection occurs, fitnesses constant and equal in the two sexes

2), and the relative fitnesses (viabilities) of *AA*, *Aa*, and *aa* are denoted w_{11}, w_{12}, and w_{22}, respectively. By definition, zygotes survive in the ratio $w_{11}:w_{12}:w_{22}$, so the ratio of *AA*:*Aa*:*aa* among adults is

$$p^2w_{11}:2pqw_{12}:q^2w_{22}$$

The ratio of *A*:*a* in gametes is therefore

$$p^2w_{11} + \frac{1}{2}(2pqw_{12}):\frac{1}{2}(2pqw_{12}) + q^2w_{22}$$

(The 1/2's enter the ratio because *Aa* heterozygotes produce 1/2 *A* and 1/2 *a* gametes due to Mendelian segregation.) The ratio of *A*:*a* in gametes readily simplifies to

$$p(pw_{11} + qw_{12}):q(pw_{12} + qw_{22})$$

To obtain the gametic frequencies, one must divide each value in the above ratio by the overall sum, which is $p(pw_{11} + qw_{12}) + q(pw_{12} + qw_{22}) = p^2w_{11} + 2pqw_{12} + q^2w_{22}$. This sum, usually denoted $\overline{w}$, is simply the AVERAGE FITNESS of the population. Dividing the gametic ratios by $\overline{w}$ gives the gametic frequencies in generation $t + 1$, namely $p_{t+1} = p(pw_{11} + qw_{12})/\overline{w}$ and $q_{t+1} = q(pw_{12} + qw_{22})/\overline{w}$, which are denoted p' and q' in Table VIII. Thus, in one generation of selection, the allele frequency of *A* changes from p to

$$p' = \frac{p(pw_{11} + qw_{12})}{\overline{w}}$$

TABLE VIII. Change in allele frequency over one generation of selection[a].

Generation	Life-history stage	Frequencies				
t	Gametes	A				a
	Gametic frequencies ($p + q = 1$)	$p_t(= p)$				$q_t(= q)$
$t + 1$	Genotypes	AA		Aa		aa
	Genotype frequencies (zygotes)	p^2		$2pq$		q^2
	Fitness (viability)	w_{11}		w_{12}		w_{22}
	Ratio of genotypes (adults)	p^2w_{11}	:	$2pqw_{12}$	:	q^2w_{22}
$t + 1$	Gametes	A				a
	Ratio of gametic frequencies	$p^2w_{11} + (1/2)(2pqw_{12})$		:		$(1/2)(2pqw_{12}) + q^2w_{22}$
		$p(pw_{11} + qw_{12})$		:		$q(pw_{12} + qw_{22})$
	Average fitness ($\overline{w}$)	$\overline{w} = p(pw_{11} + qw_{12}) + q(pw_{12} + qw_{22})$ $= p^2w_{11} + 2pqw_{12} + q^2w_{22}$				
	Gametic *frequencies* ($p' + q' = 1$)	A				a
		$p_{t+1} = p'$				$q_{t+1} = q'$
		$p' = p(pw_{11} + qw_{12})/\overline{w}$				$q' = q(pw_{12} + qw_{22})/\overline{w}$
	Change in allele frequency of A (Δp)	$\Delta p = p' - p = pq[p(w_{11} - w_{12}) + q(w_{12} - w_{22})]/\overline{w}$				

[a] The relative viabilities of AA, Aa, and aa are denoted w_{11}, w_{12}, and w_{22}, respectively.

It is often useful to know $p' - p$, often symbolized Δp, the difference in allele frequency caused by one generation of selection; in this case

$$\Delta p = \frac{pq[p(w_{11} - w_{12}) + q(w_{12} - w_{22})]}{\overline{w}}$$

(Derivation of Δp is an exercise in Box G, which also includes several applications of the formula.)

At this point, a numerical example illustrating use of the above equations might be useful. Let us consider an extreme example in which genotypes *AA*, *Aa*, and *aa* survive in the ratio $w_{11} = 1$, $w_{12} = .5$, and $w_{22} = 0$ (i.e., *aa* is lethal, and *Aa* has a probability of survival only half as great as *AA*). Suppose that initially the allele frequency of *A* among gametes is $p = .1$. What will the value of p be among gametes in the next generation? (It is natural to wonder how the allele frequency of *A* could be a mere .1 considering the large amount of selection against *a*. Such a situation can easily be created experimentally, as the investigator can set the initial allele frequency at any predetermined value by appropriate choice of genotypes in the founding population; the situation can also arise in nature when a change of environment causes previously favored genotypes to become disfavored.) To calculate p among gametes in the next generation (p'),

---

G Time Required for a Given Change in Allele Frequency

The expression $\Delta p = pq[p(w_{11} - w_{12}) + q(w_{12} - w_{22})]/\overline{w}$ is of fundamental importance in the population genetics of natural selection. Remarkably, derivation of the expression involves nothing more than simple algebra, which is provided as an exercise in part (a). Part (b) provides several important cases in which the expression for Δp is put to use to derive equations that will be applied in Chapters 4 and 5.

a. From the expression $p' = p(pw_{11} + qw_{12})/\overline{w}$ derived in Table VIII, show that the change in allele frequency in one generation,

$p' - p$ (usually denoted Δp), obeys $\Delta p = pq[p(w_{11} - w_{12}) + q(w_{12} - w_{22})]/\bar{w}$. [Note: a useful identity is $1 - 2p = (1 - p) - p = q - p$.] G

b. This part of the problem is for students who have been required to study calculus but complain that they rarely have a chance to use it. Students familiar with calculus will appreciate the observation that, if w_{11}, w_{12}, and w_{22} are not too different, then Δp will be so small that $\Delta p \approx dp/dt$ to a good approximation. In such a case, an expression for p_t can be obtained. If we agree to measure relative fitnesses so that w_{11}, w_{12}, and w_{22} are all rather close to 1, then $\bar{w} \approx 1$ and $dp/dt = pq[p(w_{11} - w_{12}) + q(w_{12} - w_{22})]$ to a good approximation. There are a number of important cases:

(1) Dominant favored: $w_{11} = w_{12} = 1$ and $w_{22} = 1 - s$, with s small. Then $dp/dt = pq^2s$ and $dp/pq^2 = sdt$, so $\int_{p_0}^{p_t}(1/pq^2)dp = \int_0^t sdt$. Now use the fact that $\int[1/x(1 - x)^2]dx = 1/(1 - x) - \ln[(1 - x)/x]$ to derive $\int_{p_0}^{p_t}[1/p(1 - p)^2]dp = 1/(1 - p_t) - \ln[(1 - p_t)/p_t] - 1/(1 - p_0) + \ln[(1 - p_0)/p_0] = \ln[p_t(1 - p_0)/p_0(1 - p_t)] + [1/(1 - p_t)] - [1/(1 - p_0)] = \ln(p_tq_0/p_0q_t) + (1/q_t) - (1/q_0)$. Then use $\int dx = x$ to obtain $\int_0^t sdt = s(t - 0) = st$. Finally, set the two integrals equal to each other to show that $st = \ln(p_tq_0/p_0q_t) + (1/q_t) - (1/q_0)$.

(2) Additive alleles, A favored: $w_{11} = 1$, $w_{12} = 1 - s/2$, and $w_{22} = 1 - s$, with s small. Then $dp/dt = pqs/2$. Use $\int[1/x(1 - x)]dx = -\ln[(1 - x)/x]$ to show $st = 2\ln(p_tq_0/p_0q_t)$.

(3) Recessive favored: $w_{11} = 1$, $w_{12} = 1 - s$, and $w_{22} = 1 - s$, with s small. Here $dp/dt = p^2qs$. Use $\int[1/x^2(1 - x)]dx = -(1/x) - \ln[(1 - x)/x]$ to show $st = \ln(p_tq_0/p_0q_t) - (1/p_t) + (1/p_0)$.

(4) Asexual, haploid, or selection acting on gametes rather than on zygotes; A favored. In this case, the fitnesses of A and a are w_{11} and w_{22}, and there are no heterozygotes. From later Box I, $p_t = p_{t-1}w_{11}/\bar{w}$, where $\bar{w} = p_{t-1}w_{11} + q_{t-1}w_{22}$, so $\Delta p = pq(w_{11} - w_{22})/\bar{w}$. Let $w_{11} = 1$ and $w_{22} = 1 - s$, with s small. Then $\bar{w} \approx 1$, and $dp/dt = pqs$. It follows as in (2) that $st = \ln(p_tq_0/p_0q_t)$.

[Interestingly, the approximations (1) to (4) are more accurate for populations with overlapping generations than for ones with nonoverlapping generations.]

c. For $s = .01$, calculate the number of generations required to change the frequency of a favored allele from $p_0 = .01$ to $p_t = .99$ in cases (1) to (4).

first calculate $\overline{w}$ as $\overline{w} = p^2w_{11} + 2pqw_{12} + q^2w_{22} = (.1)^2(1) + 2(.1)(.9)(.5) + (.9)^2(0) = .10$; then use $p' = p(pw_{11} + qw_{12})/\overline{w}$ to obtain $p' = (.1)[(.1)(1) + (.9)(.5)]/(.10) = .55$. In this case, $\Delta p = .55 - .1 = .45$, which could have been calculated directly from the equation $\Delta p = pq[p(w_{11} - w_{12}) + q(w_{12} - w_{22})]/\overline{w} = (.1)(.9)[(.1)(.5) + (.9)(.5)]/(.10) = .45$. (A change of allele frequency of .45 in one generation of selection is extreme because of the high degree of selection in this example; values of Δp for realistic examples are typically much smaller.)

The theory in Table VIII provides an equation for p_t in terms of p_{t-1}. From this equation we would ideally wish to find an expression for p_t in terms of p_0. Unfortunately, such a general expression for p_t cannot be obtained for selection except in a few special cases (some of these cases are provided as exercises in Box H, and see Box I for an interesting application to bacterial chemostats). The long-term effects of selection when the relative fitnesses are constant can be determined rather easily, however; they are summarized in Table IX. The first two cases in Table IX follow immediately from the expression for Δp in Table VIII. First, note that there are always at least two equilibria (values of p_t such that $p_{t+1} = p_t$, or $\Delta p = 0$), namely $p = 0$ and $p = 1$. In the first instance, $\Delta p = 0$ because $p = 0$; in the second instance $\Delta p = 0$ because $q = 0$. Then, too, $\overline{w}$ must always be positive, and p and q must certainly not be negative. Therefore, if w_{11}

TABLE IX. Long-term effects of constant selection[a].

(1) $w_{11} > w_{12} > w_{22}$: *A* becomes fixed ($p \to 1$)
(2) $w_{11} < w_{12} < w_{22}$: *A* becomes lost ($p \to 0$)
(3) $w_{11} < w_{12}$, $w_{22} < w_{12}$: STABLE EQUILIBRIUM at $\hat{p} = (w_{12} - w_{22})/(2w_{12} - w_{11} - w_{22})$, and $p \to \hat{p}$
(4) $w_{11} > w_{12}$, $w_{22} > w_{12}$: UNSTABLE EQUILIBRIUM at $\hat{p} = (w_{12} - w_{22})/(2w_{12} - w_{11} - w_{22})$; if $p_0 < \hat{p}$, then $p \to 0$, whereas if $p_0 > \hat{p}$, then $p \to 1$

[a] Principles apply to a locus with two alleles and random mating. The fitnesses are assumed to be constant.

Mutation, Migration, Selection: Some Special Cases

H

The theory of population genetics frequently leads to equations in which the allele frequency in generation t, called p_t, is expressed in terms of the allele frequency in the previous generation, p_{t-1}. The solution to such an equation is an expression for p_t in terms of p_0, the allele frequency in the initial generation. In some cases, the solution can be stated immediately: (1) if the expression for p_t in terms of p_{t-1} can be written in the form $p_t - a = (p_{t-1} - a)b$, where a and b are any constants, then

$$p_t - a = (p_0 - a)b^t$$

(2) if p_t in terms of p_{t-1} can be written as $p_t = p_{t-1} + a$, where a is any constant, then

$$p_t = p_0 + at$$

This box provides several cases in which (1) or (2) above can be used.

a. For reversible mutation, $p_t = p_{t-1}(1 - \mu) + (1 - p_{t-1})\nu$. Show that $[p_t - \nu/(\mu + \nu)] = [p_{t-1} - \nu/(\mu + \nu)](1 - \mu - \nu)$, and then use (1) above to show that $[p_t - \nu/(\mu + \nu)] = [p_0 - \nu/(\mu + \nu)](1 - \mu - \nu)^t$. For $\mu = 98 \times 10^{-6}$ and $\nu = 2 \times 10^{-6}$, what is the value of p at equilibrium? If $p_0 = 1$, how many generations are required for p to go halfway to equilibrium? (Optional: prove that it requires $t = \ln(.5)/\ln(1 - \mu - \nu)$ generations for p to go halfway to equilibrium, *whatever* the value of p_0.)

b. For the migration model illustrated in Figure 21, $p_t = p_{t-1}(1 - m) + \bar{p}m$. Show that $(p_t - \bar{p}) = (p_{t-1} - \bar{p})(1 - m)$ so that, from (1) above, $p_t - \bar{p} = (p_0 - \bar{p})(1 - m)^t$. For $m = .01$ and $p_0/\bar{p} = 2$, what will be the value of $p_t/\bar{p}$ after 69 generations?

c. For the selection model in Table VIII, $p_t = p_{t-1}(p_{t-1}w_{11} + q_{t-1}w_{12})/\bar{w}$ and $q_t = q_{t-1}(p_{t-1}w_{12} + q_{t-1}w_{22})/\bar{w}$. Suppose the a allele represents a recessive lethal so that $w_{11} = w_{12} = 1$ and $w_{22} = 0$. Use the expression for q_t to show that $1/q_t = (1/q_{t-1}) + 1$, and then use (2) above to show $1/q_t = (1/q_0) + t$. Because random mating is assumed, note that the frequency of homozygous *aa* genotypes in generation t is q_t^2. Suppose the frequency of the lethal homozygote is $q_0^2 = .0016$ (so $q_0 = \sqrt{.0016} = .04$). How many generations would be required to reduce the frequency of the lethal *gene* to .01 (i.e., to 1/4 of its initial value); how many generations would be required

H to reduce the frequency of the lethal *genotype* to .0004 (i.e., to 1/4 of its initial value)?

d. Again using the selection model in Table VIII, if the fitnesses of *AA*, *Aa*, and *aa* are such that $w_{12} = \sqrt{w_{11}w_{22}}$, then the relative fitnesses $w_{11}:w_{12}:w_{22}$ may be written as $w_{11}/w_{22}:w_{12}/w_{22}:w_{22}/w_{22}$, or more succinctly as $w^2:w:1$, where $w = \sqrt{w_{11}/w_{22}}$. Use the expression for p_t in part (c) above to show that $p_t = p_{t-1}w/(p_{t-1}w + q_{t-1})$ and the expression for q_t to show that $q_t = q_{t-1}/(p_{t-1}w + q_{t-1})$. Combine these equations to obtain $(p_t/q_t) = (p_{t-1}/q_{t-1})w$ and use (1) above to conclude $p_t/q_t = (p_0/q_0)w^t$. Then show that $p_t = p_0w^t/(p_0w^t + q_0)$. Suppose $w_{11} = .9604$, $w_{12} = .98$, and $w_{22} = 1.0$, so that $w_{12} = \sqrt{w_{11}w_{22}}$ exactly, and $w = \sqrt{.9604/1.0} = .98$. (Note that, in this example, *A* is a somewhat detrimental allele that is very nearly additive; that is, $w_{12} = .98 \approx (w_{11} + w_{22})/2 = (.9604 + 1.0)/2 = .9802$.) If $p_0 = .99$, what is p_{50}? If $p_0 = .60$, what is p_{50}?

I Estimating Selection in Bacterial Chemostats

The figure in this box presents data from Dykhuizen (unpublished) in which two strains of *Escherichia coli*, call the strains *A* and *a*, were grown together in a bacterial chemostat. Prior to the experiment, strain *A* had been allowed to evolve for 160 generations in a chemostat, and strain *a* had evolved for 40 generations in a separate chemostat. Note that the graph of $\ln(p/q)$ against t is linear, where p represents the frequency of strain *A* and $q(= 1 - p)$ that of strain *a*. The reason for the linearity is that, in a haploid

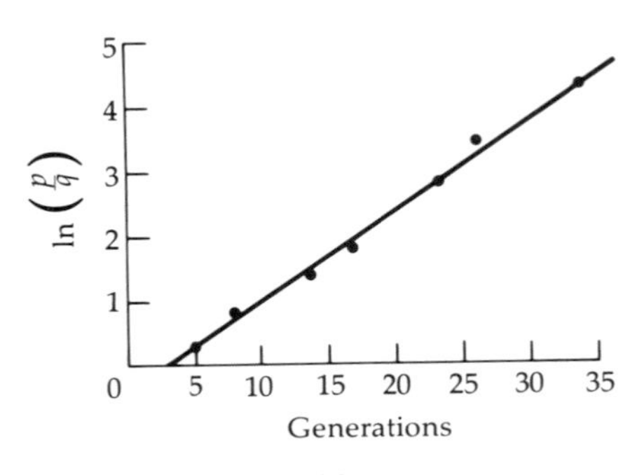

I

population with alleles A and a at frequencies p_t and q_t in generation t, the changes in allele frequency under selection can be shown by arguments analogous to those in Table VIII to be $p_t = p_{t-1}w_{11}/\bar{w}$ and $q_t = q_{t-1}w_{22}/\bar{w}$, where w_{11} and w_{22} are the fitnesses of A and a, and $\bar{w} = p_{t-1}w_{11} + q_{t-1}w_{22}$. Let $w = w_{11}/w_{22}$. Then w represents the fitness of A relative to a value of 1 for a. Show that $p_t/q_t = (p_{t-1}/q_{t-1})w$, and then recall from part (d) of Box H that $p_t/q_t = (p_0/q_0)w^t$; taking logarithms of both sides of the latter equation leads to $\ln(p_t/q_t) = \ln(p_0/q_0) + t\ln w$. Thus, a plot of $\ln(p_t/q_t)$ against t should be linear with a slope equal to $\ln w$. The line shown in the figure is the best fitting line, calculated by the method of linear regression. The line fits several of the points exactly, however; when $t = 5$ generations, $p = .561$, and when $t = 24$ generations, $p = .9462$. From these two points, calculate the line $\ln(p_t/q_t) = \ln(p_0/q_0) + t\ln w$. [Note: given two points with coordinates (x_0, y_0) and (x_1, y_1) on the straight line $y = a + bx$, the slope is given by $b = (y_1 - y_0)/(x_1 - x_0)$ and the intercept is given by $a = y_0 - bx_0$.] What is the value of w? What was the initial value of p (p_0) in the experiment? Dykhuizen's strains A and a were originally equal in fitness, but strain A had undergone $160 - 40 = 120$ additional generations of evolution in a chemostat. What was the average increase in fitness per generation of strain A relative to strain a?

$> w_{12} > w_{22}$, then Δp will always be positive because $p(w_{11} - w_{12}) + q(w_{12} - w_{22})$ will always be positive; but $\Delta p > 0$ means that p always increases, so p must eventually go to $p = 1$. Conversely, if $w_{11} < w_{12} < w_{22}$, then Δp must always be negative, so $p \rightarrow 0$. Examples of these two cases are shown in Figure 25: in panel A, $p \rightarrow 1$ for any initial frequency p_0 because $w_{11} > w_{12} > w_{22}$; in panel B, $p \rightarrow 0$ for any p_0 because $w_{11} < w_{12} < w_{22}$. Note in both cases that the most rapid changes in allele frequency occur when the allele frequency is around .5.

There is a special terminology to describe certain types of equilibria. Observe in Figure 25A that, when p is close to the equilibrium at $p = 0$, p moves farther away from the equilibrium in subsequent generations; such an equilibrium is

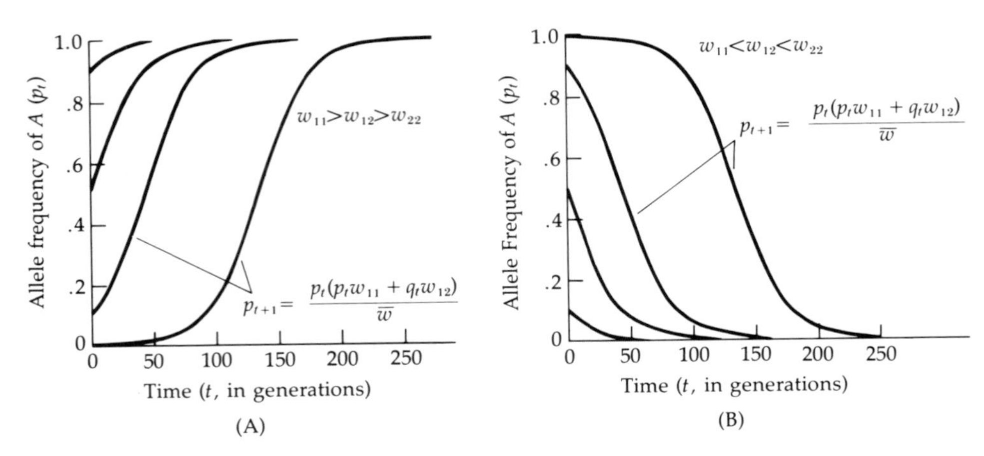

FIGURE 25. Change of allele frequency against time when (A) allele A is favored, and (B) allele a is favored. Fitnesses assumed are $w_{11} = 1$, $w_{12} = .95$, $w_{22} = .9025$ in (A) and $w_{11} = .9025$, $w_{12} = .95$, $w_{22} = 1$ in (B). In both cases $w_{12} = \sqrt{w_{11}w_{22}}$, which allows an explicit expression for p_t to be obtained (Box H).

said to be UNSTABLE. When p is close to the equilibrium at $p = 1$ in Figure 25A, on the other hand, then p moves closer to the equilibrium in subsequent generations; such an equilibrium is said to be LOCALLY STABLE. Actually, the equilibrium at $p = 1$ in Figure 25A is GLOBALLY STABLE, which means that $p \rightarrow 1$ irrespective of the starting value of p. Any globally stable equilibrium must also be locally stable, but a locally stable equilibrium need not be globally stable. In Figure 25B, $p = 0$ is globally stable, whereas $p = 1$ is unstable.

Figure 26 shows the results of selection favoring the A allele when, in the curve on the left, A is dominant to a (i.e., $w_{11} = w_{12}$ and $w_{12} > w_{22}$); in the curve on the right, A is recessive to a (i.e., $w_{11} > w_{12}$ and $w_{12} = w_{22}$). When the favored allele is dominant and at high frequency, allele frequency changes very slowly. To say the same thing in another way, when a disfavored allele is recessive and rare, allele frequency changes very slowly. For this reason, an increase in selection against rare homozygous recessive genotypes has almost no effect in changing allele frequency. Thus, the forced sterilization of rare homozygous recessive individuals, a procedure advocated in a number of naive eugenic programs to "improve" the "genetic quality" of human beings, is not only morally and ethically questionable, it is genetically unsound. Figure 27 shows the observed course of selection involving an allele in *Drosophila melanogaster.* The theoretical curve fits rather well, considering the simple nature of the model.

OVERDOMINANCE

In case 3 of Table IX, $w_{11} < w_{12}$ and $w_{22} < w_{12}$, so the heterozygote is superior in fitness to both homozygotes, a condition known as OVERDOMINANCE. When overdominance occurs, there will be a third equilibrium in addition to $p = 0$ and $p = 1$ because $p(w_{11} - w_{12}) + q(w_{12} - w_{22})$ can equal 0. The equilibrium allele frequency, $\hat{p}$, is found by solving $\hat{p}(w_{11} - w_{12}) + (1 - \hat{p})(w_{12} - w_{22}) = 0$, from which a little algebra gives $\hat{p} = (w_{12} - w_{22})/(2w_{12} - w_{11} - w_{22})$. This equilibrium is globally stable, whereas the equilibria at $p = 0$ and $p = 1$ are unstable as shown in Figure 28A. Figure 28B shows

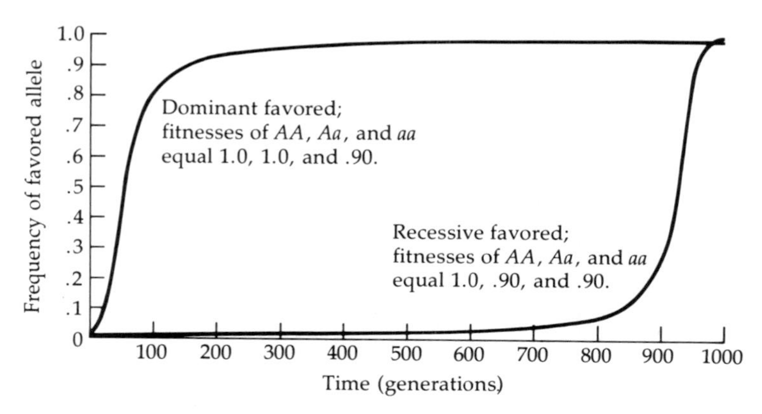

FIGURE 26. Change in frequency against time of a favored dominant (left curve) and a favored recessive (right curve). Note that the frequency of a favored dominant changes most slowly when the allele is common, but the frequency of a favored recessive changes most slowly when the allele is rare.

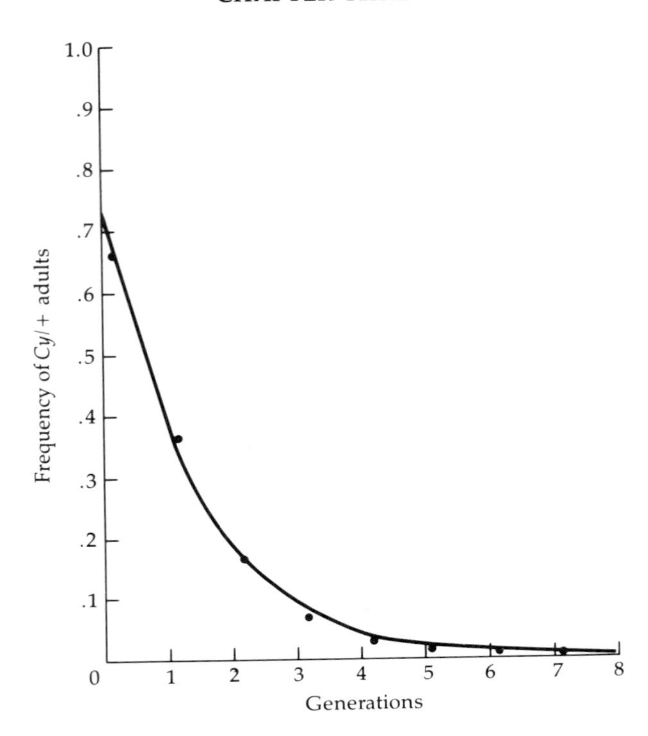

FIGURE 27. Change in frequency of adult *Drosophila melanogaster* heterozygous for the dominant mutation *Cy* (*Curly* wings) in an experimental population. *Cy/Cy* homozygotes are inviable, and the curve represents the theoretical change in frequency when the ratio of viabilities of *Cy*/+ to +/+ is 0.5:1. (Data from Teissier, 1942; the fitness value of 0.5 was calculated by Wright, 1977.)

Δp versus p; $\Delta p > 0$ when $p < \hat{p}$, so p increases toward $\hat{p}$, whereas $\Delta p < 0$ when $p > \hat{p}$, so p decreases toward $\hat{p}$. It is the shape of this curve that determines that $\hat{p}$ will be globally stable. Figure 28C shows $\overline{w}$ versus $\hat{p}$. Note that $\overline{w}$, the average fitness of the population, is maximal when $p = \hat{p}$. Maximization of average fitness is a frequent outcome of selection in random-mating populations with constant fitnesses, although there are many exceptions when more than one locus is involved.

Overdominance frequently leads to polymorphism in the sense discussed earlier in connection with electrophoresis

(Chapter 2); that is to say, overdominance frequently leads to an equilibrium in which $\hat{p}$ is between .01 and .99. For this reason, overdominance has been much discussed as a possible explanation for such polymorphisms as those involving

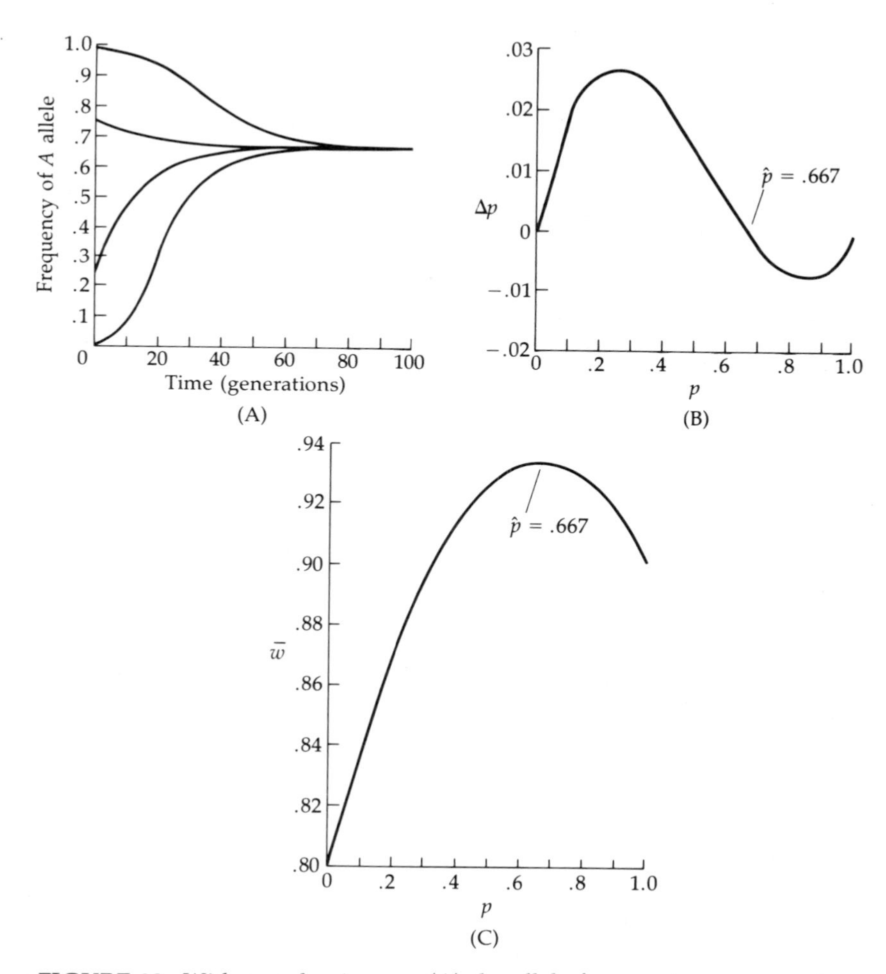

FIGURE 28. With overdominance, (A) the allele frequencies converge to an equilibrium value irrespective of the initial frequency; in this case, $w_{11} = .9$, $w_{12} = 1$, $w_{22} = .8$, so the equilibrium frequency of the *A* allele is .667. (B) Δp against p for the same example as in A, showing that $\Delta p < 0$ when $p > \hat{p}$, but $\Delta p > 0$ when $p < \hat{p}$. (C) Average fitness ($\bar{w}$) against p for the same example as in A; note that in this simple example $\bar{w}$ is a maximum at equilibrium.

allozymes. On the other hand, few proven cases of overdominance are known. The best known case involves two alleles coding for the β chain of human hemoglobin, $Hb\beta^+$, the normal allele, and $Hb\beta^S$, the allele leading to an amino acid substitution that causes sickle-cell anemia. Homozygous $Hb\beta^S/Hb\beta^S$ individuals are severely anemic due to sickling of red blood cells; $Hb\beta^+/Hb\beta^+$ individuals have normal hemoglobin but are sensitive to the type of malaria caused by the mosquito-borne protozoan parasite *Plasmodium falciparum*; $Hb\beta^S/Hb\beta^+$ heterozygotes are mildly anemic but tend to be resistant to falciparum malaria, apparently because red blood cells infested with the parasite undergo sickling and are removed from circulation. The viabilities of $Hb\beta^S/Hb\beta^S$, $Hb\beta^+/Hb\beta^S$, and $Hb\beta^+/Hb\beta^+$ individuals in high-malaria regions have been estimated as $w_{11} = 0$, $w_{12} = 1$, and $w_{22} = .85$ (see Allison, 1964), leading to a predicted equilibrium allele frequency for $Hb\beta^S$ of $(w_{12} - w_{22})/(2w_{12} - w_{11} - w_{22}) = (1 - .85)/(2 - 0 - .85) = .13$; this is quite close to the average allele frequency of .10 observed in West Africa, but there is considerable variation in allele frequency among local populations. The relationship between resistance to malaria and the $Hb\beta^S$ allele is supported by the extensive overlap in their geographical distributions (Figure 29).

Although overdominance is certainly capable of maintaining genetic polymorphisms involving two alleles (except perhaps in very small populations — see Robertson, 1962), the maintenance of polymorphisms involving three or more alleles by overdominance is much less likely because, as Lewontin et al. (1978) have pointed out, the superiority in fitness of each heterozygote over its corresponding homozygotes is by no means sufficient for overdominance to maintain multiple alleles; in most multiallelic situations, natural selection will eliminate most of the alleles, maintaining only those two or three that maximize the average fitness of the population at equilibrium. Thus, multiallelic polymorphisms in natural populations are unlikely to be maintained by simple overdominance.

On the other hand, there is one situation in which simple overdominance is probably of great importance; this situa-

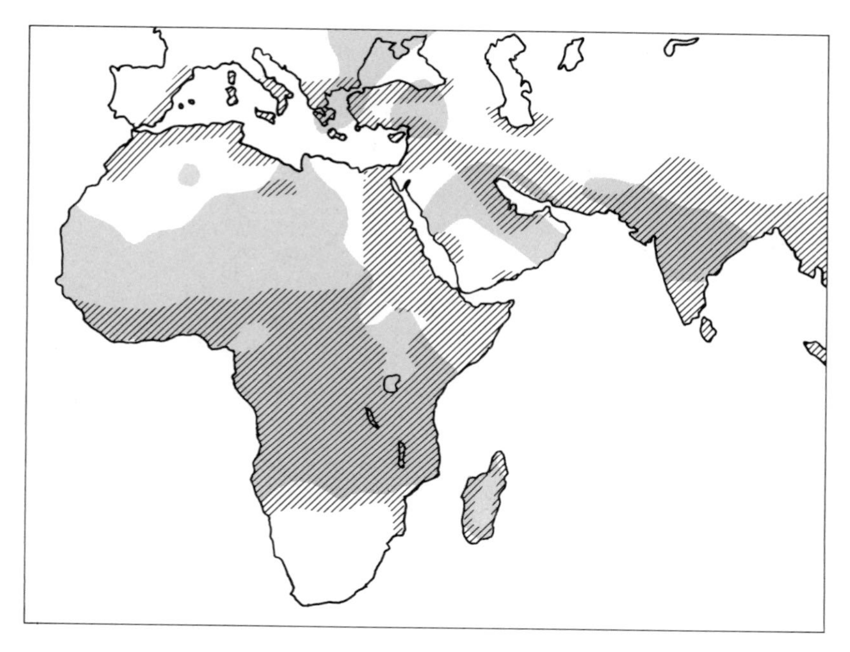

FIGURE 29. The hatched areas show the incidence of falciparum malaria around the 1920's (before control programs were implemented) in Africa, the Middle East, and southern Europe; the shaded regions designate areas of high incidence of sickle-cell hemoglobin disease. The extensive overlap in the distributions was an early indication that there might be some causal connection. (After Cavalli-Sforza, 1974.)

tion involves the polymorphic paracentric inversions found in many species of *Drosophila.* As discussed in Chapter 2, these inversions constitute coadapted "supergenes," and they frequently exhibit heterozygote superiority (Dobzhansky, 1970).

HETEROZYGOTE INFERIORITY

Case 4 in Table IX is HETEROZYGOTE INFERIORITY, in which $w_{11} > w_{12}$ and $w_{22} > w_{12}$. Again there is an equilibrium at $\hat{p} = (w_{12} - w_{22})/(2w_{12} - w_{11} - w_{22})$, but in this case the equilibrium is unstable (Figure 30A and B). The equilibria at $p = 0$ and $p = 1$ are both locally (but not globally) stable. When

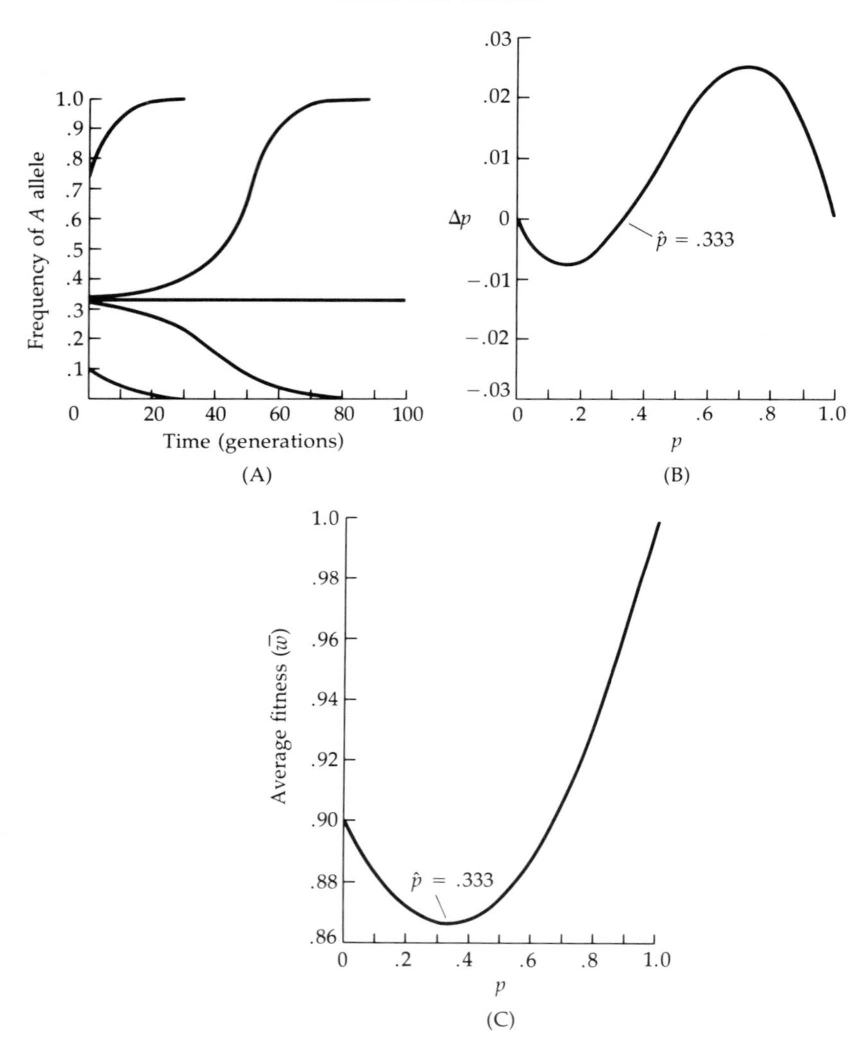

FIGURE 30. With heterozygote inferiority, (A) the allele frequency goes to 0 or 1, depending on the initial frequency; in this case $w_{11} = 1$, $w_{12} = .8$, $w_{22} = .9$, so there is an unstable equilibrium when the frequency of the *A* allele is $p = .333$; an infinite population with $p = 1/3$ will maintain that frequency, but any slight upward change in the frequency of *A* will lead to fixation of *A*, and any slight downward change in the frequency of *A* will lead to loss of *A*. (B) Δp against p for same example as in A showing that $p < \hat{p}$ implies $\Delta p < 0$ and $p > \hat{p}$ implies $\Delta p > 0$. (C) Average fitness ($\overline{w}$) against p for same example as in A; in this simple case, the unstable equilibrium represents the minimum of $\overline{w}$.

$p_0 < \hat{p}$, then $p \to 0$, whereas if $p_0 > \hat{p}$, then $p \to 1$. Figure 30C shows $\overline{w}$ versus p. Such a curve of $\overline{w}$ against allele (or chromosome) frequency is called an ADAPTIVE TOPOGRAPHY; the adaptive topography in Figure 30C is exceedingly simple, but for more complex genetic situations the adaptive topography is a complex surface in many dimensions with distinct "peaks" and "valleys." (Adaptive topographies are discussed in detail in Chapter 5.) In Figure 30C, the adaptive topography has two peaks (at $p = 0$ and $p = 1$) and one valley (at $p = .333$); the figure illustrates the important point that natural selection alone cannot carry allele frequency through a fitness "valley" in order to reach a higher fitness "peak." In this example, because $p = 0$ is locally stable, the *A* allele cannot increase in frequency in a population that is nearly fixed for *a*, even though fixation of *A* would increase average fitness. One way out of this dilemma is for population size to be sufficiently small that, in spite of selection, random genetic drift could occasionally carry the allele frequency of *A* across the valley. Random genetic drift can therefore be crucial in permitting a population to "explore" the full range of its adaptive topography (for further discussion and references, see Chapter 5).

Heterozygote inferiority occurs primarily in conjunction with such chromosomal abnormalities as translocations, for which heterozygotes are semisterile due to the production of aneuploid gametes (Chapter 1). Indeed, individuals carrying sufficiently complex chromosomal rearrangements produce no offspring in matings with normal individuals because the offspring have gross chromosomal imbalances and are inviable; yet two individuals that each carry the chromosomal rearrangement do produce some viable offspring, though even in this case a substantial fraction of the offspring are inviable due to chromosomal inbalance. Such chromosomal rearrangements have been proposed as a possible method of population control, especially in insects (Forster et al., 1972), the reason being that any equilibrium involving both normal individuals and those carrying the rearrangement must necessarily be unstable; thus, for a sufficiently high initial frequency of individuals carrying the rearrangement, the rearrangement ought to become fixed automatically and carry along such genes as ones conferring sensitivity to an insecticide. Specifically, the normal individuals may be looked upon as being homozygous for an "allele," say *N*/*N*, and the individuals with the rearrangement as being homozygous for an alternative "allele," say *R*/*R*. Because of the viability considerations mentioned above, the relative frequencies of *N*/*N*, *R*/*N*, and *R*/*R* individuals may be taken as 1:0:1 − *s*, where *s* measures the fraction of inviable zygotes in *R*/*R* × *R*/*R* matings. Thus, we have an extreme case of heterozygote inferiority with an unstable equilibrium at $\hat{p} = (1 - s)/(2 - s)$, where $\hat{p}$ represents the frequency of *N* among gametes. For $s = 3/4$ (a rather typical value for complex chromosomal rearrangements), $\hat{p} = .2$, or $\hat{q}$ (the frequency of *R* among gametes) is equal to .8. Thus, if an *N*/*N* population is overflooded with *R*/*R*'s at a level of more than four *R*/*R*'s for every *N*/*N*, the *R*/*R* type would be expected to go to fixation. (For another application of complex chromosomal rearrangements in population genetics, see Jungen and Hartl, 1979.)

MUTATION-SELECTION BALANCE

Recall from Chapter 2 that outcrossing species typically carry a large amount of hidden genetic variability in the form of recessive or nearly recessive deleterious alleles at low frequencies. This situation is to be expected; selection cannot completely eliminate such alleles because of their continual creation through recurrent mutation. To be specific, suppose *a* is the deleterious allele, and consider the model in Table VII, but with mutation allowed. Suppose mutation of *A* to *a* occurs at the rate μ per generation. (Since *q*, the allele frequency of *a*, will be small, reverse mutation from *a* to *A* can safely be ignored.) The calculation of p' in Table VIII is still valid, except that a proportion μ of *A* alleles will have mutated to *a*. Thus $p' = p(pw_{11} + qw_{12})(1 - \mu)/\overline{w}$. Here it is convenient to set $w_{11} = 1$, $w_{12} = 1 - hs$, and $w_{22} = 1 - s$, where $s > 0$. In this context, *s* is called the SELECTION COEFFICIENT against *aa*, and *h* is the DEGREE OF DOMINANCE; *a* is completely recessive when $h = 0$, partially dominant when

$0 < h < 1$, and dominant when $h = 1$. (In the special case $h = 1/2$, the fitness of *Aa* is exactly intermediate between that of *AA* and *aa*; and in this case, *A* and *a* are said to be ADDITIVE in their effects on fitness.) For example, in the case of the "dominant" allele for Huntington's chorea, a severe degenerative disorder of the neuromuscular system that typically appears after age 35, $w_{11} = 1$, $w_{12} = .81$, and $w_{22} = 0$ (data from a Michigan study by Reed and Neel, 1959). Thus $s = 1$ and $hs = 1 - .81 = .19$, so $h = .19$. In terms of neuromuscular degeneration, Huntington's chorea is dominant; in terms of fitness — ability to survive and reproduce — it is only partially dominant due to the late age of onset. Another example is provided by "recessive" lethals in populations of *Drosophila melanogaster,* where the average h is actually in the range .025 to .050, not 0 (Hiraizumi and Crow, 1960; Crow and Temin, 1964; Simmons and Crow, 1977).

At the equilibrium balance between mutation and selection, $p' = p = \hat{p}$, so $\hat{p}(\hat{p}w_{11} + \hat{q}w_{12})(1 - \mu)/\overline{w} = \hat{p}$, which, after substitution of $w_{11} = 1$, $w_{12} = 1 - hs$, $w_{22} = 1 - s$, and some algebra, leads to $\hat{q}hs(1 + \mu - 2\hat{q}) + \hat{q}^2s = \mu$. There are two important cases: (1) When *a* is completely recessive, then $h = 0$ and $\hat{q}^2s = \mu$, so

$$\hat{q} = \sqrt{\frac{\mu}{s}}$$

(2) When *a* is partially dominant, then $\hat{q}$ will often be small enough that $\hat{q}^2s$ can be ignored, and $1 + \mu - 2\hat{q}$ will be very close to 1, so $\hat{q}hs = \mu$, approximately, or

$$\hat{q} = \frac{\mu}{hs}$$

Huntington's chorea may be used to illustrate application of the above formulas. Recall that, for this disorder, $s = 1$ and $h = .19$. From the expression $\hat{q}hs = \mu$, we could calculate $\hat{q}$ if the value of μ were known, or we could calculate μ if $\hat{q}$ were known. For the Michigan population mentioned above, $\hat{q} = .00005$, so (assuming that the population is in equilibrium) we have $\mu = (.00005)(.19)(1) = 9.5 \times 10^{-6}$.

As another example, recall from Chapter 2 that the frequency of autosomes carrying one or more recessive lethals is about 33 percent in typical populations of *Drosophila melanogaster* (Ives, 1945; Salceda, 1977). Thus, the frequency of lethal-free chromosomes is 67 percent. Suppose that the mutation rate to lethals at any locus is 5×10^{-6} and that a lethal causes a 2.5 percent reduction in fitness when heterozygous (i.e., $s = 1$ and $h = .025$). For such a locus, $\hat{q} = \mu/hs = 5 \times 10^{-6}/.025 = .0002$. For n independent loci, the probability that none will carry a lethal allele is therefore $(1 - \hat{q})^n$; setting $(1 - \hat{q})^n = .67$ leads to $n = 2002$, or roughly 2000 loci capable of mutating to lethals. (While this estimate of the number of loci on an autosome of *Drosophila* is in good agreement with estimates from other sources, the agreement might be fortuitous, as the model is greatly oversimplified in assuming a constant mutation rate and degree of dominance at each locus.)

Figure 31 shows $\hat{q}$ versus μ/s for various values of h. Note that a small amount of dominance, as measured by h, sig-

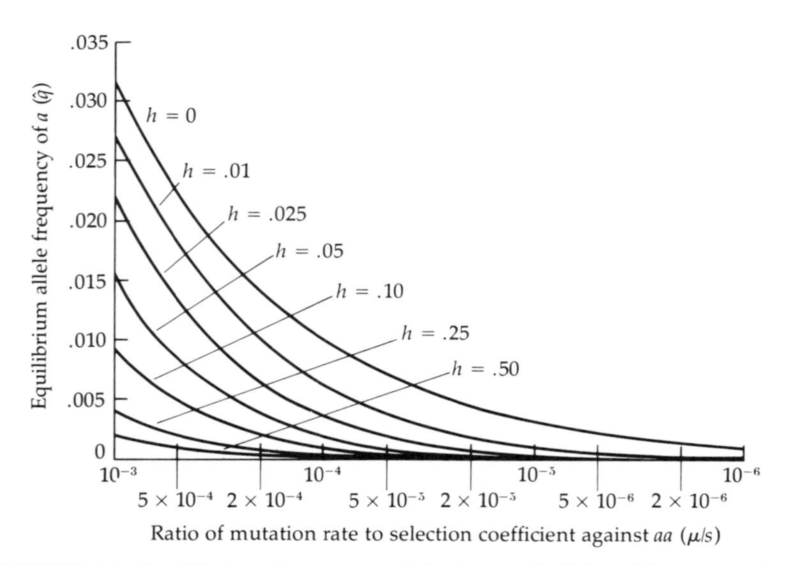

FIGURE 31. Equilibrium frequency of detrimental allele with various degrees of dominance (h), against μ/s. The fitnesses of *AA*, *Aa*, and *aa* are assumed to be 1, $1 - hs$, and $1 - s$, respectively; μ is the mutation rate of *A* to *a*; and q is the frequency of the *a* allele.

nificantly reduces the equilibrium frequency of a; this occurs because, when q is small, there will be so many more heterozygotes than aa homozygotes that even a small amount of selection against heterozygotes will have a major effect on the equilibrium frequency of a. Note also in Figure 31 that, for reasonable values of h, μ, and s, $\hat{q}$ is typically less than .01. Therefore, while mutation-selection balance satisfactorily accounts for the occurrence of low-frequency detrimentals in populations, it cannot readily account for polymorphisms in which, by definition, $q > .01$.

If there were no mutation, $\hat{q}$ would equal 0 and the average fitness of the population would equal 1. Because mutation occurs, $\hat{q}$ is not zero, and when $h = 0$ the average fitness of the population will be $1 - \hat{q}^2 s = 1 - (\mu/s)s = 1 - \mu$. The reduction in average fitness due to mutation equals $1 - (1 - \mu) = \mu$, which is called the MUTATION LOAD. (There is also a segregation load in the case of overdominance; see Box J.) When a is partially dominant, the mutation load is approximately 2μ. The effect of recurrent mutation in reducing population fitness is therefore independent of how harmful the mutation is, as measured by s. The effect of mutation on population fitness depends only on the mutation rate, so the harmful effect of an increase in mutation rate is the same, irrespective of whether the mutations produced are mildly or severely detrimental; this principle is often called the HALDANE-MULLER PRINCIPLE.

Maintenance of Genetic Polymorphisms

MORE COMPLEX TYPES OF SELECTION

The maintenance of genetic polymorphisms in populations is a difficult issue. Some population geneticists believe polymorphisms are maintained by a balance between the occurrence of new selectively neutral mutations and the extinction (or fixation) of these mutations by random genetic drift; this is the neutrality hypothesis discussed earlier in this chapter. Other population geneticists believe that selection best accounts for genetic polymorphism, a hypothesis called the SELECTIONIST HYPOTHESIS (for discussions, see Wills, 1973;

J Genetic Load

A population's GENETIC LOAD (L) due to any process is defined as the proportionate reduction in equilibrium average fitness as compared to that of a hypothetical population, otherwise identical, in which the process does not occur. Let $\overline{w}$ represent the average fitness of the population in question at equilibrium, and let w^* represent that of the hypothetical population. Then $L = (w^* - \overline{w})/w^*$. For recurrent mutation to a detrimental recessive, for example, $\overline{w} = 1 - \hat{q}^2 s$, where $\hat{q} = \sqrt{\mu/s}$, and $w^* = 1$. Thus $L(\text{mutation}) = [1 - (1 - \hat{q}^2 s)]/1 = \hat{q}^2 s = (\mu/s)s = \mu$. For a locus with overdominance, we can calculate a segregation load. Suppose the fitnesses of AA, Aa, and aa are given by $1 - s$, 1, and $1 - t$, and that the allele frequency of A is p. At equilibrium, $\hat{p} = t/(s + t)$ and $\hat{q} = s/(s + t)$. Then $\overline{w}$ at equilibrium is $1 - \hat{p}^2 s - \hat{q}^2 t = 1 - [st/(s + t)]$. In this case, the hypothetical population consists exclusively of heterozygotes, so $w^* = 1$ and $L(\text{segregation}) = st/(s + t)$. Calculate $\hat{q}$ and $L(\text{mutation})$ for a detrimental recessive with $\mu = 2 \times 10^{-6}$ and $s = .0052$; calculate $\hat{q}$ and $L(\text{segregation})$ for an overdominant locus with $s = 1.04 \times 10^{-4}$ and $t = .0052$. Note that the segregation load is much larger — in fact, $L(\text{segregation})/L(\text{mutation}) = 51$; why? (For more on theoretical and experimental aspects of genetic load, see Muller, 1950; Dobzhansky and Spassky, 1963; Brues, 1969; Crow and Kimura, 1970; Wallace, 1970; Hoenigsberg et al., 1977.)

Lewontin, 1974a; Ayala, 1976). Even the simplest model of selection, that defined in Table VII, permits polymorphism to be maintained if there is overdominance. Many other types of selection can also maintain polymorphisms, as shown in Table X. Most of these types of selection involve some sort of balance in which AA is favored under some conditions but disfavored under others. Even when the heterozygote is always intermediate in fitness between the homozygotes, polymorphisms will often be established. If genetic polymorphisms are maintained by selection, the most likely general mechanisms are spatially heterogeneous environments (Case 2 in Table X), temporally varying en-

TABLE X. Mechanisms that can maintain genetic polymorphism.

(1) Overdominance: heterozygote favored.

(2) Spatially heterogeneous environments: allele favored in some environments, disfavored in others (Levene, 1953; Hedrick et al., 1976; Bryant, 1976; Gillespie, 1978).

(3) Temporally varying environments: allele favored at some times, disfavored at others (Dempster, 1955; Haldane and Jayakar, 1963; Gillespie, 1972; Hartl and Cook, 1973; Karlin and Lieberman, 1974; Karlin and Levikson, 1974; Turelli, 1977).

(4) Epistasis: includes nonallelic interactions, modifiers, coadaptation, etc., in which an allele can be favored in some genetic backgrounds, disfavored in others (Karlin and Feldman, 1970; Feldman et al., 1974).

(5) Balance involving life-cycle stages: allele favored at one stage of life cycle, disfavored at another (Scudo, 1967).

(6) Balance of fitness components: allele favored in viability, disfavored in fertility.

(7) Density-dependent selection: allele favored at high population density, disfavored at low (Clarke, 1972; Templeton, 1974; Wallace, 1975).

(8) Frequency-dependent selection: allele favored at low frequency, disfavored at high (Clarke and O'Donald, 1964; Kojima, 1971; Ayala and Campbell, 1974; Gromko, 1977).

(9) Balance involving gametic selection: allele favored in zygotes, disfavored in gametes (Hiraizumi, 1964; Scudo, 1967; Mulcahy and Kaplan, 1979).

(10) Balance between the sexes: allele favored in one sex, disfavored in the other (Anderson, 1969; Mandel, 1971; O'Donald, 1977; Kidwell et al., 1977).

(11) Balance involving non-Mendelian segregation: allele disfavored in zygotes, but heterozygotes produce more than 50 percent of functional gametes bearing the allele (Sandler and Novitski, 1957; Charlesworth and Hartl, 1978; Lyttle, 1979).

(12) Group selection: allele neutral or detrimental to individual, but enhances survival of subpopulation; possibly important for "altruistic traits" — see Chapter 5.

(13) Hitchhiking: allele neutral, but affected by selection at linked locus (Thomson, 1977).

Thirteen mechanisms are listed above, but there are many more (especially if combinations of mechanisms are taken into account). Not all of the phenomena need necessarily lead to polymorphism, but they frequently do. The references are by no means exhaustive, nor do they imply scientific priority for the ideas; they are simply representative examples to serve as points of departure for further reading.

vironments (Case 3), and epistasis (Case 4). The other mechanisms in Table X are likely to be important only in some individual cases. (For one remarkable mechanism that prevents self-fertilization in some plants and at the same time leads to polymorphisms involving many alleles, see Box K.)

EVIDENCE FOR MAINTENANCE OF POLYMORPHISM

Statistical tests of neutrality. Earlier in this chapter observed data concerning genetic polymorphisms were compared with expectations of the neutrality hypothesis and the results found to be ambiguous. Ambiguity in such tests is by no means unusual. Indeed, considerable ambiguity is not surprising because the expectations of the neutrality hypothesis are often very similar to those from selection models, particularly selection models involving spatially heterogeneous or temporally varying environments. To date, no overall statistical test has provided convincing evidence either for or against the neutrality hypothesis. (For examples of such approaches and further discussion see Ewens, 1972, 1977, 1979; Ewens and Feldman, 1976; Weir et al., 1976; Kingman, 1977; Watterson, 1977; Schaffer et al., 1977.)

Clines. A CLINE is a regular trend of increase (or decrease) in allele frequency accompanying a change in an environmental variable such as temperature, salinity (for marine organisms), population density, or some combination of environmental variables. Clines are not at all unusual. Figure 32 shows a cline in the frequency of the *hemoglobin-*I^1 allele in the eelpout fish *Zoarces viviparus* from the North Sea to the Baltic Sea. In human aboriginal populations, there is a cline of increasing frequency of the I^B allele in the ABO blood groups from Southwest to Northeast Europe. Although clines can result from selection — one genotype being favored at one extreme of the environmental gradient, disfavored at the other extreme — clines can also result from, for example, migration coupled with founder effects at the extremes of the environmental gradient. Clines are therefore not necessarily evidence of selection.

Self-Sterility Alleles in Plants

K

Some plants, such as clover (Atwood, 1947), have systems of self-sterility alleles that prevent germination of pollen grains on the style of plants that carry the same allele as the pollen; thus a plant cannot fertilize itself, nor can homozygotes for self-sterility alleles ordinarily be formed. (The theory of such alleles has been studied by Wright, 1964, and Yokoyama and Nei, 1979.) Suppose there are n such alleles: $S_1, S_2, \ldots, S_n$. (In actual cases, n can be in the hundreds.) Let p_i be the frequency of S_i in zygotes, with $\Sigma p_i = 1$, and let $\theta = 1 - \Sigma p_i^2$. With random mating, the frequency of plants carrying any allele, say S_1, will be $2p_1(1 - p_1)/\theta$, and therefore the probability of an S_1 pollen grain landing on a compatible style is $1 - [2p_1(1 - p_1)/\theta]$, or $[\theta - 2p_1(1 - p_1)]/\theta$. Thus the relative fitnesses of $S_1, S_2, \ldots, S_n$ pollen are $w_1 = \theta - 2p_1(1 - p_1):w_2 = \theta - 2p_2(1 - p_2): \ldots :w_n = \theta - 2p_n(1 - p_n)$. In the next generation, the frequency of S_i will be $p_i' = p_i w_i/\bar{w}$ where $\bar{w} = \Sigma p_i w_i$, and at equilibrium $w_i = \bar{w}$ for all i.

a. From the above considerations, show that at equilibrium $p_i(1 - p_i) = p_j(1 - p_j)$ for any alleles S_i and S_j, and therefore that at equilibrium all alleles are equally frequent, i.e., $p_i = 1/n$ for $i = 1, 2, \ldots, n$. In this case $\bar{w}$ is the probability that a randomly chosen pollen grain settles on a compatible style. Show that at equilibrium $\bar{w} = (n - 2)(n - 1)/n^2$ and calculate $\bar{w}$ for $n = 3$, 30, and 300.

b. In an equilibrium population, the initial selection coefficient in favor of a newly arising (and therefore extremely rare) mutation to a novel allele is very nearly $1 - \bar{w}$. Show that this initial advantage equals $(3n - 2)/n^2$ and calculate its value for $n = 3$, 30, and 300. Does this consideration of initial selective advantage help explain why there are so many alleles at self-sterility loci?

The intensity of selection involved in the establishment and maintenance of clines of allele frequency can be surprisingly small, even when selection is involved; this point can be appreciated by considering a model of Fisher (1950), which provides a useful theory for the analysis of clines. Imagine a locus with two alleles, A and a, spread out uniformly along a line extending to negative infinity in one

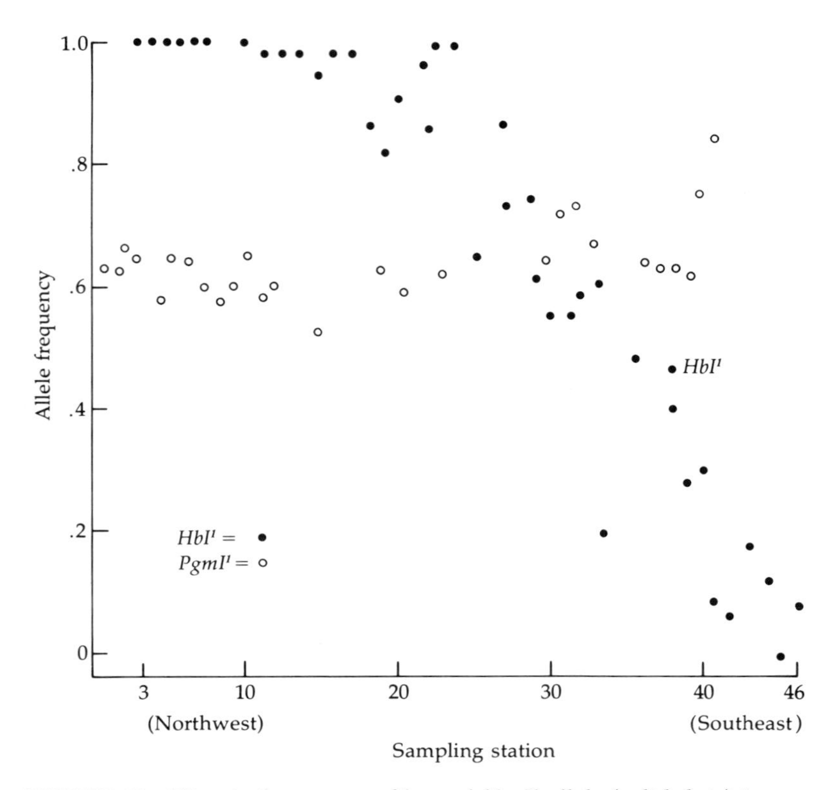

FIGURE 32. Cline in frequency of *hemoglobin-I*[1] allele (solid dots) in populations of the eelpout (*Zoarces viviparus*) in the seas northwest of Denmark and southeast to the Baltic. The frequency of the *phosphoglucomutase-I*[1] allele (open circles) changes little if at all over the same distance. (From Christiansen and Frydenberg, 1974.)

direction and positive infinity in the other. Fisher considers a model with overlapping generations in which births and deaths occur continuously in time; if the heterozygote has a fitness exactly intermediate between the fitnesses of the homozygotes, and if $s(x)$ measures the magnitude of selection at any point, x, along the line, then at any instant in time the rate of change of the natural logarithm of the ratio of allele frequencies at position x due to selection will be given by $s(x)$. (There is unfortunately no perfect analogy for this situation in terms of nonoverlapping generations;

roughly speaking, however, if $s(x)$ is small, the relative fitnesses of *AA, Aa,* and *aa* in an analogous nonoverlapping-generation model would be approximately $e^{s(x)}$, $e^{s(x)/2}$, and 1, respectively.) Partially offsetting the selection is migration, measured by σ^2, which is the variance in the distance along the line between the birthplaces of parents and offspring. To simplify matters further, Fisher assumes that $s(x)$ increases linearly with x; that is to say, $s(x) = gx$, where g is the slope of the line. As time goes on in Fisher's model, the allele frequencies eventually form a stable (i.e., permanent) cline. The allele frequencies in the cline do not change linearly with x, but Fisher found numbers related to allele frequency that do change linearly with x; these numbers he called LEGITS.

Table XI is a table of legits, considerably abridged from Fisher (1950). The column on the left gives the first digits of allele frequency, the row across the top gives the last digit, and the body of the table gives the corresponding legit. Thus, for example, the legit of $p = .13$ is legit $(.13) = .876$, and legit $(.006) = 2.087$. (The legit of $p = .50$, not included in the table, is 0.) For any allele frequency p, greater than .50, legit $(p) = -$ legit $(1 - p)$, so, for example, legit $(.65) = -$ legit $(.35) = -.294$.

Use of Table XI can be illustrated with data of Dobzhansky (1948) on the frequency of the *Standard* gene arrangement (an inversion) of the third chromosome of *Drosophila pseudoobscura* at altitudes of 259, 914, 1402, 1890, 2438, 2621, and 3018 meters in the Yosemite region of the Sierra Nevada in California. The data are laid out in Figure 33, with the pair of numbers near each point giving the frequency of *Standard* and the number of flies examined at each altitude. (These populations are highly polymorphic for three inversions of the third chromosome — *Standard, Arrowhead,* and *Chiricahua* — though four other inversions occur at aggregate frequencies of up to 10 percent.) In a much later study in the same area, Dobzhansky and Powell (1974) estimated dispersion as $\sigma^2 = 14{,}500$ m^2 per day (see also Crumpacker and Williams, 1973; Powell et al., 1976); assuming an average of seven days between emergence of an adult fly and deposi-

TABLE XI. Legits of allele frequency[a].

First digit	Second digit									
	0	1	2	3	4	5	6	7	8	9
.4	.193	.173	.154	.134	.115	.096	.076	.057	.038	.019
.3	.401	.379	.357	.336	.315	.294	.273	.253	.233	.213
.2	.648	.621	.594	.568	.542	.517	.493	.469	.446	.423
.1	1.002	.957	.915	.876	.838	.803	.770	.737	.707	.697
.0	—	1.913	1.665	1.511	1.398	1.306	1.230	1.163	1.104	1.051
.00	—	2.652	2.441	2.313	2.221	2.148	2.087	2.036	1.990	1.950

[a] The first digits of allele frequency are read from the column at the left and the last digit from the row across the top; the body of the table gives the corresponding legit. The legit of .5 equals 0, and legit (p) = −legit ($1 - p$). (Abridged from Fisher, 1950.)

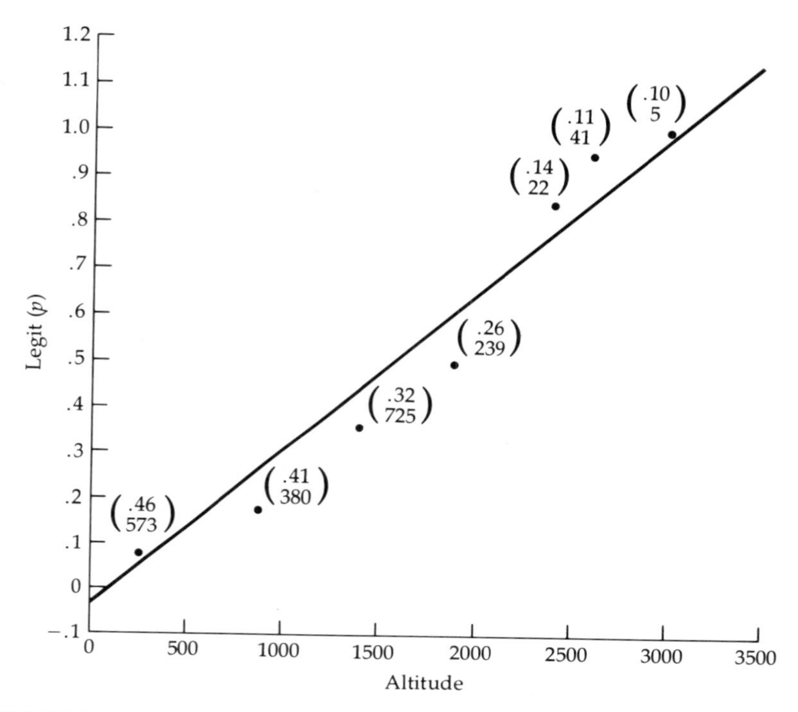

FIGURE 33. Legit of the frequency of the *Standard* gene arrangement in *Drosophila pseudoobscura*, against altitude in the Sierra Nevada. The numbers near each point give the frequency of *Standard* and the number of flies examined. (Data from Dobzhansky, 1948.)

tion of a fertilized egg (Wright, 1978), we may take $\sigma^2 = 7 \times 14{,}500 \text{ m}^2 = 101{,}500 \text{ m}^2$.

Analysis of the data proceeds as follows:

1. Find the legit corresponding to each frequency. (These are the points plotted in Figure 33.)

2. Carry out a weighted linear regression (as in Box D of Chapter 1) of legit (y-axis) on geographical position (x-axis), the weights being twice the number of animals examined at each point; in this case $y = -.037088 + .000291x$, which is the line shown. (Fisher gives a more elaborate and exact line-fitting procedure, but the procedure using weighted linear regression is adequate for most purposes.)

3. The reciprocal of the slope of the regression line gives the distance along the cline corresponding to an allele-frequency change of one legit; in this case the distance, call it a, is $a = (.000291/\mathrm{m})^{-1} = 3436.43$ m.

4. The slope, g, of $s(x) = gx$ is given according to Fisher's theory as $g = 4\sigma^2/a^3 = 4(101{,}500\ \mathrm{m}^2)/(3436.43\ \mathrm{m})^3 = 1.000469 \times 10^{-5}/\mathrm{m}$.

5. The change in selection across the entire distance $(3018 - 259 = 2759$ m) is thus $2759g = 2759\ \mathrm{m} \times 1.000469 \times 10^{-5}/\mathrm{m} = 0.0276$. (Fisher's more exact procedure gives this number as 0.0236.) Thus, a difference in selection of less than 3 percent from 259 to 3018 m can account for a cline with an allele frequency of .46 at the lowest altitude to .10 at the highest. (For more on clines and their interpretation, see Haldane, 1948; Clarke, 1966; Christiansen and Frydenberg, 1974; May et al., 1975; Slatkin and Maruyama, 1975; Endler, 1977; Mettler et al., 1977; Nagylaki, 1978; for some exercises involving Fisher's theory, see Box L.)

L Analysis of clines

This box provides two exercises involving Fisher's (1950) theory of clines. The analysis proceeds as illustrated in the text for the *Standard* gene arrangement in *Drosophila pseudoobscura.*

a. In the same samples as in Figure 33, the frequency of the *Arrowhead* inversion from lowest altitude to highest was .25, .35, .37, .44, .45, .55, and .50. (Of course, the frequencies of *Standard* and *Arrowhead* are not independent.) Carry out an analysis of the *Arrowhead* cline. (Note: the slope of the regression line turns out to be negative, so change its sign when calculating a.)

b. The frequency of the I^A allele in the human ABO blood groups increases along the Japanese archipelago from $p = .24$ in the northeast to $p = .30$ in the southwest, a distance of about 2000 kilometers (Nei and Imaizumi, 1966). Assuming $\sigma^2 = 10\ \mathrm{km}^2$ for Japan (Cavalli-Sforza and Bodmer, 1971, from data of Yasuda, 1968), use Fisher's theory to calculate the difference in selection on I^A across the 2000 km in order to account for the cline, and be prepared for a surprise! [Note: the slope of a straight line passing through points (x_0, y_0) and (x_1, y_1) is $(y_1 - y_0)/(x_1 - x_0)$.]

In vitro enzyme studies. A large number of studies have shown differences among allozymes in the amount, activity, or thermostability of the enzymes (for examples, see Gibson, 1972; Harper and Armstrong, 1973; Day et al., 1974a, 1974b; Koehn, 1978; Somero, 1978). Although such studies indicate that selection *could* be acting on the allozymes, specifically on the observed differences between them, such studies do not prove that selection *is* acting, because the differences that are observed may not effect overall fitness. Thus, *in vitro* studies are necessarily ambiguous unless combined with other types of experiments.

Physiological stress. A promising approach to determining whether individuals with different allozymes have different fitnesses involves subjecting a polymorphic population to conditions that place a physiological stress on the enzyme in question. Such conditions are likely to amplify any fitness differences that might exist. Growth media containing various concentrations of alcohol or various types of alcohol subject the enzyme alcohol dehydrogenase to physiological stress, for example, and the enzyme amylase is "stressed" on growth media containing starch, the substrate of amylase. Such experiments conducted on *Drosophila melanogaster* with alcohol dehydrogenase allozymes (Figure 34; see also van Delden et al., 1978) and amylase allozymes (De Jong and Scharloo, 1976) suggest that selection occurs at both loci, but the evidence is not yet conclusive (see, for example, Yardley, 1978). There is, of course the additional problem that environments associated with selection in the laboratory may only rarely, or never, occur in nature.

All in all, then, whether neutrality or selection best accounts for the majority of genetic polymorphisms is still controversial. If selection is involved, the type of selection at work is also unclear. This is a very unsatisfactory state of affairs, of course, especially since the selection-neutrality issue has been hotly debated for more than ten years. Unfortunately, a compelling body of evidence favoring one hypothesis over the other has not yet accumulated. Indeed, it may very well be the case that the selective mechanisms

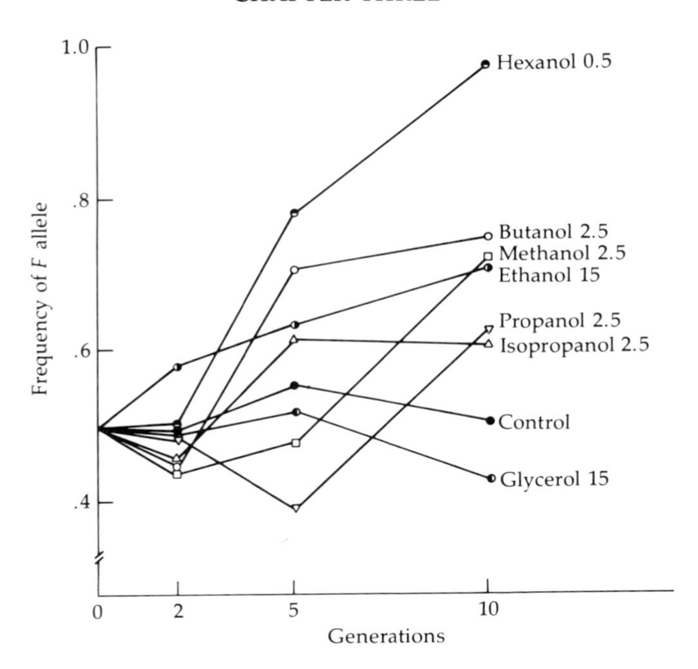

FIGURE 34. Change in frequency of alcohol dehydrogenase fast (*F*) allele in experimental populations of *Drosophila melanogaster* maintained on media containing various alcohols at the volume percentages indicated by the numbers. (From van Delden et al., 1975.)

by which polymorphisms are maintained differ from locus to locus and that some polymorphic loci experience so little selection as to be effectively neutral. To make matters even more complex, certain alleles at individual loci may be subject to selection, whereas other alleles may be neutral. Resolution of the selection-neutrality issue will require detailed studies of many loci in a variety of organisms.

The models of selection discussed in this chapter are simple ones involving one locus with constant selection coefficients. More elaborate models have been studied, but their mathematical analysis can be very complex, especially when there are multiple loci and when such forces as random genetic drift, mutation, and migration are taken into account. (For examples, see the references in Table X, those at the end of this chapter, and also Feldman et al., 1975; Karlin, 1975, 1979a; Felsenstein, 1976; and Li, 1978.) One relatively simple model involving multiple loci has been studied extensively, however; this model involves loci that influence quantitative traits such as height or weight, and it is an especially useful model in plant and animal breeding. The model in question and its implications is one of the subjects of the next chapter.

Problems

1. Suppose an experimenter maintains a large number of small populations, all of the same size and with an equal number of males and females. At one point the experimenter replaces one organism in each of the populations with an organism known to be heterozygous for a neutral mutation, and the resulting populations are maintained at constant size until the neutral allele is fixed or lost. Suppose fixation occurred in 5 percent of the populations. What was the actual size of the populations?
2. What is the average effective population number of a population that, in five consecutive generations, has effective numbers of 500, 1500, 10, 50, and 1000?
3. Suppose a locus has eight alleles at frequencies .55, .20, .09, .06, .04, .03, .02, and .01. What is the effective number of alleles? What would the effective number be if each allele had the same frequency, namely, .125?
4. For a population of effective size N, the variance in allele frequency after one generation of random genetic drift is $p(1 - p)/2N$. Kerr and Wright (1954) studied experimental populations of *Drosophila melanogaster* containing the mutation *aristapedia*. In one case, the population consisted of 4 females and 4 males with an initial allele frequency of $p = .5$. The variance in allele frequency in the next generation was .01707. The actual population size was 8; what was the effective size? Note: effective size based on the variance in allele frequency is called the VARIANCE EFFECTIVE SIZE; that based on the rate of increase in F, as in the text, is called the INBREEDING EFFECTIVE SIZE. In many cases, the two effective sizes agree, but not always. For discussion see Crow and Kimura, 1970.

5. Calculate the autozygosity, F, after 200 generations in a random-mating population of effective size $N = 50$.

6. For two alleles at a locus, if the mutation rate from A to a is 3×10^{-5} and that of a to A is 7×10^{-7}, what is the equilibrium frequency of A?

7. Consider a locus with two alleles, A and a, in each of two random-mating populations of the same size. In one of the populations, the frequency of the A allele is $p = .4$; in the other, it is $p = .1$. Calculate H_S, H_T, and σ^2, where σ^2 represents the variance in allele frequency. Then calculate F_{ST} as $(H_T - H_S)/H_T$ and as $\sigma^2/\bar{p}\bar{q}$.

8. For the populations in Problem 7, suppose a is a recessive allele. Calculate the average frequency of homozygous recessives in the two populations. Suppose the populations are of equal size and undergo fusion followed by one generation of random mating. Calculate the frequency of homozygous recessives in the fused population, and verify Wahlund's formula: $\overline{q^2} = \bar{q}^2 + \sigma^2$.

9. Suppose that the populations in Problem 7 exchange migrants at the rate $m = .10$ for one generation. What will be the frequency of A in the two populations? If the migration continues for a long time, what will the allele frequencies eventually become?

10. Suppose genotypes AA, Aa, and aa have frequencies in zygotes of .16, .48, and .36, respectively, and relative viabilities of $w_{11} = 1.0$, $w_{12} = .8$, and $w_{22} = .6$, respectively. Calculate the genotype frequencies in zygotes of the next generation.

11. If genotypes AA and Aa have relative viabilities of .98 and 1, respectively, what would the relative viability of aa have to be for there to be overdominance with an equilibrium frequency of the A allele of $p = .8$?

12. If the genotype aa has a fitness of 0.2 relative to a value of 1 for AA, what is the equilibrium frequency of a if a is a complete recessive and the mutation rate from A to a is 5×10^{-6}? What is the equilibrium frequency if a is partially dominant with a degree of dominance $h = .035$?

13. Consider a locus with two neutral alleles, A and a, at frequencies .4 and .6, respectively, in a large random-mating population. Suppose the population is split up into subpopulations of effective size $N = 30$, and the subpopulations

are so maintained with random mating within each. Calculate the average frequency of *AA, Aa,* and *aa* genotypes in the subpopulations after 50 generations. One subpopulation has an allele frequency of *A* of .2; what are the genotype frequencies in that population?

14. For the populations in Problem 13, what is the variance in allele frequency among subpopulations after 10, 50, and 100 generations?

15. Consider a protein of 120 amino acids that evolves at a rate of 0.5×10^{-9} amino acid-changing nucleotide substitutions per nucleotide site per year. What number of amino acid differences would you expect to find when comparing this protein in two species that diverged from their common ancestor 180 million years ago? (Remember that there are two lineages emanating from the common ancestor, and ignore the possibility that the same amino acid might change twice along a lineage.)

Further Readings

Ayala, F. J. (ed.). 1976. *Molecular Evolution.* Sinauer Associates, Sunderland, Massachusetts.

Cavalli-Sforza, L. L. and W. F. Bodmer. 1971. *The Genetics of Human Populations.* W. H. Freeman, San Francisco.

Christiansen, F. B. and T. M. Fenchel (eds.). 1977. *Measuring Selection in Natural Populations.* Springer-Verlag, New York.

Crow, J. F. and M. Kimura. 1970. *An Introduction to Population Genetics Theory.* Harper & Row, New York.

Endler, J. A. 1977. *Geographic Variation, Speciation, and Clines.* Princeton Univ. Press, Princeton, New Jersey.

Ewens, W. J. 1968. *Population Genetics.* Methuen, London.

Fisher, R. A. 1930. *The Genetical Theory of Natural Selection.* Clarendon, Oxford.

Futuyma, D. J. 1979. *Evolutionary Biology.* Sinauer Associates, Sunderland, Massachusetts.

Jacquard, A. 1974. *The Genetic Structure of Populations.* (Trans. by D. and B. Charlesworth.) Springer-Verlag, New York.

Karlin, S. and E. Nevo (eds.). 1976. *Population Genetics and Ecology.* Academic Press, New York.

Kimura, M. 1979. The neutral theory of molecular evolution. *Scientific American* 241:98–126.

Kimura, M. and T. Ohta. 1971. *Theoretical Aspects of Population Genetics.* Princeton Univ. Press, Princeton, New Jersey.

Kojima, K. (ed.). 1970. *Mathematical Topics in Population Genetics.* Springer-Verlag, New York.

Lewontin, R. C. 1974. *The Genetic Basis of Evolutionary Change.* Columbia Univ. Press, New York.

Malécot, G. 1969. *The Mathematics of Heredity.* (Trans. by D. M. Yermanos.) W. H. Freeman, San Francisco.

Maruyama, T. 1977. *Stochastic Problems in Population Genetics.* Springer-Verlag, New York.

Moran, P. A. P. 1962. *The Statistical Processes of Evolutionary Theory.* Clarendon, Oxford.

Nagylaki, T. 1977. *Selection in One- and Two-Locus Systems.* Springer-Verlag, New York.

Nei, M. 1975. *Molecular Population Genetics and Evolution.* American Elsevier, New York.

Spiess, E. B. 1977. *Genes in Populations.* John Wiley and Sons, New York.

Wallace, B. 1968. *Topics in Population Genetics.* W. W. Norton, New York.

Wright, S. 1977. *Evolution and the Genetics of Populations,* Vol. 3: *Experimental Results and Evolutionary Deductions.* Univ. of Chicago Press, Chicago.

Wright, S. 1978. *Evolution and the Genetics of Populations,* Vol. 4: *Variability within and among Natural Populations.* Univ. of Chicago Press, Chicago.

· 4 ·

Polygenic Inheritance

Genetic variation affects such diverse traits as allozymes, blood groups, and visible genetic abnormalities. Such traits have a relatively simple pattern of inheritance because they are influenced by relatively few genes: for example, the human glandular disease cystic fibrosis is due to a single recessive mutation, and the ABO blood groups are controlled by the alleles at a single locus. In this chapter, we consider more complex traits that are influenced by alleles at several or many loci; so they are said to be POLYGENIC (or MULTIFACTORIAL), and they are usually more or less strongly influenced by environment. Thus, for example, variation in human weight is partly due to genetic differences among individuals and partly due to such environmental factors as exercise and level of nutrition.

Three principal types of multifactorial traits may be distinguished. First there are CONTINUOUS TRAITS (often called QUANTITATIVE TRAITS) — traits for which there is a continuum of phenotypes; examples of quantitative traits include height, weight, milk yield, and growth rate. A second category of polygenic traits consists of MERISTIC TRAITS — traits for which the phenotype is expressed in discrete classes; examples of meristic traits include number of offspring (i.e., litter size), number of ears on a stalk of corn, number of flowers on a petal, and number of bristles on a fruit fly. A third category of polygenic traits consists of THRESHOLD TRAITS — traits that are either present or absent in any one individual; examples of threshold traits in humans include diabetes and schizophrenia.

Polygenic traits are of utmost importance to plant and animal breeders, since such agriculturally important characteristics as yield of grain, egg production, milk production, efficiency of food utilization by domesticated animals, and meat quality are all polygenic. QUANTITATIVE GENETICS may be considered as the study of genetic principles underlying the inheritance of multifactorial traits. In addition to being essential ingredients in modern plant and animal improvement programs, the principles of quantitative genetics have been applied, sometimes wrongly, to the analysis of polygenic traits in humans. This chapter concerns the nature and uses of genetic variation affecting multifactorial traits.

Quantitative Traits: The Normal Distribution

The first problem that arises in dealing with continuous quantitative traits is to find a concise description of a population. Variation in quantitative traits is continuous, which means that clear-cut phenotypic classes do not exist. Nevertheless, in dealing with continuous traits, it is usually convenient to group phenotypes arbitrarily into categories, as is done for the example in Figure 1; in this figure, all men with heights between 64.50 and 65.49 inches have been grouped and treated as if each had a height of 65 inches. Such grouping produces a set of phenotypic measurements

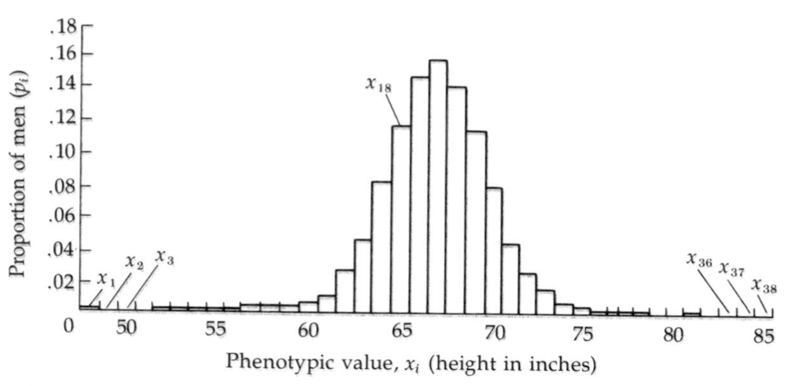

FIGURE 1. Distribution of height among 91,163 English soldiers called for military duty in 1939. Height classes are denoted x_i (so, for example, x_{18} denotes a class of heights for men 64.50 to 65.49 inches tall); the proportion of men in height class x_i is denoted p_i. (After Harrison et al., 1964.)

(called PHENOTYPIC VALUES) denoted $x_1, x_2, \ldots, x_n$ with respective frequencies $p_1, p_2, \ldots, p_n$. Here p_i is the proportion of individuals that have a phenotypic value of x_i, and $p_1 + p_2 + \ldots + p_n = 1$. For the data in Figure 1, for example, $x_1 = 48$ inches and $p_1 = .001$; $x_{18} = 65$ inches and $p_{18} = .116$; and $x_{38} = 85$ inches and $p_{38} = 0$. The MEAN (or average) of x, denoted $\bar{x}$, is simply $p_1x_1 + p_2x_2 + \ldots + p_nx_n$. Equal in importance to the mean in summarizing quantitative traits is the VARIANCE, denoted σ_x^2, a measure of how closely the individual observations cluster around the mean. To calculate the variance of x, one must first calculate the DEVIATION of each value from the population mean by subtracting the mean from each individual value — the deviations are therefore $x_1 - \bar{x}, x_2 - \bar{x}, \ldots, x_n - \bar{x}$ — then square each of these deviations and take the average. The variance of x is therefore equal to $\sigma_x^2 = p_1(x_1 - \bar{x})^2 + p_2(x_2 - \bar{x})^2 + \ldots + p_n(x_n - \bar{x})^2$. In words, the variance is the average squared deviation from the population mean. Table I illustrates calculation of the mean (column 1) and variance (column 3) for a small set of numbers. (See also Box A for further examples and definitions of some important related quantities.) The STANDARD DEVIATION of x, denoted σ_x, is defined as the square root of the variance (Table I). For the data graphed in Figure 1, $\bar{x}$ works out to be 67.5 and σ_x^2 works out as 6.86.

Use of only a mean and variance to summarize a quantitative trait in a large population might seem like gross ov-

TABLE I. Calculation of mean, variance, and standard deviation.

Frequency	x (1)	$x - \bar{x}$ (2)	$(x - \bar{x})^2$ (3)
1/4	10	−10	100
1/2	20	0	0
1/4	30	10	100

Mean ($\bar{x}$) = (1/4)(10) + (1/2)(20) + (1/4)(30) = 20 [column (1)]
Variance (σ_x^2) = (1/4)(100) + (1/2)(0) + (1/4)(100) = 50 [column (3)]
Standard deviation (σ_x) = $\sqrt{\sigma_x^2} = \sqrt{50} = 7.07$

Statistical Tools Used in Quantitative Genetics A

For those lacking prior experience with statistical operations used in biology, a summary of the principal statistical quantities used in quantitative genetics might be useful. Suppose we have n pairs of measurements, $(x_1, y_1), (x_2, y_2), \ldots, (x_n, y_n)$; in an actual example, x might be the clean weight of a sheep's wool and y the number of crimps per inch in the fibers, or x might be the clean wool weight of a ewe and y that of her daughter. As elsewhere, the symbol Σ denotes summation and a superior bar denotes a mean. Thus $\bar{x} = (\Sigma x)/n$ is the mean of x, the mean of the squares of x is $\overline{x^2} = (\Sigma x^2)/n$, and the mean of the products of x and y is $\overline{xy} = (\Sigma xy)/n$. The principal quantities that will be used are:

(1) $\bar{x} = (\Sigma x)/n$ (the mean of x).

(2) $\bar{y} = (\Sigma y)/n$ (the mean of y).

(3) σ_x^2 (variance of x) $= [\Sigma(x - \bar{x})^2]/n$, which equals $\overline{x^2} - \bar{x}^2$; the standard deviation of x, denoted σ_x, is the square root of the variance.

(4) σ_y^2 (variance of y) $= [\Sigma(y - \bar{y})^2]/n$, which equals $\overline{y^2} - \bar{y}^2$; the standard deviation of y, σ_y, is the square root of the variance.

(5) σ_{xy} (covariance of x and y) $= [\Sigma(x - \bar{x})(y - \bar{y})]/n$, which equals $(\overline{xy}) - (\bar{x})(\bar{y})$.

(6) r (correlation coefficient of x and y) $= \sigma_{xy}/\sigma_x\sigma_y$.

(7) b (regression coefficient of y on x) $= \sigma_{xy}/\sigma_x^2$.

The covariance and the correlation coefficient are convenient measures of the degree of association between x and y. If x and y are independent, then σ_{xy} and r are both zero. The regression coefficient of y on x is the slope of the regression line (see Box D of Chapter 1). It should be noted that the expressions (3) through (5) are "large sample" estimates of variance and covariance; for estimation of variances and covariances from small samples, the n in formulas (3) through (5) should be replaced with $n - 1$.

a. Calculate quantities (1) through (7) for the data $(x, y) = (11, 6), (15, 1), (13, 4), (8, 8), (10, 7)$, and $(16, 2)$.

In addition to the product-moment correlation coefficient defined in expression (6) above, another kind of correlation coefficient is extremely important in quantitative genetics because it can be used to compare groups of measurements rather than only pairs; this coefficient is called the INTRACLASS CORRELATION COEFFICIENT (which we will denote as t) and its calculation will be illustrated using data in the table presented in this box.

A

Group 1	Group 2	Group 3
$x_{11} = 1$	$x_{21} = 4$	$x_{31} = 7$
$x_{12} = 2$	$x_{22} = 5$	$x_{32} = 8$
$x_{13} = 3$	$x_{23} = 6$	$x_{33} = 9$
		$x_{34} = 10$
		$x_{35} = 11$

Quantities							Grand Totals
$\Sigma_j x_{ij}$	6	+	15	+	45	=	66
n_i	3	+	3	+	5	=	11
$\bar{x}_{i.}$	2		5		9		$(\bar{x}_{..} = 66/11 = 6)$
n_i^2	9	+	9	+	25	=	43
$\Sigma_j(x_{ij} - \bar{x}_{i.})^2$	2	+	2	+	10	=	14 (*WSS*)
$(\bar{x}_{i.} - \bar{x}_{..})^2$	16		1		9		
$n_i(\bar{x}_{i.} - \bar{x}_{..})^2$	48	+	3	+	45	=	96 (*BSS*)
$\Sigma_j(x_{ij} - \bar{x}_{..})^2$	50	+	5	+	55	=	110 (*TSS*)
$\Sigma_j x_{ij}^2$	14	+	77	+	415	=	506
$n_i \bar{x}_{i.}^2$	12	+	75	+	405	=	492

$m = 3$

$\Sigma n_i - m = 11 - 3 = 8$

$\tilde{n} = [1/(m - 1)][\Sigma n_i - (\Sigma n_i^2/\Sigma n_i)] = (1/2)[11 - (43/11)] = 3.545$

In the table, the data fall into $m = 3$ groups, which may, for example, represent the offspring of different parents. Individual observations within groups are denoted by double subscripts, so, for instance, x_{12} refers to the 2nd individual in group 1. Various quantities, whose calculation is straightforward but tedious, are shown below the data, where $\bar{x}_{i.}$ denotes the "group mean" (the mean of group i), and $\bar{x}_{..}$ denotes the "grand mean" (the mean of the entire population). [That is to say, $\bar{x}_{i.} = (\Sigma_j x_{ij})/n_i$, where n_i is the number of individuals in the ith group and where the j beneath the summation sign means that the summation includes all possible values of j. Similarly, $\bar{x}_{..} = (\Sigma_i \Sigma_j x_{ij})/(\Sigma_i n_i)$, where $\Sigma_i n_i$ is the total number of individuals and where the double summation over i and j means summation over all possible values of both i and j.] The calculations are carried out to extract the three quantities designated *WSS*, *BSS*, and *TSS* under the column headed "grand totals." *TSS* stands for TOTAL SUM OF SQUARES, and it equals the sum of the squares of the deviation of each observation from the grand mean. *WSS* stands for WITHIN-GROUP SUM OF SQUARES, and it equals the sum of squares of the deviation of each observation from its respective group mean. *BSS* stands for BETWEEN-GROUP (often called AMONG-GROUP) SUM OF SQUARES, and it equals the sum of squares of the deviation of the group mean from the grand mean, each multiplied by the number of observations in the group. A

Note that $TSS = WSS + BSS$. Since the *TSS* and *BSS* are easier to calculate for large sets of data than is the *WSS*, the latter is usually obtained by subtraction: $WSS = TSS - BSS$. [Indeed, for computational convenience, two other formulas are usually used: $TSS = \Sigma_i \Sigma_j x_{ij}^2 - (\Sigma_i n_i)\bar{x}_{..}^2$ and $BSS = \Sigma_i n_i \bar{x}_{i.}^2 - (\Sigma_i n_i)\bar{x}_{..}^2$; in the case shown in the table, $TSS = 506 - (11)(6^2) = 506 - 396 = 110$ and $BSS = 492 - 396 = 96$.]

Now calculate the BETWEEN-GROUP MEAN SQUARE (often called the AMONG-GROUP MEAN SQUARE), abbreviated *BMS*, in the following manner: $BMS = BSS/(m - 1) = 96/2 = 47.5$ because $m = 3$ is the number of groups. Similarly, *WMS* stands for the WITHIN-GROUP MEAN SQUARE and is calculated as $WMS = WSS/(\Sigma_i n_i - m) = 14/8 = 1.75$.

A sort of "average" group size, $\tilde{n}$, is now calculated as $\tilde{n} = [1/(m - 1)][\Sigma_i n_i - (\Sigma_i n_i^2/\Sigma_i n_i)] = 3.545$. If all groups have the same size, say $\bar{n}$, then $\tilde{n} = \bar{n}$. We next proceed by calculating V_B, which is an estimate of the variance among observations due to differences between groups, as $V_B = (BMS - WMS)/\tilde{n} = (47.5 - 1.75)/3.545 = 12.906$. Next we set $V_W = WMS = 1.75$; V_W is an estimate of the variance among observations due to differences among individuals within groups.

Finally, we are ready to calculate the intraclass correlation coefficient as $t = V_B/(V_B + V_W) = 12.906/(12.906 + 1.75) = .881$. As you can see, calculation of t is lengthy when done by hand, but these days such computations are almost always carried out on high-speed computers.

Remember that the intraclass correlation coefficient conveys how much of the total variation in a set of data is due to differences *between* groups as opposed to differences *within* groups. If $V_W \approx 0$, for example, then $t \approx 1$, so the correlation between individuals within the same group is very high as compared to the correlation between individuals in different groups; if, on the other hand, $V_B \approx 0$, then $t \approx 0$, and the correlation between individuals within the same group is not much greater than that between individuals in different groups.

b. Calculate t as above for the following data: $x_{11} = 6$, $x_{12} = 4$, $x_{13} = 2$, $x_{14} = 3$, $x_{21} = 5$, $x_{22} = 5$, $x_{23} = 3$, $x_{31} = 7$, $x_{32} = 6$, $x_{33} = 4$, $x_{34} = 5$, $x_{35} = 4$.

ersimplification, but these two numbers are usually adequate because many quantitative traits have a so-called NORMAL DISTRIBUTION. That is to say, when the data are graphed in a histogram like the one in Figure 1, the overall shape of the histogram is often closely approximated by a bell-shaped curve (see Figure 2) whose height (y) at phenotypic value x is $y = (1/\sqrt{2\pi}\sigma)\, e^{-(x-\mu)^2/2\sigma^2}$, where μ represents the mean of the population and σ^2 the variance. (As usual, $\pi = 3.14159$ and $e = 2.71828$ are constants.) This curve is called the NORMAL DENSITY in statistics, but the term normal distribution is often used in quantitative genetics as a synonym. The true mean (μ) and variance (σ^2) of a large population are unknown, but if the sample of individuals from the population is sufficiently large and representative, then $\bar{x}$ is expected to be close to μ and σ_x^2 close to σ^2. If a quantitative trait is normally distributed, therefore, $\bar{x}$ and σ_x^2 provide a complete description of the population because the entire normal density can be calculated from these two quantities. In Figure 1, recall, $\bar{x} = 67.5$ and $\sigma_x^2 = 6.86$; the smooth curve in Figure 2 is the normal density calculated from the formula given above with $\mu = 67.5$ and $\sigma^2 = 6.86$.

The comment above that the true mean (μ) and variance (σ^2) of a distribution are unknown is extremely important. For example, Figure 1 is not the distribution of height among all English men in 1939; it is rather the distribution of height among a sample of 91,163 English men. The true mean and variance of the distribution of height among all English men could be found only by measuring the height of every man in the population. Nevertheless, as was done in the paragraph above, we can use the sample to estimate μ and σ^2. Specifically, if a sample consists of N individuals (in Figure 1, $N = 91{,}163$), the sample mean, $\bar{x}$, is $\bar{x} = \Sigma x_i/N$, where x_i is the phenotypic value (in this case height) of the ith individual and Σ denotes summation over all individuals in the sample; likewise, the sample variance, σ_x^2, is $\sigma_x^2 = \Sigma(x_i - \bar{x})^2/N$. Having calculated $\bar{x}$ and σ_x^2 (67.5 and 6.86 for the sample in Figure 1), we use $\bar{x}$ and σ_x^2 as estimates of μ and σ^2, respectively. For large samples, these estimates of μ and σ^2 are perfectly adequate, but for small samples, the estimate

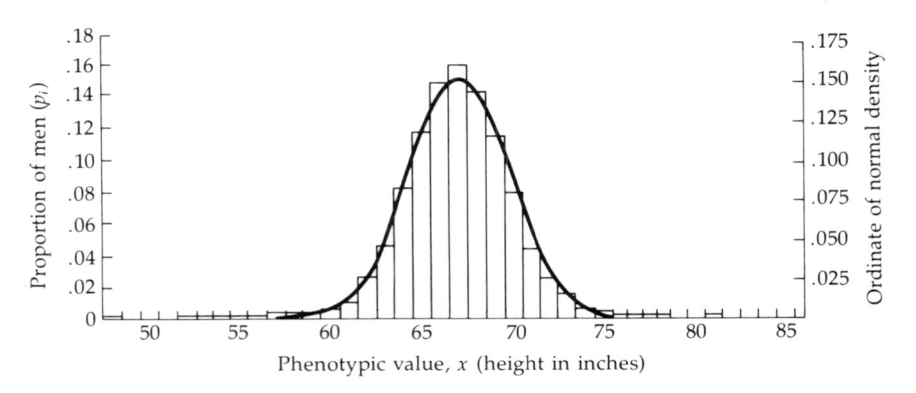

FIGURE 2. The normal curve best approximating the height distribution in Figure 1. The normal curve has a mean of 67.5 inches and variance 6.86. The right-hand ordinate is the ordinate of the normal density, $y = 1/(\sqrt{2\pi}\sigma)\exp[-(x - \mu)^2/2\sigma^2]$, where $\exp(\cdot)$ means $e^{(\cdot)}$ and where μ and σ^2 are the mean and variance of the density. [The $(\cdot)$ in $\exp(\cdot)$ can stand for anything; here it stands for $-(x - \mu)^2/2\sigma^2$.]

of σ^2 should be modified. In particular, for small samples, the best estimate of the true population variance is not $\Sigma(x_i - \bar{x})^2/N$ but rather $\Sigma(x_i - \bar{x})^2/(N - 1)$. Of course, use of $N - 1$ instead of N in the denominator as a correction for small samples makes very little difference unless N is relatively small to begin with; for $N = 91{,}163$, the correction is negligible.

The normal distribution is extremely important in quantitative genetics because the distribution of many traits is normal and because the entire distribution of such traits is conveyed by a mere two numbers — μ and σ^2. Moreover, the normal distribution is of value not only for continuous traits but also for many meristic traits. That is to say, the distribution of phenotypes for many meristic traits is very nearly a normal distribution, especially for traits like number of bristles on the abdomen of fruit flies or number of eggs laid per year in chickens in which the numbers involved are quite large.

Armed with the normal distribution as an adequate description of the phenotypic distribution of many continuous or meristic traits, we can approach the issue of how a population is expected to change genetically and phenotypically when certain phenotypes are favored by selection.

Artificial Selection

ARTIFICIAL SELECTION is the deliberate choice of a select group of individuals to be used for breeding. The most common type of artificial selection is directional selection, in which phenotypically superior animals or plants are chosen for breeding (see Figure 24A in Chapter 3). Although artificial selection has been practiced successfully for thousands of years (for example, in the body size of domesticated dogs), only during this century have the genetic principles underlying its successes become clear (for reviews, see Robertson, 1967; Sprague, 1967; Comstock, 1977, 1978; and Harris, 1977). Moreover, application of genetic principles permits prediction of how rapidly and by how much a population can be altered by artificial selection in any particular generation or small number of generations.

Artificial selection in outcrossing, genetically heterogeneous populations is virtually always successful in the sense that the mean phenotype of the population changes over generations in the direction of selection, provided the population has not previously been subjected to long-term artificial selection for the trait in question. In experimental animals, the mean of almost any quantitative trait can be altered in whatever direction desired by artificial selection; thus, in *Drosophila,* body size, wing size, bristle number, growth rate, egg production, insecticide resistance, and many other traits can be increased or decreased by selection (for examples, see F. W. Robertson, 1955; Clayton et al., 1956a, 1956b; Clayton and Robertson, 1956; Milkman, 1970; and Bell, 1974). In domesticated animals and plants, birth weight, growth rate, milk production, egg production, grain yield, and countless other traits respond to selection. (Figure 3 shows the results of a famous long-term selection program involving oil content in corn.) On the other hand, artificial selection within inbred and essentially homozygous lines consistently results in failure (provided the inbred lines are

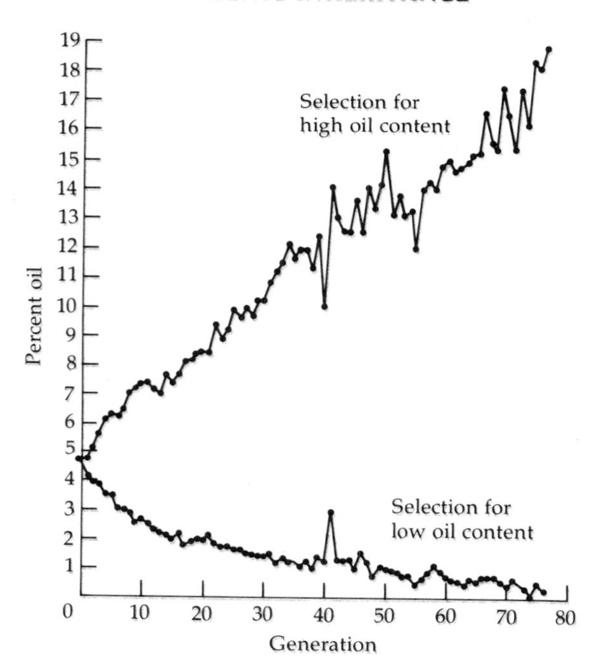

FIGURE 3. Results of a famous long-term experiment selecting for high and low oil content in corn seeds. Begun in 1896, the experiment has the longest duration of any on record and still continues at the University of Illinois. Note the steady, linear rise in oil content shown by the upper curve. The lower curve started on a roughly linear path and continued so for about ten generations, but then the response tapered off, presumably because 0 percent oil is an absolute lower limit for the trait. (After Dudley, 1977.)

truly homozygous); mean phenotype cannot be changed by artificial selection in genetically uniform populations because genetic variation is essential to progress under artificial selection. The general success of artificial selection in outcrossing species indicates that a wealth of genetic variation affecting quantitative traits exists.

A Prediction Equation

Figure 4 illustrates a type of artificial selection called TRUNCATION SELECTION. The curve in (A) represents the normal

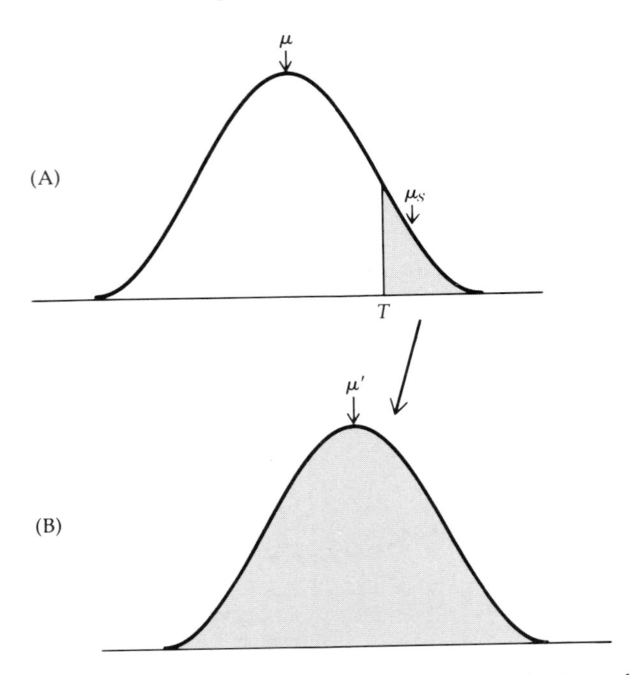

FIGURE 4. Diagram of truncation selection. (A) Distribution of phenotypes in parental population, mean μ. Those individuals with phenotypes above the truncation point (T) are saved for breeding the next generation. The selected parents are denoted by the shading, and their mean phenotype by μ_S. (B) Distribution of phenotypes in offspring generation derived from the selected parents. The mean phenotype is denoted μ'. Note that μ' is greater than μ but less than μ_S.

density of a quantitative trait in a population, and the shaded part of the density to the right of the phenotypic value denoted T indicates those individuals selected for breeding. (T is called the TRUNCATION POINT). The mean phenotype in the whole population is denoted μ, that of the selected parents is denoted μ_S. When the selected parents are mated at random, their offspring will have the phenotypic distribution shown in (B), where the mean phenotype is denoted μ'. An actual example involving seed weight in edible beans is shown in Figure 5. Here T = 650 mg, μ = 403.5 mg, μ_S = 691.7 mg, and μ' = 609.1 mg. Note that

μ' is greater than μ but less than μ_S, a result that is typical of truncation selection. The reason μ' is greater than μ is that some of the selected parents have favorable genotypes and therefore pass favorable alleles on to their offspring. At the same time, μ' is generally less than μ_S for two reasons:

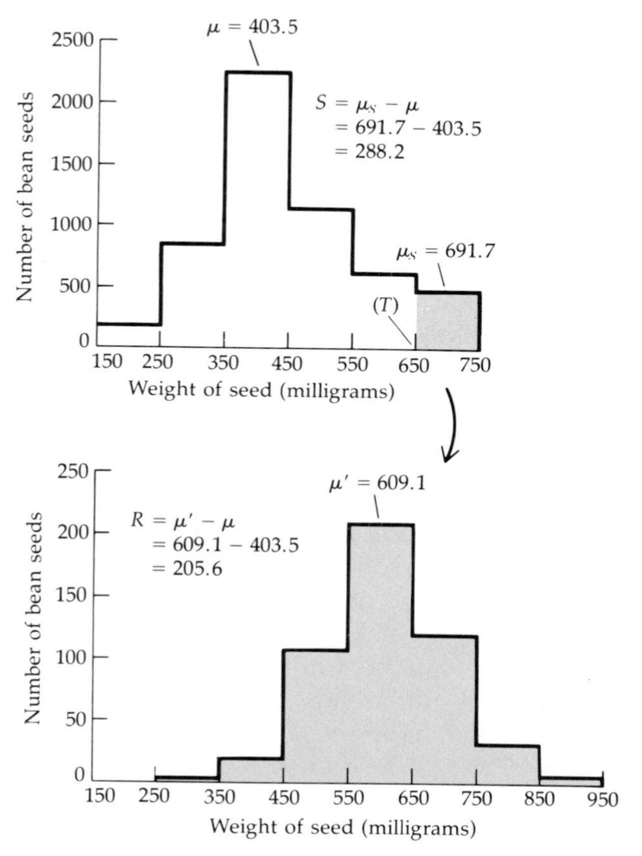

FIGURE 5. Truncation selection experiment involving seed weight in edible beans of the genus *Phaseolus*, laid out as in Figure 4. The truncation point (T) is 650 mg. S is the difference in means between the selected parents and the whole population and is called the selection differential. R is the difference in means between the progeny generation and the whole population in the previous generation, and it is called the response. The quantity R/S is called the realized heritability. (Data from Johannsen, 1903.)

first, because some selected parents do not have favorable genotypes — their phenotypes result from chance exposure to exceptionally favorable environments — and, second, because alleles, not genotypes, are transmitted to the offspring, and exceptionally favorable genotypes are therefore disrupted by Mendelian segregation and recombination. The difference in mean phenotype between the selected parents and the whole parental population is called the SELECTION DIFFERENTIAL and is symbolized S; thus $S = \mu_S - \mu$. The difference in mean phenotype between the progeny generation and the previous generation is called the RESPONSE TO SELECTION and is symbolized R; thus $R = \mu' - \mu$. In Figure 5, $S = 288.2$ mg and $R = 205.6$ mg.

In quantitative genetics, any equation that defines the relationship between the selection differential (S) and the response to selection (R) is called a PREDICTION EQUATION. Since selection can be applied to a population in many different ways (some will be discussed later in this chapter), there are many different prediction equations corresponding to the different modes of selection (see A. Robertson, 1955, and Gardner, 1978, for examples). For truncation selection, which is the type of selection illustrated in Figure 4, the prediction equation reads $R = h^2S$, where h^2 is a quantity called the HERITABILITY of the trait. (For historical reasons, heritability is symbolized as h^2, not as h. For a history of the heritability concept, which is due to Lush, 1937, see Bell, 1977.) Heritability can be understood at several different levels. At one level, heritability can be interpreted as a mere description of what happens under selection. In Figure 5, for example, $S = 288.2$ and $R = 205.6$, so $h^2 = R/S = 205.6/288.2 = 71.3$ percent; when estimated in this manner, h^2 is called the REALIZED HERITABILITY. It is important to understand that the concept of realized heritability has no genetic content whatsoever; realized heritability is merely a shorthand description of an observed result.

On the other hand, the equation $R = h^2S$, where h^2 is the realized heritability, is not much of a "prediction equation" inasmuch as the equation merely describes what has already happened in one generation of selection. Of course, the equation could be used to predict the result of the next generation of selection, but artificial selection in many domesticated plants and animals is time consuming and expensive, so it would be useful if one could estimate heritability without actually performing any artificial selection. If the heritability, h^2, could be estimated in such a manner, then $R = h^2S$ would be a true prediction equation in the sense that the response, R, could be predicted for any selection differential, S, based on the estimated value of h^2. Such an estimate of h^2 is indeed possible, but it involves an understanding of heritability at a level that includes the underlying genetic basis of the quantitative trait. Such a level of understanding is also of importance in human genetics because many unjustified assertions about the genetic basis of such human quantitative traits as I.Q. score have been based on misunderstandings of the concept of heritability. An understanding of the genetics behind the equation $R = h^2S$ requires three items: (1) a concept of how the alleles at an individual locus affect a quantitative trait; (2) a determination of how selection changes allele frequency at the locus; and (3) a calculation of how much the mean of the trait increases as a result of the change in allele frequency. Some detail is required to establish these three items, but the detail is necessary in order to understand the genetic meaning of heritability.

GENETIC BASIS OF QUANTITATIVE TRAITS

Figure 6 shows the normal density of a trait in a hypothetical randomly mating population; its mean is denoted μ and its variance is σ^2. In truncation selection, all individuals with phenotypes above the truncation point (T) are saved for breeding, and the shaded area (B) of the density represents the proportion of the population selected. (The total area under any normal density equals 1.) The height of the normal density at the point T is denoted Z, and, as before, the mean phenotype among the selected individuals is called μ_S. One of the special properties of the normal density to be used below is that $(\mu_S - \mu)/\sigma^2 = Z/B$; the derivation of this is provided later as an exercise in part (a) of Box C.

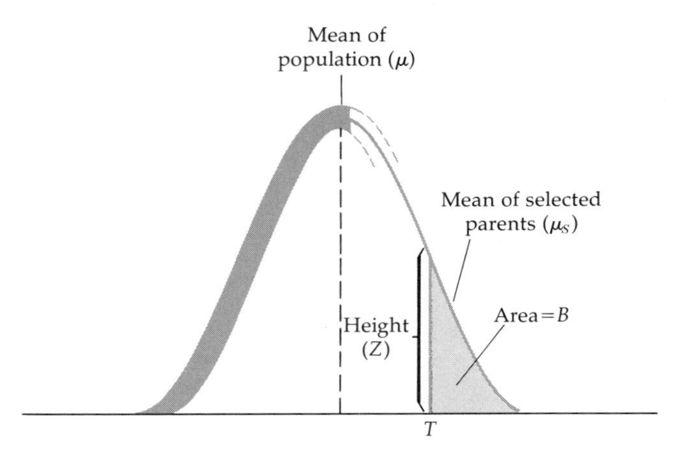

FIGURE 6. Normal distribution of a quantitative trait in a hypothetical population, showing some important symbols used in quantitative genetics. Here μ is the mean of the population, T the truncation point, Z the height (ordinate) of the normal density at the point T, B (shaded) the area under the normal curve to the right of T, and μ_S the mean among selected parents.

To determine the amount of increase in mean phenotype of a population caused by one generation of truncation selection, we first imagine a locus that affects the trait in question and that has alleles A and A' at respective allele frequencies p and q. Because of random mating, genotypes AA, AA', and $A'A'$ will be present in the population with frequencies p^2, $2pq$, and q^2, but the individual genotypes cannot be identified by phenotype because of the variation in phenotype caused by environmental factors and genetic differences at other loci. If the genotypes could be identified, their phenotypic distributions might appear as shown in Figure 7. Each density is normal and has the same variance, but their means are very slightly different. The mean phenotypes of AA, AA', and $A'A'$ genotypes are denoted $\mu^* + a$, $\mu^* + d$, and $\mu^* - a$. The symbols a and d serve as convenient representations of the effects of the alleles in question on the quantitative trait. The difference between means of homozygotes is $(\mu^* + a) - (\mu^* - a) = 2a$, and d/a

serves as a measure of dominance. The relationship $d = a$ means that A is dominant, $d = 0$ means that heterozygotes are exactly intermediate in phenotype between the homozygotes (in which case the alleles are said to be ADDITIVE), and $d = -a$ means that A' is dominant. (Use of a and d in this manner simplifies some of the subsequent formulae.) Calculation of a and d for an actual example involving two alleles that affect coat coloration in guinea pigs is illustrated in Table II. In this case, $a = .1775$ and $d/a = .1235/.1775 = .696$, so the c^k allele is partially dominant. In Figure 7, the mean phenotype of the whole population is $p^2(\mu^* + a) + 2pq(\mu^* + d) + q^2(\mu^* - a)$, which we have called μ.

CHANGE IN ALLELE FREQUENCY

Suppose for the moment that we were practicing artificial selection for increased amount of black coat coloration in the guinea pigs in Table II. Selection for black coat coloration in a population containing both the c^k (i.e., A) and c^d (i.e., A') alleles would be successful in increasing the allele frequency

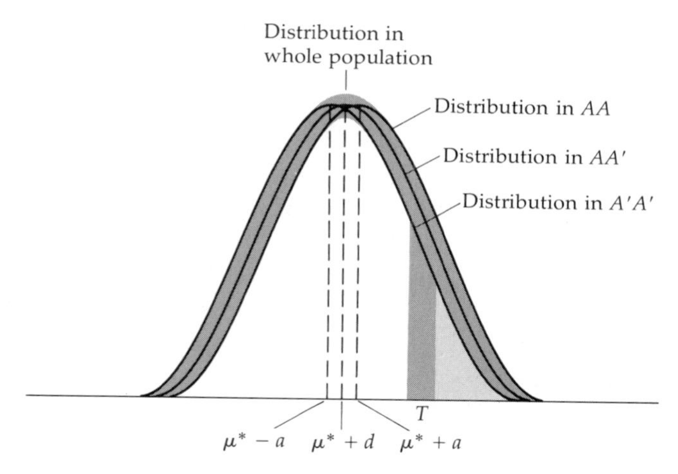

FIGURE 7. Same distribution as in Figure 6, showing the slightly different phenotypic distributions of the three genotypes (AA, AA', and $A'A'$) for a locus with two alleles that contributes to the quantitative trait. The means of the distributions of AA, AA', and $A'A'$ are symbolized $\mu^* + a$, $\mu^* + d$, and $\mu^* - a$, respectively.

TABLE II. Calculation of μ^*, a, and d for alleles at a locus affecting coat coloration in guinea pigs.[a,c]

Genotype	Amount of black coloration[b]		
c^kc^k (AA)	1.303	=	$\mu^* + a = 1.1255 + .1775$
c^kc^d (AA')	1.249	=	$\mu^* + d = 1.1255 + .1235$
c^dc^d ($A'A'$)	.948	=	$\mu^* - a = 1.1255 - .1775$

$$\mu^* = (1.303 + .948)/2 = 1.1255$$
$$a = 1.303 - 1.1255 = .1775$$
$$d = 1.249 - 1.1255 = .1235$$

[a] The calculations to be carried out first are those beneath the data; then the right-hand column is completed.
[b] Here the amount of black coloration is measured as arcsin $\sqrt{x}$, where x is the percentage of black coloration on the animal. For c^kc^k, c^kc^d, and c^dc^d genotypes, the corresponding x values are 93 percent, 90 percent, and 66 percent.
[c] Data from Wright (1968).

of A, so the average amount of black coloration among individuals of the next generation would also increase. Therefore, in order to calculate the expected increase in black coloration in one generation of selection, we must first calculate the corresponding change in the allele frequency of A. Equations for changes in allele frequency with natural selection were derived earlier in Table VIII of Chapter 3; these equations are equally valid for artificial selection if we agree to interpret the "fitness" of an individual as the probability that the individual will be included among the group selected as parents of the next generation. With this interpretation of fitness, differences in fitness (i.e., reproductive success) of AA, AA', and $A'A'$ genotypes correspond to the differences in area to the right of the truncation point in Figure 7, because only those individuals in the shaded area are allowed to reproduce. The differences in area are easy to calculate if you shift or slide each curve until its mean coincides with μ^*; thus the $A'A'$ curve must slide a units to the right, and the AA' and AA curves must slide d and a units to the left. This shifting brings the densities into coincidence, but it slides the truncation points slightly out of register, as shown in Figure 8. The difference in "fitness" between AA and AA′, denoted $w_{11} - w_{12}$ (as in Table VIII of Chapter 3), is equal to the small area indicated in Figure 8, as is the difference in fitness between AA' and $A'A'$, denoted $w_{12} - w_{22}$. The areas corresponding to $w_{11} - w_{12}$ and $w_{12} - w_{22}$ are approximately rectangles, and the area of a rectangle is the product of its base and its height. Therefore, since Z represents the height of the normal curve at the point T, $w_{11} - w_{12} = Z[(T + \mathrm{a}) - (T + \mathrm{d})] = Z(a - d)$ and $w_{12} - w_{22} = Z[(T + d) - (T - a)] = Z(a + d)$. The average fitness of the whole population, $\overline{w}$, simply equals B, the proportion of the population saved for breeding. From Table VIII in Chapter 3, we know that $\Delta p = pq[p(w_{11} - w_{12}) + q(w_{12} - w_{22})]/\overline{w}$, where Δp represents the change in allele frequency of A in one generation of selection. Substituting

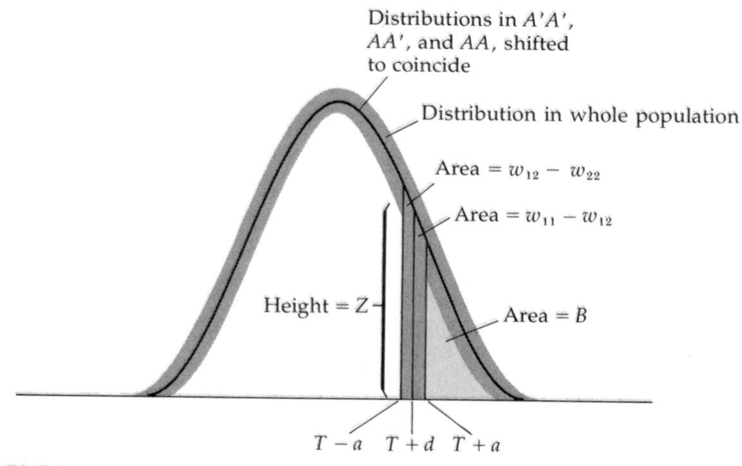

FIGURE 8. Same distribution as in Figures 6 and 7, but with the distributions of AA, AA' and $A'A'$ shifted laterally so that they coincide. Shifting the distributions slides the truncation points slightly out of register, so the truncation points for AA, AA', and $A'A'$ are now $T - a$, $T + d$, and $T + a$, respectively. The small area denoted $w_{12} - w_{22}$ is the difference between the proportions of AA' and $A'A'$ genotypes that are included among the selected parents, and the area $w_{11} - w_{12}$ is the difference in the proportion of AA and AA' genotypes included among the selected parents.

for $w_{11} - w_{12}$, $w_{12} - w_{22}$, and $\overline{w}$ leads to $\Delta p = pq[pZ(a - d) + qZ(a + d)]/B = (Z/B)pq[a + (q - p)d]$ since $p + q = 1$. A corresponding expression could be obtained for any locus affecting the trait, of course, but the values of p, a, and d would differ for each locus. (Box B provides some important equivalent expressions for Δp and relates them to the selection coefficients discussed in Chapter 3.)

B Effect of a Constituent Locus on a Quantitative Trait

In Chapter 3 we developed an important equation for the change in allele frequency at a locus due to natural selection. In this chapter we developed an analogous equation for the change in allele frequency at a locus due to truncation selection. The purpose of this box is to show the relationship between the two equations. Part (a) introduces a new quantity, use of which simplifies certain other formulas. Part (b) demonstrates the connection between the equations of Chapter 3 and Chapter 4 mentioned above. And part (c) shows that the amount of selection on a locus that affects a quantitative trait increases with the heritability of the trait but decreases with the number of loci.

a. The quantity $\alpha = a + (q - p)d$ is called the AVERAGE EFFECT OF AN ALLELE SUBSTITUTION (in this case, a substitution of A for A'), because if an A' allele is chosen at random from the population and replaced with A, the average change in phenotype will be α. An A' allele will be chosen from an AA' individual a proportion p of the time and from an $A'A'$ individual a proportion q of the time. In the former case, the genotype will change from AA' to AA, so the effect will be to change phenotype by an amount $(\mu + a) - (\mu + d) = a - d$; in the latter case, the genotype will change from $A'A'$ to AA', leading to a phenotypic change of $(\mu + d) - (\mu - a) = d + a$. Altogether, the average effect of substituting A for A' will be $\alpha = p(a - d) + q(a + d)$, whence $\alpha = a + (q - p)d$. Show that the average effect of substituting A' for A is $-\alpha$, and show that $\sigma_a^2 = 2pq\alpha^2$.

b. The text shows that, for any locus affecting a quantitative trait, $\Delta p = (Z/B)pq[a + (q - p)d] = (Z/B)pq\alpha$. This expression can be simplified somewhat by using the quantity i, which is called the intensity of selection. Intensity of selection will be discussed in detail in Box C later in this chapter, but for the moment, we need only the relationship $i = \sigma Z/B$, where σ is the phenotypic standard deviation. Thus we can write $\Delta p = (\sigma Z/B)(pq\alpha/\sigma) = ipq\alpha/\sigma$. In Chapter 3, we showed $\Delta p = pq[p(w_{11} - w_{12}) + q(w_{12} - w_{22})]$, assuming $\overline{w} \approx 1$, and comparison of the two expressions for Δp shows how selection of a quantitative trait alters allele frequency at a single locus that affects the trait. For example, if the alleles at the locus are additive, we can write $w_{11} = 1$, $w_{12} = 1 - s/2$, and $w_{22} = 1 - s$, so $\Delta p = pqs/2$, where s is the selection coefficient; at the same time, for additive alleles, $\alpha = a + (q - p)d = a$ because $d = 0$, so $\Delta p = pq(ia/\sigma)$; thus $s = 2ia/\sigma$ for additive alleles. In other words, the selection coefficient for an allele that influences a quantitative trait increases with the intensity of selection (i) and the effect of the allele (measured by a), but it decreases as the phenotypic standard deviation (σ) increases. Show that $s = 2ia/\sigma$ also holds when the favored allele is dominant as well as when the favored allele is recessive. (This sort of relationship between s and a/σ is quite general; see Milkman, 1978.) **B**

c. For n completely equivalent and additive loci, $\sigma_a^2 = 2npqa^2$, thus from $h^2 = \sigma_a^2/\sigma^2$, we obtain $\sigma^2 = 2npqa^2/h^2$ or $\sigma = (\sqrt{2npq/h^2})a$, assuming $h^2 \neq 0$. The selection coefficient per locus is approximately $s = 2ia/\sigma$, so $s = 2ih/\sqrt{2npq}$, where h is the square root of the heritability. For $p = q = 1/2$ (as would be true if the selected population were the F_1 of a cross between highly inbred lines), then $s = 2ih\sqrt{2/n}$. Calculate s for all combinations of $h^2 = .09$ and $.16$; $n = 8$, 18, and 98; and $i = .80$ and 2.06.

CHANGE IN MEAN PHENOTYPE

We now have an expression for Δp and can apply it to a hypothetical example involving the guinea pigs in Table II, in which $a = .1775$ and $d = .1235$. Suppose a guinea pig population has an allele frequency of A (i.e., c^k) of $p = .25$ and that only this one locus affects coat color (an unreasonable assumption, but worthwhile for purposes of illustration); suppose further that there are such environmental effects on coat color that the overall distribution of coat color in the population is normal with variance $\sigma^2 = .16$ (again a questionable assumption made only for purposes of illustration). Since $p = .25$, the mean coat-color phenotype in the population can be calculated as $\mu = (.25)^2(1.303) + 2(.25)(.75)(1.249) + (.75)^2(.948) = 1.083$.

Suppose now that truncation selection is practiced in such a way that the darkest 30 percent of the animals are used for breeding; thus $B = .30$, for which the corresponding value of Z/B is 2.9. (Calculation of Z/B for a given value of B is discussed in detail in Box C.) Thus we have $Z/B = 2.9$, $p = .25$, $a = .1775$, and $d = .1235$. Hence the change in allele frequency of A in one generation of selection is $\Delta p = (Z/B)pq[a + (q - p)d] = (2.9)(.25)(.75)[.1775 + (.75 - .25)(.1235)] = .13$. In the next generation, therefore, the allele frequency of A will be $p' = p + \Delta p$ and that of A' will be $q' = q - \Delta p$. In the guinea pig example, $p' = .25 + .13 = .38$ and $q' = .75 - .13 = .62$.

C Intensity of Selection

For ease of reference, here is a short glossary of some of the symbols we have defined in connection with truncation selection, assuming a normal distribution of phenotypes:

μ is the mean of the whole population,
σ^2 is the phenotypic variance of the population,
T is the truncation point,
Z is the ordinate of the normal density at the point T,
B is the proportion of the population selected for breeding,
μ_S is the mean of the breeding population.

In the text, the expression $(\mu_S - \mu)/\sigma^2 = Z/B$ appears several times without proof. Students willing to accept the relationship on faith may wish to skip part (a) of this box, but skeptical students will find in part (a) some guidelines on how to go about proving it for themselves. Part (b) introduces the concept of intensity of selection; this concept is widely used in quantitative genetics because it allows relatively easy comparison of selection programs involving different traits or even different populations. Part (c) deals with the prediction of response to selection in the important case when the selection intensity differs in sires and dams.

C

a. Students familiar with integral calculus can derive a fundamental relationship used in quantitative genetics, namely $(\mu_S - \mu)/\sigma^2 = Z/B$. By definition, $Z = (1/\sqrt{2\pi}\sigma)\exp[-(T - \mu)^2/2\sigma^2]$, $B = (1/\sqrt{2\pi}\sigma)\int_T^\infty \exp[-(x - \mu)^2/2\sigma^2]\, dx$, and $\mu_S = (1/B\sqrt{2\pi}\sigma)\int_T^\infty x \exp[-(x - \mu)^2/2\sigma^2]dx$, where $\exp(\cdot)$ means $e^{(\cdot)}$. Use the definition of μ_S and the integration formula $\int x \exp[-(x - \mu)^2/2\sigma^2]dx = -\sigma^2\exp[-(x - \mu)^2/2\sigma^2] + \mu\int \exp[-(x - \mu)^2/2\sigma^2]dx$ to show that $\mu_S = (\sigma^2Z/B) + \mu$, or $(\mu_S - \mu)/\sigma^2 = Z/B$.

b. Let $N(\mu,\sigma^2) = (1/\sqrt{2\pi}\sigma)\exp[-(x - \mu)^2/2\sigma^2]$ represent a normal distribution. Usually it is convenient to deal with the so-called STANDARD NORMAL DISTRIBUTION, which is $N(0,1) = (1/\sqrt{2\pi})\exp(-x^2/2)$, because ordinates and areas of the standard normal distribution are tabulated in mathematical handbooks. The table included here gives values of t and z of the standard normal distribution for various values of B. (Note that we use the lower case symbols t and z with reference to the standard normal distribution.) The number t satisfies $\int_t^\infty N(0,1)dx = B$, and z equals $N(0,1)$ evaluated at $x = t$; thus, $t = (T - \mu)/\sigma$ and $z = \sigma Z$. To obtain t for a value of $B > .50$, take the negative of t corresponding to $1 - B$, and to obtain z for $B > .50$, simply use the z for $1 - B$; for $B = .70$, for example, $1 - B = .30$, so $t = -.52$ and $z = .34769$. The quantity i, which is called the INTENSITY OF SELECTION, is calculated as $i = z/B$; i is a convenient number by which to compare different selection programs because i depends only on B, the proportion saved for breeding. (For some simple approximations for i, see Simmonds, 1977.)

B	t	z	i
.001	3.09	.00337	3.37
.005	2.58	.01446	2.89
.01	2.33	.02665	2.66
.05	1.64	.10314	2.06
.10	1.28	.17550	1.76
.15	1.04	.23316	1.55
.20	.84	.27996	1.40
.25	.67	.31778	1.27
.30	.52	.34769	1.16
.35	.39	.37040	1.06
.40	.25	.38634	.97
.45	.13	.39580	.88
.50	0	.39894	.80

C Since $R = h^2S$ and since $S = \mu_S - \mu = \sigma^2Z/B$, it follows that $S = \sigma(\sigma Z)/B = \sigma\, z/B = \sigma i$, so the prediction equation for truncation selection can be written as $R = i\sigma h^2$.

As is evident from the table, i changes much more slowly than B. Indeed, suppose a breeder converts from a breeding program with $B = .10$ to one with $B = .01$ (so B is decreased by a factor of 10); by what factor is i increased?

c. A breeder cannot choose i with complete freedom but must take reproductive considerations into account. A cattle breeder must typically save $B = 70$ percent of the cows in order to maintain herd size, whereas with artificial insemination, only 1 percent of the bulls need be saved. Because of the greater reproductive capacity of swine and chickens, a breeder can typically save 10 percent of sows and hens as compared to 1 percent of boars and cockerels. Using the table in (b), what is i for $B = .70$ (cows), $B = .10$ (sows and hens), and $B = .01$ (bulls, boars, and cockerels)?

When the intensity of selection differs in the sexes, the prediction equation becomes $R = \bar{i}\sigma h^2$ where $\bar{i} = (i_S + i_D)/2$, that is, the arithmetic average of the selection intensities on sires and dams. The proportion of the expected response due to sire selection is thus $i_S/\bar{i}$ and that due to dam selection is $i_D/\bar{i}$. Calculate the proportion of the expected response due to sire and dam selection using the above values for cattle, swine, and chickens.

To predict the mean coat-color phenotype after one generation of selection, all we have to do is calculate the mean phenotype of a population with allele frequencies of A and A' of $p + \Delta p$ and $q - \Delta p$, respectively. For the guinea pig example, the mean of the offspring would be expected to equal $\mu' = (.38)^2(1.303) + 2(.38)(.62)(1.249) + (.62)^2(.948)$, which works out to $\mu' = 1.141$. More generally, using the symbols in Table II, $\mu' = (p + \Delta p)^2(\mu^* + a) + 2(p + \Delta p)(q - \Delta p)(\mu^* + d) + (q - \Delta p)^2(\mu^* - a)$. When the right-hand side of the expression for μ' is multiplied out and terms in $(\Delta p)^2$ are ignored because Δp will usually be small, then μ' is found to be $\mu' = \mu + 2[a + (q - p)d]\Delta p$. (The algebra for this approximation is suggested as an exercise in part (a) of a later Box G.) The approximation is rather good even for the relatively large value of Δp in the guinea pig example; in that case, recall, the exact value of μ' was calculated as 1.141, whereas the approximation yields $\mu' = 1.145$.

The equation $\mu' = \mu + 2[a + (q - p)d]\Delta p$ deserves a little more development as it will yield the prediction equation $R = h^2S$ and provide an expression for h^2 in terms of parameters (a, d, and p) that can be interpreted genetically. First, rewrite the equation for μ' as $\mu' - \mu = 2[a + (q - p)d]\Delta p$. Then substitute for Δp the expression derived earlier, namely $\Delta p = (Z/B)pq[a + (q - p)d]$; this substitution yields $\mu' - \mu = (Z/B)2pq[a + (q - p)d]^2$. However, as noted in connection with Figure 6, $Z/B = (\mu_S - \mu)/\sigma^2$, so we may substitute for Z/B to obtain $\mu' - \mu = (\mu_S - \mu)2pq[a + (q - p)d]^2/\sigma^2$.

Now, $\mu' - \mu$ is the selection response (R) and $\mu_S - \mu$ is the selection differential (S). Thus $R = (S)2pq[a + (q - p)d]^2/\sigma^2$. But $R = h^2S$ also, so $h^2 = 2pq[a + (q - p)d]^2/\sigma^2$. This expression for h^2 is the one we were after, as it defines the heritability in terms of p, q, a, and d — each of which has a genetic meaning. Returning to the guinea pig example, $h^2 = 2(.25)(.75)[.1775 + (.75 - .25)(.1235)]^2/.16 = .134$. For $Z/B = 2.9$, S, which equals $\mu_S - \mu$, works out to be .464. [That is to say, $\mu_S = .464 + 1.083 = 1.547$, which brings up the paradox that the average coat-color score among selected parents is greater than the mean of AA homozygotes (1.303, recall); this is possible because of the substantial environmental contribution to the trait that we have assumed for purposes of illustration — because of the large environmental effects, some individuals in all genotypic classes will have phenotypes substantially different from the overall average of their genotypic class.] From $R = h^2S$, we calculate $R = (.134)(.464) = .062$, which, since $R = \mu' - \mu$, implies $\mu' = .062 + 1.083 = 1.145$ — the same as given above. (Box C provides an important alternative expression for R and some additional examples.)

The expression $h^2 = 2pq[a + (q - p)d]^2/\sigma^2$ is valid when a single locus affects the trait in question. However, when many loci affect the trait, $2pq[a + (q - p)d]^2/\sigma^2$ is replaced by a sum of such terms, one for each locus (see Griffing, 1960, for more complex types of models). That is to say, for many loci, $R = h^2S$ where

$$h^2 = \frac{\Sigma\, 2pq[a + (q - p)d]^2}{\sigma^2}$$

the summation being over all loci that affect the trait. (Of course, each locus may have different values of a, d, p, and q.) As will be noted later, the quantity $\Sigma 2pq[a + (q - p)d]^2$ is called the ADDITIVE GENETIC VARIANCE of the trait, and while the individual components in the sum are difficult to identify except in contrived examples like the one involving guinea pigs, the collective effects (represented by the summation) can be estimated.

The Meaning of Heritability

The above theoretical formula for heritability is important in showing that heritability says virtually nothing about the actual mode of inheritance of a quantitative trait, useful as the concept is in predicting response to selection. The heritability of a trait represents the cumulative effect of all loci affecting the trait. The number of loci involved can be determined only with elaborate and specially designed experiments (as in Box D; and see Seyffert and Forkman, 1976). Even for a single locus, heritability depends in a complex manner on the values of p, a, and d (Figure 9), and these individual components cannot be disentangled. (The values of p, a, and d are said to be STATISTICALLY CONFOUNDED.) As noted above, for more than one locus, each with two alleles, the heritability is the sum of quantities like $2pq[a + (q - p)d]^2/\sigma^2$, one term for each locus, each term having its own particular values of p, q, a, and d. Here, precisely, is the problem: for a quantitative trait determined by, say, 10 loci, there would be 30 quantities involved in heritability — 10 allele frequencies, 10 values of a, and 10 values of d. Heritability is but a single number that gives the combined effect of all 30 quantities; it says nothing about any of them. With more loci affecting the trait or with multiple alleles, matters would be even worse. Also, we have assumed that all loci affecting the trait act independently of one another and are

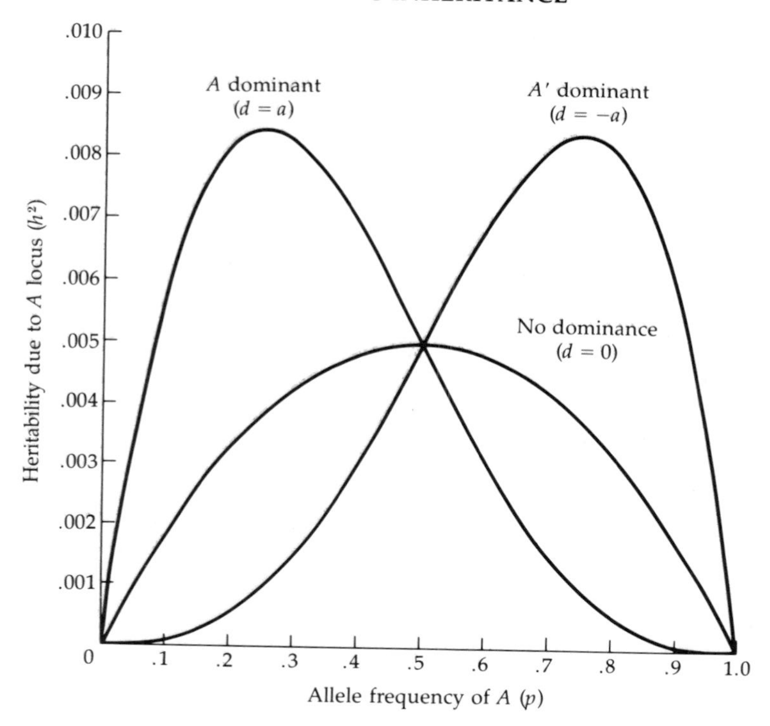

FIGURE 9. Heritability (h^2) due to a single locus with two alleles (A and A') as a function of p, the allele frequency of A. In general, for one locus, $h^2 = 2pq[a + (q - p)d]^2/\sigma^2$, where σ^2 is the total phenotypic variance. The curves correspond to $a = .1$ and $d = .1$ (A dominant), $d = 0$ (no dominance), and $d = -.1$ (A' dominant).

unlinked; in actual cases, loci often interact and can be linked. Moreover, we have assumed that environmental effects are the same for all genotypes, but this assumption can easily be violated, even in some well-designed breeding programs. All in all, while heritability, especially realized heritability, is an indispensable aid to plant and animal breeders, it lends itself to no easy interpretation in simple genetic terms. These cautions should be kept in mind especially when heritabilities of traits in humans are involved.

There is another problem in interpreting heritability — heritability depends on the range of environments involved.

D Number of Loci for Quantitative Traits

Determining the actual number of loci that influence a quantitative trait is extremely difficult, but it is rather easy to get an order-of-magnitude approximation. Suppose we make a cross of two highly inbred lines and practice "up" and "down" selection for the trait of interest, as was done for oil content in corn in the experiment shown in Figure 3. Suppose further that the trait is influenced by n unlinked loci having identical effects. Then for any one locus (say locus 1), the mean phenotype of A_1A_1, A_1A_1', and $A_1'A_1'$ genotypes may be written (as in the text) as $\mu^* + a$, $\mu^* + d$, and $\mu^* - a$, where A_1 is the favorable allele. Were the favorable allele fixed at each locus, the genotype $A_1A_1A_2A_2A_3A_3 \ldots$ would have mean $\mu^* + na$, whereas, were the unfavorable alleles fixed, the genotype $A_1'A_1'A_2'A_2'A_3'A_3' \ldots$ would have mean $\mu^* - na$. Thus, if selection were continued on a large population (large to minimize the effects of random genetic drift) until all favorable alleles were fixed in the "up" line and all unfavorable ones in the "down" line, the difference in mean phenotype between the "up" and "down" lines, call it $U - D$, would be $U - D = \mu^* + na - (\mu^* - na) = 2na$.

Moreover, for any one locus, $\sigma_a^2 = 2pq[a + (q - p)d]^2$, and for n equivalent loci $\sigma_a^2 = 2npq[a + (q - p)d]^2$. The base population arising from the cross of inbreds must have $p = q = 1/2$ for all segregating loci, so in the base population $\sigma_a^2 = 2n(1/2)(1/2)[a + (1/2 - 1/2)d]^2 = na^2/2$. (Note how $p = q = 1/2$ makes the term involving d disappear.) Actually, the inbred lines would be likely to be fixed for different alleles only if the alleles were present at substantial frequency in the foundation population from which the inbreds were derived. Thus, the following procedure for estimating the number of loci influencing the trait actually estimates the minimum number.

From an experiment such as that described above, we have two observable quantities: $U - D$ (the difference between the "up" and "down" lines) and σ_a^2 (which may, for example, be estimated as four times the between-sire component of variance in the paternal half-sib experimental design outlined later in Box H). Under our simplifying assumption, $U - D = 2na$ and $\sigma_a^2 = na^2/2$. Thus $(U - D)^2 = 4n^2a^2 = 8n\sigma_a^2$, so $n = (U - D)^2/8\sigma_a^2$ estimates the number of loci. (When n is estimated in this manner, it is often called the EFFECTIVE NUMBER OF LOCI or the SEGREGATION INDEX in order to emphasize the simplifying assumptions.)

D

For a single locus, the difference between means of homozygotes is $2a = (U - D)/n = \sqrt{8n\sigma_a^2}/n = 2\sigma_a\sqrt{2/n}$. In units of σ — the phenotypic standard deviation of the trait — the effect of a single locus is $2a/\sigma = 2(\sigma_a/\sigma)\sqrt{2/n} = 2h\sqrt{2/n}$, where h is the square root of the heritability. The quantity $2a/\sigma$ is often called the PROPORTIONATE EFFECT of a locus.

As an example of the use of these formulas, we may use data of Clayton and Robertson (1957) on the number of abdominal bristles in *Drosophila melanogaster*. In these experiments, the total response was $U - D = 20\sigma$ (i.e., 20 times the phenotypic standard deviation in the base population), and $\sigma_a^2 = .51\sigma^2$. Thus $n = (20\sigma)^2/8(.51\sigma^2) = 400/4.08 = 98$ was the effective number of loci; $h^2 = \sigma_a^2/\sigma^2 = .51$, so $2a/\sigma = 2\sqrt{.51}\sqrt{2/98} = .20$ is the proportionate effect (i.e., the difference between means of homozygotes at any one locus is about one-fifth of a phenotypic standard deviation).

The effective number of loci for abdominal bristle number is rather large, about 100. However, a locus has its greatest effect on a quantitative trait when its alleles are at intermediate frequencies, and since the base population here arises from the cross of inbreds, all loci are initially at intermediate frequencies and contribute to the response. In a heterogeneous population, such as a natural population, only a handful of loci affecting the trait may have alleles at intermediate frequencies, and these few loci will contribute most of the genetic variation and response to selection. Thus, we have a situation where a great many loci may potentially influence a quantitative trait, but in a heterogeneous population only a few of these will actually be of great importance at any one time. (For more on the detection and measurement of the effects of individual genes involved in quantitative traits, see Wehrhahn and Allard, 1965; McMillan and Robertson, 1974; Jinks and Towey, 1976; and Hill and Avery, 1978.)

a. For length of thorax in *Drosophila melanogaster*, F. W. Robertson (1955) found $U - D = 12\sigma$ and $\sigma_a^2 = .3\sigma^2$. For weight at six weeks in mice, Falconer (1955) found $U - D = 8\sigma$ and $\sigma_a^2 = .25\sigma^2$. Calculate the effective number of loci and the proportionate effect of a locus for each of these traits.

b. For birth weight in Hereford steers, the data in Box H imply that $\sigma_a^2 = 4V_B = 4(11.9) = 47.6$ and $\sigma^2 = V_B + V_W = 11.9 + 55 = 66.9$. Assuming that the effective number of loci is $n = 50$ and that all the other simplifying assumptions above are valid (which they surely are not), calculate what $(U - D)/\sigma$ would be after long continued selection for higher and lower birth weight.

For a single locus, recall, $h^2 = 2pq[a + (q - p)d]^2/\sigma^2$, where σ^2 is the total phenotypic variance in the population. Because the total phenotypic variance depends partly on the variance due to environmental differences among individuals (as will be detailed in a later section of this chapter) increasing the variation in environment will decrease h^2. How heritability depends on environment is an even more subtle problem, however, because the values of a and d also depend on the environment. A hypothetical example is illustrated in Figure 10, which shows the norms of reaction of AA, AA', and $A'A'$ genotypes. (The NORM OF REACTION of any genotype is the relationship between phenotype and environment for that genotype.) If we were dealing with the range of environments denoted E_1 in Figure 10, A would be the favored allele and A would be nearly dominant to A'; if we were dealing with the range of environments denoted E_2 on the other hand, A' would be the favored allele and

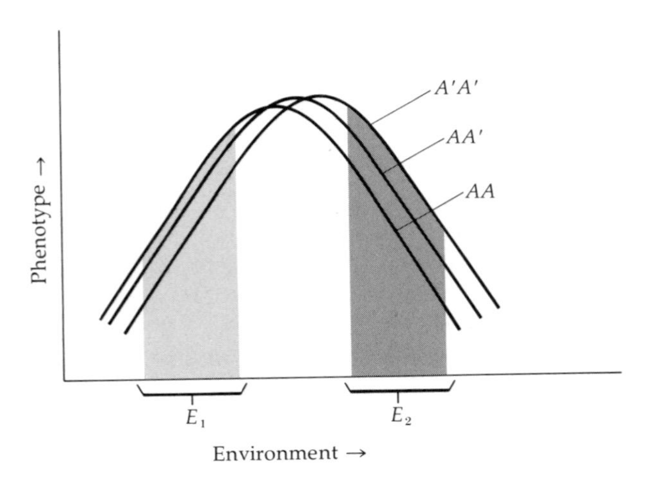

FIGURE 10. Hypothetical norms of reaction — relationships between phenotype and environment — for genotypes AA, AA', and $A'A'$. In the range of environments denoted E_1, A is very nearly dominant to A' (i.e., AA and AA' have nearly the same phenotype); in the range E_2, A and A' are very nearly additive (no dominance). Thus the heritability due to this locus will be different depending on whether it is measured in E_1 environments or E_2 environments.

there would be essentially no dominance. Thus, switching a population from E_1 to E_2 would change the values of a and d and substantially alter the heritability of the trait, even though the total phenotypic variance of the population might remain the same. In short, the heritability of a trait has reference only to a particular population at a particular time in a particular range of environments. On the other hand, practical breeders are usually interested in improving a particular population at a particular time in a particular range of environments, so the concept of heritability is of great utility when used for its intended purpose — the prediction of response to selection.

Recall again that $R = h^2S$, where, for one locus, $h^2 = 2pq[a + (q - p)d]^2/\sigma^2$. (An interesting application of this expression to the trait "fitness" can be found in Box E.) Because the allele frequencies (p and q) are expected to change during the course of selection, the heritability (h^2) is also expected to change. In practice, however, the heritability changes sufficiently slowly that over the course of a few generations h^2 can be regarded as a constant.

Fundamental Theorem of Natural Selection — E

The selection differential, S, can be measured as the average deviation from the population mean of the breeding population; that is, $S = (1/n)\Sigma(x_i - \bar{x})$, where the summation extends over all of n selection individuals (n is assumed to be very large), x_i is the phenotype of the ith selected individual, and $\bar{x}$ is the population mean. Thus $S = (1/n)\Sigma x_i - \bar{x}(1/n)\Sigma 1 = \mu_S - \mu(1/n)n = \mu_S - \mu$, in the symbols of the text.

This formulation of S assumes that all individuals in the breeding population have the same fitness; if they differ in fitness, then it is necessary and proper to calculate S as a weighted average, namely $S = (1/\Sigma w_i)\Sigma w_i(x_i - \bar{x})$, where w_i is the fitness of the ith selected individual.

An important conclusion now emerges if no individuals are selected but the trait represented by x is interpreted as fitness itself, for then $S = (1/\Sigma w_i)\Sigma w_i(w_i - \bar{w}) = (n/\Sigma w_i)[(1/n)\Sigma w_i^2 -$

E $\bar{w}(1/n)\Sigma w_i] = (1/\bar{w})(\overline{w^2} - \bar{w}^2) = \sigma_p^2/\bar{w}$, since $(1/n)\Sigma w_i = \bar{w}$. The quantity σ_p^2 is, of course, the total phenotypic variance in fitness.

Now, $R = \Delta\bar{w}$ is the change in fitness in one generation of selection (in this case, natural selection), and since $R = h^2S = (\sigma_a^2/\sigma_p^2)S$, we can substitute to obtain $\Delta\bar{w} = (\sigma_a^2/\sigma_p^2)(\sigma_p^2/\bar{w}) = \sigma_a^2/\bar{w}$; that is to say, the change in average fitness in one generation of natural selection is proportional to the additive genetic variance in fitness. This result, first obtained by Fisher (1930), is known as the FUNDAMENTAL THEOREM OF NATURAL SELECTION, and much more general versions of it can be derived (see, for example, Kimura, 1958; Edwards, 1967; Li, 1967b; Turner, 1970; Hartl, 1972; and Pollak, 1978).

a. The expression $R = h^2S$ was derived under the assumption that Δp is small enough that $(\Delta p)^2$ can be neglected, so $\Delta\bar{w} = \sigma_a^2/\bar{w}$ is an approximation whose validity depends on changes in allele frequency being small. Consider the overdominance model

Genotype	AA	AA'	$A'A'$
Frequency	p^2	$2pq$	q^2
Fitness	$w_{11} = 1 - s$	$w_{12} = 1$	$w_{22} = 1 - t$

and let $s = .20$ and $t = .60$. For $p = .4$, calculate $\bar{w}$. Then calculate p' (the prime denotes the value after one generation of selection), and from p' calculate $\bar{w}'$. Then compare $\Delta\bar{w} = \bar{w}' - \bar{w}$ with $\sigma_a^2/\bar{w}$, where $\sigma_a^2 = 2pq[a + (q - p)d]^2$ and, in this case, $a = (t - s)/2$ and $d = (t + s)/2$. The approximation, you will note, is about 15 percent too large, the reason being that the change in allele frequency is too great to justify the assumption $(\Delta p)^2 \approx 0$. Verify numerically in this case that $\Delta\bar{w}$ is given exactly by $\Delta\bar{w} = (\sigma_a^2/\bar{w})[1 + (pq/2\bar{w})(w_{11} - 2w_{12} + w_{22})]$.

b. In the same model as above, but with $s = .02$ and $t = .06$, calculate $\Delta\bar{w}$ and $\sigma_a^2/\bar{w}$ for $p = .4$. In this case, because Δp is suitably small, the approximation is off by less than 1 percent.

c. Overdominance is the clearest example of the important distinction between genetic variance and additive genetic variance. At equilibrium, $\Delta\bar{w} = 0$, which, from the fundamental theorem, implies $\sigma_a^2 = 0$. There can be ample genetic variance in a population, but if all of it is nonadditive variance that natural selection cannot use to alter allele frequency, the population will be at equilibrium. In the case of overdominance, all of the equilibrium genetic variance will be dominance variance. For the model in (a), $\sigma_a^2 = 2pq[a + (q - p)d]^2$ and $\sigma_d^2 = (2pqd)^2$, where, as noted, $a = (t - s)/2$ and $d = (t + s)/2$. Verify that $\sigma_a^2 = 0$ at equilibrium, and calculate σ_d^2 at equilibrium for the numerical examples in (a) and (b). E

d. The fundamental theorem seems to imply that, because a variance cannot be negative, fitness will always increase to a maximum under natural selection. This is indeed the case in simple models not involving such complications as multiple interacting loci, nonrandom mating, frequency-dependent selection, non-Mendelian segregation, and many others. For the overdominant model in (a), $\overline{w} = 1 - p^2s - q^2t = 1 - p^2s - (1 - p)^2t$. Show that the first derivative $d\overline{w}/dp = 0$ when p is at its equilibrium value and that the second derivative $d^2\overline{w}/dp^2 < 0$ at that point, indicating that $\overline{w}$ is a maximum at equilibrium. (To refresh your memory of calculus, the maxima or minima of $\overline{w}$ are found by solving $d\overline{w}/dp = 0$ for p. Such a value of p represents a maximum for $\overline{w}$ if $d^2\overline{w}/dp^2 < 0$ and a minimum if $d^2\overline{w}/dp^2 > 0$.)

The approximate constancy of heritability has a twofold cause: first, if a particular locus accounts for only a small proportion of the total phenotypic variance in a quantitative trait, then the allele frequency at the locus will not change very rapidly; second, the values of a and d will remain nearly constant provided that the environment does not change drastically from one generation to the next. Thus, at least for the first ten generations or so, heritability will usually remain approximately constant and can be used as a constant in the prediction equation. To be precise, suppose h^2 is constant and let μ_t and S_t represent the mean of the population and the selection differential in the tth generation. From the prediction equation, we can write $\mu_t - \mu_{t-1} = h^2S_{t-1}$ and $\mu_{t-1} - \mu_{t-2} = h^2S_{t-2}$; adding these two equations yields $\mu_t - \mu_{t-2} = h^2(S_{t-1} + S_{t-2})$. Continuing in exactly this manner all the way back to the first generation of selection, we finally obtain $\mu_t - \mu_0 = h^2(S_0 + S_1 + \ldots + S_{t-1})$. The quantity $\mu_t - \mu_0$ is the TOTAL RESPONSE to selection, and $S_0 + S_1 + \ldots + S_{t-1}$ is called the CUMULATIVE SELECTION DIFFERENTIAL. For the length of time during which h^2 is approximately constant, therefore, a plot of μ_t against cumulative selection differential is expected to yield a straight line with slope equal to h^2 (Figure 11).

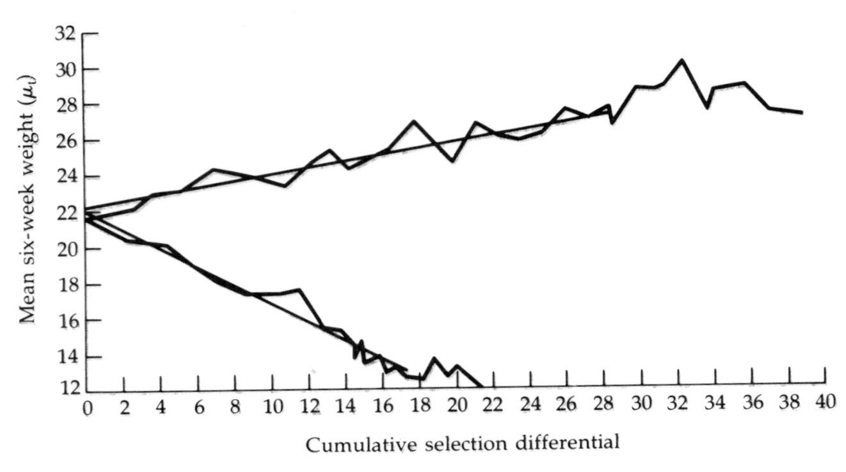

FIGURE 11. Linearity in response against cumulative selection differential for body weight in mice at age six weeks. Linearity in the "up" (high-weight) direction continues for about twice as long as it does in the "down" (low-weight) direction. (After Falconer, 1955.)

Modes of Selection

A plant or animal breeder must choose among many alternatives in the design of breeding programs, and heritability is a useful aid in making these choices. For example, virtually all breeding programs attempt to improve performance for several traits simultaneously; whether an individual is to be saved for breeding or not depends on its phenotypic value for each of the traits. Faced with the problem of simultaneous selection of, for example, two traits, a breeder has basically three options. First, selection can be in favor of those animals that are superior with respect to both traits simultaneously; this mode of selection is called INDEPENDENT CULLING. Second, selection could alternate between the traits, one trait being selected in one generation and the other trait in the next; this is known as TANDEM SELECTION. Finally, selection could be practiced on both traits simultaneously, with the overall phenotypic value of any individual calculated according to a so-called SELECTION INDEX — an arithmetic combination of the measurements of all relevant traits in the individual. (Box F compares the efficiencies of these three modes of selection.) In designing an optimal selection index, heritability is one of the components used to decide which traits should receive the most emphasis in the overall selection index (see Box F for details).

Another use of heritability is in selection for traits that cannot be evaluated in certain individuals — for example, egg production in roosters, milk production in bulls, and litter size in boars. In such cases, the selection criterion for males must be based on the performance of their female

Simultaneous Selection for Multiple Traits

F

The value of an animal or plant is determined by many traits. The value of a poultry flock, for example, is determined in part by growth rate of the birds, efficiency of food utilization, age at first egg laying, lifetime egg production, and various aspects of carcass quality. Consequently breeders must usually try to improve several traits simultaneously. Here we consider just two traits, denoted X_1 and X_2.

1. TANDEM SELECTION. The breeder may select X_1 in one generation (ignoring X_2), select X_2 in the next generation (ignoring X_1), and so on, alternating the trait selected with each generation.

2. INDEPENDENT CULLING LEVELS. The breeder may cull (i.e., discard) a certain proportion of individuals that are inferior for each trait without regard to the other. If $B = .10$ of all individuals are to be saved, and if X_1 and X_2 are uncorrelated and equal in importance, then a proportion $\sqrt{B} \approx .32$ of individuals should be saved with respect to each trait. That is to say, if the superior 32 percent for trait X_1 are saved and the superior 32 percent for trait X_2 are saved, the overall proportion saved will be $.32 \times .32 \approx .10$. Independent culling decreases the selection intensity on each trait; for $B = .10$, $i = 1.76$ (from Box C), whereas for $B = .32$, $i = 1.12$.

3. SELECTION INDEX. The breeder may select individuals based on a calculated score, or index, that, in complex cases, takes into account the heritabilities and genetic correlations of the traits as well as their relative economic value (Hazel and Lush, 1942; and see Berger, 1977, for a more recent discussion). If the traits are uncorrelated, the selection index takes the form $I = a_1h_1^2x_1 + a_2h_2^2x_2 + \ldots + a_nh_n^2x_n$, where, for the ith trait, x_i represents the

F phenotype of the individual, h_i^2 is the heritability, and a_i is the relative economic value of the trait (measured as the relative profit expected from an improvement of one unit in the trait). For two traits of equal economic value and heritability, $I = ah^2(x_1 + x_2)$, so the index value on which selection is practiced is proportional to the sum of the phenotypic measurements of the two traits.

Theoretically, independent culling is superior to tandem selection, and index selection is superior to both tandem selection and independent culling. Indeed, for n uncorrelated traits of equal heritability and economic importance, the expected gains from tandem selection, independent culling, and index selection are in the ratios $i{:}ni'{:}i\sqrt{n}$, where i is the selection intensity corresponding to a proportion saved of B and i' is the selection intensity corresponding to a proportion saved of $\sqrt[n]{B}$. When $n = 2$ traits and $B = .10$, $i = 1.76$, as noted above; and for a proportion saved of $\sqrt{B} = .32$, $i' = 1.12$; thus the selection gains are in the ratio $1.76{:}(2)(1.12){:}(1.76)(\sqrt{2})$, or 1.76:2.24:2.49. (It should be noted, however, that the theoretical superiority of index selection does not usually apply in practice, one reason being that the efficiency of index selection is rather sensitive to errors in the estimates of heritability and economic value.)

a. For $n = 3$ uncorrelated and equally heritable and important traits, calculate the relative gains under tandem selection, independent culling, and index selection, for $B = .10$ and $B = .01$. (Use the approximations $\sqrt[3]{.10} \approx .45$ and $\sqrt[3]{.01} \approx .20$ so that the table in Box C can be used.)

b. Dunlop and Young (1960, cited in Turner and Young, 1969) devised an index for selection of clean wool weight (x_1) and number of crimps per inch (x_2) in Merino sheep. For these traits $h_1^2 = .47$ and $h_2^2 = .47$, and the relative economic weights were estimated as $a_1 = 81.5$ and $a_2 = 20.1$. Calculate the appropriate selection index from the formula in (3) above. (Note: the formula in (3) ignores any genetically based correlation that may exist between the traits. In the present example, there is actually a rather large correlation, and the correlation is negative, which means that crimps per inch is negatively correlated with clean wool weight. However, clean wool weight has a higher economic value than crimps per inch, so the best selection index gives crimps per inch less importance than the index calculated according to the formula in (3). Indeed, the best index calculated by Dunlop and Young, which takes the genetically based correlation into account, is $I = 38.3x_1 + 4.7x_2$.)

relatives, often sisters, in which case the selection procedure is called SIB SELECTION (discussed later in part (b) of Box J). The heritability of a trait can be used to calculate how many sisters of each male should be evaluated.

Heritability is also useful in deciding what mode of selection will best achieve progress. Many modes of selection are possible, some of which are summarized in Table III. Up to

TABLE III. Summary of some selection methods.

I. Intrapopulation selection systems.
   A. Selection based on individual merit (called mass selection or individual selection; bulk population breeding is a mass-selection method in plants in which genetically heterogeneous seeds are mixed in bulk and planted, seed for each succeeding generation being obtained from the previous harvest).
   B. Selection based on family mean (family selection).
      1. Full-sib families (full-sib family selection).
      2. Half-sib families (half-sib family selection).
   C. Selection based on deviation of individual merit from family mean (within-family selection).
      1. Full-sib families (full-sib within-family selection).
      2. Half-sib families (half-sib within-family selection).
   D. Selection based on both family mean and deviation of individual merit from family mean (combined selection; Lush, 1947a,b). A type of combined selection in corn breeding uses the ear-to-row planting procedure, in which rows in plots are grown from the seed of a single ear (Lonnquist, 1964).
   E. Selection based on progeny merit (called progeny testing).
      1. Progeny result from random mating within population.
      2. Progeny result from one generation of self-fertilization or sib mating.
      3. Progeny result from two generations of self-fertilization or sib mating.

II. Interpopulation selection systems.
   A. Selection based on combining ability, i.e., the ability to produce superior progeny in matings to members of another population, called the tester population (recurrent selection for combining ability; Hull, 1945). The trait of interest may be *general combining ability* (which refers to the average performance of crossbreds with a genetically heterogeneous tester or with a series of genetically uniform testers), or the trait of interest may be *specific combining ability* (which refers to the performance of crossbreds with a single, usually genetically uniform, tester).
      1. Tester population is a random-mating population.
      2. Tester population is a hybrid.

TABLE III. *(Continued)*

3. Tester population is an inbred line (called "topcrossing" if inbred parent is male or pollen parent).

B. Selection based on specific combining ability, i.e., on best performance of offspring of individual crosses between members of population A and population B (reciprocal recurrent selection; Comstock et al., 1949).

C. Selection based on mean of families resulting from interpopulation crosses between A and B.
   1. Full-sib families (full-sib reciprocal recurrent selection; Lonnquist and Williams, 1967).
   2. Half-sib families (half-sib reciprocal recurrent selection; Calhoon and Bohren, 1974).

D. Selection based on mean of full-sib families of cross of A with an inbred tester line derived from population B, and vice versa (Russell and Eberhart, 1975).

III. Improvement of inbred lines (primarily plants).

A. Cross elite inbred lines to produce F_1; self-fertilize F_1 to produce F_2; select superior F_2's for self-fertilization to produce new elite lines (pedigree selection).

B. Cross elite inbred lines to produce F_1; self-fertilize F_1 to produce F_2; each plant in F_2 and subsequent generations propagated by a single self-fertilized seed; select superior plants in F_5 or F_6 generation as new elite lines (single-seed descent).

C. Cross elite inbred lines to produce F_1; backcross F_1 to one of the parental inbreds; select superior offspring for self-fertilization or further backcrossing (backcross selection).

The most appropriate method to be used in any given case depends on the organism, on management considerations, on whether selection is practiced on a single trait or on an index (Box F), and on whether the method has proven successful in the past. For practical reasons, selection methods are usually not used in a "pure" manner, but rather are used as a combination of two or more methods (e.g., in reciprocal recurrent selection, one can eliminate obviously inferior individuals prior to testcrossing; this would constitute a combination of individual selection and reciprocal recurrent selection). Prediction equations for most selection methods, as well as more discussion, can be found in Lush (1945), A. Robertson (1955), Falconer (1960), Allard (1960), Turner and Young (1969), and Gardner (1978).

now we have discussed primarily INDIVIDUAL SELECTION, in which individuals are selected for breeding or not according to their own phenotypic value (Figure 12A). Individual selection is most useful for traits with high heritabilities — h^2 = .20 or greater. For traits with lower heritability, other modes of selection are often used, particularly FAMILY SELECTION, in which individuals are saved or culled according to the value of the family mean (Figure 12B). Family selection is useful for traits with low heritability when environmental effects on the trait are large but independent from individual to individual because, by averaging among members of a family, the environmental effects tend to cancel out. A breeder may in some cases apply WITHIN-FAMILY SELECTION — a selection procedure in which individuals are saved or culled according to the deviation of their own phenotypic value from the family mean (Figure 12C). Within-family selection is most useful when environmental effects on the trait are large but common to members of a family (preweaning weight in pigs and mice are examples of traits with a large but familial component of variation, which in these cases is due to the nurturing ability of the mother); selection of individuals with large deviations from their family mean has the effect of eliminating the nongenetic familial effects because the selection is among individuals whose familial effects are in common. (The classic papers on family and within-family selection are by Lush, 1947a, 1947b.)

Individual selection, family selection, and within-family selection are often used to improve the performance of a single "purebred" population. In commercial application, superior agricultural performance often involves crossbreeding, as in the case of hybrid corn. In such cases, selection is not based upon the phenotypic value of an individual but upon the individual's COMBINING ABILITY for the trait in question, measured as the average phenotypic value of the offspring formed by crossbreeding the individual with members of another population. As noted in Table III, selection may be for GENERAL COMBINING ABILITY, which refers to the average performance of crossbreds with a genetically heterogeneous "tester" population or with a series of genetically uniform testers; or selection may be for SPECIFIC COMBINING ABILITY, which refers to the performance of crossbreds with a single, usually genetically uniform, tester. Two principal modes of selection for combining ability are in widespread use. One mode is called RECURRENT SELECTION (Hull, 1945), in which the selection criterion is the

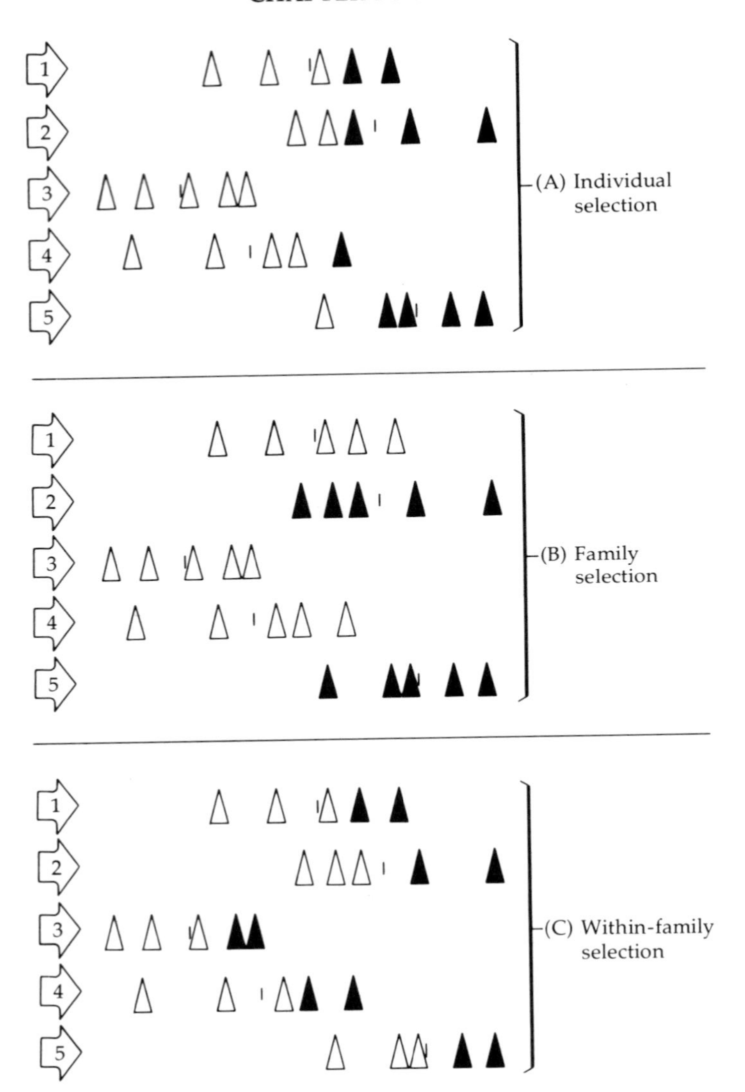

FIGURE 12. Five individuals (triangles) from each of five families (arrows) illustrating (A) individual, (B) family, and (C) within-family selection. In each case, the abscissa represents phenotype, and each individual is positioned along the abscissa according to its phenotype; the vertical line in each family represents that family's mean. In each case, ten individuals are to be selected. In individual selection (A), the ten best individuals (shaded) are selected, irrespective of the families to which they belong. In family selection (B), entire families are selected on the basis of the

performance of the offspring obtained from matings between an individual and members of an unselected (and usually genetically uniform) tester strain; the other mode is called RECIPROCAL RECURRENT SELECTION (Comstock et al., 1949), in which case two populations are selected simultaneously, and the selection criterion for each population is the performance of the offspring obtained from matings with members of the other population. Recurrent selection and reciprocal recurrent selection are particularly useful for traits in which a major part of the genetic variation is due to nonadditive gene action, such as overdominance. The modes of selection in Table III represent only a sample of the many possibilities open to breeders, and decisions about the mode of selection to be applied in any individual case must be based not only on heritability but also on such factors as population size and economic and management considerations.

Resemblance between Relatives

PARENT-OFFSPRING RESEMBLANCE

To carry out the artificial selection required for the estimation of realized heritability is in some cases impractical and, in humans, impossible. Alternative methods of estimating heritability are based on the similarity between relatives. There are of course many degrees of relationship — parent-offspring, sibling, half-sibling, first-cousin, and so on — but in genetics one of the most important relationships is that between parents and their offspring. Figure 13 shows a plot of the mean of male offspring for a quantitative trait (y values) against the phenotypic value of the father (x values). The

family mean, and the selected individuals are those belonging to these selected families; in this case, families 2 and 5 (which have the greatest means) are selected. In within-family selection (C), the superior individuals in each family are selected irrespective of their own phenotype or that of the family mean. In general, individual selection is most effective for traits having high heritability; the other modes of selection are most useful for low-heritability traits.

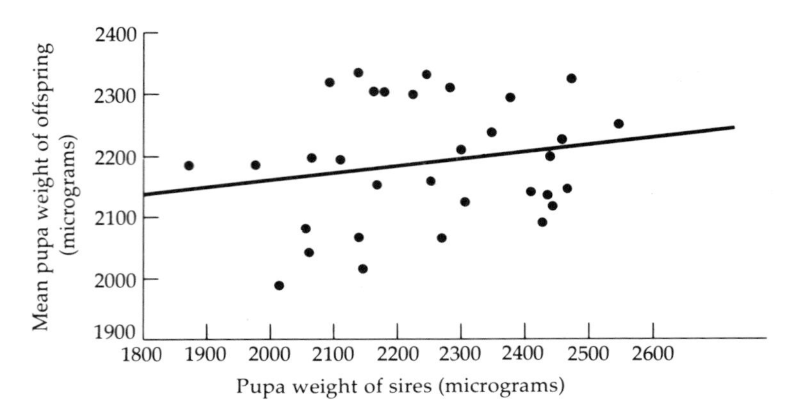

FIGURE 13. Mean weight of male pupae of the flour beetle, *Tribolium castaneum*, against pupa weight of father (sire). Each point is the mean of about eight male offspring. The regression coefficient of male offspring weight on sire's weight is $b = .11$, and h^2 is estimated as $2b$. (Courtesy of F. D. Enfield.)

line is the best fitting straight line, called the REGRESSION LINE, of offspring on parent. For reasons that will become clear in a moment, we are interested in the slope of this line, which is most easily expressed in terms of the covariance of x and y, as defined below. When values occur in pairs as they do here for parents and their offspring, the covariance can always be calculated. The required calculations are illustrated for a simple example in Table IV. First the means $\bar{x}$ and $\bar{y}$ are calculated from columns 1 and 2. Then the deviation (difference) of each x and y from its respective mean is calculated as in columns 3 $(x - \bar{x})$ and 4 $(y - \bar{y})$. Multiplying $x - \bar{x}$ times $y - \bar{y}$ as in column 5 of Table IV yields products of the deviations. The covariance of x and y is defined as the mean of the products of the deviations (see Box A); that is, the covariance equals the mean of the $(x - \bar{x})(y - \bar{y})$ values in column 5.

Returning now to Figure 13, if σ_{xy} represents the covariance between phenotypic values of fathers (sires) and those of their male offspring, and σ_x^2 represents the variance of phenotypic values of the fathers, then the slope of the regression line is simply σ_{xy}/σ_x^2. This can be seen as follows:

TABLE IV. Calculation of regression coefficient of y on x.

Frequency in population	x (1)	y (2)	$(x - \bar{x})$ (3)	$(y - \bar{y})$ (4)	$(x - \bar{x})(y - \bar{y})$ (5)
1/4	10	14	−10	−4	40
1/2	20	19	0	1	0
1/4	30	20	10	2	20

Mean of x $(\bar{x})$ = 20 [Table I]
Variance of x (σ_x^2) = 50 [Table I]
Mean of y $(\bar{y})$ = (1/4)(14) + (1/2)(19) + (1/4)(20) = 18 [column (2)]
Covariance of x and y (σ_{xy}) = (1/4)(40) + (1/2)(0) + (1/4)(20) = 15 [column (5)]
Regression coefficient of y on x $(b = \sigma_{xy}/\sigma_x^2)$ = 15/50 = .30

suppose the equation of the line is represented as $y = c + bx$, where c and b are constants, b being the slope. Taking means of both sides yields $\bar{y} = c + b\bar{x}$. (Here we have used the fact that the mean of $c + bx$ equals $c + b\bar{x}$, as illustrated in Table V.) Subtracting $\bar{y} = c + b\bar{x}$ from $y = c + bx$ yields $y - \bar{y} = c + bx - c - b\bar{x} = b(x - \bar{x})$. Now multiply both sides by $x - \bar{x}$ to obtain $(x - \bar{x})(y - \bar{y}) = b(x - \bar{x})^2$. Taking means of both sides produces $\sigma_{xy} = b\sigma_x^2$. In other words, the slope of the regression line is $b = \sigma_{xy}/\sigma_x^2$; this slope is called the regression coefficient of offspring on one parent (also defined in Box A).

TABLE V. Mean of $c + bx = c + b\bar{x}$.

Frequency in population	x	$c + bx$
1/4	10	$c + 10b$
1/2	20	$c + 20b$
1/4	30	$c + 30b$

c and b are constants.
Mean of x $(\bar{x})$ = (1/4)(10) + (1/2)(20) + (1/4)(30) = 20.
Mean of $c + bx$ = $(1/4)(c + 10b) + (1/2)(c + 20b) + (1/4)(c + 30b) = c + 20b = c + b\bar{x}$. This relationship is always true.

The procedure for estimating b — the regression coefficient of offspring on one parent — is clearly related to the procedure for estimating realized heritability because data pertaining to parents and their offspring are involved. The regression coefficient of offspring on one parent (b) can be calculated for any random-mating population, however, and the heritability (h^2) can be estimated from the relationship $b = h^2/2$. (The 1/2 enters because the regression involves only a single parent — the father, in the case of Figure 13 — and only half the alleles from any one parent are passed on to the offspring.) In Figure 13, $b = .11$, so $h^2 = .22$. Notice the considerable scatter among the points in the figure, which represents data from 32 families. Because this sort of scatter is typical, heritability estimates tend to be quite imprecise unless based on data from several hundred families. One further point about Figure 13: the regression is better performed on the father's phenotype, rather than on the mother's, in order to avoid potential bias in the estimate of heritability caused by such maternal effects as nurturing ability. (See Meyer and Enfield, 1975, and Katz and Enfield, 1977, for some complications that can arise in estimation of heritability by means of parent-offspring regression.)

ANALYSIS OF GENETIC VARIATION

The variance of a quantitative trait can be split into various components representing different causes of variation. Similarity between relatives is conveniently expressed in terms of these variance components, but variance partitioning is also of interest in its own right. Recall that the rate of change of a trait under selection depends on the amount of genetic variation affecting the trait; if there is no genetic variation, there will obviously be no response to selection. What is not so obvious is that some genetic variation cannot be acted upon by some kinds of selection; in other words, certain populations can have ample genetic variation and yet fail to respond to selection. The part of the genetic variation amenable to selection can be clarified by partitioning the variance.

Table VI shows the kind of reasoning involved in parti-

TABLE VI. Phenotypes of various genotypes as the sum of μ, G, and E.[a]

Genotype	Phenotypic value
AA	$\mu + G_1 + E_1$
AA	$\mu + G_1 + E_2$
AA	$\mu + G_1 + E_3$
AA'	$\mu + G_2 + E_4$
AA'	$\mu + G_2 + E_5$
AA'	$\mu + G_2 + E_6$
$A'A'$	$\mu + G_3 + E_7$
$A'A'$	$\mu + G_3 + E_8$
$A'A'$	$\mu + G_3 + E_9$

[a] μ is the population mean. G is a contribution due to genotype, different for each genotype. E is a contribution due to environment, different for each individual.

tioning the variance. The actual phenotype of any individual is represented as a sum of three components: (1) the mean, μ, of the entire population, (2) a deviation from the population mean due to the specific genotype of the individual in question — these genotypic deviations are symbolized in Table VI as G_1, G_2, and G_3 for AA, AA', and $A'A'$ genotypes — and (3) a deviation from the population mean due to the specific environment of the individual in question — these environmental deviations are unique to each individual and are represented in Table VI as $E_1, E_2, \ldots, E_9$. It is important to note that the G's and E's are not directly observable; nevertheless, as will be shown, the total *variance* in phenotype can be partitioned into a component due to *variation* among the G's and another component due to *variation* among the E's. Table VI can be summarized by writing $P = \mu + G + E$, where P represents the phenotypic value of any individual and G and E represent the genotypic and environmental deviations pertaining to that individual.

To connect the above symbols with actual numbers, recall the earlier example of coat color in guinea pigs based on Table II, where we assumed an allele frequency of A of p = .25 in a hypothetical population. The mean of the population, recall, was earlier calculated as $\mu = 1.083$. Thus the G_1 deviation for AA genotypes in this population is $1.303 - 1.083 = .220$, the G_2 deviation for AA' genotypes is $1.249 - 1.083 = .167$, and the G_3 deviation for $A'A'$ genotypes is $.948 - 1.083 = -.135$. For a particular animal of genotype AA whose actual coat-color score is, for example, 1.403, the corresponding value of E for the animal would be $1.403 = 1.083 + .220 + E$ because of the relationship $P = \mu + G + E$; thus, for this animal, $E = .100$. Similarly, a particular animal of genotype AA' with an actual phenotype of P = 1.05 would have a value of E given by $1.05 = 1.083 + .167 + E$, or $E = -.200$. Because the E's are measured as deviations from their mean, the average of E's for any genotype will be 0. Likewise, since the G's are measured as deviations from their mean, the mean of the G's will be 0, as can be verified for the guinea pig example because $(.25)^2(.220) + 2(.25)(.75)(.167) + (.75)^2(-.135) = 0$.

The equation $P = \mu + G + E$ is appropriate if the effects of genotype and environment are additive; that is to say, if the deviation of the phenotype of any particular individual from the population mean ($P - \mu$) can be written as the sum of an effect due to the genotype of that individual and a separate effect due to the environment of that individual. This additivity will be true whenever the ratio of G_1:G_2:G_3 is the same in each of the relevant environments. For the genotypes in Figure 10, for example, if the actual range of environments is the range designated E_1, then the genetic and environmental effects will be additive because the ratio G_1:G_2:G_3 is the same for any particular environment in E_1; for the same reason, the genetic and environmental effects will be additive if the actual range of environments is E_2. However, if the actual range of environments includes both E_1 and E_2, then the ratio G_1:G_2:G_3 will depend on the particular environment, so the genetic and environmental effects will not be additive. Nonadditivity of genetic and environmental effects is called GENOTYPE–ENVIRONMENT INTERACTION, and in writing $P = \mu + G + E$, we are assuming that there is no genotype–environment interaction. (For one example in which genotype–environment interaction does occur, see Orozco, 1976.)

Now, when $P = \mu + G + E$, the total phenotypic variance in the population, call it σ_p^2, is simply the mean of $(P - \mu)^2$. But $(P - \mu)^2$ equals $(\mu + G + E - \mu)^2$, which is just $(G + E)^2 = G^2 + 2GE + E^2$. Because G and E are already deviations from the mean, the mean of G^2 is the phenotypic variance in the population that is due to differences in genotype, and the mean of E^2 is the phenotypic variance in the population that is due to differences in environment. The mean of G^2 is called the GENOTYPIC VARIANCE and denoted σ_g^2; the mean of E^2 is called the ENVIRONMENTAL VARIANCE and denoted σ_e^2. The remaining term — the mean of $2GE$ — is twice the GENOTYPE–ENVIRONMENT COVARIANCE. If the genotypic and environmental deviations are independent — that is to say, if there is no systematic association between genotype and environment — then there is said to be no GENOTYPE–ENVIRONMENT ASSOCIATION and the mean of $2GE$ equals zero. When there is no genotype-environment association, therefore, $\sigma_p^2 = \sigma_g^2 + \sigma_e^2$. The assumption that genotype-environment association is negligible is frequently a valid assumption in animal and plant breeding where, because breeders have a degree of control not available to, for example, human geneticists, experiments can be intentionally designed in such a way as to minimize genotype-environment association. However, genotype-environment association can occur even in animal and plant breeding. For example, dairy farmers routinely provide more feed supplements to cows that produce more milk; because milk-producing ability is partly due to genotype, this feed regimen will provide superior environments (i.e., better feed) to those cows that have superior genotypes to begin with, so there will be a genotype-environment association.

The biological meaning of the expression $\sigma_p^2 = \sigma_g^2 + \sigma_e^2$ is shown for one locus in Figure 14. The solid curves represent the phenotypic distributions in the genotypes AA, AA', and

$A'A'$ with means denoted G_1, G_2, and G_3, and the dashed curve represents the phenotypic distribution in the whole population. The total phenotypic variance, σ_p^2, is the variance of the dashed distribution; the genotypic variance, σ_g^2, is the variance among the G's (i.e., $\sigma_g^2 = p^2G_1^2 + 2pqG_2^2 + q^2G_3^2$, where p is the allele frequency of A); the environmental variance, σ_e^2, is obtained by subtraction: $\sigma_e^2 = \sigma_p^2 - \sigma_g^2$. Of course, the G's are not generally known, but in a genetically uniform population, σ_g^2 must equal zero. The observed variance of a randomly bred population, therefore, provides an estimate of $\sigma_g^2 + \sigma_e^2$, whereas the observed variance of a genetically uniform population provides an estimate of σ_e^2. The estimate of σ_g^2 is obtained by subtraction as shown for the data on *Drosophila* in Table VII; in this case, genetic variation among individuals in the randomly bred population accounts for about 0.180/0.366 = 49.2 percent of the phenotypic variance. Genetically uniform populations such as inbred lines or crosses between inbreds are not available in human populations, of course, but identical twins are often used instead because of the identical genotypes of the

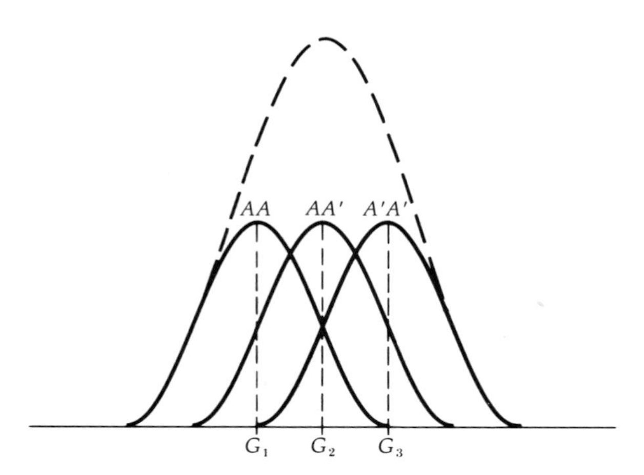

FIGURE 14. Phenotypic distribution (dashed curve) of a quantitative trait in a hypothetical population showing distributions (solid curves) of three constituent genotypes for two alleles at a locus. The means of AA, AA', and $A'A'$ are here denoted as G_1, G_2, and G_3, respectively.

TABLE VII. Calculation of genotypic variance (σ_g^2) and environmental variance (σ_e^2).[a]

Variance	Populations	
	Random-bred	Uniform
Theoretical	$\sigma_g^2 + \sigma_e^2$	σ_e^2
Observed	0.366	0.186

$\sigma_e^2 = 0.186$
$\sigma_g^2 = (\sigma_g^2 + \sigma_e^2) - \sigma_e^2 = 0.366 - 0.186 = 0.180$

[a] Trait is length of thorax in *Drosophila melanogaster* (in units of 10^{-2} mm). Data from Robertson (1957).

twins. As will be discussed later in this chapter, such a use of twins presents many serious complications, and twin studies must therefore be carried out and interpreted with utmost caution.

COMPONENTS OF GENOTYPIC VARIATION

So far, the phenotypic variance has been partitioned into the genotypic variance and the environmental variance according to the formula $\sigma_p^2 = \sigma_g^2 + \sigma_e^2$. The genotypic variance can be partitioned further into terms that are particularly important for interpreting the resemblance between relatives. The appropriate setup is shown in Table VIII, where the means of AA, AA', and $A'A'$ genotypes are denoted in column 1 by $\mu^* + a$, $\mu^* + d$, and $\mu^* - a$ as they were earlier in Figure 7. To obtain the G values, the mean of each genotype must be expressed as a deviation from the population mean, which is $\mu = \mu^* + (p - q)a + 2pqd$, and the deviations are shown in column 2 of Table VIII. (The algebra is suggested as an exercise in Box G.) The genotypic variance, σ_g^2, equals $p^2G_1^2 + 2pqG_2^2 + q^2G_3^2$, which works out in Box G to be $2pq[a + (q - p)d]^2 + (2pqd)^2$. The quantity $2pq[a + (q - p)d]^2$ is called the ADDITIVE GENETIC VARIANCE and symbolized σ_a^2; the quantity $(2pqd)^2$ is called the DOMINANCE VARIANCE and symbolized σ_d^2. Thus we have $\sigma_g^2 = \sigma_a^2 + \sigma_d^2$. This expression allows us to write the total phenotypic

TABLE VIII. Expressions for population mean and genotypic deviations.

Genotype	Frequency	Mean phenotype (1)	Genotypic deviation from population mean (G)[a] (2)
AA	p^2	$\mu^* + a$	$G_1 = \mu^* + a - \mu = 2q[a + (q - p)d] - 2q^2d$
AA'	$2pq$	$\mu^* + d$	$G_2 = \mu^* + d - \mu = (q - p)[a + (q - p)d] + 2pqd$
$A'A'$	q^2	$\mu^* - a$	$G_3 = \mu^* - a - \mu = -2p[a + (q - p)d] - 2p^2d$

$$\begin{aligned}\text{Population mean } (\mu) &= p^2(\mu^* + a) + 2pq(\mu^* + d) + q^2(\mu^* - a)\\ &= (p^2 + 2pq + q^2)\mu^* + (p^2 - q^2)a + 2pqd\\ &= (p + q)^2\mu^* + (p - q)(p + q)a + 2pqd\\ &= \mu^* + (p - q)a + 2pqd\end{aligned}$$

[a] Derivation of the equations in column (2) is an exercise in part (b) of Box G.

variance as the sum of three terms: $\sigma_p^2 = \sigma_a^2 + \sigma_d^2 + \sigma_e^2$.

To apply these concepts in a particular case, consider again the (by now familiar) hypothetical population of guinea pigs where we assumed the values $a = .1775$ and $d = .1235$ from Table II, an allele frequency of A of $p = .25$, and a total phenotypic variance in coat-color score of $\sigma_p^2 = .16$. Earlier we had calculated that for AA, AA', and $A'A'$ in this population, $G_1 = .220$, $G_2 = .167$, and $G_3 = -.135$, respectively. The total genotypic variance in the population is therefore $\sigma_g^2 = (.25)^2(.220)^2 + 2(.25)(.75)(.167)^2 + (.75)^2(-.135)^2 = .024$. Thus the environmental variance is $\sigma_e^2 = \sigma_p^2 - \sigma_g^2 = .16 - .024 = .136$. The additive genetic variance in the population is $\sigma_a^2 = 2(.25)(.75)[.1775 + (.75 - .25)(.1235)]^2 = .022$, and the dominance variance is $\sigma_d^2 = [2(.25)(.75)(.1235)]^2 = .002$. Note that $\sigma_g^2 = \sigma_a^2 + \sigma_d^2$ (i.e., $.024 = .022 + .002$) and $\sigma_p^2 = \sigma_g^2 + \sigma_e^2$ (i.e., $.16 = .024 + .136$).

Recall now that the heritability, h^2, is given by $h^2 = 2pq[a + (q - p)d]^2/\sigma_p^2$, so we may write equivalently $h^2 = \sigma_a^2/\sigma_p^2$. This is an important result because it says that the heritability depends only on the additive genetic variance and not on the dominance variance. Thus, if all the genetic variance in a population is dominance variance (i.e., $\sigma_a^2 = 0$), then the population will not respond to individual selection because h^2 will equal zero. To say the same thing in another way, the dominance variance (σ_d^2) represents that part of the genetic variance that is not acted upon by individual selection. (See part (c) of Box E for an example in which the additive variance equals zero but the dominance variance does not.)

The equation $h^2 = \sigma_a^2/\sigma_p^2$ means that the heritability of a

---

Average Effect of an Allele Substitution

G

The quantity $a + (q - p)d$ arises so often in quantitative genetics that it is given a special name — the average effect of an allele substitution (see Box B). This box deals with two cases in which the quantity $a + (q - p)d$ appears naturally. The first exercise involves the approximation used in the text, namely, that $\mu' = \mu + 2[a + (q - p)d]\Delta p$ is the mean phenotype of a population after one generation of truncation selection; part (a) below shows how the approximation can be derived. The second exercise, part (b), involves derivation of the expressions in column 2 of Table VIII, in which the deviation of each genotype from the population mean is expressed in terms of $a + (q - p)d$. (Students who wish only to verify that the expressions are correct will find the answers worked out in full at the end of the book.)

a. For a change in allele frequency Δp, the mean phenotype in the next generation will be $\mu' = (p + \Delta p)^2(\mu^* + a) + 2(p + \Delta p)(q - \Delta p)(\mu^* + d) + (q - \Delta p)^2(\mu^* - a)$. Multiply out the right-hand side, ignore terms in $(\Delta p)^2$, and use the fact that $\mu = \mu^* + (p - q)a + 2pqd$ from Table VIII to show that $\mu' = \mu + 2[a + (q - p)d]\Delta p$.

b. In Table VIII we showed that $\mu = \mu^* + (p - q)a + 2pqd$. Here the values in column 2 of the table are to be derived. First use the above expression for μ to show that

(1) $\mu^* + a - \mu = 2q(a - pd)$
(2) $\mu^* + d - \mu = (q - p)a + (1 - 2pq)d$
(3) $\mu^* - a - \mu = -2p(a + qd)$.

Then show that

(4) $2q(a - pd) = 2q[a + (q - p)d] - 2q^2d$
(5) $(q - p)a + (1 - 2pq)d = (q - p)[a + (q - p)d] + 2pqd$
(6) $-2p(a + qd) = -2p[a + (q - p)d] - 2p^2d$.

(Hint: the easiest way to proceed for (4) through (6) is to multiply out the right-hand sides and then show that they reduce to the left-hand sides.)

---

trait is the ratio of the additive genetic variance to the total phenotypic variance. Sometimes the word "heritability" is also used in reference to a different variance ratio, namely the ratio of the total genotypic variance to the total phenotypic variance (i.e., σ_g^2/σ_p^2). To avoid confusion, the ratio σ_a^2/σ_p^2 is often called the HERITABILITY IN THE NARROW SENSE, and the ratio σ_g^2/σ_p^2 is called the HERITABILITY IN THE BROAD SENSE. For the guinea pig example discussed earlier, we had assumed $\sigma_p^2 = .16$ and calculated $\sigma_a^2 = .022$ and $\sigma_g^2 = .024$. Thus, in this hypothetical population, the narrow-sense heritability is $\sigma_a^2/\sigma_p^2 = .022/.16 = .1375$, and the broad-sense heritability is $\sigma_g^2/\sigma_p^2 = .024/.16 = .1500$. Generally speaking, narrow-sense heritability is the more important with individual selection (or any mode of selection that capitalizes primarily on the additive genetic variance), whereas broad-sense heritability is the more important when selection is practiced among clones (a CLONE is a group of genetically identical individuals), inbred lines, or varieties. In this book, the word "heritability" will be used to mean "narrow-sense heritability" unless otherwise stated.

As mentioned earlier, heritability has no easy interpretation in simple genetic terms. The same is true of the variance components σ_a^2 and σ_d^2. Even for a single locus, the variance components depend on allele frequency and on the particular values of a and d (Figure 15). For many loci, σ_a^2 represents a sum of the values of $2pq[a + (q - p)d]^2$ for each locus affecting the trait, and σ_d^2 represents a sum of the values of $(2pqd)^2$ for each locus. Indeed, for more than one locus there must be an additional term included in the genotypic variance that pertains to interaction among loci; this term is called the INTERACTION VARIANCE or the EPISTATIC VARIANCE and is symbolized σ_i^2; with the interaction variance included, the genotypic variance becomes partitioned as $\sigma_g^2 = \sigma_a^2 + \sigma_d^2 + \sigma_i^2$. The important point to remember about the components of genotypic variance is that they represent the cumulative, statistical effect of all loci affecting the trait; few inferences about the actual mode of inheritance of the trait are possible from the variance components, particularly concerning the number of loci involved and their individual effects.

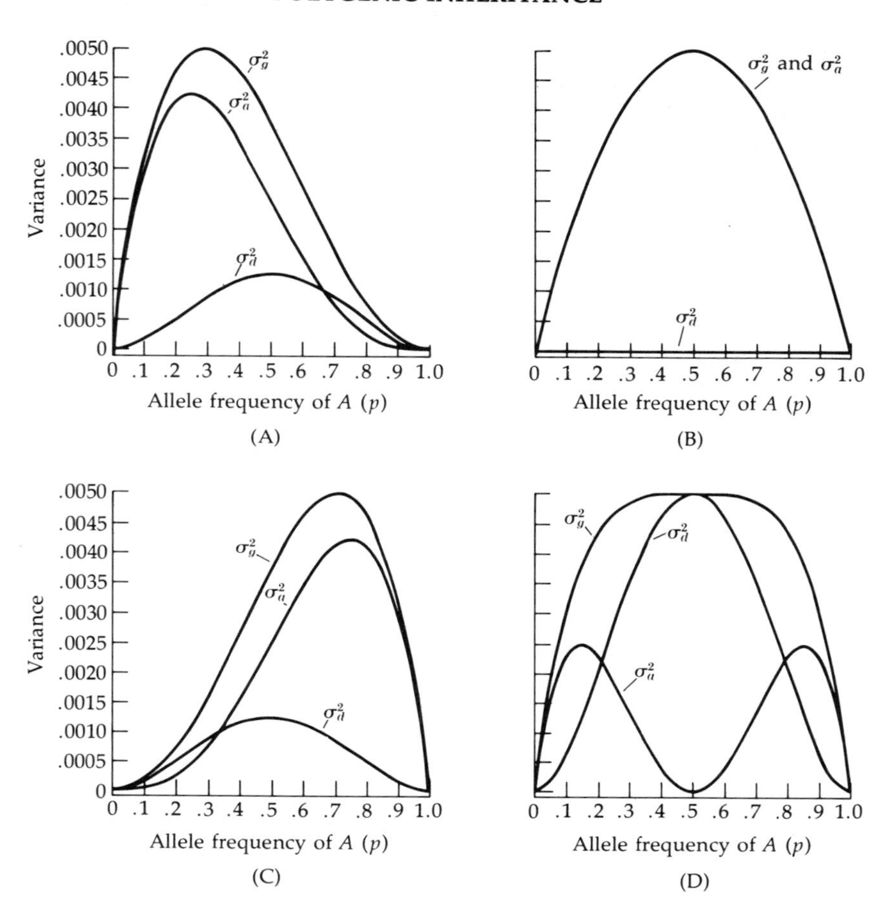

FIGURE 15. Total genetic variance (σ_g^2), additive genetic variance (σ_a^2), and dominance variance (σ_d^2) for a locus with two alleles (A and A'), against allele frequency of A (p). The mean phenotypes of AA, AA', and $A'A'$ are denoted $\mu^* + a$, $\mu^* + d$, and $\mu^* - a$. In all cases $\sigma_a^2 = 2pq[a + (q-p)d]^2$, $\sigma_d^2 = (2pqd)^2$, and $\sigma_g^2 = \sigma_a^2 + \sigma_d^2$. (A) $a = d = .07071$ (A dominant to A'); (B) $a = .1$, $d = 0$ (no dominance); (C) $-a = d = .07071$ (A' dominant to A); (D) $a = 0$, $d = .14142$ (overdominance). For ease of comparison, the values of a and d have been chosen to make the maximum of σ_g^2 equal to .005 in each case.

COVARIANCE BETWEEN RELATIVES

Components of genetic variation are important because they may be used to express the phenotypic covariance between relatives. The algebra involved in deriving the covariances is lengthy and tedious, but the most important results are summarized in Table IX. As can be seen, the additive genetic variance can be estimated directly either from parent–offspring covariance or from half-sib covariance; the full-sib covariance includes a term due to dominance, however. The expressions in Table IX are correct as long as there are no complications such as genotype–environment associations or other nonrandom environmental effects such as full sibs sharing environmental factors common to the whole family but not shared by other families. Once σ_a^2 is estimated from

TABLE IX. Theoretical covariance in phenotype between relatives.[a]

Degree of relationship	Covariance
Offspring and one parent	$\sigma_a^2/2$
Offspring and average of parents (midparent)	$\sigma_a^2/2$
Half siblings	$\sigma_a^2/4$
Full siblings	$(\sigma_a^2/2) + (\sigma_d^2/4)$
Monozygotic twins	$\sigma_a^2 + \sigma_d^2$
Nephew and uncle	$\sigma_a^2/4$
First cousins[b]	$\sigma_a^2/8$
Double first cousins	$(\sigma_a^2/4) + (\sigma_d^2/16)$

[a] Variance terms due to interaction between loci (epistasis) have been ignored.
[b] First cousins are the offspring of matings between siblings and unrelated individuals; double first cousins are the offspring of matings between siblings from two different families.

the covariance between relatives, the heritability is estimated as $h^2 = \sigma_a^2/\sigma_p^2$. The first three relationships in Table IX are the most useful ones in quantitative genetics and are commonly used in animal and plant breeding; the other relationships are used mainly in human quantitative genetics. (A procedure frequently used for heritability estimation in animal breeding is outlined in Box H. For more detail on partitioning of genotypic variance and covariances between relatives, see Cockerham, 1954, 1956, 1959; Hayman, 1954; Kearsey and Jinks, 1968; and Kempthorne, 1969.)

Estimation of Heritability by Half-Sib Analysis of Variance

H

In the practice of animal breeding, heritability is often estimated by a procedure called analysis of variance. The data are typically generated by mating each of a number of males (sires) with a number of females (dams), and measurements are made on the offspring. In much the same manner that the total variance of observations in Box A could be partitioned into a between-group component and a within-group component, the total variance of the offspring can be partitioned into components that can be related to quantities of genetic interest, for example, the additive genetic variance (σ_a^2), the dominance variance (σ_d^2), and so on.

An analysis of variance can take many forms, some quite complex, depending on the particular experimental design and the variance components to be estimated. A particularly simple design involving paternal half sibs is often used in animal breeding, however. In this design, a number of sires are each mated to a number of dams and a single offspring from each mating is measured to provide the raw data. The data can be laid out as in the table in Box A, where each group now consists of the progeny of a single sire and each entry within a group is the offspring of the corresponding sire with a different dam. Using the symbols of Box A, and letting σ_a^2, σ_d^2, σ_i^2, σ_e^2 refer, respectively, to the additive genetic, dominance, interaction (or epistatic), and environmental variance components, the expected values of the *BMS* and *WMS* can be shown to be

$$BMS = \left(\frac{3}{4} + \frac{\tilde{n}}{4}\right) \sigma_a^2 + \sigma_d^2 + \sigma_i^2 + \sigma_e^2$$

$$WMS = \frac{3}{4} \sigma_a^2 + \sigma_d^2 + \sigma_i^2 + \sigma_e^2$$

Thus, $V_B = (BMS - WMS)/\tilde{n}$ estimates $(1/4)\sigma_a^2$; and, since $V_W = WMS$, $V_B + V_W$ estimates $\sigma_a^2 + \sigma_d^2 + \sigma_i^2 + \sigma_e^2 = \sigma^2$, where σ^2 is the total phenotypic variance. The intraclass correlation coefficient between half sibs is $t = V_B/(V_B + V_W)$, which in this case is $t = (1/4)(\sigma_a^2/\sigma^2)$. Since $\sigma_a^2/\sigma^2 = h^2$ by definition, we can estimate heritability as $h^2 = 4t$.

Data from Shelby et al. (1955), pertaining to various economic characteristics of 635 Hereford steers in half-sib families from 88

H sires, may be used to illustrate the analysis. For birth weight, $BMS = 140$, $WMS = 55$, and $\tilde{n} = 7.14$. Here $V_B = (140 - 55)/7.14 = 11.90$ and $V_W = 55$, so $t = 11.90/(55 + 11.90) = .178$. Thus, $h^2 = 4(.178) = .71$, or 71 percent, is the estimate of heritability of birth weight.

Calculate the heritability of carcass grade ($BMS = 7.6$, $WMS = 5.9$, $\tilde{n} = 6.81$) and thickness of fat ($BMS = 16.8$, $WMS = 10.4$, $\tilde{n} = 6.60$).

(For more information on this method and other methods of estimating heritability, see Turner and Young, 1969; Ponzoni and James, 1978; and Hill, 1978.)

Figure 16 presents heritabilities of various traits in farm animals estimated from the correlation between relatives. The data are presented merely to show the values of heritability that breeders typically must deal with; it is important to keep in mind that the heritabilities in Figure 16 pertain to one population in one type of environment at one particular time. The same trait in a different population or in a different environment could have a different heritability. Generally speaking, traits that are closely related to fitness (such as calving interval in cattle or eggs per hen in poultry) tend to have rather low heritability.

Long-Term Artificial Selection

There are a few items concerning the long-term effects of artificial selection that are important enough to warrant mention, and we shall mention them very briefly.

GENETIC DIVERGENCE IN SMALL POPULATIONS

In many cases the size of the population subjected to artificial selection must be limited for reasons of space or expense, especially in breeding programs involving large animals such as cattle, hogs, and sheep. Small population size implies that random genetic drift may become important, and alleles that are favorable for the trait might be lost from a small population by chance, as discussed in Chapter 3. (A

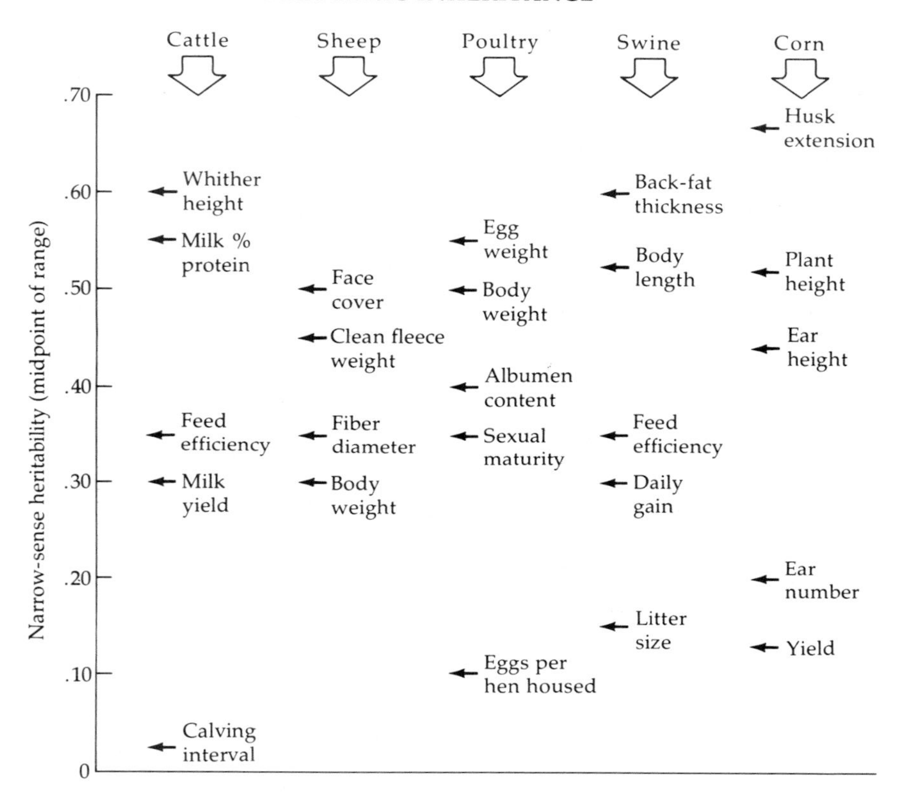

FIGURE 16. Narrow-sense heritabilities for traits in various plants and animals. Note that traits closely related to fitness (calving interval, eggs per hen, litter size of swine, yield and ear number of corn) tend to have rather low heritabilities. (Animal data from Pirchner, 1969, who gives the range of heritabilities in various studies; the midpoint of the range is plotted here. Corn data from Robinson et al., 1949.)

theoretical description of the effects of random genetic drift in small populations undergoing inbreeding is provided in Box I, which employs the concept of identity by descent.) Eventually, after long-continued artificial selection, a population may be expected to reach a SELECTION LIMIT or PLATEAU — a phenotypic limit at which response to selection no longer occurs. A selection limit will obviously be reached whenever all alleles that affect the trait have become either

I Population Subdivision with Inbreeding

Breeders often subdivide a large population into smaller subpopulations and carry out mating between relatives within subpopulations. In such a scheme the overall inbreeding coefficient (probability of autozygosity) steadily increases, not only because of nonrandom mating within subpopulations but also because of random genetic drift due to the restricted size of the subpopulations. With population subdivision and inbreeding, there are three levels of complexity to be dealt with — individuals (I), subpopulations (S), and the total population (T) — and the levels are "hierarchical" because each individual is part of some subpopulation and each subpopulation is part of the total population.

Theoretical analysis of hierarchical populations is somewhat complex, but two levels of the hierarchy have already been discussed. The inbreeding coefficient in Chapter 2 is a convenient measure of the effects of nonrandom mating within subpopulations; this inbreeding coefficient will be symbolized as F_{IS} to emphasize that it is the inbreeding coefficient of an individual (I) relative to its subpopulation (S), and F_{IS} can be calculated from pedigrees as in Chapter 2. The fixation index, F_{ST}, of Chapter 3 is a measure of the effect of population subdivision by itself on the probability of autozygosity; F_{ST}, recall, is the probability that two alleles chosen at random from within the same subpopulation are identical by descent. A breeder is naturally interested in the overall inbreeding coefficient of an individual, which takes into account not only the nonrandom mating within subpopulations but also the effects of population subdivision; this overall inbreeding coefficient is denoted F_{IT} to emphasize that it is the inbreeding coefficient of an individual (I) relative to the total population (T). (These concepts are due primarily to Wright, 1969, 1977, 1978, and earlier.)

As in earlier chapters, the values of F_{IS}, F_{ST}, and F_{IT} (which are called F STATISTICS) are related to amounts of heterozygosity at various levels in the hierarchy. Moreover, with population subdivision, the genetic variance in the total population becomes partitioned into a variance *within* subpopulations and a variance *between* subpopulations, and this effect on the variance is conveniently expressed in terms of F statistics.

This box discusses the theoretical relationships between genetic variance, heterozygosity, and F statistics. The material is rather abstract, but a practical example showing its usefulness is provided

I

in part (d). (Further application of the concepts to the analysis of natural populations will be carried out in Box F of Chapter 5.) For the sake of simplicity, we consider the following model of a locus with two additive alleles:

Genotype:	AA	AA'	$A'A'$
Frequency:	$\bar{p}^2(1 - F_{IT}) + \bar{p}F_{IT}$	$2\bar{p}\bar{q}(1 - F_{IT})$	$\bar{q}^2(1 - F_{IT}) + \bar{q}F_{IT}$
Phenotype:	2	1	0

Here $\bar{p}$ and $\bar{q}$ are the average frequencies of A and A' among subpopulations, F_{IT} is the overall probability of autozygosity, and the phenotypes are measured on an arbitrary scale.

a. Show that the mean phenotype is $M = 2\bar{p}$ and the variance in phenotype is $V_{IT} = 2\bar{p}\bar{q}(1 + F_{IT})$.

Note that the mean is independent of F_{IT}, which is true only in the case of additive alleles. Moreover, if the population were not subdivided and there were no inbreeding, F_{IT} would equal 0 and the phenotypic variance would be $V_0 = 2\bar{p}\bar{q}$. Thus we can write the total phenotypic variance of the population as $V_{IT} = (1 + F_{IT})V_0$.

A subpopulation with allele frequency p_i has a mean phenotype of $2p_i$. The variance between subpopulation means is therefore $V_{ST} = \overline{(2p_i)^2} - (2\bar{p}_i)^2 = 4\overline{p_i^2} - 4\bar{p}_i^2 = 4(\overline{p_i^2} - \bar{p}_i^2) = 4\sigma_p^2$, where σ_p^2 represents the variance of allele frequency between subpopulations. In Chapter 3, we showed that for subdivided populations $\sigma_p^2 = \bar{p}\bar{q}F_{ST}$, so $V_{ST} = 4\bar{p}\bar{q}F_{ST} = 2\,F_{ST}V_0$.

b. The total phenotypic variance, V_{IT}, is the sum of the variance between population means, V_{ST}, and the variance within populations, V_{IS}; that is, $V_{IT} = V_{IS} + V_{ST}$. Show that $V_{IS} = (1 + F_{IT} - 2\,F_{ST})V_0$.

c. Turning now to the heterozygosity, the overall heterozygosity in the population is $H_I = 2\bar{p}\bar{q}(1 - F_{IT})$. If the population were subdivided but mating were random within subpopulations, the heterozygosity would be $H_S = 2\bar{p}\bar{q}(1 - F_{ST})$, as shown in Chapter 3. If the population were one large randomly mating unit, the heterozygosity would be $H_T = 2\bar{p}\bar{q}$. F_{ST} therefore measures the decrease in heterozygosity due to population subdivision; that is $F_{ST} = (H_T - H_S)/H_T$. Similarly, F_{IT} measures the decrease in heterozygosity due to both subdivision and inbreeding within subdivisions, that is, $F_{IT} = (H_T - H_I)/H_T$. Finally, F_{IS} measures the decrease in heterozygosity due only to inbreeding within subdivisions, that is, $F_{IS} = (H_S - H_I)/H_S$. Use the expressions involving heterozygosity to show that $(1 - F_{IS})(1 - F_{ST}) = (1 - F_{IT})$.

I **d.** Suppose a population of chickens is divided into flocks of effective size $N = 20$ and maintained by random mating within flocks for $t = 9$ generations, at which time a generation of sib mating is carried out to produce individuals with an inbreeding coefficient of $F_{IS} = 0.25$ (appropriate for the offspring of sibs). Use the formula $1 - F_{ST} = [1 - (1/2N)]^t$ from Chapter 3 to calculate F_{ST}, and combine F_{IS} and F_{ST} to obtain F_{IT} — the total probability of autozygosity after the 10 generations.

---

fixed or lost, because in such cases, the trait will have no remaining genotypic variance. (However, complete fixation or loss of all relevant alleles is not the most common reason that populations reach selection limits, except perhaps in very small populations; other, more typical, reasons for selection limits will be discussed in a few pages.) In any case, small populations undergo genetic differentiation due to random genetic drift even when they are also subjected to artificial selection; and small populations that have reached their selection limit may nevertheless differ in genotype, some favorable alleles having been fixed by selection in some populations but lost by chance in others. Additional progress can sometimes be made by crossing the populations and practicing selection in the population produced by the cross (or crosses), but the first offspring of such crosses may be phenotypically inferior to the parents; this temporary fallback is the price to be paid in order to obtain a selected population that may eventually surpass the original selection limit (for an example, see Roberts, 1967).

INBREEDING DEPRESSION

A second problem that arises in artificial selection because of small population size is a steady increase in the inbreeding coefficient in the selected population and a consequent reduction in fitness due to inbreeding depression (see Figure 20 in Chapter 2 for an example). Traits that are closely related to fitness typically show the greatest inbreeding depression; although there is considerable variation in the amount of inbreeding depression, there is often about a 5-10 percent reduction in phenotypic value for each 10 percent increase in the inbreeding coefficient (Falconer, 1960). Because of inbreeding depression, breeders usually try to arrange matings so as to avoid inbreeding as much as possible.

CORRELATED RESPONSES

Genes are pleiotropic. Every gene potentially affects every trait in the organism, either as a primary effect or as a secondary, indirect effect. Thus, the alleles that are favorable for one quantitative trait may have unfavorable effects on another quantitative trait, and as these alleles are increased in frequency by artificial selection, thereby improving the phenotypic value with respect to the selected quantitative trait, the very same alleles may bring about a deterioration of some other aspect of performance. Pleiotropy is one cause of CORRELATED RESPONSE — a change in phenotypic value of one trait that accompanies response to selection of a different trait. A second possible cause of correlated responses is linkage disequilibrium (Chapter 2); a favorable allele for one trait that increases in frequency under selection may drag along with it an allele at another, tightly linked locus that has a detrimental effect on an unselected trait.

Correlated responses are quite common in artificial selection and often, but not always, involve a deterioration in reproductive performance. In one case in Leghorn chickens, for example, twelve generations of selection for increased shank length reduced the egg hatchability by nearly half (Lerner, 1958). In turkeys, to take another example, there was intense selection during the period 1944-1964 for growth rate, body conformation, and body size; but there was also a steady decline in such aspects of reproductive fitness as fertility, egg production, and egg hatchability (Nordskog and Giesbrecht, 1964; Johansson and Rendel, 1968). On the other hand, correlated responses can sometimes be useful. For example, selection for larger mature body size will often increase litter size in mice and swine. If a trait has a low heritability or is difficult to measure, it is sometimes possible to practice selection for another, correlated trait, obtaining progress in the trait of interest by correlated response.

From a theoretical point of view, the covariance between two quantitative traits can be partitioned in a manner analogous to the partitioning of the variance for one trait outlined in an earlier section; the covariance can thus be partitioned into an additive covariance, a dominance covariance, an environmental covariance, and so on. The most important theoretical result is that the amount of correlated response with individual selection depends only on the additive covariance, much as the direct response to individual selection depends only on the additive variance. The components of covariance between traits can be estimated from the resemblance between relatives, but often it is preferable to estimate the correlated response by direct observation in a manner analogous to the determination of realized heritability. (Plant and animal breeders often use the term GENETIC CORRELATION, which is defined as the ratio $Cov_a(X, Y)/\sqrt{Var_a(X) \times Var_a(Y)}$, where $Cov_a(X, Y)$ denotes the additive genetic covariance between the traits X and Y, and $Var_a(X)$ and $Var_a(Y)$ are the additive genetic variances of X and Y. Use of the concept is illustrated in part (c) of Box J.)

SELECTION LIMITS

Progress under artificial selection cannot go on forever, of course. As noted earlier, the population will eventually reach a selection limit, or plateau, after which it will no longer respond to selection. One of the obvious reasons why a population would eventually reach a plateau is exhaustion of additive genetic variance; all alleles affecting the selected trait may have become fixed, lost, or have attained a stable equilibrium. With no additive genetic variance, no progress under individual selection can be achieved. However, many experimental populations that have reached a selection limit readily respond to REVERSE SELECTION (selection in the reverse direction of that originally applied), so additive genetic variance affecting the trait is still present. Indeed, in such populations, the phenotype may change in the direction of

Aids to Selection

J

Here we consider three methods that breeders sometimes use to circumvent certain problems in selection. Part (a) involves traits that may vary within the same individual; part (b) involves traits that can be measured in only one sex; and part (c) involves correlated response to selection.

a. Certain traits are susceptible to environmental fluctuations, and repeated measurements on the same individual will, therefore, yield different values. For example, "time to finish" in racehorses will vary from race to race for the same horse. The REPEATABILITY of a trait, denoted t, is defined as the intraclass correlation coefficient of repeated measurements on the same individual (see Box A); thus, t measures the proportion of the phenotypic variation in the trait caused by permanent differences between individuals as opposed to temporary fluctuations. If, instead of using a single performance record to evaluate an individual's phenotype, we use the mean of a number of records (say, k of them), the expected gain with selection can be shown to be $R = h^2\sigma\sqrt{K}i$, where σ is the phenotypic standard deviation, i is the standardized selection intensity (see Box C), and $K = k/[1 + (k - 1)t]$.

The repeatability of litter size in sows is about $t = .15$. Calculate the expected gain for $k = 1, 2, 3, 4$, and 5 records (expressing the gain as multiples of $h^2\sigma i$). Note how the increased gain falls off for each additional record. Indeed, considering the time and expense often involved in collecting performance records, it is seldom worthwhile to use more than two or three records per individual.

b. Some traits can only be measured in one sex, such as egg production, milk production, and so on. In such cases, the breeder assesses a male's "performance" based on the records of his female relatives. Although any degree of relationship can be taken into account, we shall consider only the male's full sisters or half sisters. (This method of selection is called sib selection.) Suppose the male's "performance" is based on the records of k sisters. Then the expected gain is $R = h^2\sigma r\sqrt{K}i$, where σ and i are as in (a), and $K = k/[1 + (k - 1)t]$. For full sisters, $r = 1/2$, and for half sisters, $r = 1/4$, and t represents the intraclass correlation between full or half sisters, respectively. For a single full sister, $K = 1$ and $r = 1/2$, so $R = h^2\sigma i/2$, a formula that is also correct when evaluation of a male is based on the performance of his mother.

[J] Show that the maximum expected gain from sib selection is $R = h^2\sigma ri/\sqrt{t}$. (Hint: examine what happens to K as k becomes very large.)

c. Sometimes it is possible to achieve gains in one trait as a correlated response to selection for another trait (INDIRECT SELECTION) rather than by direct selection. Indirect selection can be profitable if the trait of interest has a low heritability but a secondary trait is highly correlated with the primary trait and has a high heritability, or if the secondary trait can be measured in both sexes so that higher selection intensities can be achieved for the secondary trait. To be specific, let a subscript 1 denote the trait of interest and a subscript 2 denote the secondary trait. Direct selection on trait 1 leads to an expected response of $R_1 = ih_1^2\sigma_1$; the indirect response of trait 1 (call it R_1') arising from selection for trait 2 can be shown to be $R_1' = ih_1h_2r_A\sigma_1$, where h_1 and h_2 denote the square roots of the corresponding heritabilities of the traits and where r_A is a quantity called the ADDITIVE GENETIC CORRELATION (or just GENETIC CORRELATION). The genetic correlation can be estimated by a procedure known as analysis of covariance, much as the additive genetic variance can be estimated by analysis of variance.

Suppose trait 1 represents body weight in poultry (for which $h_1^2 = .20$) and trait 2 represents age at laying of first egg ($h_2^2 = .50$). For these traits, $r_A = .29$. Suppose a breeder selects for age at first egg with $i = 2.0$. What is R_1'/σ_1 (i.e., the correlated response of body weight expressed in units of the phenotypic standard deviation)? If body weight were selected directly with intensity i^*, say, what would i^* have to be to obtain the same response?

The additive genetic correlation between milk yield and percent butterfat in dairy cattle is about $-.20$. What implication does this have for percent butterfat when there is direct selection for milk yield?

---

its original value if continuing artificial selection is simply suspended (RELAXED SELECTION). The consequences of relaxed selection for a case in Drosophila are illustrated in Figure 17.

One frequent reason for the occurrence of selection limits in populations with considerable additive genetic variation

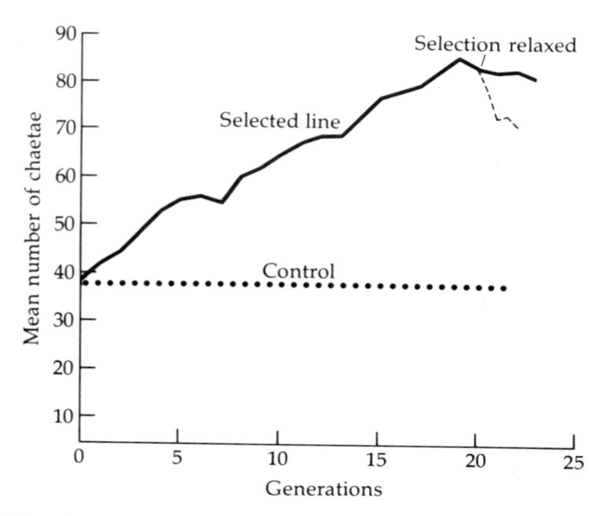

FIGURE 17. Response to selection for number of abdominal chaetae bristles in females of *Drosophila melanogaster* (solid line) and control (no selection applied: dotted line). Note the fallback in bristle number (dashed line) in the selected line upon relaxation of selection. (After Clayton et al., 1953.)

is that artificial selection may be opposed by natural selection. In mice, for example, selection for small body size ultimately ceases because small animals are less fertile than larger ones and the smallest animals are sterile (Falconer, 1960). Selection for small body size gradually becomes less effective due to the opposing effects of natural selection until, eventually, no further progress is possible.

A final point about selection limits: in most genetically heterogeneous populations, artificial selection can change phenotype well beyond the range of variation found in the original population. Pertinent data for populations of mice are presented in Table X. As can be seen, a total selection response of three to five times the original phenotypic standard deviation is not unusual, and for selection to change a population of effective size N halfway to its selection limit typically requires about $0.5N$ generations. (For more on selection limits, see Robertson, 1960, 1970; Roberts, 1966; Wilson et al., 1971; Falconer, 1971, 1977.)

TABLE X. Selection limits and duration of response for various traits in laboratory mice.[a]

Character selected	Direction of selection	Total response[b]	Half-life of response[c]
Weight (in strain N)	Up	$3.4\sigma_p$	$.6N$
	Down	$5.6\sigma_p$	$.6N$
Weight (in strain Q)	Up	$3.9\sigma_p$	$.2N$
	Down	$3.6\sigma_p$	$.4N$
Growth rate	Up	$2.0\sigma_p$	$.3N$
	Down	$4.5\sigma_p$	$.5N$
Litter size	Up	$1.2\sigma_p$	$.5N$
	Down	$0.5\sigma_p$	$.5N$

[a] From Falconer, 1977.
[b] Total response is expressed as a multiple of the initial phenotypic standard deviation, σ_p.
[c] Half-life of response is the number of generations taken to progress half way to the selection limit; here the half-life is expressed in multiples of effective population number (N).

Some of the methods that breeders use as aids to selection to maximize progress are discussed in Box J.

Human Quantitative Genetics

METHODS: HERITABILITY IN THE BROAD SENSE

The principles of quantitative genetics, as outlined in this chapter, can be applied to human populations as well as to populations of other animals or plants. For the estimation of heritability, human quantitative geneticists can study individuals with almost any degree of relationship employed by animal breeders, so estimation of heritability in humans would seem to be a simple matter. The issue is not entirely straightforward, however. The human quantitative geneticist must live with a number of serious problems that the animal breeder can circumvent with proper experimental design. First among these problems is bias; the breeder can ensure that all relevant individuals in the population are studied, but the human geneticist may unwittingly study a biased sample of people, such as a group of hospital patients and their relatives, who may not fairly represent the entire population.

A second serious problem in human quantitative genetics is genotype–environment association, discussed earlier in this chapter. The breeder may, by appropriate experimental design, ensure that environments will be associated nearly randomly with genotypes; plant breeders can assign individual plants to field plots at random, for example. The human geneticist has no such experimental control, however, and in some instances serious genotype–environment association may occur. In the case of human I.Q. score, for example, certain genotypes may provide home environments that are conducive to higher-than-average I.Q.'s among their children. Still a third problem in human quantitative genetics is assortative mating — that is, mating in which the phenotypes of mates are more similar or dissimilar than would be expected with random mating: the breeder can ensure that mating be random, but the human geneticist must live with a mating system as it is. Assortative mating increases the covariance between relatives and thereby inflates estimates of heritability (see Box K). That assortative mating in humans is rather common is shown by the data for Caucasians in Table XI, although the actual amount of assortative mating varies considerably among populations.

TABLE XI. Average correlations between spouses for various traits in Caucasians.[a]

Trait	Average correlation
Intelligence quotient	0.40
Hair color	0.23
Height	0.20
Weight	0.20
Head circumference	0.09
Tooth size	0.15
Eye color	0.19
Total finger ridge count	0.05

[a] After Spuhler, 1968.

That the correlations in Table XI are positive implies that mates are phenotypically more similar than would be expected with random mating.

Despite potential sources of error in estimating the heritability of quantitative traits in humans, estimates for many traits have been made. The estimates for humans frequently use the notion of broad-sense heritability mentioned earlier. Recall that the narrow-sense heritability of a trait is defined as $h^2 = \sigma_a^2/\sigma_p^2$, where σ_a^2 represents the additive genetic variance in the trait and where σ_p^2 represents the phenotypic variance in the whole population. In human genetics, the quantity σ_g^2/σ_p^2 is often called "heritability," but it is more properly called the broad-sense heritability; σ_g^2 represents

K Assortative Mating for a Quantitative Trait

The theory of assortative mating is, in general, quite complex, but if one assumes that the assortative mating is based on phenotypic resemblance for a quantitative trait, the major results can be summarized rather easily (see Crow and Felsenstein, 1968; Crow and Kimura, 1970). To be specific, suppose assortative mating has been going on for so long that the population is in equilibrium. (Actually, much of an assortative mating effect appears in early generations.) For the quantitative trait in question, let r be the correlation coefficient (see Box A) between the phenotypes of mates, and let $\hat{\sigma}_a^2$ and $\hat{\sigma}_p^2$ be the additive genetic and total phenotypic variance; the heritability of the trait in the equilibrium population is, of course $\hat{h}^2 = \hat{\sigma}_a^2/\hat{\sigma}_p^2$. (The circumflexes over the symbols are added to emphasize the assumption that the population is in equilibrium.)

The inbreeding coefficient (probability of autozygosity) in the equilibrium population is

$$\hat{F} = \frac{r}{2n_e(1 - r) + r}$$

where n_e is called the EFFECTIVE NUMBER OF LOCI. The quantity n_e equals the actual number of loci if the loci are in linkage equilibrium and all contribute equally to the variance, otherwise n_e is less than the actual number of loci. Note that unless n_e is very small or r is close to 1, $\hat{F} \approx 0$, which implies that assortative mating, in contrast to inbreeding, does not increase the homozygosity by much. K

Moreover, although assortative mating inflates the total phenotypic variance, it disproportionately inflates the additive genetic variance, so the heritability of the trait is increased. Indeed, $\hat{\sigma}_a^2$, $\hat{\sigma}_p^2$ and $\hat{h}^2$ are rather simply related to what the corresponding quantities would be were the population undergoing random mating; we'll use the symbols σ_a^2, σ_p^2, and h^2 to denote the random-mating quantities. Suppose that n_e is large enough that $1/2n_e$ is small compared to 1. Then

$$\hat{\sigma}_a^2 = \sigma_a^2 \left(\frac{1}{1 - A}\right) \approx \sigma_a^2(1 + A)$$

$$\hat{\sigma}_p^2 = \sigma_p^2 \left[1 + \frac{\hat{h}^2 A}{1 - \hat{h}^2 A}\right] \approx \sigma_p^2(1 + \hat{h}^2 A)$$

$$h^2 = \hat{h}^2 \left[\frac{1 - A}{1 - \hat{h}^2 A}\right] \approx \hat{h}^2[1 - (1 - \hat{h}^2)A]$$

where the symbol A stands for $r\hat{h}^2$ (that is, A is the product of the heritability and the phenotypic correlation between mates). The last approximations given after the $\approx$ signs assume that A is small enough that A^2 and higher powers can be ignored.

a. Caucasians undergo assortative mating for height with a phenotypic correlation coefficient between mates of $r = .28$ (data from Pearson and Lee, 1903, cited in Crow and Kimura, 1970). The heritability of height in these data is $\hat{h}^2 = .79$, which we will assume represents the equilibrium value. What would h^2 be were the population not undergoing assortative mating? By what fraction are the additive variance, the total variance, and the heritability increased by assortative mating? (Answer the questions twice, once using the exact expressions above and once the approximations.)

b. Use the above expression for h^2 and the relation $\hat{h}^2 = A/r$ to show that A must satisfy $(1 - h^2)A^2 - A + h^2 r = 0$. (This equation has two solutions for A; here the appropriate one for assortative mating is the one that lies between 0 and 1.)

c. Consider a quantitative trait controlled by $n_e = 20$ loci in a randomly mating population with $\sigma_p^2 = 60$ and $\sigma_a^2 = 15$. If the population initiates assortative mating for this trait with $r = .5$, what would be the equilibrium values of $\hat{F}$, $\hat{\sigma}_p^2$, $\hat{\sigma}_a^2$ and $\hat{h}^2$. (Hint: use part (b) to find the appropriate value of A, and then substitute into the equations and solve for the quantities of interest.)

the total genotypic variance in the trait, and σ_g^2 includes not only the additive genetic variance but also the dominance variance, the interaction variance, and it includes in addition another variance component due to assortative mating. Broad-sense heritability measures that fraction of the total phenotypic variance in the population caused by genetic differences among individuals. There is a question whether heritability, either in the broad sense or in the narrow sense, but most especially in the broad sense, is a quantity worth estimating for human quantitative traits because heritability has almost nothing to do with the actual mode of genetic determination of a trait; this is an issue to be discussed in a moment.

TWINS

Identical twins are among the favorite degrees of relationship for study in human genetics; because identical twins are genetically identical, phenotypic differences between identical twins would seem to be a straightforward reflection of how much phenotypic variance is caused by environment. Twin studies raise their own unique problems, however, and the results must be interpreted with caution (see Harris et al., 1979, for a recent study). Before discussing the use of twins in quantitative genetics, we should back up a few steps and first discuss the phenomenon of twinning itself.

Twins are relatively frequent among human births, though the rate of twinning varies from population to population. Among Caucasians in the United States, for example, about one in 88 births are twins; among Japanese in Japan, the rate is about one in 145 births (Crow, 1966). Two kinds of twins actually occur. IDENTICAL TWINS, often called MONOZYGOTIC or ONE-EGG twins, arise from a single zygote that very early in embryonic development splits into two distinct clumps of cells, each clump thereafter undergoing its own embryonic development. Because they arise from a single zygote, identical twins are necessarily genetically identical. The other kind of twins are called FRATERNAL TWINS, DIZYGOTIC twins, or TWO-EGG twins. Fraternal twins arise from a double ovulation in the mother, each egg being fertilized by a different sperm. Because of their mode of origin, fraternal twins are related genetically as siblings. Most of the variation in twinning rates in humans is due to variation in the rate of dizygotic twinning. For example, the rates of monozygotic twinning among Caucasians in the United States and among Japanese in Japan are one in 256 and one in 238, respectively, whereas the rates of dizygotic twinning are one in 135 and one in 370 (Crow, 1966).

For studies in quantitative genetics, identical twins are often compared with same-sex fraternal twins in order to discount the effects of common intrauterine environments. Such an approach is only partially successful, as identical twins often share embryonic membranes *in utero* that are not usually shared by fraternal twins. Moreover, because identical twins often have astonishingly similar facial features (Figure 18), they may be treated more similarly by parents, teachers, and peers than are fraternal twins. Some of these problems can be overcome by studying twins that are raised

FIGURE 18. Identical twins showing a striking resemblance in facial features. (Courtesy of J. C. Christian.)

apart (in different households), but data of this sort are very limited (Shields, 1962), making estimates of heritability highly imprecise. In any case, if r_{MZ} and r_{DZ} represent the correlation coefficients of a quantitative trait among monozygotic and dizygotic twins, then $2(r_{MZ} - r_{DZ})$ provides a rough estimate of the broad-sense heritability of the trait. (See Smith, 1975, for a good discussion. Recall that the correlation coefficient, r, in a set of pairs of numbers (x, y) is defined as $r = \sigma_{xy}/\sigma_x\sigma_y$, where σ_{xy} is the covariance between x and y and where σ_x and σ_y represent the standard deviations of x and y. See Box A.)

Figure 19 shows estimated broad-sense heritabilities derived from studies of twins (T) or correlations between relatives (R) for a number of quantitative traits in Caucasians. Broad-sense heritabilities vary widely for different traits, as they do in other species.

HERITABILITY: WHAT IT DOESN'T MEAN

Figure 19 again raises the issue of how useful knowledge of broad-sense heritability may be in human genetics. Because the broad-sense heritability of a trait means very little in terms of the genetic basis of the trait, some geneticists have argued that estimation of broad-sense heritabilities for human traits is a meaningless exercise (Lewontin, 1974b; Feldman and Lewontin, 1975). On the other hand, the broad-sense heritability does convey how much of the total phenotypic variation in a trait is due to genetic variation, combining together the additive, dominance, interaction, and assortative mating effects. This is not much information, to be sure, but it is something, and a little knowledge is probably to be preferred over total ignorance (for an excellent discussion, see Kempthorne, 1978). Where the main problem arises is in pushing the interpretation of broad-sense (or narrow-sense) heritability beyond what is justified. Thus, for example, the heritability of a trait is often taken as a measure of how important "heredity" is in comparison with "environment"; such an interpretation is justified only in a very restricted sense — broad-sense heritability measures the relative importance of heredity versus environment on

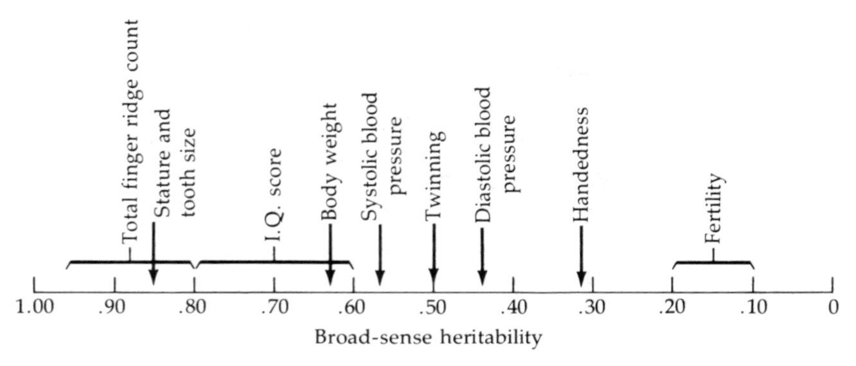

FIGURE 19. Broad-sense heritabilities (and ranges of heritabilities) for various traits in humans. Note the low heritability of fertility, a trait that is obviously closely related to fitness. (Data from Smith, 1975.)

the *variance* of the trait, and it applies to only one population in one set of environments at one particular time.

A related misconception about heritability is to imagine that a high heritability implies that a trait is relatively insensitive to environmental change. This idea is wholly false. A herd of cattle maintained on a substandard diet could have a very high heritability for growth rate, for instance, but the easiest and fastest way to improve the average growth rate would be to supply adequate feed. For this reason, heritabilities of human cognitive traits are of no significant value in setting or justifying educational policies.

Finally, it may be wrongly assumed that high heritabilities for the same quantitative trait in two different populations imply that any difference in the means of the two populations is hereditary or largely hereditary. This again is wholly false. The high heritabilities mean only that, *within* each population, much of the phenotypic variance is attributable to genetic differences among individuals. By themselves, the high heritabilities are meaningless in comparing the two populations. Thus, attributing the observed 15-point average I.Q. difference between blacks and whites in the United States to hereditary differences because the trait is "highly heritable" is fallacious and harmful; such an argument uses within-population heritability to compare different populations. (For further discussion see McClearn and DeFries, 1973; Lewontin, 1975; and Loehlin et al., 1975.) In addition, there is also a question of how high the heritability of I.Q. really is; most I.Q. studies can be disputed on the purely technical grounds of inadequate experimental design or improper controls.

Threshold Traits

There is a class of polygenic traits that does not exhibit continuous variation. Although in these cases an individual either expresses the trait or does not, making the variation discontinuous, the trait is nevertheless influenced by more than one locus and also by environment. Such traits are called THRESHOLD TRAITS. A human example is diabetes, an abnormality in sugar metabolism that affects one or two percent of the Caucasian population. In a sense, diabetes is a continuous trait because the severity of the disease varies from nearly undetectable to extremely severe. On the other hand, diabetes can also be considered a threshold trait because all individuals can be classified according to whether or not they are so severely affected that clinical treatment is required. With such a classification, there are only two phenotypes, "affected" and "not affected," even though there is phenotypic variation within each category. That there is a genetic influence on the trait is evidenced by an enhanced risk of diabetes in relatives of affected individuals, yet such environmental factors as diet are also important in determining whether high-risk genotypes will actually develop the disease. Years ago many threshold traits were "explained" by postulating a simple genetic mechanism (such as a single recessive allele in the case of diabetes) and invoking "incomplete penetrance" to account for the poor fit of pedigree data to a hypothesis. Nowadays it is possible to consider threshold traits as *bona fide* polygenic traits and to calculate heritabilities as for any other quantitative trait. In most cases, however, it is simply not known whether the genetic influence on the trait is due to a single gene or is polygenic.

HERITABILITY OF LIABILITY

The basic idea behind the modern treatment of threshold traits is illustrated in Figure 20. The normal curve in (A) represents the (unobservable) distribution of a hypothetical LIABILITY (or risk) toward the threshold trait, measured on a scale such that the mean value is 0 and the variance is 1. It is assumed that individuals whose liability is above a certain threshold (*T*) will actually express the trait. Thus, the

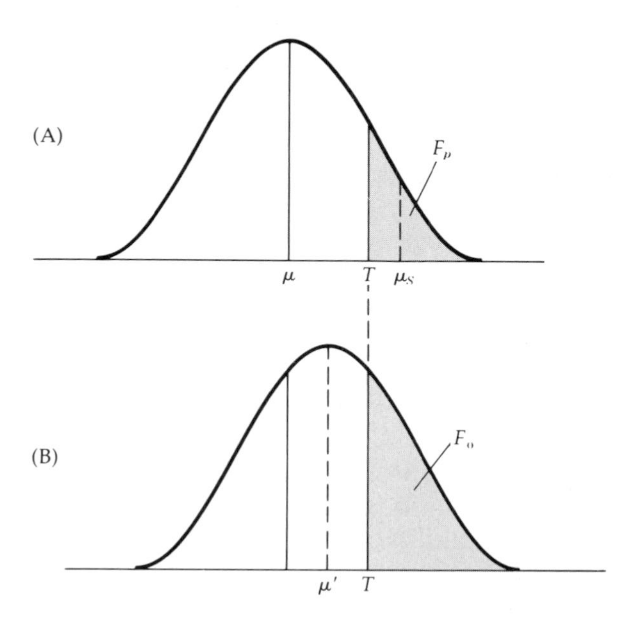

FIGURE 20. (A) Distribution of liability assumed for a threshold trait in a hypothetical population. The shaded area denotes individuals having liability above a certain threshold (*T*) and therefore affected with the trait. F_p is the frequency of affected individuals in the whole population, μ is the mean liability of individuals in the population, and μ_S the mean liability of affected individuals. (B) Distribution of liability among offspring who have one parent affected with the trait. μ' denotes the mean liability among offspring, and F_o is the proportion of affected offspring. Note similarity to Figure 4. In Figure 4, however, the quantitative trait could be measured directly in each individual. In this case, liability cannot be measured directly, and inferences about the distribution of liability must be made from F_p and F_o.

shaded area in Figure 20A represents the proportion of affected individuals in the population (F_p), and the mean liability among affected individuals is denoted μ_S. Figure 20B gives the (again unobservable) distribution of liability among offspring of affected individuals; its mean is denoted μ', and the proportion of offspring above the threshold is denoted F_o. The setup here is like that in the earlier section of this chapter in which we calculated the regression coefficient of offspring on one parent. Here the regression coefficient, b, is $b = \mu'/\mu_S$, and the appropriate estimate of the HERITABILITY OF LIABILITY is $h^2 = 2b = 2\mu'/\mu_S$. Given F_p (the only observable quantity pertaining to Figure 20A), one can find T and μ_S from standard tables of the normal curve. (See Box L for details.) Having found T, one then uses the observed value of F_o to determine μ' (again from standard tables, as outlined in Box L). With μ' and μ_S in hand, $h^2 = 2\mu'/\mu_S$ is the estimate of heritability of liability.

Methods similar to those outlined above have been developed for comparing the frequency of affected individuals among all important degrees of relationship, not merely parent-offspring, and heritability of liability can be calculated in either the narrow sense or the broad sense (Dempster and Lerner, 1950; Falconer, 1965; Gottesman and Shields, 1967; Kidd and Cavalli-Sforza, 1973; Smith, 1975; Karlin, 1979b). All the problems involved in the correct interpretation of heritability of any quantitative trait apply to the heritability of liability as well, perhaps even more severely inasmuch as estimates of the latter tend to be very imprecise because the distribution of liability cannot be ascertained directly but must be inferred from the proportion of affected individuals.

TWIN CONCORDANCE

Twins are frequently used in human quantitative genetics for the study of threshold traits. For threshold traits, twin data are best expressed in terms of concordance. The CONCORDANCE of a trait in a population of twins is the proportion of affected twins that have affected co-twins. For example, suppose that 100 affected individuals are found to be twins

Heritability of Liability: Threshold Traits

L

Here we illustrate calculation of the heritability of liability using data from Carter (1961) on pyloric stenosis, which is an obstruction of the opening at the lower end of the stomach. The incidence of this condition differs in males and females; we will use the incidences in fathers and their sons. For pyloric stenosis, the incidence among males in the general population is $F_p = .005$, and the incidence among sons of affected males is $F_o = .05$. We assume that "liability" in the general population has mean $\mu = 0$ and variance $\sigma^2 = 1$, and to find μ_S (the mean liability of affected parents), we refer to the table in Box C under the entry $B = .005$ (which corresponds to F_p) to find $t = 2.58$ and $z = .01446$. Since $\mu_S - \mu = \sigma z/B$ and $\mu = 0$ and $\sigma = 1$, we have $\mu_S = z/B = .01446/.005 = 2.89$.

The mean liability of sons of affected fathers is denoted μ'. If we slide the distribution μ' units to the left, the new mean will be 0 and the truncation point corresponding to $t = 2.58$ will become $t - \mu' = 2.58 - \mu'$. Reference to the table in Box C under the entry $B = .05$ (which corresponds to F_o) gives us the offspring truncation point, t', as $t' = 1.64$. Thus, since $t - \mu' = t'$, $2.58 - \mu' = 1.64$, so $\mu' = 2.58 - 1.64 = .94$.

The heritability of liability for pyloric stenosis based on father-son incidences is therefore $h^2 = 2\mu'/\mu_S = 2(.94)/(2.89) = 65$ percent.

a. For schizophrenia, the population incidence is about 1 percent, and the incidence of the condition among children of an affected parent is about 10 percent. Estimate h^2 of the liability toward schizophrenia.

b. What is the probability that a child of a parent who has a trait with a population incidence of 0.1 percent and a heritability of liability of 86.0 percent will also have the trait?

---

and that in 35 cases the co-twin is also affected. The concordance rate is then $35/100 = 35$ percent. From the concordance rates for monozygotic and dizygotic twins and the incidence of the trait in the population, the correlations in liability between monozygotic twins (r_{MZ}) and dizygotic twins (r_{DZ}) can be calculated (Figure 21). The broad-sense

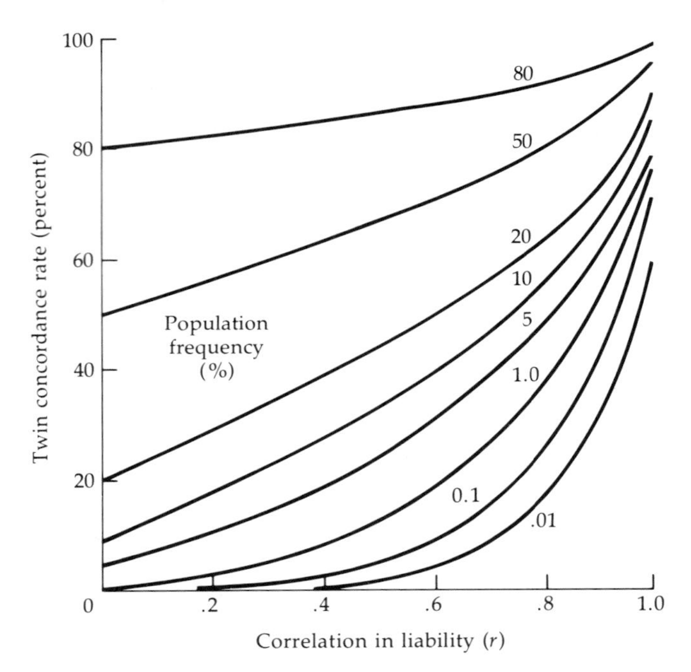

FIGURE 21. Theoretical concordance rates expected in monozygotic twins, against the correlation in liability toward the trait and the population incidence of the trait. (From Smith, 1975.)

heritability is then estimated by $2(r_{MZ} - r_{DZ})$, as discussed earlier. As with quantitative traits in general, twin data are more reliable if the twins are reared apart, but it is seldom possible in practice to obtain a sufficient number of such twin pairs. Figure 22 shows broad-sense heritabilities of various threshold traits in Caucasians as estimated from twin (T) or family (F) data.

FIGURE 22. Broad-sense heritabilities estimated for various threshold traits in humans. Ranges are given for some of the traits. Thus, for example, the population incidence of schizophrenia is one or two percent, and the broad-sense heritability estimates range from .70 to .80. Estimates based on family data are designated (F), those based on twins are designated (T). (Data from Smith, 1975.)

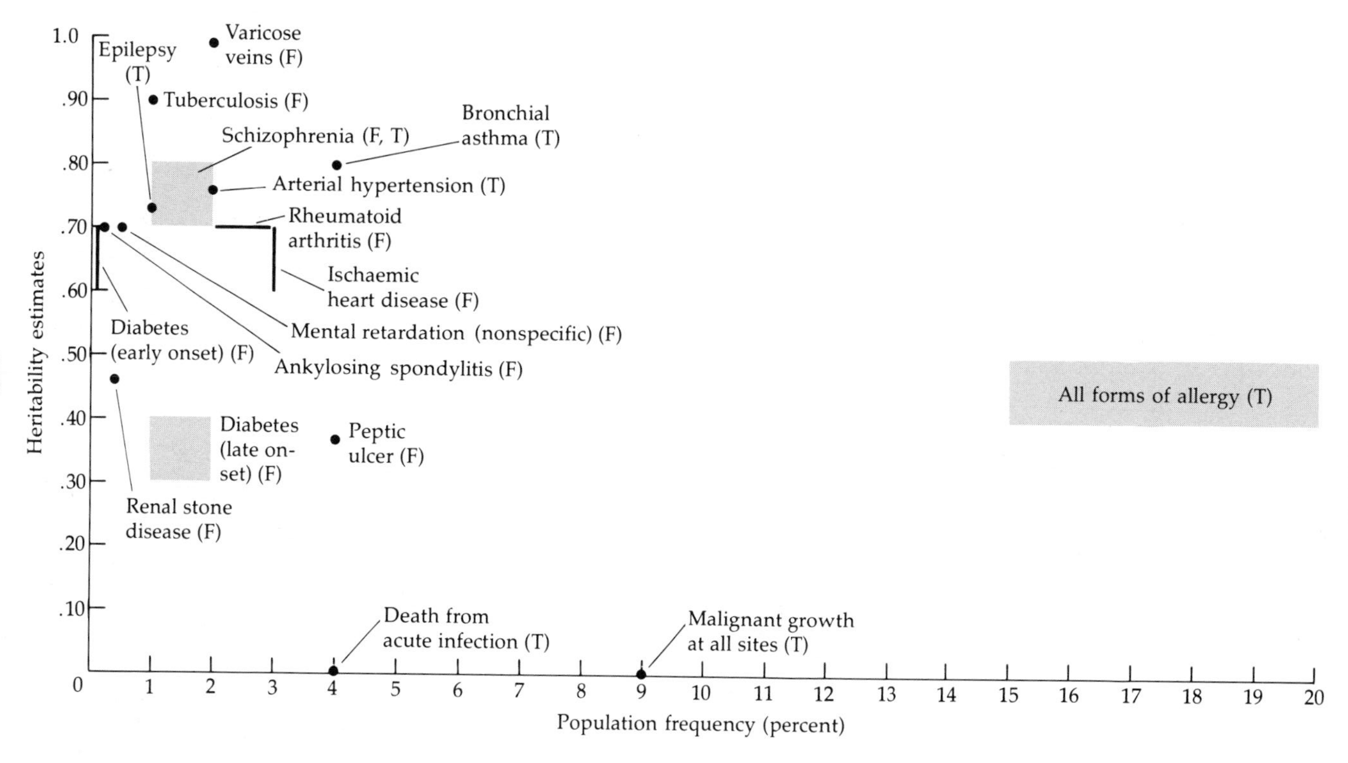
Heritability estimates
1.0
.90
.80
.70
.60
.50
.40
.30
.20
.10
0
1
2
3
4
5
6
7
8
9
10
11
12
13
14
15
16
17
18
19
20
Population frequency (percent)
Epilepsy (T)
Varicose veins (F)
Tuberculosis (F)
Schizophrenia (F, T)
Bronchial asthma (T)
Arterial hypertension (T)
Rheumatoid arthritis (F)
Ischaemic heart disease (F)
Mental retardation (nonspecific) (F)
Diabetes (early onset) (F)
Ankylosing spondylitis (F)
Diabetes (late on-set) (F)
Peptic ulcer (F)
Renal stone disease (F)
Death from acute infection (T)
Malignant growth at all sites (T)
All forms of allergy (T)

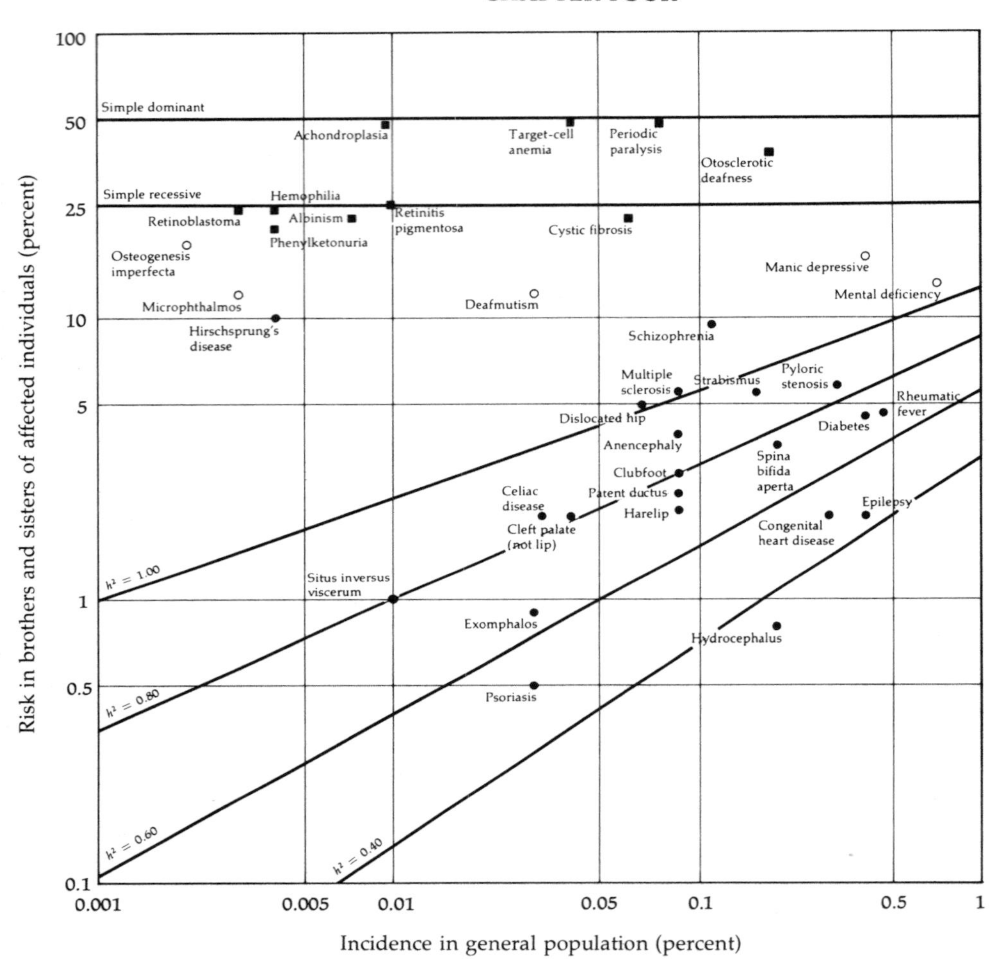

EMPIRICAL RECURRENCE RISKS

One of the uses of broad-sense heritability in threshold traits is the prediction of RECURRENCE RISK — the risk of a couple having an affected child when they have already had an affected one. Although such theoretical predictions are extremely useful for comparing different threshold traits, actual genetic counseling is often based on an EMPIRICAL RECURRENCE RISK — that is, on the recurrence risk actually

FIGURE 23. Empirical recurrence risks for various conditions in Caucasians, against population incidence. The heavy lines denote theoretical recurrence risks for simple dominant mutations, simple recessives, or quantitative threshold traits with the broad-sense heritabilities as noted. Traits with a recurrence risk of 20 percent or greater are represented by filled squares; traits with a recurrence risk of 10 percent or less are represented by filled circles. (From Hartl, 1977.)

observed in a large number of families. Such empirical recurrence risks for many of the most common conditions in Caucasians are shown in Figure 23. Generally speaking, conditions that are due to simple recessive or dominant alleles have high recurrence risks (25 or 50 percent), whereas most polygenic threshold traits have rather low recurrence risks (usually 10 percent or less). In Figure 23, high-risk conditions are represented by filled squares and low-risk conditions by filled circles.

As emphasized throughout this chapter, the principles of quantitative genetics are particularly useful in plant and animal breeding, and many of the examples were drawn from those fields. (For examples of alternative approaches to the theory and analysis of quantitative traits, see Morton and MacLean, 1974; Lande, 1976, 1977; Cavalli-Sforza and Feldman, 1976, 1978; Feldman and Cavalli-Sforza, 1976; and Reich et al., 1979.) In the next chapter, we turn again to natural populations, particularly to cases of evolution in action and to a consideration of the process of speciation in the light of population genetics.

Problems

1. From a population having a mean of 80 units for a quantitative trait, individuals having a mean of 110 were selected to be parents of the next generation; their offspring had a mean of 90. Calculate the realized heritability.
2. From a population having a mean of 140 units for a quantitative trait, individuals having a mean of 165 units were selected to be parents of the next generation. The narrow-sense heritability of the trait is 40 percent. What is the ex-

pected mean of the population in the next generation? What is the answer if the 165 applies only to females, the males being chosen at random from the entire population?

3. For a quantitative trait with a narrow-sense heritability of 12 percent in a population with mean 60, what is the expected mean of offspring when their parents' mean is 48?

4. If 25 is the mean of the offspring of parents having a mean of 35 for a quantitative trait with a narrow-sense heritability of 40 percent, what was the mean of the whole population in the parental generation?

5. A quantitative trait has a mean of 110 and a narrow-sense heritability of 25 percent. Calculate the mean of the population after five generations of selection with a cumulative selection differential of 80.

6. For a quantitative trait in a certain randomly mating population, the mean is 100, the variance is 240, and the phenotypic covariance between paternal half sibs is 15. Truncation selection is practiced with a selection differential of 32. What is the expected mean of the population in the next generation?

7. Suppose the phenotypic variance of a quantitative trait is 240 in a random-bred population and 190 in a genetically uniform population. Calculate the broad-sense heritability in the random-bred population.

8. If a trait has a broad-sense heritability of 70 percent, what are the maximum and minimum possible values for the narrow-sense heritability of the trait in that same population?

9. Jones and Smith are both poultry producers. The heritability of an index of carcass grade in Jones' flock is .24; in Smith's flock it is .12. Yet Jones and Smith use the same ration and housing procedures, and, indeed, the total phenotypic variance of the carcass-grade index in both flocks is 4. What is the additive genetic variance in Jones' flock? In Smith's?

10. Suppose that a quantitative trait in a population is measured in many individuals and found to be nonnormally distributed, but that the logarithm of the measurement is normally distributed. That is to say, the direct measurement, x, is nonnormal, but $y = \ln x$ is normal. For a locus with two alleles, suppose the mean phenotype of AA, AA', and $A'A'$ is $x = 21$, 15, and 9, respectively. Calculate μ^*, a, and d as

in Table II for x and $y = \ln x$. Which values are most appropriate to use in calculation of heritability, and why? (Note and hint: the transformations $y = \ln x$, $y = \sqrt{x}$, and $y = \arcsin\sqrt{x}$ are often useful in finding a scale of measurement in which the trait satisfies the assumption of being normally distributed; $y = \arcsin\sqrt{x}$ is particularly useful for transforming percentages, as in Table II.)

11. Consider a locus with genotypes AA, AA', $A'A'$ whose contribution to a quantitative trait has $a = .6$ and $d = .2$; consider also another locus with genotypes BB, BB', $B'B'$ whose contribution to the same trait has $a = .4$ and $d = 0$. Assuming that the loci are unlinked and that they contribute additively to the trait (i.e., no epistasis), calculate the heritability of the trait when the allele frequency of A is $p = .5$ and the allele frequency of B is $p = .7$. (Use the symbol σ^2 for the total phenotypic variance.)

12. For the overdominance model in Figure 15D, $\sigma_a^2 = 0$ when the allele frequencies at the locus are in equilibrium. Why?

13. For a locus with $p = .3$, $a = .10$, and $d = .08$, calculate σ_a^2, σ_d^2, and σ_g^2.

14. What would happen to allele frequencies at a locus after long-continued selection of a quantitative trait for which the locus has $a = .10$ and $d = .12$?

15. Suppose one has two inbred lines differing at n completely equivalent, unlinked, and additive loci (i.e., for n loci, one of the inbred lines is genetically AA, the other is $A'A'$, and for each locus $d = 0$). Suppose further that the phenotypic variance in the inbred lines is the same and equal to σ^2. The lines are crossed to produce an F_1 generation, and the F_1's are randomly mated to produce an F_2 generation. If the number of F_2 individuals is large enough that all genotypes are represented, then the difference between the extreme phenotypes will be $R = na$ [representing homozygosity for all n favorable alleles] minus $(-na)$ [representing homozygosity for all n unfavorable alleles] $= 2na$. The genotypes AA, AA', $A'A'$ in the F_2 are represented in the proportions 1/4 AA, 1/2 AA', and 1/4 $A'A'$ for each locus, and their respective contributions to phenotype (measured as deviations from the mean) will be a, 0 [because $d = 0$], and $-a$. Show that the phenotypic variance in the F_2 is $\sigma_2^2 = (na^2/2) + \sigma^2$. Then show that n can be estimated as $n = R^2/8(\sigma_2^2 - \sigma^2)$.

Further Readings

Allard, R. W. 1960. *Principles of Plant Breeding.* John Wiley and Sons, New York.

Brewbaker, J. 1964. *Agricultural Genetics.* Prentice-Hall, Englewood Cliffs, New Jersey.

Bulmer, M. G. 1980. *The Mathematical Theory of Quantitative Genetics*. Clarendon, Oxford.

Cavalli-Sforza, L. L. and W. F. Bodmer. 1971. *The Genetics of Human Populations.* W. H. Freeman, San Francisco.

Falconer, D. S. 1960. *Introduction to Quantitative Genetics.* Ronald Press, New York.

First World Congress on Genetics Applied to Livestock Production. 1974. Graficas Orbe, Madrid, Spain.

Haldane, J. B. S. et al. 1960. *Papers on Quantitative Genetics and Related Topics.* North Carolina State Univ. Press, Raleigh.

Hanson, W. D. and H. F. Robinson (eds.). 1963. *Statistical Genetics and Plant Breeding.* Natl. Acad. Sci.-Natl. Res. Council Publ. 982, Washington, D.C.

Hayes, H. K., F. R. Immer and D. C. Smith. 1955. *Methods of Plant Breeding.* McGraw-Hill, New York.

Johansson, I. and J. Rendel. 1968. *Genetics and Animal Breeding* (trans. by M. Taylor). W. H. Freeman, San Francisco.

Kempthorne, O. 1969. *An Introduction to Genetic Statistics.* Iowa State Univ. Press, Ames.

Lasley, J. 1978. *Genetics of Livestock Improvement,* 3rd ed. Prentice-Hall, Englewood Cliffs, New Jersey.

Lerner, I. M. 1958. *The Genetic Basis of Selection.* John Wiley and Sons, New York.

Lerner, I. M. and H. P. Donald. 1966. *Modern Developments in Animal Breeding.* Academic Press, New York.

Lush, J. L. 1945. *Animal Breeding Plans,* 3rd ed. Iowa State Univ. Press, Ames.

McClearn, G. E. and J. C. DeFries. 1973. *Introduction to Behavioral Genetics.* W. H. Freeman, San Francisco.

Mather, K. and J. L. Jinks. 1971. *Biometrical Genetics.* Cornell Univ. Press, Ithaca, New York.

Pirchner, F. 1969. *Population Genetics in Animal Breeding.* W. H. Freeman, San Francisco.

Pollak, E., O. Kempthorne and T. B. Bailey (eds.). 1977. *International Conference on Quantitative Genetics.* Iowa State Univ. Press, Ames.

Rendel, J. M. 1967. *Canalization and Gene Control.* Academic Press, New York.

Robertson, A. 1967. The nature of quantitative genetic variation. *In* Brink, R. A. (ed.). *Heritage from Mendel.* Univ. of Wisconsin Press, Madison.

Smith, C. 1975. Quantitative inheritance. *In* G. Fraser and O. Mayo (eds.). *Textbook of Human Genetics.* Blackwell, Oxford.

Thompson, J. M., Jr. and J. M. Thoday (eds.). 1979. *Quantitative Genetic Variation.* Academic Press, New York.

Turner, H. N. and S. S. Y. Young. 1969. *Quantitative Genetics in Sheep Breeding.* Cornell Univ. Press, Ithaca, New York.

Walden, D. B. (ed.). 1978. *Maize Breeding and Genetics.* John Wiley and Sons, New York.

Warwick, E. J. and J. E. LeGates. 1979. *Breeding and Improvement of Farm Animals,* 7th ed. McGraw Hill, New York.

Wright, S. 1958. *Systems of Mating and Other Papers.* Iowa State College Press, Ames.

·5·

Population Genetics, Ecology, and Evolution

The theme of this book has been genetic variation — its origin, maintenance, and significance. Chapter 1 dealt with the statistical rules of inheritance, the biochemical makeup of the gene, and the nature of genetic variation. Chapter 2 considered how genetic variation is organized into genotypes under such mating systems as random mating and inbreeding. Chapter 3 was concerned with the effects of the four principal evolutionary forces: selection, migration, mutation, and random genetic drift. Then Chapter 4 focused on the analysis of genetic variation affecting polygenic traits by examining the resemblance between relatives.

Up to now the approach has been analytic; each phenomenon discussed has been reduced to the simplest terms possible in order that a situation could be represented by a relatively uncomplicated mathematical model, but one still retaining a semblance of biological reasonableness. Examples, too, have been chosen because they illustrate specific points with minimal complications. This analytic approach, emphasizing simplification, can be extremely useful in that it permits precise definition and study of individual evolutionary forces and processes, uncomplicated by other phenomena. A danger of simplification, however, is that an overall view of the evolutionary process and its ecological setting may be lost. For example, the intensity of selection on genes influencing the manner in which a vertebrate predator searches for prey may depend on the number of competitors that hunt the same prey, and this number may vary from generation to generation due to relatively unpredictable factors, such as weather. Thus, any model in which the intensity of selection on such genes is assumed to be constant is an oversimplified model, one that ignores an essential element of the situation.

In this chapter, the approach is synthetic, integrated: an attempt to put the pieces of the previous chapters together into a coherent overview. First there is discussion of several general, comprehensive theories of the evolutionary process, which contrast in the roles and relative importance ascribed to such factors as a population's geographical structure, selection, migration, and random genetic drift. Then natural selection, as it may act on entire groups of individuals, is considered; this subject includes group selection, kin selection, and attendant issues in behavioral genetics and sociobiology. The third major subject in this chapter is evolution in action and is exemplified by several well-studied cases that depict the complexities brought about by the interaction of evolutionary forces in an ecological setting. Finally, we push to the frontiers of present knowledge and consider the contribution of population genetics to our current understanding of the grand macroevolutionary process of speciation.

The Shifting Balance Theory of Evolution

The major forces of evolution are undoubtedly natural selection, mutation, migration, and random genetic drift. Two questions about these forces inevitably arise: first, what roles do the various forces play in the overall evolutionary process, and, second, how important are the forces relative to one another? Neither of these questions has been answered satisfactorily at the present time. There are at least two broad views and many intermediate shades of opinion, but perhaps the most far-reaching hypothesis is the SHIFTING BALANCE THEORY of evolution, developed by Sewall Wright (1977, 1978, and earlier references cited in these books).

The shifting balance theory assumes four major premises concerning polymorphism, pleiotropy, the relation between genotype and fitness, and population structure. Each of these premises warrants a brief discussion.

POLYMORPHISM

The conventional definition of polymorphism was given in Chapter 2: polymorphic loci are those at which the most common allele has a frequency of less than .99. Widespread polymorphism is a key ingredient in the shifting balance theory because the theory assumes that genetic variation is nearly universal, affecting virtually all loci and every trait. Experimental documentation of the prevalence of genetic polymorphism has been outlined in Chapter 2, where it was emphasized that the 25 to 50 percent of loci found to be polymorphic by conventional electrophoresis may be an underestimate (although the point is highly uncertain — see the discussion in Chapter 2), and in Chapter 4, where the nearly universal occurrence of genetic variation affecting quantitative traits was pointed out. Natural populations do, in any case, seem to contain the substantial genetic variation required by the shifting balance theory.

PLEIOTROPY

Recall from Chapter 1 that pleiotropic alleles are alleles that affect more than one trait. In the shifting balance theory, it is assumed that most alleles have pleiotropic effects on fitness. The probable validity of this assumption is evidenced by the almost universal occurrence of pleiotropic effects that visible, laboratory-derived mutations have on such fitness components as viability or fertility and also by the frequency with which fitness components deteriorate as a correlated response in long-term directional selection programs for almost any specific characteristic. There is a question whether typical allozyme alleles at polymorphic loci have effects on fitness (this is precisely what the selection *versus* neutrality controversy in population genetics is all about), but it is clear, in any case, that pleiotropy is a common feature of gene action. The premise of pleiotropy in the shifting balance theory therefore has experimental support.

RELATION BETWEEN GENOTYPE AND FITNESS

The genetic underpinning of fitness is extremely complex, involving alleles at many loci that influence fitness in two ways — one through the main effect of each locus (which typically may be an effect on some quantitative trait such as body size, growth rate, or developmental time) and the other through pleiotropic effects on other traits related to fitness.

It is often the case that quantitative traits have intermediate optima for fitness; that is to say, individuals having an intermediate phenotype of the quantitative trait will have the highest fitness. This concept of an intermediate optimum is illustrated in Table I for two loci, each with two alleles. (In an actual case there would, of course, be many more loci and many more alleles.) The second column in Table I gives the zygotic frequencies of the nine genotypes expected under random mating and linkage equilibrium (see Chapter 2); p_1 and q_1 represent the allele frequencies of *A* and *a*, p_2 and q_2 those of *B* and *b*. The third column gives the phenotypic measurement of a quantitative trait in arbitrary units, which may represent, for example, the deviation of each genotype from a standard value in centimeters (for height), kilograms (for weight), or weeks (for developmental time). The model of gene action assumed is extremely simple — each "favorable" allele (*A* or *B*) in the genotype adds one unit to the measurement of the quantitative trait.

The contribution of these two loci to fitness through their main effects on the quantitative trait, assuming an intermediate optimum, is shown in the fourth column. Here we assume that a phenotype of 2 represents the optimum phenotype for the quantitative trait and that each unit of phenotype above or below the optimum subtracts a constant quantity (α) from the maximum fitness.

Pleiotropic effects on fitness, which arise from secondary effects of the alleles on traits other than the trait in question, are illustrated in the fifth column. In this case we assume that, because of pleiotropy, each *A* allele in a genotype adds an increment (ϵ) to fitness. The overall fitness of each of the nine genotypes is shown in the last column of Table I, calculated for the case α = .2 and ϵ = .05.

Any correspondence between genotype and fitness such as that exemplified in Table I implies that the fitness of a genotype at a locus will depend on allele frequencies at other loci. This frequency-dependence of fitness is seen, for the example of Table I, in Figure 1, which shows the fitnesses of *BB*, *Bb*, and *bb* as functions of p_1, the allele frequency of *A*. (See Box A for the manner of calculation.) For the range

TABLE I. Two-locus model with intermediate optimum for fitness.

Genotype	Frequency	Phenotype of quantitative trait (arbitrary units)	Fitness: Contribution due to deviation from optimum		Fitness: Contribution due to pleiotropy	Total
AABB	$p_1^2p_2^2$	4	$1 - 2\alpha$	+	2ϵ	= .70
AABb	$p_1^2(2p_2q_2)$	3	$1 - \alpha$	+	2ϵ	= .90
AAbb	$p_1^2q_2^2$	2	1	+	2ϵ	= 1.10
AaBB	$(2p_1q_1)p_2^2$	3	$1 - \alpha$	+	ϵ	= .85
AaBb	$(2p_1q_1)(2p_2q_2)$	2	1	+	ϵ	= 1.05
Aabb	$(2p_1q_1)q_2^2$	1	$1 - \alpha$	+	ϵ	= .85
aaBB	$q_1^2p_2^2$	2	1			= 1.00
aaBb	$q_1^2(2p_2q_2)$	1	$1 - \alpha$			= .80
aabb	$q_1^2q_2^2$	0	$1 - 2\alpha$			= .60

The phenotype for a quantitative trait is assumed to depend on the additive contributions of two loci, one phenotypic unit being added for each *A* or *B* allele in the genotype. Fitness is assumed to be maximum for the intermediate value of the quantitative trait and to decrease by an amount (α) for every unit that the phenotype deviates from the optimum. The *A* allele is also assumed to have a pleiotropic effect of fitness due to its contribution to other fitness-related traits; in this case, the pleiotropic effect is the addition of an amount (ϵ) to fitness for each *A* allele in the genotype. The numerical values in the last column have been calculated for $\alpha = .2$ and $\epsilon = .05$.

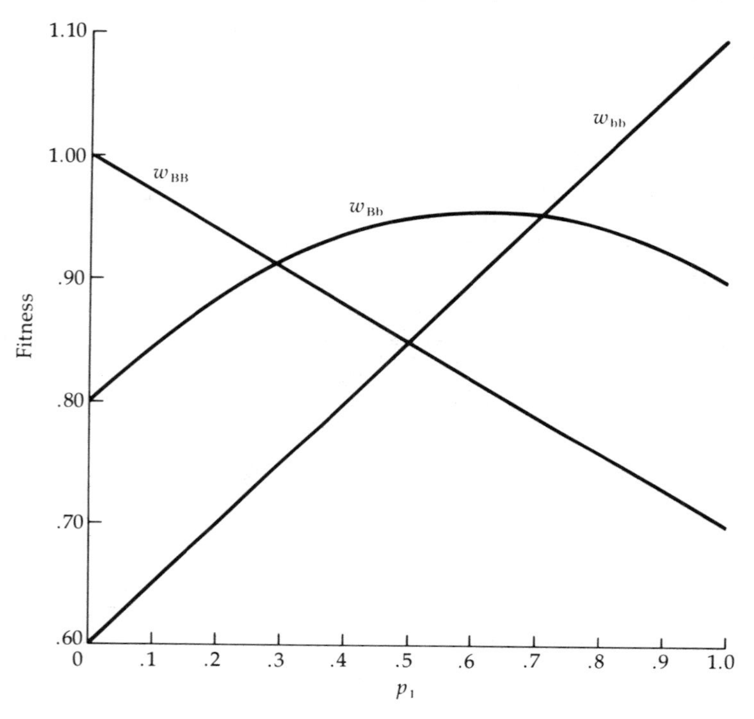

FIGURE 1. Fitnesses of *BB, Bb,* and *bb* genotypes for the model in Table I, against p_1, the allele frequency of *A*. Formulas for w_{BB}, w_{Bb}, and w_{bb} are derived in Box A.

$p_1 = 0$ to $p_1 = .293$, the fitness relationships at the *B* locus are $w_{bb} < w_{Bb} < w_{BB}$; for $p_1 = .293$ to $.500$, we have $w_{bb} < w_{BB} < w_{Bb}$; for $p_1 = .500$ to $.707$, the relationships are $w_{BB} < w_{bb} < w_{Bb}$; and for $p_1 = .707$ to 1.0, $w_{BB} < w_{Bb} < w_{bb}$. Thus, unless the allele frequency at the *A* locus is specified, one cannot even say which genotype at the *B* locus has the highest fitness.

For any values of p_1 and p_2 (remember that $q_1 = 1 - p_1$ and $q_2 = 1 - p_2$), the average fitness of a population ($\overline{w}$) can be calculated from Table I as the sum of the crossproducts of the second and sixth columns; that is, $\overline{w} = p_1^2p_2^2(.70) + p_1^2(2p_2q_2)(.90) + p_1^2q_2^2(1.10) + \cdots$. Calculations of $\overline{w}$ over the whole range of p_1 and p_2 result in a fitness surface called an

A A Type of Frequency-Dependent Selection

Here we consider the model of Table I in order to show that the fitnesses of genotypes at one locus depend on the allele frequencies at the other locus. (This is only one type of frequency-dependent selection; in other models, the fitnesses of genotypes at a locus may depend on the allele frequencies at that same locus.)

For genotype *BB*, say, the fitness is the average fitness of *AABB*, *AaBB*, and *aaBB*, which is

(1) $w_{BB} = p_1^2(1 - 2\alpha + 2\epsilon) + 2p_1q_1(1 - \alpha + \epsilon) + q_1^2$

Similarly,

(2) $w_{Bb} = p_1^2(1 - \alpha + 2\epsilon) + 2p_1q_1(1 + \epsilon) + q_1^2(1 - \alpha)$

(3) $w_{bb} = p_1^2(1 + 2\epsilon) + 2p_1q_1(1 - \alpha + \epsilon) + q_1^2(1 - 2\alpha)$.

Show that

a. $w_{BB} = 1 - 2p_1(\alpha - \epsilon)$

b. $w_{Bb} = 1 - (p_1^2 + q_1^2)\alpha + 2p_1\epsilon$

c. $w_{bb} = 1 - 2q_1\alpha + 2p_1\epsilon$

d. $w_{AA} = 1 + 2\epsilon - 2p_2\alpha$

e. $w_{Aa} = 1 + \epsilon - (p_2^2 + q_2^2)\alpha$

f. $w_{aa} = 1 - 2q_2\alpha$

g. $\overline{w} = p_2^2w_{BB} + 2p_2q_2w_{Bb} + q_2^2w_{bb} = p_1^2w_{AA} + 2p_1q_1w_{Aa} + q_1^2w_{aa} = 1 - 2\alpha[1 - p_1 - p_2 + 2p_1p_2(p_1 + p_2 - p_1p_2)] + 2p_1\epsilon$.

As emphasized in the text, it is this sort of interaction between loci that leads to the complex kinds of adaptive topographies exemplified in Figure 2. Such complex adaptive topographies are central to the shifting balance theory of evolution.

ADAPTIVE TOPOGRAPHY, as shown in Figure 2. Notice that there are two "peaks" of fitness (corresponding to fixation of *AAbb* or *aaBB*), two "pits" of fitness (corresponding to fixation of *AABB* or *aabb*), and a "saddle" at approximately $p_1 = .45$, $p_2 = .65$. Although, in this particular illustration, the peaks and pits correspond to fixation and the saddle to a two-locus polymorphism, in other models the peaks and pits could well occur at interior points on the surface, not at the corners. In any case, the shifting balance theory assumes that, because of intermediate optima and pleiotropy, adaptive topographies will typically exhibit multiple peaks, pits, and saddles, as illustrated in the simplified example in Figure 2. (For more discussion and examples of complex adaptive topographies, see Lewontin and White, 1960; Moran, 1964; Wright, 1967; Nassar, 1972; Turner, 1972; and Semeonoff, 1977.)

The evolutionary consequences of the adaptive topography in Figure 2 are apparent in Figure 3, which depicts the fitness surface in two rather different ways. Figure 3A is a contour map showing the FITNESS ISOCLINES (curves that connect points corresponding to populations that have equal average fitness). Under the influence of natural selection, allele frequencies will tend to change in such a way as to increase average fitness; that is to say, populations will evolve toward isoclines of higher fitness. The trajectories along which allele frequencies change under natural selec-

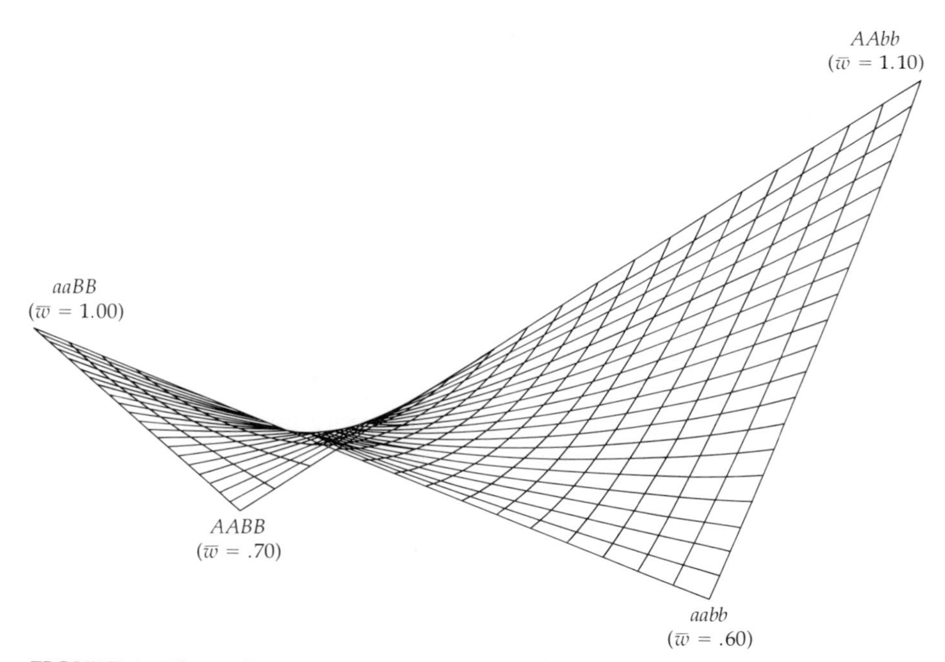

FIGURE 2. Three-dimensional fitness surface (adaptive topography) for the model in Table I. The corners of the surface represent the homozygous populations shown. Points on the fitness surface give the average fitness of populations that are polymorphic for (*A, a*) or (*B, b*) or both. Note the "saddle" on the fitness surface, which indicates an unstable equilibrium.

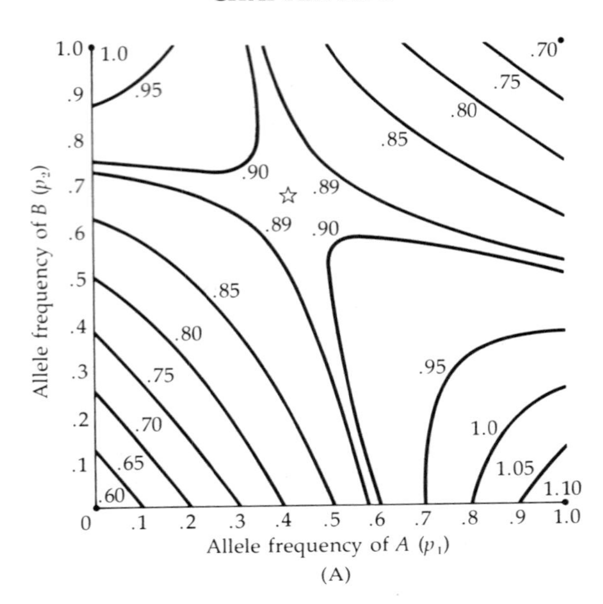

FIGURE 3. (A) Alternative representation of fitness surface in Figure 2. Each curve connects allele frequencies in populations that have the same average fitness, indicated for each curve. (Such curves are called fitness isoclines.) The region marked by the star corresponds to the "saddle" in Figure 2. (B) Changes in allele frequency for the two-locus model in Table I. The base of each arrow represents presumed allele frequencies of *A* and

tion are illustrated in Figure 3B. Note that populations in the region of the lower right quadrant of Figure 3B will evolve toward fixation of *AAbb* and ultimately become fixed at that corner, where the average fitness is $\overline{w} = 1.10$; on the other hand, populations near the upper left corner of Figure 3B will evolve toward the *aaBB* corner and eventually become fixed there, where the average fitness is $\overline{w} = 1.00$. The two fitness peaks at fixation of *AAbb* and *aaBB* have distinct domains over which they attract populations, and this is a central feature of the shifting balance hypothesis, for the genotypes corresponding to the fitness peaks can be termed "balanced" genotypes. One could as well convey the essential situation in Figures 2 and 3 by saying that genotype *AA* interacts "harmoniously" with *bb* (this symphonic metaphor

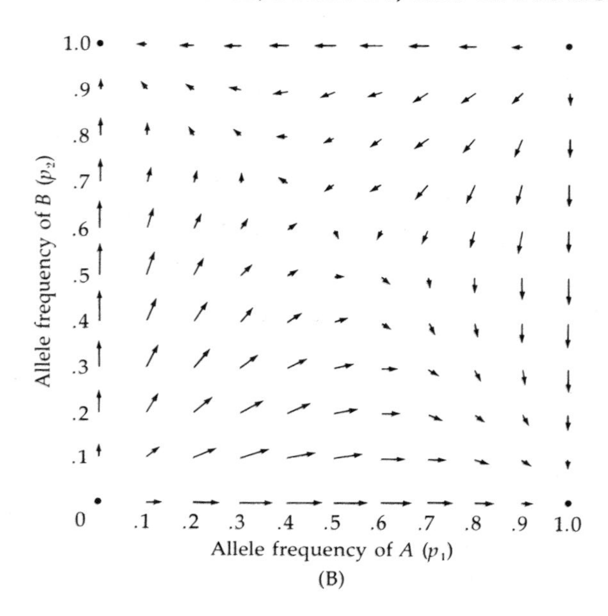

B, and the head of the arrow shows how these frequencies would change after one generation of selection (assuming linkage equilibrium). The scale is the same as in part (A). Note how populations to the upper left of the "saddle" evolve to the *aaBB* corner; those to the lower right of the "saddle" evolve to the *AAbb* corner.

is rather common in evolutionary biology) or that the *AAbb* genotype is "balanced"; in the same sense, one could say that the *aaBB* genotype is "balanced."

In terms of average fitness, of course, a population would be better off to be fixed for *AAbb* than to be fixed for *aaBB*. Nevertheless, any population in the domain of attraction of *aaBB* will become fixed for *aaBB* because to evolve in any other way would entail a temporary reduction in average fitness, and such a reduction cannot occur by natural selection in this sort of situation. Thus, the significant implication of a complex adaptive topography with many fitness peaks is that a population may very well evolve toward a peak that, while higher than its immediate surroundings on the fitness surface, is not the highest that exists. Any population on a submaximal fitness peak is destined to stay there because natural selection will not carry the population down into any of the surrounding valleys and thus, perhaps, into the domain of attraction of a higher fitness peak. The population is simply "stuck."

There are several ways out of the dilemma for populations "stranded" on submaximal fitness peaks. The environment may change, for example, causing changes in fitness of the various genotypes and a consequent alteration in the adaptive topography surface. What was a peak may become a valley and any population in such a valley will evolve; if, after awhile, the environment should revert back to what it was originally, the population may have evolved far enough from where it was originally to find itself in the domain of attraction of a higher fitness peak. To assume that such fortuitous changes of environment will regularly occur is, of course, overoptimistic. There is, however, another process by which a population stranded on a submaximal fitness peak can move into a nearby valley — random genetic drift; it is in this context that the shifting balance theory ascribes a special and very important evolutionary role to chance changes in allele frequency caused by random genetic drift (Wright, 1970).

POPULATION STRUCTURE

As stated above, random genetic drift is a fundamental process in the shifting balance hypothesis because random drift can shift the constellation of allele frequencies from a submaximal fitness peak into a nearby valley; from there, natural selection can predominate and carry the population to a higher fitness peak. As emphasized in Chapter 3, random genetic drift increases in importance as the effective population number decreases, so a key assumption in the shifting balance hypothesis is that effective population numbers are sufficiently small that appreciable random changes in allele frequency can occur; that is to say, effective population numbers should be on the order of 100 to 200, or less, rather than 500 or more.

In species that are relatively rare and occupy a limited

geographic range with low population density, it is likely that effective numbers are sufficiently small for random genetic drift to play the role ascribed to it. The situation with regard to abundant species spread over a large geographic area is by no means clear. As mentioned in Chapter 1, most species have a kind of clustered distribution brought about by geographical or ecological barriers or by the limited range of movement of the organisms themselves (or their pollen in the case of plants); each local population or DEME constitutes a locally interbreeding group that, depending on the magnitude of migration, is more or less isolated from other such groups. The effective population number in demes is the primary issue in evaluating the evolutionary role of random genetic drift.

Effective deme sizes are by no means uniform in a species. Local populations at the edges of a species' range are smaller than those in the center, and a deme can be reduced drastically in numbers by disease, predation, competition, or inhospitable weather. Even in abundant species with a large average effective deme size, such bottlenecks of population size can reduce the effective number in a deme far below the average and thus provide the opportunity for significant random genetic drift. In *Drosophila melanogaster,* for example, the effective deme size in the summer months is at least 1000 and sometimes much larger (Mukai, 1977; Choi, 1978); in the winter months, when population density drops precipitously, the effective number is surely much smaller (Begon, 1977). In the prairie deer mouse, *Peromyscus maniculatus,* in southern Michigan, effective deme sizes are in the range 80 to 120 (Dice and Howard, 1951; discussed in Wright, 1978); and the effective number for the prairie flower *Phlox pilosa* in Illinois is about 400 (Levin and Kerster, 1968). These numbers actually refer to "neighborhood size," estimated as $N = 4\pi\sigma^2 d$ where σ^2 is the variance in the distance between the birthplaces of parents and offspring and d is population density. As discussed in Box F of Chapter 3, the neighborhood size is thought to be the number most strictly comparable to deme size when a species is distributed relatively evenly in space. These representative estimates of neighborhood size, which pertain to very abundant species, suggest that random genetic drift can play the role envisaged in the shifting balance hypothesis, especially when geographic, seasonal, and accidental bottlenecks are taken into account.

SUMMARY OF THE SHIFTING BALANCE PROCESS

Having discussed the assumptions underlying the shifting balance hypothesis, it is appropriate to summarize the overall process in terms of its three major phases. There is first the phase of random genetic drift, during which local populations randomly "explore" their nearby surroundings on

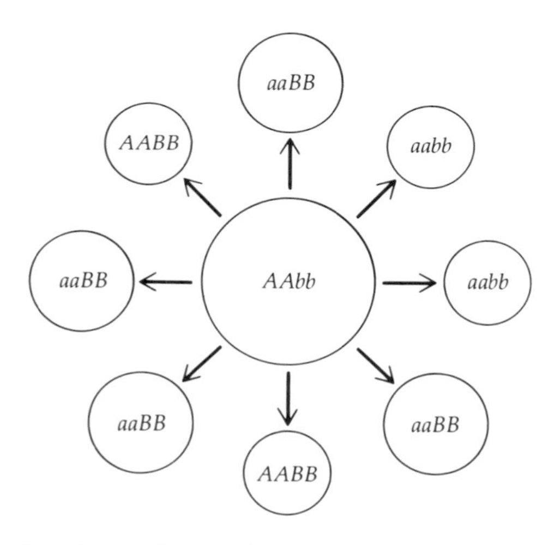

FIGURE 4. Role of interdeme selection in the shifting balance theory. The circles represent populations fixed for the alleles shown, and their differing sizes represent differing average fitnesses, based on the model in Table I. In the center is a population that, by chance, has crossed the fitness "saddle" and, by mass selection, has become fixed for *A* and *b*. This population will expand in size and so, by the pressure of population density, send out substantial numbers of emigrants to the surrounding populations (arrows). With sufficient immigration into the surrounding populations, their allele frequencies will become so altered that these populations, too, will cross the fitness saddle and eventually become fixed for *A* and *b*. The figure artificially separates fixation of *A* and *b* in the central population from the emigration phase; both processes, of course, occur simultaneously.

the adaptive topography, occasionally, by chance, producing a new balanced genotype in sufficient frequency that the population enters the domain of attraction of a higher fitness peak. There follows the phase of MASS SELECTION in which a peak-shift occurs; natural selection in favor of the new balanced genotype overwhelms the effects of random drift and carries the population to the summit of a higher fitness peak.

Accompanying the phase of mass selection, or subsequent to it, a third phase of evolution occurs, namely INTERDEME SELECTION. In the phase of interdeme selection, the population attaining the higher fitness peak increases in numbers and by excess dispersion shifts the allele frequencies in nearby local populations until they, too, have crossed the valley and come under the influence of the higher fitness peak; this process spreads the favorable genotype throughout the species in ever-widening concentric circles. Should the region of spread from two such centers overlap, a new and yet more favorable interaction system may arise in the region of overlap and itself become a center for interdeme selection (Figure 4). In this manner, virtually the whole of the adaptive topography can be explored, and there is a continual shifting of control by one adaptive peak to control by a superior one.

Alternatives to the Shifting Balance Theory

The shifting balance theory is Wright's bold attempt to synthesize the major forces of evolution into a coherent picture that ascribes to each force a distinct evolutionary role and importance. We do not know that the shifting balance hypothesis is true, only that it could be true, and it therefore becomes necessary to examine some of the possible alternatives. We here focus on contemporary alternatives — as opposed to theories popular in the early years of the twentieth century that imagined, for example, a major feature of evolution to be some sort of internal driving and guiding force in species or that emphasized major visible mutations ("hopeful monsters") as being of prime importance. (See Wright, 1977, for an excellent discussion of modern alternatives to the shifting balance theory.)

SELECTIVE NEUTRALITY

One alternative to the shifting balance theory arises from doubts about the evolutionary relevance of polymorphism, taking the point of view that most polymorphisms observed are selectively neutral or nearly neutral. This hypothesis, which assumes that the alleles at polymorphic loci have no effects on fitness, is the hypothesis of selective neutrality discussed earlier in Chapter 3. According to the hypothesis of selective neutrality, random genetic drift accounts for the constellations of allele frequencies at polymorphic loci, and real evolutionary advance in the sense of increased adaptation depends on the occurrence of rare favorable mutations. Because favorable mutations are assumed to be rare, much of the natural selection that usually occurs in a species is a kind of cleansing or purifying action that serves mainly to keep recurrent deleterious mutations at low frequencies.

SINGLE SELECTIVE PEAKS

Another alternative to the shifting balance model is the supposition that the relationship between genotype and fitness

TABLE II. Two-locus model with additive contributions to fitness.

Genotype	Frequency	Fitness[a]
AABB	$p_1^2p_2^2$	$1 + 4\alpha = 2.00$
AABb	$p_1^2(2p_2q_2)$	$1 + 3\alpha = 1.75$
AAbb	$p_1^2q_2^2$	$1 + 2\alpha = 1.50$
AaBB	$(2p_1q_1)p_2^2$	$1 + 3\alpha = 1.75$
AaBb	$(2p_1q_1)(2p_2q_2)$	$1 + 2\alpha = 1.50$
Aabb	$(2p_1q_1)q_2^2$	$1 + \alpha = 1.25$
aaBB	$q_1^2p_2^2$	$1 + 2\alpha = 1.50$
aaBb	$q_1^2(2p_2q_2)$	$1 + \alpha = 1.25$
aabb	$q_1^2q_2^2$	$1 = 1$

[a] Each *A* or *B* allele adds an amount $\alpha = .25$ to fitness.

is more like that in Table II than that in Table I. The adaptive topography pertaining to Table II has a single peak (Figure 5A), and the trajectories of allele frequency change all lead

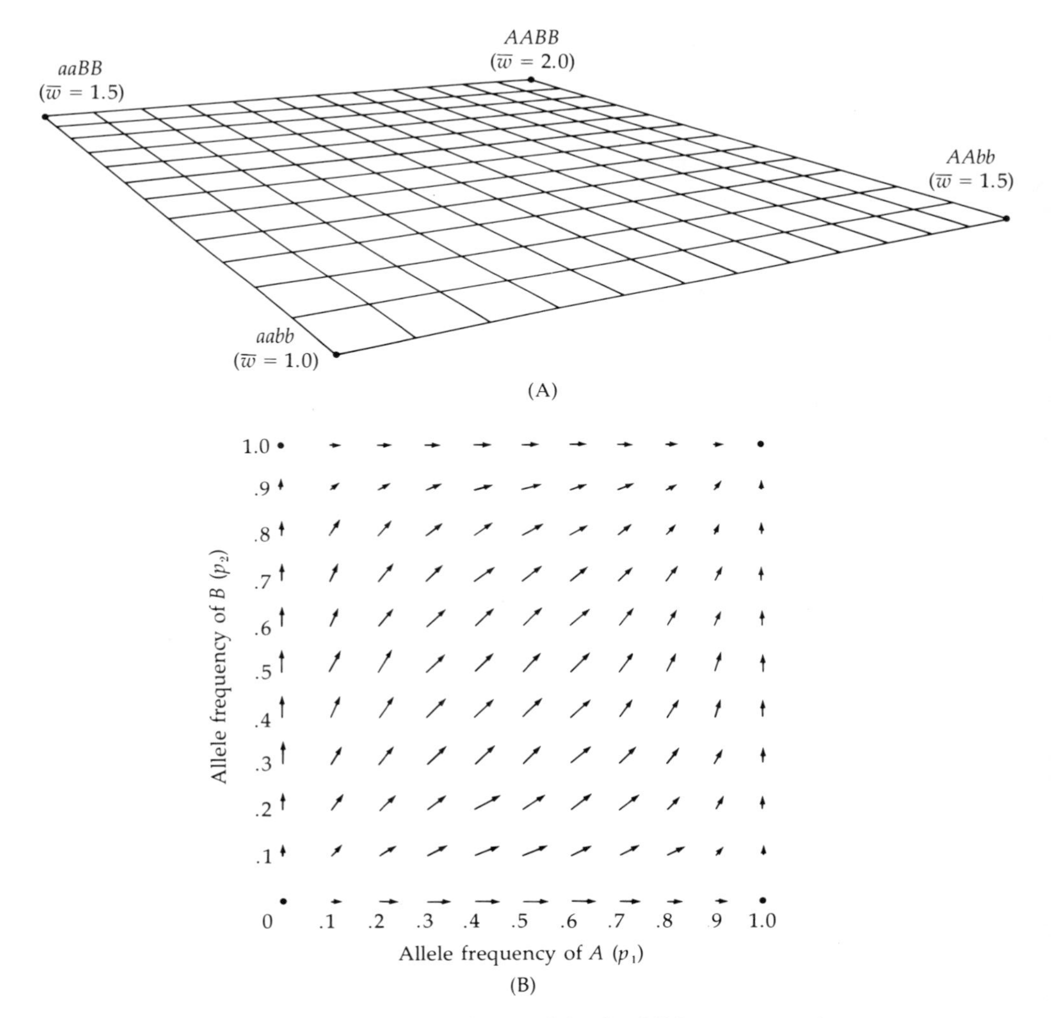

FIGURE 5. (A) Fitness surface for the model of additive gene action shown in Table II. The fitness surface is a plane tilted upward at the *AABB* corner; there are no fitness "saddles." (B) Changes in allele frequency in one generation of selection for the model in Table II, represented as in Figure 3B. All populations segregating for (A, a) and (B, b) will evolve to fixation at the *AABB* corner.

to that one peak (Figure 5B). (Although in this example the fitness peak entails fixation of *AABB*, in other examples the peak could involve polymorphism of one or both loci.) When the adaptive topography is dominated by one fitness peak, as in Figure 5, the geographical structure of the population becomes irrelevant to evolutionary advance because all populations tend to evolve toward that peak irrespective of random genetic drift.

MASS SELECTION

A third alternative to the shifting balance theory is based on doubts about the premise pertaining to population structure. The assumption made, instead, is that effective deme sizes are so large or the amount of migration between demes so extensive that random drift becomes an insignificant process, one able to alter allele frequencies only trivially. The contrast between this model and the shifting balance hypothesis is depicted in Figure 6, where the size of the circles represents deme size and the width of the migration corridors represents the amount of migration. The overall geographical structure of the population is the same in Figure 6A (shifting balance model) and Figure 6B (mass selection model), except that in the latter case there can arise little random differentiation among demes. (Chapter 3 contains calculations showing what a small amount of migration is required to effectively prevent random genetic differentiation of demes.) In Figure 6B, all demes evolve in unison as if they formed a single, large, essentially unstructured population.

In the case of mass selection, natural selection always predominates, whether the adaptive topography has a single peak or many. Because demes are expected to undergo little genetic differentiation except that caused by their response to differing local selective conditions, interdeme selection becomes a process of little importance.

Overview of Evolutionary Theories

We do not know which of the possible modes of evolution are the most important in actual evolving populations. The

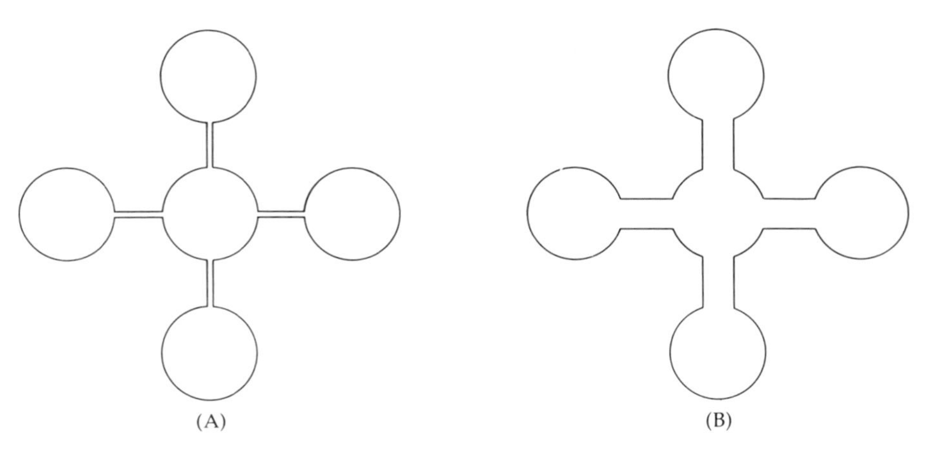

FIGURE 6. The amount of migration between populations is critical in the shifting balance theory. In (A), the amount of migration (represented by the "corridors" connecting the circular populations) is assumed to be sufficiently small that substantial genetic differentiation between demes can occur, some evolving toward certain fitness peaks and others to different peaks; each population, itself, is sufficiently small that considerable "exploration" of the fitness surface by random genetic drift can occur. In (B), the amount of migration (indicated by the corridor width) is so large that local genetic differentiation is swamped; the entire group of populations behaves effectively as a single, randomly mating unit, and random genetic drift plays no significant role. The critical amount of migration determining whether (A) or (B) is the most appropriate model for an actual population is not known. It depends on population structure, patterns of migration, and other factors, but it is nevertheless quite small, on the order of a few individuals or less per generation.

shifting balance hypothesis has been examined most extensively because it is the most complex and comprehensive theory and because the other theories are conveniently discussed as contrasts to the shifting balance hypothesis. Perhaps some species evolve predominantly according to one model of evolution and other species evolve according to another model; or perhaps during certain periods of time a species evolves in accordance with one hypothesis and at other times according to another. Critical information from a wide range of plant and animal species pertaining to the relationships between genotype and fitness, effective numbers of individuals within demes, and migration rates is needed to decide the issue. At the present time, such crucial information is lacking.

Ecological Opportunity

It is important to point out that, judged on the basis of the fossil record, observed rates of evolution vary enormously. Different groups of organisms are observed to evolve at very different rates; a single group of organisms can evolve more rapidly at one time and less rapidly at another; and different traits can evolve at different rates even within a single group of organisms over a given period. (Figure 7 provides an example of variation in the rate of evolution in lungfish.) Such variation in evolutionary rates implies that the limiting factor is not the actual mode of evolution — whether evolution proceeds mainly by the shifting balance process, the mass selection process, or whatever. The rate of evolution is limited by something else, and this something else can be called ECOLOGICAL OPPORTUNITY.

Most species, most of the time, are constrained by their physical environment and by other organisms. (In this context, the word "constraint" means "limitation.") Physical constraints include, for example, seasonal weather; the vast majority of plants and animals find winter inhospitable and have evolved behavioral or physiological adaptations to survive the cold times. Actual physical barriers may also be physical constraints; for fish and water lilies, for example, the shoreline of a lake is such a barrier. Physical constraints also include availability of essential nutrients in soil or water.

Biological constraints are those associated with the presence of other organisms. Predation of the young is a limiting factor in fish population growth; excess shade caused by a forest canopy limits sunlight available for germination and growth of other green plants in the undergrowth; infectious disease often decimates populations that have become too dense.

The evolution of most species, most of the time, merely keeps pace with their ever changing physical and biological constraints. Such offsetting evolutionary changes are often

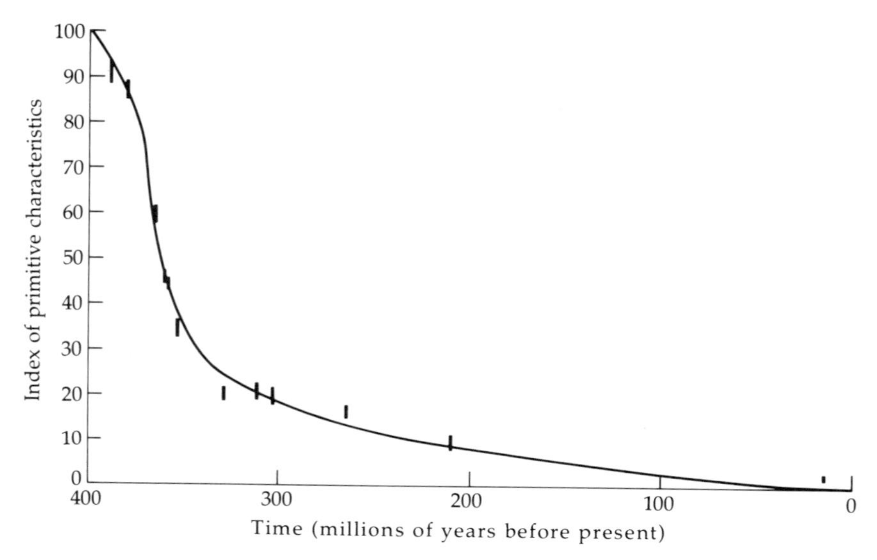

FIGURE 7. Variation in rates of evolution in lungfish (*Dipnoi*). The ordinate is an index of the number of primitive characteristics exhibited by the organisms, and the vertical lines on the graph show the range in this index for various lungfish genera existing at the times shown. Note the extremely rapid rate of evolution about 375 to 350 million years ago. (After Westoll, 1949.)

called "treadmill evolution"; more colorfully this situation is sometimes called the "Red Queen" hypothesis after the Queen of Hearts in Lewis Carroll's *Through the Looking Glass,* who tells Alice, "Now, *here,* you see, it takes all the running you can do, to keep in the same place" (Van Valen, 1973). Note that a constraint must be changing — intensifying — for treadmill evolution to occur; a steady-state constraint may eventually be overcome by adaptations resulting from natural selection, provided that the necessary genetic variation becomes available through mutation.

Occasionally an opportunity for rapid evolution becomes available. A geographical constraint may disappear or be surmounted, as by the chance colonization of a new favorable habitat such as an island. Or a biological constraint may disappear, as when a major predator, competitor, or parasite becomes extinct. Or, in less striking situations, the rate of intensification of a constraint may ease to a level insufficient to balance natural selection. Removal or easing of such constraints partly releases a species from its treadmill, and the species becomes free to evolve so as to expand its ecological niche or even evolve into new niches. In extreme cases, evolution can be very rapid and result in a spectacular adaptive radiation of many newly formed species that occupy available niches; one exceptionally well-studied case — the explosive speciation of *Drosophilia* flies in the Hawaiian islands — resulted in the evolution of at least 300 endemic species in the several million years since the formation of the archipelago by volcanic action. (See Carson et al., 1970; Richardson, 1974; Carson and Kaneshiro, 1976; Templeton, 1979.) Thus, most species have ample reserves of genetic variation to enable rapid evolution in new directions; their rate of evolution is limited by the absence of ecological opportunity. (Box B gives some simple genetic implications of the widely used Lotka-Volterra model of competition between populations.)

Selection Acting on Entire Populations

Interdeme selection, discussed in connection with the shifting balance theory, is only one of several kinds of natural selection that involve entire populations rather than single individuals. The purpose of this section is to define and discuss varieties of group selection that can occur, or that have been postulated to occur. It is convenient to discuss these modes of selection in contrast to individual selection.

INDIVIDUAL SELECTION

The term INDIVIDUAL SELECTION refers to natural selection that operates on individual phenotypes and not on properties of groups. Individual selection is the usual form of selection envisaged in most of population genetics and evolutionary biology. In considering individual selection, we imagine that each individual in a population has a certain phenotype that determines its fitness and that the survival and reproduction of each individual is determined solely by

Intensity of Selection for Ecological Parameters r, K, and α

B

Here we combine simple theories of quantitative traits from Chapter 4 with models of population growth from Chapter 1 to derive some approximations for the intensity of selection on traits of interest to evolutionary ecologists.

Consider a locus with two alleles that affects a trait, x, as well as fitness, w. From Box B of Chapter 4, we can write $\Delta p = i_w pq\alpha_w/\sigma_w$, where the subscript w refers to fitness. Moreover, in Box E of Chapter 4 we showed that, for fitness, $i_w = \sigma_w/\overline{w}$; thus we have $\Delta p = pq\alpha_w/\overline{w}$.

Now, since allele frequencies change because of natural selection for fitness, the trait x will change. Indeed, were x changing as a direct response to selection rather than as a correlated response to selection on another trait, we could write $\Delta p = i_x pq\alpha_x/\sigma_x$. Setting this expression for Δp equal to the one in the paragraph above gives us a measure of i_x — the "effective" intensity of selection on the trait x. Equating the two expressions for Δp produces $i_x = (\sigma_x\alpha_w)/(\overline{w}\alpha_x)$.

If $w = f(x)$ is the equation relating w and x, and if there is not too much variation among the x's and w's, then it can be shown that, approximately

$$\alpha_w = \left(\frac{df(x)}{dx}\right)\alpha_x, \text{ where } \frac{df(x)}{dx}$$

is the first derivative and is to be evaluated at $x = \bar{x}$. Furthermore, it can also be shown that

$$\frac{df(x)}{\overline{w}dx} = \frac{d\ln f(x)}{dx}$$

approximately, which we can write as $d\ln\overline{w}/dx$. Substituting the expression for α_w into that for i_x and replacing $df(x)/\overline{w}dx$ with $d\ln\overline{w}/dx$ yields $i_x = \sigma_x(d\ln\overline{w}/dx)$.

The change in x in one generation is $R_x = i_x\sigma_x h_x^2$ (see Box C of Chapter 4). Substituting for i_x and noting $h_x^2 = \sigma_{ax}^2/\sigma_x^2$ yields

$$R_x = \left(\frac{d\ln\overline{w}}{dx}\right)\sigma_{ax}^2$$

(If $d\ln\overline{w}/dx$ is negative, of course, it means that x decreases with selection.)

B Now, from Box E of Chapter 1, the Lotka-Volterra competition equation for the growth of species 1 is

$$\frac{d\ln N_1}{dt} = r - \frac{rN_1}{K} - \frac{r\alpha N_2}{K}$$

where $r = r_1$, $K = K_1$, $\alpha = \alpha_{21}$, and $d\ln N_1/dt = (1/N_1)(dN_1/dt)$.

We can define "fitness" as that quantity that obeys the fundamental theorem of natural selection (Box E in Chapter 4), i.e., $\Delta\bar{w} = \sigma^2_{aw}/\bar{w}$. The most appropriate quantity in this case is $\bar{w} = exp(d\ln N_1/dt)$, or $\ln\bar{w} = d\ln N_1/dt$.

Taking the derivatives $d\ln\bar{w}/dx$ for $x = r$, K, and α produces

$$\frac{R_r}{\sigma^2_{ar}} = 1 - \frac{(N_1 + \alpha N_2)}{K} \quad \textbf{(1)}$$

$$\frac{R_K}{\sigma^2_{aK}} = \frac{r(N_1 + \alpha N_2)}{K^2} \quad \textbf{(2)}$$

$$\frac{R_\alpha}{\sigma^2_{a\alpha}} = \frac{-rN_2}{K} \quad \textbf{(3)}$$

Calculate these quantities at equilibrium for the numerical example in part (b) of Box E in Chapter 1, with $r = r_1 = r_2 = .05$.

The expressions on the right-hand sides of (1) to (3) above are equal to the selection intensities for r, K, and α, respectively, provided that the signs of the quantities are disregarded. An alternative derivation of the selection intensities can be found in Emlen (1973), who points out several important implications of the formulas. First, from equation (1), if $N_2 = 0$ (i.e., the competitor is absent) and N_1 is much less than K (i.e., if the number of individuals of species 1 is held well below its carrying capacity by, for example, inclement weather), then the intensity of selection for increased r is large (in fact, it equals $1 - N_1/K$). However, if both species are present and at equilibrium then $N_1 + \alpha N_2 = K$, so the intensity of selection for r equals 0. Second, from equation (2), if both species are present and at equilibrium, the intensity of selection for K is r/K. Third, from equation (3), selection always acts so as to decrease competition (i.e., to decrease α), and the selection intensity for the avoidance of competition is most intense when the competitor's population size (N_2) is large relative to the carrying capacity of species 1 (K). (For more discussion see Emlen, 1973.)

its own fitness. Most of the models of selection in Chapters 3 and 4 of this book are based on individual selection, as is the mass selection phase of the shifting balance theory discussed earlier in this chapter. In considering varieties of group selection, the assumption is that an individual's own fitness does not provide sufficient information to predict its evolutionary success; rather it is assumed that certain properties of the population to which the individual belongs must also be taken into account.

GROUP SELECTION IN THE STRICT SENSE

The term GROUP SELECTION has been used in a loose sense to describe almost any kind of selection other than individual selection; that is to say, group selection is often used to denote any kind of genetic change brought about by the differential extinction or proliferation of populations. In this section, it will be convenient to use the term in a more restricted sense — to refer to selection that increases the frequencies of genes beneficial to the group as a whole but detrimental (or at best selectively neutral) to the individuals that carry them. This more restricted definition of group selection we will call GROUP SELECTION IN THE STRICT SENSE. (For discussions of the varieties of group selection, see Brown, 1966; Levin and Kilmer, 1975; Leigh, 1977; and Wade, 1978.)

Considerations of group selection in the strict sense arise in the context of the regulation of population growth. In many species, excessive population growth of a deme can cause extensive damage to the deme's environment; this leads to an inevitable population crash and perhaps even extinction of the deme. Such might be the fate of a predator population that excessively depleted its prey, for example; and North American deer, whose natural predators have been largely removed by human activity, can increase dramatically in population size, only to be cut back ruthlessly by winter starvation, especially of fawns (for discussion see Smith, 1974). Some species, particularly among birds and mammals, have evolved elaborate behavioral mechanisms such as pecking orders and other social hierarchies or the establishment and defense of territories that, in times of population excess, serve to prevent reproduction of certain members of the population. The basic idea behind group selection in the strict sense is that genes for mechanisms of population regulation are selected because, in preventing overpopulation, they benefit the entire group, even though the genes are surely detrimental to the fitness of those individuals excluded from reproduction. When there are many demes, there may be genetic variation among demes in the extent of population regulation. Those demes with genes that regulate population size at an optimal level will prosper at the expense of those demes that overpopulate the environment and become extinct or at the expense of those demes that reproduce insufficiently to perpetuate themselves. In this manner the genes for population regulation are spread throughout the species (Wynne-Edwards, 1962).

Many population geneticists doubt that group selection in the strict sense used here can occur except perhaps under unusual circumstances. The main problem is in accounting for the evolution of population regulation within demes — how can a gene that causes its bearers to refrain voluntarily from reproduction increase in frequency? Since population self-regulation and other traits beneficial to groups have evolved [in fact, repeatedly (Wade, 1976, 1977)] skeptics of group selection in the strict sense have searched for alternatives to account for the evolution, alternatives based either on individual selection of some kind or on such modes of selection as kin selection (discussed below) that are not seemingly so anti-Darwinist. One simple alternative is to suppose that the self-regulating mechanisms of populations may not always be selected directly but rather may arise as a correlated response to individual selection for some other trait or complex of traits. An even simpler alternative is an explanation based directly on individual selection. For example, individual birds that establish and defend a territory surely enhance the transmission of their own genes; the genes for territorial behavior will therefore increase in frequency due solely to individual selection. That such behavior is ultimately beneficial to the group is not the essential cause for its evolution but rather a happy coincidence. (Optimal population size is only one trait that has been hypothesized to evolve according to some mode of group selection. Other traits include rates and patterns of migration, mating, competition, feeding, and predation, and especially patterns of behavior of individuals toward other members of the same group. The mode of group selection envisaged is not always group selection in the strict sense, however. See D. S. Wilson, 1975, for further discussion.)

INTERDEME SELECTION

Interdeme selection is one form of selection involving groups, and it plays a central role in the shifting balance theory of evolution. Like group selection in the strict sense, interdeme selection is based on genetic differences among demes; unlike group selection in the strict sense, however, the genetic variation that arises among demes is based on conventional principles of individual selection. In the shifting balance hypothesis, genetic variation among demes arises only because there are many different selective peaks; within each deme, selection is always in favor of those individuals in the population that have the highest fitness. It may happen that some demes flourish at the expense of others because they have evolved traits beneficial to the group; nevertheless, the ultimate cause for the evolution of these traits is individual selection within the deme.

Two forms of interdeme selection may conveniently be distinguished: (1) extinction and recolonization and (2) overflooding. Demes are often small and, therefore, impermanent, subject to the vagaries of bad weather, infectious disease, overpredation, and many other causes of decimation. Consequently, by chance, and sometimes perhaps for genetic reasons, a deme may become extinct. Eventually the vacant habitat will be repopulated by colonists from other, more successful demes (for one sort of theoretical model, see Slatkin, 1977). If the number of colonists is few, then there will be a founder effect (Chapter 3), and perhaps the initial allele frequencies among the colonists will be such as to lead the resulting population to a new selective peak.

When the number of colonists is large, on the other hand, the genetic structure of the source population may be preserved nearly intact.

A successful deme that reaches a new, higher selective peak will increase in size, and the resulting population pressure will increase the number of emigrants. Nearby demes will therefore be subject to increased immigration from the successful deme, a situation that may be called OVERFLOODING, and the rate of immigration may be large enough that the recipient population itself undergoes a peak shift. In this manner, favorable genotypic combinations can be spread throughout the species in ever-widening concentric circles.

The important distinction between interdeme selection based on extinction and recolonization as opposed to overflooding is that, in the former case, local populations are completely replaced, whereas, in the latter case, they are not. In the case of overflooding, therefore, rare alleles unique to a deme can be retained even if the deme is overflooded from a nearby population and undergoes a peak shift.

KIN SELECTION

Kin selection is a special kind of group selection that has been postulated in attempts to account for the evolution of altruistic behavior. KIN SELECTION refers to indirect selection for alleles that occurs through the relatives of carriers of those alleles rather than by direct selection through an enhanced fitness of the carriers themselves. ALTRUISM consists of behavioral traits or other attributes that increase the fitness of other individuals at the expense of one's own fitness. Altruistic behavior is exhibited most dramatically by such social insects as termites, ants, and bees, in which certain worker castes exert their labors for the care, protection, and reproduction of the queen and her offspring but do not reproduce themselves. Other less dramatic examples of altruistic behavior include phenomena such as parental care. As an example of parental care, consider a female deer, who will often flaunt herself before a wolf or other predator when the predator is too near her fawn. The display distracts the predator, who may then pursue the mother; the mother thus places herself at risk in order to protect the fawn, so the parental behavior in this case is altruistic.

Evolution of Altruism. A central consideration in kin selection is that relatives have genes in common; thus, a gene that causes altruistic behavior can increase in frequency if the increase in the recipient's fitness as a result of altruism is sufficiently large to offset the decrease in the altruist's own fitness. The essentials of the situation can be made clear by considering the case of identical twins. Because identical twins are genetically identical, the reproduction of one's twin is genetically equivalent to reproduction by oneself. Thus, it makes no difference if an altruistic individual decreases its own fitness for the sake of an equal increase in fitness of an identical twin; from an evolutionary point of view, it is an even trade because the combined number of offspring from both twins remains unchanged. By the same token, if an altruistic act decreases the fitness of an individual by an amount less than the increase gained by an identical twin, then the altruism results is a net increase in the combined number of offspring. One would, therefore, expect altruism between identical twins to be favored by natural selection as long as the risk to the altruist is no greater than the benefit to the recipient.

The above reasoning can be made quantitative and extended to other degrees of relationship by referring to Figure 8. In Figure 8A, the circle labeled I represents an individual that is heterozygous for the *A* allele. (For ease of argument, we suppose that the *A* allele is rare, so that *AA* homozygotes can be neglected, and it is also assumed that there is no inbreeding.) The circle labeled R in Figure 8A represents a relative (genotype unknown) of the *Aa* individual, and the degree of relationship between the two individuals is measured by a quantity denoted r in the figure. The quantity r is the probability that R carries the *A* allele when we know that I carries that allele. In the absence of inbreeding, r is called the COEFFICIENT OF RELATIONSHIP and equals two times the probability that a randomly chosen gamete from

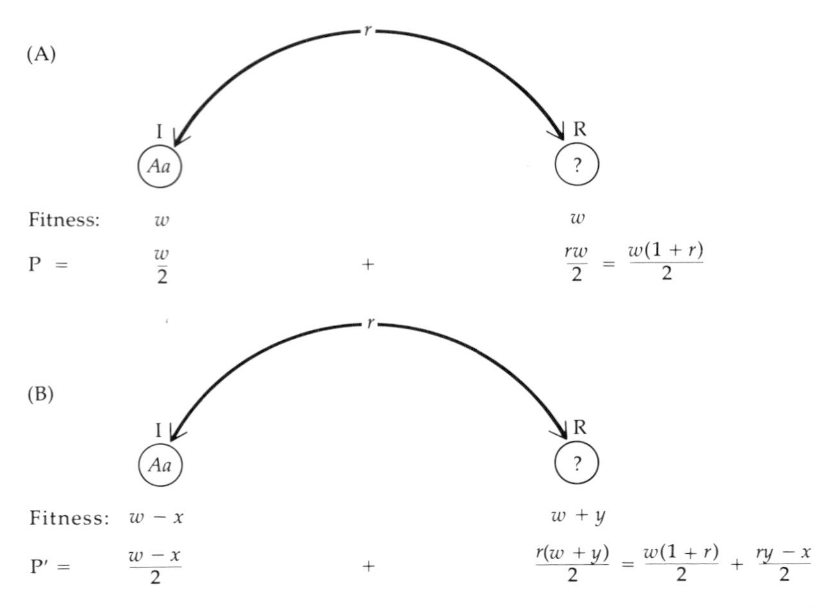

FIGURE 8. Calculation of inclusive fitness of individual I based on I's own fitness (w), that of a relative (R), and on the coefficient of relationship between I and R. (A) No altruism. (B) I is an altruist who increases R's fitness by an amount y at a cost of x units in fitness to himself.

I and a randomly chosen gamete from R carry alleles that are identical by descent; the latter probability is, by definition, the inbreeding coefficient of a hypothetical offspring of I and R. (See Box C; the concepts of identity by descent and inbreeding are discussed in Chapter 2.)

Now, what is the overall number of A alleles contributed to the next generation by the individuals in Figure 8A? The fitness of each individual in the figure is given as w, so the number of A alleles contributed to the next generation, P, is $P = w/2$ (from individual I, the 1/2 entering because only half the offspring receive the A allele) + $rw/2$ (from relative R, recalling that r is the probability that R carries the A allele, given that I does; here again the 1/2 enters because, when R carries A, the probability that it will be included in a particular gamete is 1/2). Such fitness calculations that take into account not only an individual's own fitness but also

Measurement of Degrees of Genetic Relationship

C

The text defines r as the probability that a relative (R) carries a specific allele, given that the individual in question (I) carries that allele. In a diploid organism, assuming no inbreeding, $r = 2F_{IR}$, where F_{IR} denotes the inbreeding coefficient of a hypothetical offspring of I and R. (The offspring must be hypothetical, of course, as both I and R could be of the same sex.) The relationship $r = 2F_{IR}$ can be shown as follows. Let the genotype of I be denoted $\alpha_1\alpha_2$ and that of R be denoted $\alpha_3\alpha_4$. Since I is assumed to be non-inbred, $\alpha_1 \neq \alpha_2$ (i.e., α_1 is not identical by descent with α_2); similarly, since R is assumed to be non-inbred, $\alpha_3 \neq \alpha_4$. The hypothetical offspring I and R could have genotypes $\alpha_1\alpha_3$, $\alpha_1\alpha_4$, $\alpha_2\alpha_3$, and $\alpha_2\alpha_4$, each with probability 1/4; thus the inbreeding coefficient of the hypothetical offspring will be $F_{IR} = (1/4)[Pr(\alpha_1 = \alpha_3) + Pr(\alpha_1 = \alpha_4) + Pr(\alpha_2 = \alpha_3) + Pr(\alpha_2 = \alpha_4)]$, where $Pr(\alpha_1 = \alpha_3)$ is the probability that α_1 is identical by descent with α_3, and likewise for $Pr(\alpha_1 = \alpha_4)$ and the other symbols.

Now, r is the probability that R carries a specific allele, given that I carries it. Thus $r = Pr(\text{R carries } \alpha_1 | \text{I carries } \alpha_1) = Pr(\alpha_3 = \alpha_1) + Pr(\alpha_4 = \alpha_1)$; similarly, $r = Pr(\text{R carries } \alpha_2 | \text{I carries } \alpha_2) = Pr(\alpha_3 = \alpha_2) + Pr(\alpha_4 = \alpha_2)$. Substituting these into the expression for F_{IR} yields $F_{IR} = (1/4)(r + r)$, so $r = 2F_{IR}$.

As for terminology, F_{IR} (which is the inbreeding coefficient of a hypothetical offspring of I and R) is usually called the COEFFICIENT OF KINSHIP. Another measure of genetic relationship, which we will denote here as r^*, is called the coefficient of relationship and is defined as $r^* = 2F_{IR}/\sqrt{(1 + F_I)(1 + F_R)}$, where F_I and F_R are the inbreeding coefficients of I and R. In the absence of inbreeding ($F_I = F_R = 0$), r as defined above is the same as the coefficient of relationship.

a. Assume $F_I = F_R = 0$ and calculate the coefficient of relationship between identical twins, parent-offspring, full siblings, half siblings, first cousins, and uncle-nephew.

b. Geneticist J. B. S. Haldane is said to have quipped that he would lay down his life for two brothers, four nephews, or eight cousins. Can his statement be justified on the grounds of kin selection? Hint: consider the conclusion from Figure 9 in light of the values for r calculated in part (a).

the fitness of relatives (other than direct descendants) constitute what is often called the INCLUSIVE FITNESS of the individual; inclusive fitness therefore represents the total effect of both individual and kin selection.

Figure 8A depicts a situation in which no altruistic behavior is involved. Altruism is represented in Figure 8B, where we assume that the *A* allele leads to an altruistic behavior that reduces I's fitness by an amount x while increasing R's fitness by an amount y. The same reasoning as before shows that, in this case, the overall number of *A* alleles contributed to the next generation, denoted P', is $P' = (w - x)/2 + [r(w + y)]/2$.

Altruism will be favored by natural selection if $P' > P$, that is to say, if $(w - x)/2 + [r(w + y)]/2 > w/2 + rw/2$. A little manipulation reduces this inequality to the simple statement

$$\frac{x}{y} < r$$

To state the result in words: a gene for altruism toward a relative will be favored if the ratio of the loss in fitness of the altruist to the gain in fitness of the relative is less than the coefficient of relationship. (For more discussion of the evolution of altruism, see Alexander, 1974; Eshel and Cohen, 1976; Matessi and Jayakar, 1976; Yokoyama and Felsenstein, 1978; and Darlington, 1978.)

As an aside, note in Figure 8B that if x is negative and y is positive, then the *A* allele may be said to induce COOPERATIVE BEHAVIOR (i.e., behavior that increases the fitness of the individual in question as well as of the recipient). The condition $P' > P$ is still $x/y < r$, but note, since x is negative, $x/y < 0$. Thus, cooperative behavior will be favored for *any* positive value of r and, indeed, for $r = 0$, which corresponds to I and R being unrelated (Felsenstein, 1979)! An example of cooperative behavior is found in large predators such as wolves, which hunt in packs. Hunting in packs increases the probability of a successful kill, and all animals in the pack are equally benefited. This sort of cooperative behavior will be promoted by natural selection, even if the individuals involved are unrelated.

Male Haploidy. Kin selection is somewhat more complicated in species of the insect order Hymenoptera (e.g., various ants and bees) because of their peculiar mode of reproduction. In ants, bees, and most other hymenopteran insects, females originate in the normal manner by fertilization of a haploid egg by a haploid sperm; males, however, originate by parthenogenetic development of unfertilized haploid eggs (Hartl and Brown, 1970; Kerr, 1974; Crozier, 1977). This mode of reproduction is known as ARRHENOTOKY or MALE HAPLOIDY, and its genetic consequence is that all genes are inherited as if they were sex-linked (Chapter 1).

The peculiar feature of male haploidy is that a female is more closely related to her sisters than she is to her own daughters. An illustration of this point is found in Figure 9,

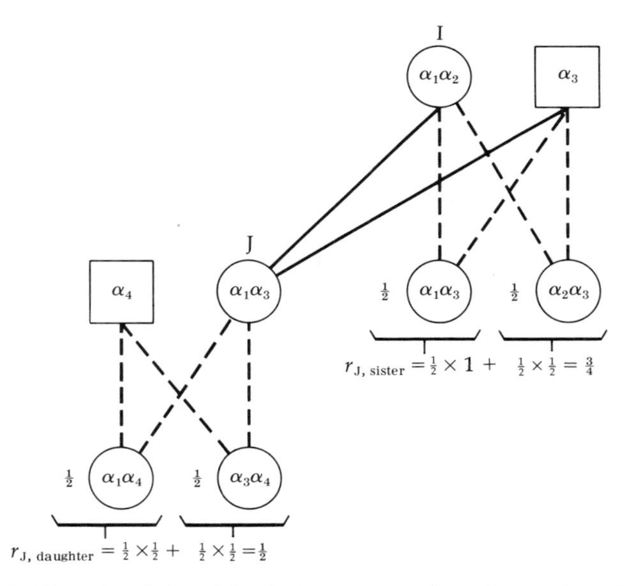

FIGURE 9. Genetic relationships between organisms in species with haploid males. The coefficient of relationship between the female J and her own daughters is 1/2, whereas the coefficient of relationship between J and her sisters is 3/4. Thus, J can bring more of her own alleles into the population by aiding her mother (I) in producing more sisters than she can by reproducing herself, provided that the mother mates with only one male.

where we make the simplifying assumptions that there is no inbreeding and that females mate with a single male. As in Chapter 2, alleles that are denoted by α's with different subscripts (α_1 and α_3, for example) are assumed not to be identical by descent. The genotype of the female (I) at the top of the pedigree is designated $\alpha_1\alpha_2$, and her daughter's genotype (J) is arbitrarily assumed to be $\alpha_1\alpha_3$. Recall that, in the absence of inbreeding, r is the coefficient of relationship between two individuals; if the two individuals have one allele in common (as for parents and offspring with autosomal inheritance or for mothers and daughters with male haploidy), $r = 1/2$ (Box C); if the individuals have two alleles in common (as for identical twins), $r = 1$ (Box C).

The coefficient of relationship between J and her daughters is calculated in Figure 9 as 1/2 (which is the probability that the daughter is genotypically $\alpha_1\alpha_4$) $\times$ 1/2 (which is the coefficient of relationship in that case) + 1/2 (the probability that the daughter is $\alpha_3\alpha_4$) $\times$ 1/2 (the coefficient of relationship in that case); $r_{\text{J, daughter}}$ therefore works out to be 1/2.

As between individual J and her sisters, the coefficient of relationship is calculated in Figure 9 as 1/2 (the probability that the sister is genotypically $\alpha_1\alpha_3$) $\times$ 1 (the coefficient of relationship in that case) + 1/2 (the probability that the sister is $\alpha_2\alpha_3$) $\times$ 1/2 (the coefficient of relationship in that case); it thus turns out that $r_{\text{J, sister}} = 1/2 + 1/4 = 3/4$.

Since the coefficient of relationship between J and her sisters is greater than that between J and her own daughters, it stands to reason that, from an evolutionary point of view, J would be better off to produce a sister than to produce a daughter. But how can J produce a sister? She can do so by promoting further reproduction by her own mother — the queen. The startling fact that hymenopteran females are more closely related to their sisters than to their own offspring may explain why true social behavior has evolved independently several times in the Hymenoptera, although it is rare elsewhere among animals. Moreover, for a hymenopteran male, the probability that his sister carries the same allele as he does is 1/2, whereas the probability that his daughter carries the same allele is 1. Thus hymenopteran males are more closely related to their daughters than they are to their sisters, and a male Hymenopteran would consequently be better off to have a daughter of his own than to promote further reproduction by the queen. This kind of reasoning may explain why, when social behavior has evolved in the Hymenoptera, the sterile worker castes that aid the queen are always female. (For more on social insects, see Kerr, 1969; Wilson, 1971; Hamilton, 1972; Trivers and Hare, 1976; and Oster et al., 1977.)

Arguments such as those above, which attempt to explain the evolution of social behavior, are important ingredients in the field of evolutionary sociobiology. SOCIOBIOLOGY itself is defined as the systematic study of all social behavior, particularly animal social behavior (E. O. Wilson, 1975). Evolutionary sociobiology is, therefore, the study of the mechanisms by which social behavior evolves. In the case of social behavior among the Hymenoptera, considerations of kin selection play a central role in understanding the evolution of sociality. Kin selection of some sort, and perhaps other modes of group selection as well, may play a role in the evolution of social behavior in other organisms, such as primates. It should be noted, however, that it is one thing to provide a theoretical demonstration that social behavior could evolve by means of kin selection (or any other mode of selection), but it is quite another thing to prove that the behavior actually did evolve according to the theoretical mechanism. The actual manner of evolution of social behavior can be determined only by experiment, observations of organisms in their natural habitats, and comparisons of social behavior among related organisms. Unfortunately, such systematic studies are only in their infancy, and a comprehensive overview of the evolution of sociality must await future studies. (See E. O. Wilson, 1975, for the definitive book on sociobiology.)

Evolution in Action

In this section, several cases of "evolution in action" are briefly summarized. From among the numerous examples that are known, four cases singled out for discussion are (1) the evolution of transferable antibiotic resistance factors in bacteria (important because of its obvious medical and veterinary significance and for what it illustrates about the complex mode of evolution in prokaryotes); (2) the evolution of heavy metal tolerance in plants (important because it shows the profound effect of strong selection on population differentiation); (3) the evolution of industrial melanism in moths (a classic and well-studied case illustrating the spread of favorable alleles and the importance of migration as a cohesive force between populations); and (4) the evolution of Batesian mimicry among butterflies (important because it involves the evolution of a complex adaptation). Three of the four cases involve evolutionary response to environmental conditions that have been changed as a result of human activity; this coincidence occurs because the most likely place to find rapid evolution is where populations have been challenged to adapt to new and unusual environments.

RESISTANCE TRANSFER FACTORS IN BACTERIA

Plasmids are small, usually circular molecules of DNA that exist inside bacterial cells as self-replicating genetic elements not attached to the bacterial chromosome (see Wolstenholme and O'Connor, 1969; Clowes, 1972; Meynell, 1972; Campbell, 1969; and Levin, 1977). Plasmids are heritable agents, transmitted from generation to generation as bacterial cells divide. Many plasmids are also infectious agents, transferable from one bacterial cell to another, even to cells of different bacterial species. Transfer is mediated through tube-like projections from the cell surface called pili (Figure 10), formation of which is often controlled by genes on the plasmid itself. (Bacteria have many different types of pili, only some of which are associated with the transfer of plasmids.)

Plasmids play a central role in the evolution of antibiotic resistance in bacteria, a conclusion that emerged in the late 1950's with the discovery of multiple drug-resistant bacteria. The first such strain was isolated in 1955 from a Japanese woman who developed a stubborn case of dysentery after

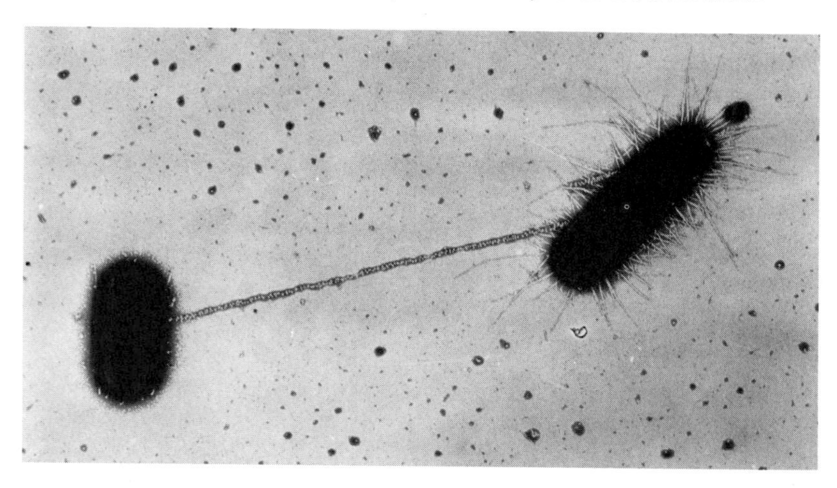

FIGURE 10. Pilus (the diagonally running structure) connecting a donor *E. coli* cell (upper right) with a recipient *E. coli* cell (lower left). The pilus has been treated with a pilus-specific phage for ease of visualization. This particular type of pilus is used in the transfer of a plasmid called the F factor and also in the transmission of certain resistance transfer factors. The numerous other appendages on the donor cell are a different type of pili not involved in transfer of genes but required for *E. coli* cells to colonize the intestine. (Courtesy of C. C. Brinton, Jr., and J. Carnahan.)

returning from vacation in Hong Kong (Watanabe, 1967). The cause was a typical dysentery bacillus of the genus *Shigella,* but in one way this *Shigella* was unique — it was resistant to four drugs: sulfonamide, streptomycin, chloramphenicol, and tetracycline. Moreover, this strain of *Shigella* was able to transfer its drug resistance to such other intestinal bacteria as *Escherichia coli.*

The troublesome *Shigella* turned out to harbor what has since become known as a RESISTANCE TRANSFER FACTOR, i.e., a plasmid that carries genes conferring resistance to one or more antibiotics as well as genes enabling its own transfer to other bacteria. Since 1955 many other resistance transfer factors have been found in a wide variety of pathogenic and nonpathogenic bacteria. These factors can confer resistance to various combinations of drugs (including sulfonamide, penicillin, streptomycin, tetracycline, chloramphenicol, ampicillin, kanamycin, neomycin, polymyxin, radiomycin, bluensomycin, spectinomycin, viomycin, and gentamycin); resistance to such heavy metal ions as mercury, cobalt, and nickel; and even resistance to ultraviolet radiation (Campbell, 1969; Davies and Smith, 1978).

Antibiotic resistance presumably evolves in response to the presence of antibiotics in the environment. Antibiotics are chemicals produced by organisms, generally fungi or bacteria, that are inhibitory to the growth of other microorganisms. Although antibiotics have been known since 1928 when Sir Alexander Fleming in England isolated penicillin from the mold *Penicillium,* the industry did not undergo rapid expansion until after the Second World War. Today several thousand types of antibiotics and their chemical derivatives are known, and use of antibiotics has revolutionized the treatment of infectious disease. Indeed, antibiotics have been misused in medicine by their administration to patients with nonspecific ailments or with virus-caused conditions such as the common cold or influenza. Misuse has been particularly brazen in animal husbandry, where antibiotics have been incorporated into livestock feed as "preventives" against pathogens. As might be expected, the upshot of the continuous presence of antibiotics in the bacterial environment has been the evolution of antibiotic-resistant strains. The spread of multiple drug resistance in bacteria is evident in Figure 11, which shows the frequency of antibiotic-resistant *Shigella* in Japanese hospital patients. Similar increases in resistance have been noted in bacterial pathogens of domestic animals (E. S. Anderson, 1969).

It is not known how the resistance genes on resistance transfer factors arise, but there are a number of clues. In the first place, the molecular mechanisms of resistance are typically quite different from those observed in antibiotic-resistant mutants selected directly in plasmid-free bacteria. The procedure for obtaining such antibiotic-resistant mutants is to spread a large number of bacterial cells on agar-solidified medium containing the antibiotic, which prevents growth of all the cells except for the tiny minority that have

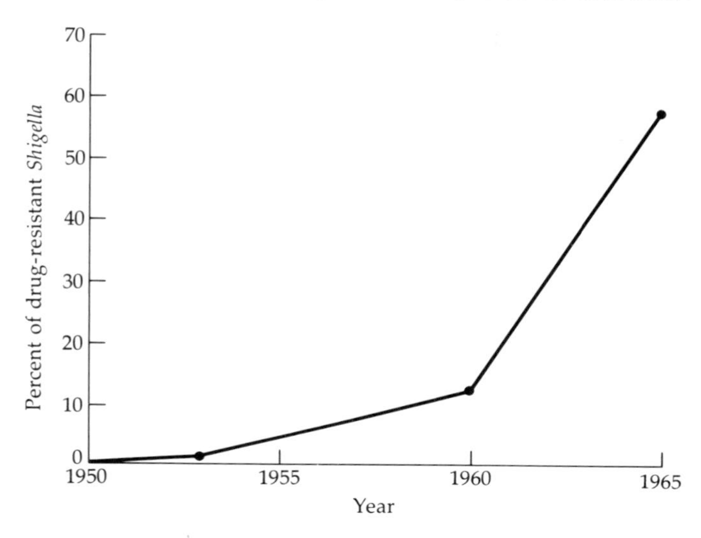

FIGURE 11. Increase in drug-resistant *Shigella* in Japanese hospital patients during the period 1950 to 1965. The increased drug resistance is due to the spread of resistance transfer factors. (Data from Campbell, 1969.)

a spontaneously occurring mutation that confers antibiotic resistance. For a number of years it was assumed that such direct selection provided a suitable experimental model of the evolution of antibiotic resistance in nature. The discovery of resistance transfer factors has shown that the experimental model is much too simple. In the case of streptomycin, for example, the antibiotic exerts its killing effect by interfering with ribosomes and thus disrupting protein synthesis. Streptomycin-resistant mutants obtained by direct selection have either defective ribosomes that are insensitive to the antibiotic or defective membranes that prevent entry of the antibiotic into the cell. The mutations conferring resistance are themselves harmful, and natural selection acts against them in antibiotic-free environments. In the case of genes for streptomycin resistance carried on resistance transfer factors, by contrast, the mechanism of resistance involves neither ribosomes nor membranes; rather, the genes code for enzymes that chemically modify the antibiotic molecule and render it innocuous (Figure 12).

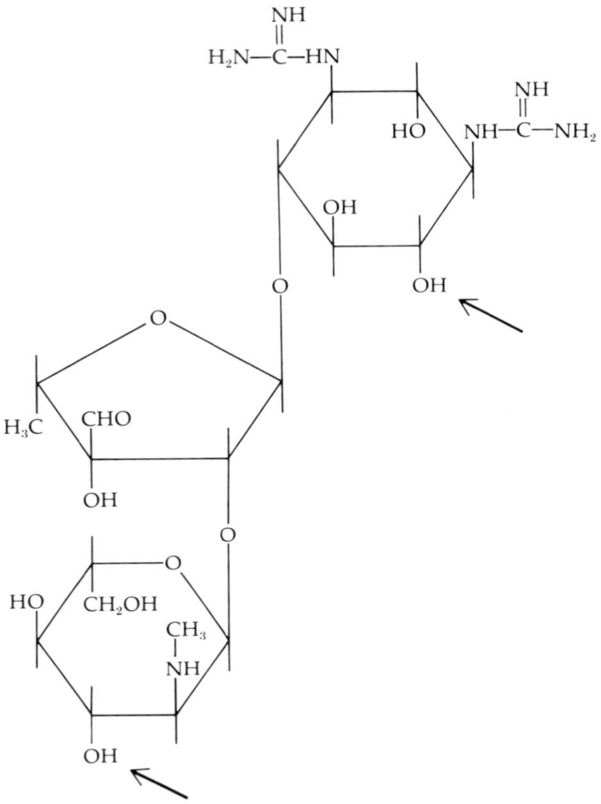

FIGURE 12. Structural formula of streptomycin, showing groups (arrows) that are chemically modified by various enzymes coded by genes on resistance transfer factors.

It seems likely that antibiotic-resistance genes evolve in natural populations of bacteria in response to naturally occurring antibiotics. Many microorganisms produce antibiotics, particularly soil-dwelling fungi and bacteria of the genera *Actinomyces* and *Streptomyces* (Benveniste and Davies, 1973). In the Solomon Islands, for example, where antibiotics have not been widely used, streptomycin and tetracycline resistance transfer factors have been isolated from both soil and stool specimens (Gardner et al., 1969). It would thus seem that the wide use of antibiotics in medicine and agriculture has simply promoted the spread of resistance transfer factors that probably already existed in nature.

The means by which resistance transfer factors can accumulate genes for resistance to several antibiotics has been clarified somewhat by examination of their molecular structure. The factors contain nontandem duplications of any of a number of special DNA sequences that can, by an as yet unknown mechanism, change position from one place in a DNA molecule to another place in the same or a different molecule, the DNA sequence between the duplications being carried along during the transposition. Such transposable elements are known as TRANSLOCATABLE ELEMENTS or TRANSPOSONS (Bukhari et al., 1977). One transposon carrying tetracycline resistance has, for example, been followed genetically as it transposed from its original position on a resistance transfer factor to the *Salmonella* phage P22, from P22 to a *Salmonella* gene, from the *Salmonella* gene to the *Escherichia coli* phage λ, from λ to an *E. coli* gene, and from the *E. coli* gene back again to λ (Figure 13; see also Berg et al., 1975, and Kleckner, 1977).

The duplicated DNA sequences involved in transpositions are fairly common and are even present and can change position in normal bacterial chromosomes. A newly evolved gene for antibiotic resistance that is or becomes flanked by such sequences therefore constitutes a new transposon that can translocate into, say, a resistance transfer factor. From there the gene can become widely disseminated by the infectious properties of the transfer factor itself. By means of such transpositions, resistance genes that evolve in several perhaps distantly related bacterial species can be brought together on a single resistance transfer factor and inherited as a unit.

Much remains to be learned about the evolution of resistance transfer factors and the origin of the resistance genes they carry. From what is already known, it is clear that transposons and resistance transfer factors provide prokaryotes with potent mechanisms for the spread and sharing between species of useful evolutionary novelties.

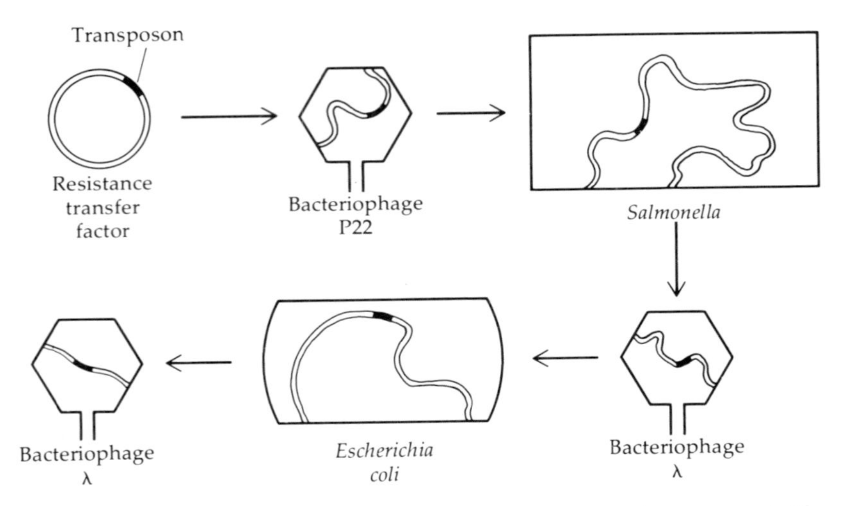

FIGURE 13. Observed transpositions of a transposon known as Tn10. Originally located on a resistance transfer factor inside a cell of the bacterial genus *Salmonella,* Tn10 moved to a P22 phage that had infected the same cell. (A phage, or bacteriophage, is a virus that infects bacterial cells.) When the P22 infected a different *Salmonella* cell, Tn10 moved to the bacterial DNA of this new host, and from there to a λ phage that had infected the same *Salmonella* cell. The Tn10-bearing λ phage then infected a cell of the bacterial species *Escherichia coli,* and the transposon moved onto the *E. coli* DNA; from there it moved onto the DNA of a different λ bacteriophage. (Data from Kleckner, 1977.)

HEAVY METAL TOLERANCE IN PLANTS

Some 38 elements having a density greater than five grams per cubic centimeter constitute the "heavy metals," metals like mercury, nickel, lead, zinc, tin, and copper. In excessive quantities, all heavy metals are lethal to most living organisms.

At various places throughout the world, soil is contaminated by toxic concentrations of one or more heavy metals. Such areas can result from the rare presence of undisturbed metal ore near the soil surface; more commonly they result from unproductive ore, tailings, or seepage near actual mining areas. In addition to heavy metal contamination, such soils are typically low in the essential nutrients nitrogen, phosphorous, and potassium. Because of their relative lack of vegetation, such areas generally have soil of poor texture and are lacking in organic matter. Any plants that do survive are subject to exposed, often sand-dunelike conditions (Figure 14).

Inhospitable as such contaminated sites are, some populations have evolved tolerance not only to the generally poor conditions but also to the toxic metals (Jain and Bradshaw, 1966; Antonovics et al., 1971). Certain bacteria and fungi are frequently found in such areas (recall that bacterial resistance transfer factors often carry genes for heavy metal tolerance), as are some algae, mosses, lichens, liverworts, and flowering plants; ferns and gymnosperms tolerant of heavy metals are rarely found, however. Mechanisms of resistance to heavy metal poisoning vary among species, ranging from those in which the metal is not taken up by the plant to those in which the metal is taken up but is sequestered in the cell walls, away from vital metabolic processes. Tolerance to

FIGURE 14. Desolate appearance of a site contaminated with a heavy metal due to mining operations, showing resistant grasses in the foreground. (Trelogan site, courtesy of J. Antonovics.)

heavy metals seems to evolve very rapidly, as mining sites only 50 to 100 years old often have well-established populations (Antonovics et al., 1971).

Populations that have evolved heavy metal tolerance often exist side by side with nontolerant populations of the same species, and gene exchange occurs. Although such gene exchange is often considered a cohesive force that acts to prevent the genetic divergence of populations, the example of heavy metal tolerance shows that selection can nevertheless overcome the effects of migration. The situation has been particularly well studied in certain grasses, among them *Agrostis tenuis* (bent grass), *Agrostis stolonifera* (creeping bent), and *Anthoxanthum odoratum* (sweet vernal). Figure 15A shows how an index of lead tolerance in *A. tenuis* changes abruptly across a mine boundary from values above 60 to ones below 40; Figure 15B shows a comparable abrupt change in copper tolerance in *A. tenuis* at another mine; and Figure 15C shows how zinc tolerance in *A. odoratum* at yet another mine declines dramatically from a value of nearly 80 to one of 15 over a distance of only about four meters across the mine boundary. Thus, in spite of gene flow, genetic differentiation associated with heavy metal tolerance at mining sites is sharp and distinct. The genetic basis of heavy metal tolerance is not known with precision, but the trait seems to be polygenic and to have a high broad-sense heritability (Antonovics et al., 1971; MacNair, 1979); moreover, tolerance to one heavy metal does not confer resistance to any other.

Selection coefficients pertaining to heavy metal tolerance in natural habitats are not known, but they have been studied under experimental conditions simulating natural habitats (Jain and Bradshaw, 1966; Cook et al., 1972). The intensity of selection involved can be very large, as shown by the data summarized in Table III. The fourth column in Table III gives estimated fitnesses of nontolerant plants relative to tolerant ones when grown in contaminated soil. With respect to heavy metal tolerance (examples 1 to 3 in the table), selection against nontolerant plants in contaminated soil is very intense, the relative fitnesses ranging from .01 to .05.

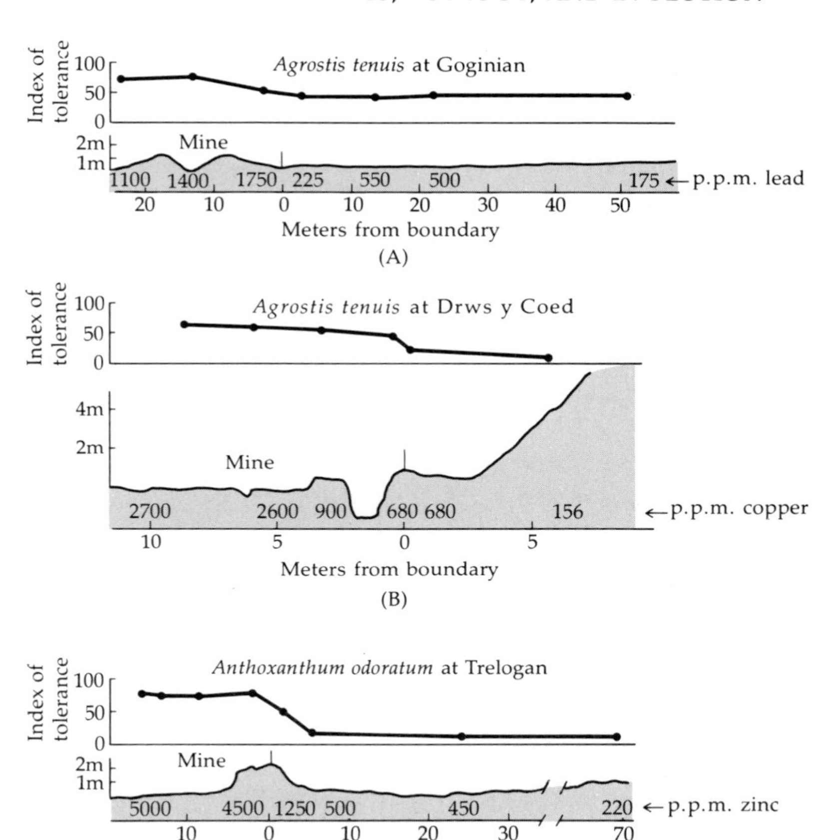

FIGURE 15. Profiles of three mine sites in Wales showing, below each profile, the amount of heavy metal contamination of the soil and, above each profile, the observed cline of heavy metal tolerance in grasses collected along the profile. Note the particularly abrupt drops in resistance at the mine boundaries in (B) and (C). (After Jain and Bradshaw, 1966.)

In practical terms, these fitnesses amount to relative survival rates of 1 percent to 5 percent, which essentially denote semilethality of the nonadapted plants. The last column of Table III gives estimated fitnesses of tolerant plants relative to nontolerant ones when grown in uncontaminated soil. Selection in this case is generally less intense than that ob-

TABLE III. Relative fitnesses of tolerant and nontolerant plants grown in contaminated and contamination-free soil.

Example	Species	Type of "contamination"	Nontolerant:tolerant type in contaminated soil	Tolerant:nontolerant type in contamination-free soil
1	*Agrostis tenuis*	Lead	0.05:1 (s = .95)	0.60:1 (s = .40)
2	*Agrostis tenuis*	Copper	0.05:1 (s = .95)	0.95:1 (s = .05)
3	*Anthoxanthum odoratum*	Zinc	0.01:1 (s = .99)	0.70:1 (s = .30)
4	*Agrostis stolonifera*	Exposure	0.20:1 (s = .80)	0.50:1 (s = .50)
5	*Anthoxanthum odoratum*	Lime	0.91:1 (s = .09)	0.73:1 (s = .27)

In each case, the fitness of plants adapted to the soil type is arbitrarily assigned the value of 1, and the relative fitness of the nonadapted type of plant is given. The selection coefficients (s) are those acting against the nonadapted type and are calculated as 1 minus the relative fitness of the nonadapted type. Examples 1 through 3 pertain to heavy metal tolerance; example 4 pertains to adaptation to growth under exposed conditions on a cliffside (in this case "contaminated soil" means "exposed conditions"); and example 5 pertains to adaptation to growth in soil treated with lime (calcium oxide). Sources of evidence are as follows. Example 1: growth of spaced plants on two soils (data of A. D. Bradshaw); example 2: growth of plants under spaced and competitive conditions on two soils (T. S. McNeilly); example 3: growth of plants under spaced and competitive conditions on two soils (P. D. Putwain); example 4: growth of spaced plants under exposed cliff and garden conditions (J. L. Aston); example 5: result of three competition experiments carried out under greenhouse conditions (R. W. Snaydon). (Adapted from Jain and Bradshaw, 1966.)

served in contaminated soil, with relative fitnesses in the case of heavy metal tolerance ranging from .60 to .95. On the other hand, even a selection coefficient of s = .05 represents very considerable selection. The last two examples of Table III give relative fitnesses of plants adapted to weather exposure on a cliffside (example 4) or to limed soil (example 5) rather than to heavy metals. Again the selection coefficients are quite large.

The amount of migration occurring between adjacent tolerant and nontolerant populations is not known with certainty (see Levin and Kerster, 1974, for a review of gene flow in seed plants). The grasses in question are wind pollinated, so pollen flow is undoubtedly more important for gene exchange than is seed transport. The pattern of pollen flow in such grasses is probably similar to that depicted in Figure 16, which pertains to ryegrass (*Lolium perenne*). Pollen flow in ryegrass is considerable over short distances, being 25 percent at a distance of about one meter, for example. The band in Figure 16 falls off rapidly, however; the amount of pollen flow at distances of five meters or more is down to

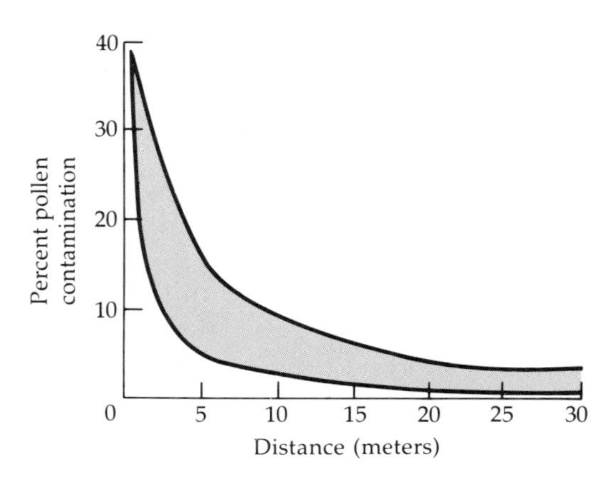

FIGURE 16. Gene flow from a source as measured by the amount of genetically marked pollen in perennial ryegrass *Lolium perenne,* a wind-pollinated species. The curve is a (gray) band rather than a line because the amount of gene flow from a source depends on direction relative to the wind. (After Griffiths, 1950.)

roughly 10 percent. On the other hand, genetic differentiation between heavy metal tolerant and nontolerant populations at mine boundaries occurs over distances of only a few meters, so it is clear that the genetic differentiation is maintained in the face of substantial gene exchange.

How wind direction influences gene exchange has been beautifully illustrated in studies of a copper mine so situated in a valley that the wind always blows in much the same direction (McNeilly, 1968). The situation is illustrated in Figure 17. The histograms in the vertical column on the left-hand side of Figure 17 show an index of copper tolerance among adult plants (gray bars) and plants grown from seed (open bars) at various positions along a transect running perpendicular to the wind. Copper tolerance in adult plants falls off rapidly across the mine boundary; plants grown from seed have a lower tolerance than the plants from which the seeds were collected, evidence of the effect of pollen flow from nontolerant plants situated upwind from the mine.

By contrast, over on the right-hand side of Figure 17 are similar histograms from a transect running parallel to the wind on the downwind side of the mine. Here the index of copper tolerance among adults falls off much more slowly than before, and plants grown from seed are *more* tolerant than the plants from which the seeds were collected, clearly illustrating pollen flow from tolerant plants on the mine site to nontolerant plants downwind. Indeed, plants grown from seed collected 160 meters downwind from the mine are almost three times as tolerant as those grown from seed collected 20 meters upwind from the mine. Considering the situation downwind from the mine, therefore, it is clear that a sufficient amount of gene flow can at least partially offset even strong selection.

As noted earlier, abandoned mine sites differ from surrounding areas not only in their heavy metal contamination but also in such other factors as lower nutrient levels and higher degrees of exposure. Natural selection for heavy metal tolerance is therefore only one aspect of evolutionary adaptation on abandoned mine sites. For *A. odoratum* at the

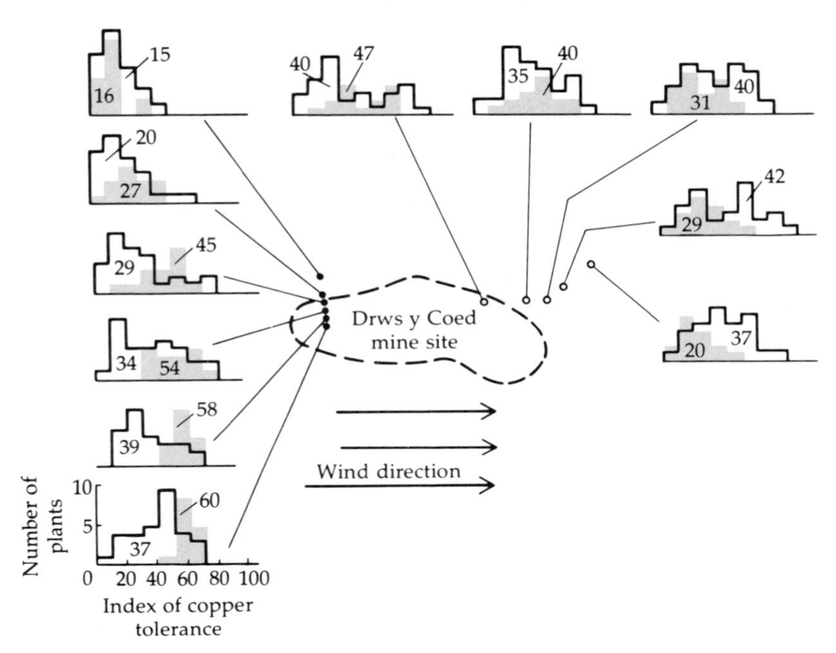

FIGURE 17. Copper tolerance among adult *Agrostis tenuis* and their seedlings in a crosswind transect (solid dots) and a downwind transect (open circles) from a mine site; prevailing wind direction indicated by arrows. Gray histograms give index of copper tolerance among 15 adults, open histograms give index among 30 seedlings grown from their seed. The numbers enclosed in the histograms are the means. Upwind from the mine, the seedlings are less resistant than their seed parents due to nonresistant pollen blown in from upwind of the mine; downwind from the mine, the seedlings are more resistant than their seed parents due to resistant pollen blown downwind from resistant plants on the mine site. (After McNeilly, 1968.)

Trelogan mine site illustrated in Figure 15C, clines for such traits as plant height and flowering time occur (Antonovics and Bradshaw, 1970), plants at the mine site being about 30 percent shorter (presumably an adaptation to exposure) and flowering several days earlier (which tends to reduce "contamination" by pollen flow from nontolerant plants). A remarkable further finding is that *A. odoratum* at the Trelogan mine undergoes self-fertilization about 1 percent of the time, in contrast to a virtual absence of self-fertilization in nearby tolerant plants; this suggests selection for reproductive isolation (Antonovics, 1968; Caisse and Antonovics, 1978).

INDUSTRIAL MELANISM IN MOTHS

INDUSTRIAL MELANISM refers to the evolution of black (melanic) color patterns in certain species of moths that accompanied progressive pollution of the environment by coal soot during the industrial revolution. (The various color forms of the moths are known as MORPHS.) Industrial melanism occurs in England, West Germany, Eastern Europe, the United States, and in other heavily industrialized areas. The species involved in industrial melanism are large moths that fly by night and rest in a sort of cataleptic state by day, often on the trunks of trees, using their cryptic black-and-white mottled color pattern for concealment from such visually cued predators as hedge sparrows, redstarts, and robins (Figure 18). Of nearly 800 species of large moths in the British Isles, where industrial melanism has been most intensively studied, about 100 species are industrial melanics. The best known of these are the peppered moth (*Biston betularia*) and the scalloped hazel moth (*Gonodontis bidentata*). (Industrial melanism has been extensively reviewed by Kettlewell, 1973, from whom much of the following account is taken.)

Industrial melanism evolves extremely rapidly, which is indicative of strong selection. Although melanic *B. betularia* were first noted in Manchester, England, in 1848, their frequency was low (a reasonable guess is 1 percent), but by 1898 the frequency of melanics in the same area was about 95 percent. Since *B. betularia* has nonoverlapping generations with a single generation per year, and since the melanic phenotype is basically due to a single dominant allele (some minor complications in the genetics are discussed below), the selection coefficient required to account for the observed increase in melanic frequency is on the order of .20 (see Haldane, 1956, and Box D for calculations). Comparable increases in the frequency of melanics as an accompaniment to industrialization have been noted elsewhere. In southeastern Michigan, for example, the frequency of melanics in the related moth *Biston cognataria* increased from 1 percent

 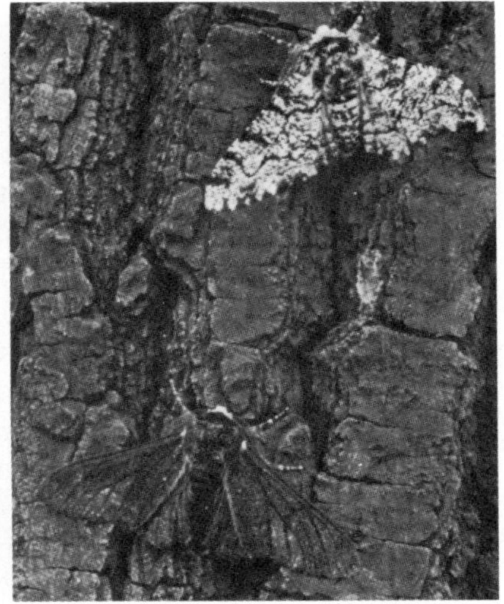

FIGURE 18. Melanic and nonmelanic moths (*Biston betularia*), showing camouflage of light moths on light background and dark moths on dark. (Courtesy of H. B. D. Kettlewell.)

Intensity of Selection for Industrial Melanism 

In Chapter 3, Box G, we showed that, for a favored dominant allele, $dp/dt = pq^2s/\bar{w} \approx pq^2s$, and from this relationship derived the expression $st = \ln(p_tq_0/p_0q_t) + (1/q_t) - (1/q_0)$, where p represents the frequency of the dominant allele and s is the selection coefficient.

The frequency of melanics among *Biston betularia* increased from 1 percent in 1848 to 95 percent in 1898, and we can use the above theory to calculate an approximate s. Let A represent the dominant allele for melanism. With random mating, the frequencies of AA, Aa, and aa will be p^2, $2pq$, and q^2, only aa being nonmelanic. Thus, the frequency of nonmelanics at any time estimates q^2. We may take 1848 as the base year (with $t = 0$), so $q_0^2 = 1 - .01 = .99$, or $q_0 = \sqrt{.99} = .995$; therefore $p_0 = 1 - q_0 = .005$. In 1898, $q_t^2 = 1 - .95 = .05$, or $q_t = \sqrt{.05} = .224$ (so $p_t = 1 - q_t = .776$). *B. betularia* has one generation per year and the generations are nonoverlapping, so $t = 1898 - 1848 = 50$ generations. Using the formula in the paragraph above, we have $s = (1/50)\{\ln[(.776 \times .995)/(.005 \times .224)] + (1/.224) - (1/.995)\} = .20$.

In southeastern Michigan, the frequency of melanics in *Biston cognataria* increased from 1 percent to 80 percent in 30 generations. Calculate s for this case.

to over 80 percent in about 30 years, again suggesting a selection coefficient of around .20 (Box D, and see Box E for an important implication of such large selection coefficients).

It has been experimentally confirmed that differential predation of melanic and nonmelanic morphs by birds is an important source of selection. First, the color morphs preferentially rest on backgrounds that make them least conspicuous. (The effect of this behavior can be seen in Figure 18.) For example, when melanic and nonmelanic morphs are released on a tree trunk that is half light and half dark because the dark half has been artificially stripped of lichens and sprayed with soot, the melanic moths tend to rest on the dark surface and the nonmelanics on the light surface (Figure 19A). Second, when marked melanic and nonmelanic moths are released and later recaptured, the proportion of melanics recovered is about twice as great as that of nonmelanics if the release was in a polluted area but only about one-third that of nonmelanics if the release was in a pollution-free area. Third, direct observations of predation

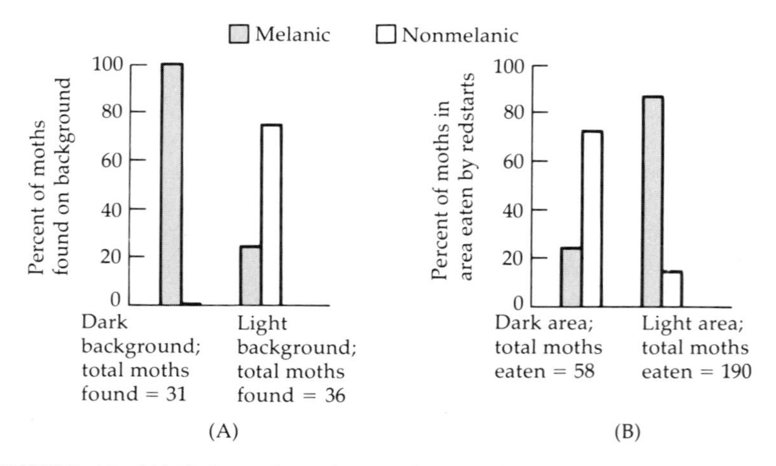

FIGURE 19. (A) Released moths tend to settle on backgrounds where they are least conspicuous; in this case, the choice was between the light half of a tree trunk and a half that had been artificially "polluted" by stripping it of lichens and spraying it with soot. (B) Birds (in this case redstarts) preferentially eat moths that have settled on backgrounds that make the moth conspicuous. (Data from Kettlewell, 1973.)

Cost of Natural Selection

E

Haldane (1957, 1960; see also Felsenstein, 1965) developed a theory for calculating how many organisms must be eliminated by selection based on viability differences in order to effect a change in allele frequency from p_0 to p_t. Consider first a dominant allele, and use the symbols in Box D. In any generation, the proportion of zygotes eliminated by selection is sq^2; the proportion of survivors is $\overline{w} = 1 - sq^2$. The ratio of "selective deaths" to survivors is therefore $sq^2/\overline{w}$, and, summing this from $t = 0$ to t produces $C = \int_0^t (sq^2/\overline{w})dt$, where C is called the COST OF NATURAL SELECTION and represents the total number of selective eliminations (relative to the number of survivors) in the course of evolution at the locus in question.

The quantity C can be expressed in terms of allele frequency. Since $dp/dt = pq^2s/\overline{w}$, $(q^2s/\overline{w})dt = (1/p)dp$. As time goes from 0 to t, p changes from p_0 to p_t. Thus we have $C = \int_0^t (sq^2/\overline{w})dt = \int_{p_0}^{p_t} (1/p)dp = \ln(p_t) - \ln(p_0)$. If $p_t \approx 1$, then $C = -\ln p_0$, approximately. The cost of fixing a selectively favored dominant allele is therefore independent of its selective advantage and depends only on its initial frequency.

a. Calculate C for *B. betularia* and *B. cognataria* using $C = \ln(p_t) - \ln(p_0)$ and the data in Box D.

b. If the alleles are additive, then $dp/dt = pqs/2\overline{w}$ (from Box G of Chapter 3) and $C = \int_0^t (sq/\overline{w})dt$. Show that $C = 2[\ln(p_t) - \ln(p_0)]$, or, if $p_t \approx 1$, $C \approx -2\ln p_0$.

c. If the favored allele is recessive, $dp/dt = p^2qs/\overline{w}$ (from Box G of Chapter 3) and $C = \int_0^t [sq(1 + p)/\overline{w}]dt$. Use $\int [(1 + x)/x^2]dx = -1/x + \ln x$ to derive $C = -1/p_t + \ln p_t + 1/p_0 - \ln p_0$, or, if $p_t \approx 1$, $C \approx -(1 + \ln p_0 - 1/p_0)$.

d. Calculate C for a favored allele with $p_t = 1$ for $p_0 = .01$ and $p_0 = .001$, for the three cases in which the favored allele is dominant, recessive, or additive.

A cost of $C = 30$, say, is the total number of selective eliminations (relative to the number of survivors) that accompanies a change in allele frequency at the locus. In a population that maintains its size at $N = 20{,}000$ individuals in each generation, an allele substitution of cost $C = 30$ entails $CN = 600{,}000$ organisms to be eliminated by selection. Haldane argued that the number of independent gene substitutions a species can undergo must be limited; otherwise there would be so many selective eliminations that

E the population could not maintain its size and would go extinct. For $C = 30$, for example, one allele substitution requires $30N$ selective eliminations; if a species can sustain only $.1N$ eliminations per generation (i.e., 10 percent of the zygotes can be lost to selection without decreasing population size), then there can be, on the average, only $.1N/30N = .0033$ allele substitutions per generation. Haldane's rather severe limitation applies only if selection at each locus is independent of the others; if there were, say, n completely equivalent additive loci, the average fitness of the population with independent selection at the loci is $\overline{w} = (1 - sq)^n$, which tends to 0 as n becomes large.

Fitness is a quantitative trait, however, and for a quantitative trait selection at the loci is not independent. For n completely equivalent additive loci, all with the same allele frequencies, $dp/dt = pqs/2$ and $s = 2i/\sqrt{2npq}$ (from Box B of Chapter 4, assuming $h^2 = 1$). Thus $dp/dt = \sqrt{pq}i/\sqrt{2n}$, or $dt = dp\sqrt{2n}/i\sqrt{pq}$. The time it takes to change the frequencies at all loci from p_0 to p_t is therefore $\int_0^t dt = t = (\sqrt{2n}/i)\int_{p_0}^{p_t} (1/\sqrt{pq})dp = (\sqrt{2n}/i)[\text{Sin}^{-1}(2p_t - 1) - \text{Sin}^{-1}(2p_0 - 1)]$. If $p_t \approx 1$ and $p_0 \approx 0$, $t = \pi\sqrt{2n}/i$, since $\text{Sin}^{-1}(1) = \pi/2$ and $\text{Sin}^{-1}(-1) = -\pi/2$. In this time, there will be n allele substitutions, so the number of substitutions per generation is $n/t = (i/\pi)\sqrt{n/2}$. The proportion of a population saved for breeding was denoted in Chapter 4 by the symbol B; for 10 percent selective eliminations per generation, therefore, $B = 1 - .10 = .90$, and Box C of Chapter 4 gives the corresponding value of i as $i = .195$.

e. Calculate the average number of substitutions per generation for a quantitative trait with $n = 10$, 30, and 100, assuming $i = .195$; note that the numbers are an order of magnitude or more greater than the .0033 calculated under the assumption of independence between loci.

As derived above, $t = \pi\sqrt{2n}/i$ is the time required to fix n equivalent additive loci. The ratio of selective eliminations to survivors in any generation is $(1 - B)/B$, and this is constant over time. Thus the total cost for n allele substitutions is $t(1 - B)/B = \pi\sqrt{2n}(1 - B)/Bi$, or the cost of each substitution is $C = \pi(1 - B)\sqrt{2}/Bi\sqrt{n}$.

f. Calculate the cost per substitution for $B = .90$ and $n = 10$, 30, and 100. Note that these values are much smaller than any of those calculated in part (d). (For more discussion of the cost of natural selection, see Van Valen, 1963; Maynard Smith, 1968; and O'Donald, 1969.)

by redstarts in light and dark areas provide the data in Figure 19B. In dark areas, proportionally more nonmelanic moths were eaten, and in light areas, proportionally more melanics were eaten, though in both types of areas the moths were released in equal numbers.

The story that emerges, then, is this: at the beginning of the industrial revolution, when coal was first burned on a large scale, melanic moths were rare because they were conspicuous on light-colored, lichen-covered trees and heavily preyed upon by birds. As soot and other pollutants belched forth from smokestacks and settled over the countryside, first the lichens (which are very sensitive to air pollution) died, then the trees began to turn black. In such blackened areas, the melanic morph was no longer so conspicuous; indeed, the nonmelanic moth, lacking light-colored areas for background camouflage, was now at a selective disadvantage. The frequency of the dominant mutation leading to melanic coloration thus began to increase, reaching, as has been noted, about 95 percent by 1898.

In the 1950s, air pollution became a matter of public concern because of its harmful effects on human health. In England, following the Beaver report of 1954, certain "smokeless zones" were established. In the smokeless zones, rains began to wash and clean the trees. The bark lightened. The lichens began to reappear on tree trunks. Suddenly the melanic form again found itself at a selective disadvantage, and its frequency began to decline — in one area from 94.8 percent in 1961 to 91.1 percent in 1964 to 89.5 percent in 1974 (Bishop and Cook, 1975).

A number of complications in this otherwise simple story have been uncovered. First, there are genetic complications. While the principal cause of melanism in *B. betularia* is a single dominant allele, in certain areas moths neither light nor dark but of intermediate shades of grey are found; these seem to be due to alleles at the same locus, but other loci may also be involved (Lees and Creed, 1977). Moreover, the melanic specimens of *B. betularia* collected in the last century are considerably lighter in color than those found today, as if in the course of evolution genetic modifiers leading to greater dominance of the melanic allele have been selected. There are also complications in the selective forces. Although bird predation is undoubtedly an important source of selection, there seem to be differences in viability (and perhaps also fertility) between melanic and nonmelanic morphs that cannot be accounted for by predation. And considering the intensity of selection involved, it is surprising that the allele for melanism never became fixed, suggesting that when the melanic morph is at a high frequency some unidentified form of selection comes into play that tends to maintain polymorphism. Yet another complication is in the mating system, as evidence has been found that the color morphs may undergo a degree of disassortative mating; that is to say, melanics sometimes may mate preferentially with nonmelanics.

Despite these complications, industrial melanism is perhaps the most widely known case of evolution in animals, demonstrating an immediate and dramatic response to changed environmental conditions. The example also illustrates how genetic differentiation between populations can be impeded by migration. This aspect of the situation is depicted in Figure 20. The peppered moth, *B. betularia,* is a species that exists in low population density, and individuals must fly relatively long distances to find a mate; as seen in Figure 20A, there is little local differentiation of populations with respect to frequency of melanics: where the frequency of melanics is high it is uniformly high in all populations,

FIGURE 20. (A) Distribution of melanic individuals of *Biston betularia* over an area including Liverpool and Manchester, as viewed from rural Wales. Note the fall-off in frequency in the nonindustrial areas toward the front of the picture. This species of moth occurs in low population densities and must fly relatively long distances to find a mate. The high rate of migration leads to little population differentiation, hence the smooth surface. (B) Distribution of melanic individuals of *Gonodontis bidentata* in a smaller area than in part (A) but viewed from the same perspective. This species occurs in high population densities and the migration rate is low, so there is substantial genetic differentiation between populations, as evidenced by the bumpy surface of the graph. (From Bishop and Cook, 1975.)

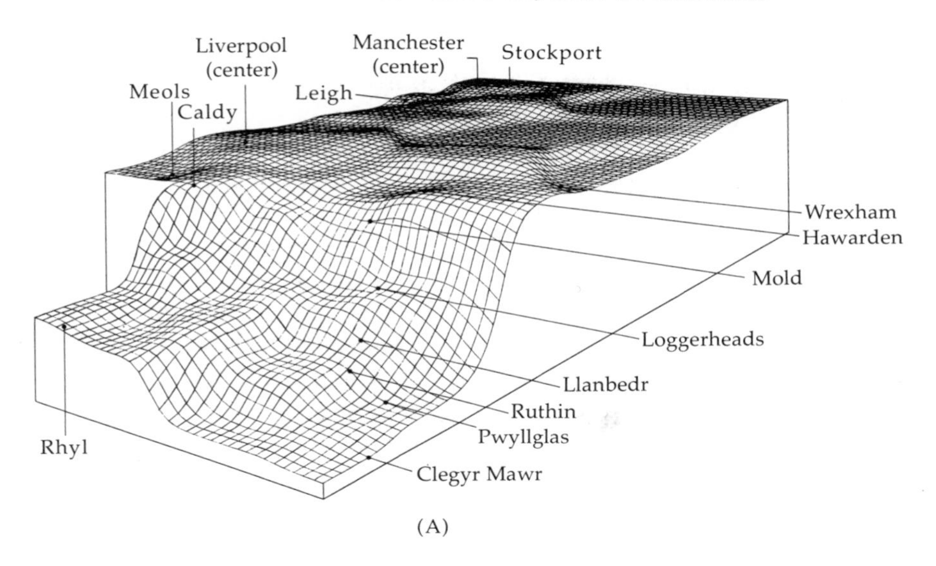
Liverpool
(center)
Manchester
(center)
Stockport
Meols
Caldy
Leigh
Wrexham
Hawarden
Mold
Loggerheads
Llanbedr
Ruthin
Pwyllglas
Rhyl
Clegyr Mawr

(A)

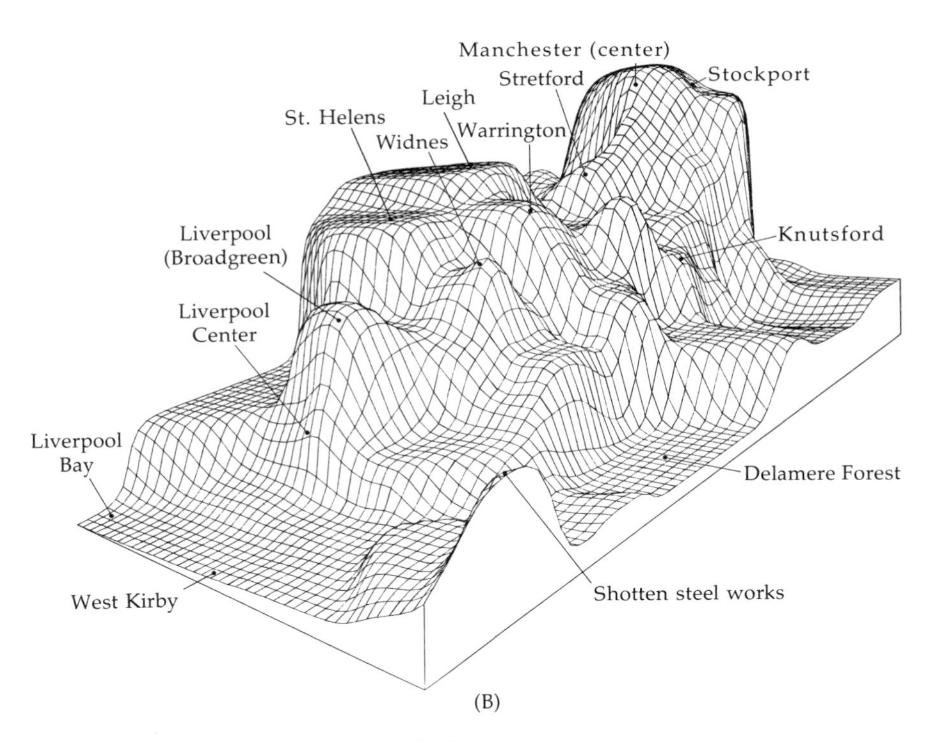
Manchester (center)
Stretford
Stockport
Leigh
St. Helens
Widnes
Warrington
Liverpool
(Broadgreen)
Knutsford
Liverpool
Center
Liverpool
Bay
Delamere Forest
West Kirby
Shotten steel works

(B)

where the frequency is low, it is almost uniformly low. By contrast, the scalloped hazel moth, *G. bidentata,* exists in much higher densities and migration is less extensive; in this case, there is substantial local differentiation of populations with respect to frequency of melanics (Figure 20B).

BATESIAN MIMICRY

Industrial melanism is an example of CRYPTIC COLORATION — coloration for inconspicuousness or camouflage — the nonmelanic form of *B. betularia* being relatively inconspicuous on clean, lichen-covered tree trunks, the melanic form being inconspicuous on soot-blackened trunks. (For a case involving shell coloration in land snails, see Jones et al., 1977.) Many insects exhibit quite an opposite kind of coloration, called WARNING COLORATION, involving bright, often gaudy, highly conspicuous colors and patterns; these insects are usually extremely distasteful to vertebrate predators such as birds and lizards that hunt for prey primarily by sight. The warning colors flaunt the presence of the insects, the predators having learned from prior unpleasant experience to leave them alone. (The distastefulness of the insects often comes from chemicals in the plants they eat as larvae.)

In the mid-nineteenth century, the English naturalist Henry W. Bates noticed a curiosity of butterfly coloration while making collections in Brazil; certain atypical species, whose near relatives were cryptically colored, exhibited the bright warning colors and patterns of quite unrelated species living in the same habitat. Bates reasoned that the atypical species were palatable MIMICS, using their coloration to disguise themselves and masquerade as members of the warningly colored, unpalatable species (the MODELS), thereby deceiving their predators (the DUPES). This phenomenon, in which a palatable species evolves the warning coloration of an unpalatable one in order to escape predation, has come to be known as BATESIAN MIMICRY (for reviews see Sheppard, 1959; Turner, 1977).

Batesian mimicry is only one of several kinds of mimicry. Certain insects mimic twigs or leaves, for example, and some orchids have flowers that mimic the abdomen of female bees in order to attract male bees for pollination. An important kind of mimicry found especially in tropical butterflies is called MÜLLERIAN MIMICRY after Fritz Müller, a nineteenth-century German zoologist (for an example see Turner et al., 1979). In Müllerian mimicry, as in Batesian mimicry, two or more unrelated species will exhibit strikingly similar warning colors and patterns; in contrast to Batesian mimicry, however, all the species involved in Müllerian mimicry are unpalatable, their mutual mimicry serving to protect them mutually from predation inasmuch as a predator need learn to avoid only a single pattern of coloration instead of many. Fewer individuals are lost to the predator's learning experience.

An important premise behind the theory of Batesian mimicry is that vertebrate predators do indeed learn to avoid noxious prey. That this is the case is abundantly clear from the data shown in Figure 21, which are the results of an experiment in which artificial cylindrical "caterpillars" composed of lard and flour and either red or blue food coloring were offered as bait to wild birds, principally blackbirds, song thrushes, house sparrows, dunnocks, and robins. As shown in Figure 21, the birds initially ate (or pecked) the red and blue "caterpillars" equally. When the red baits were made noxious by the addition of quinine sulfate, however, the birds quickly learned to avoid them; by the end of the sixth day only about 2 percent of the baits eaten or pecked were red. Moreover, when birds that had learned to avoid quinine-flavored red baits were presented with unflavored bait that was half red and half blue, they carefully ate away the blue half but left the (to them unknown) equally tasty red half behind.

For Batesian mimicry to occur, of course, the predator must not only learn to avoid the unpalatable model, it must also avoid the palatable but deceptively colored mimic. Whether predators actually are duped by mimics has been examined in a famous case of Batesian mimicry involving two large orange and black butterflies found in the eastern United States, the monarch butterfly (*Danaus plexippus*), which is the model, and the viceroy butterly (*Limenitis ar-*

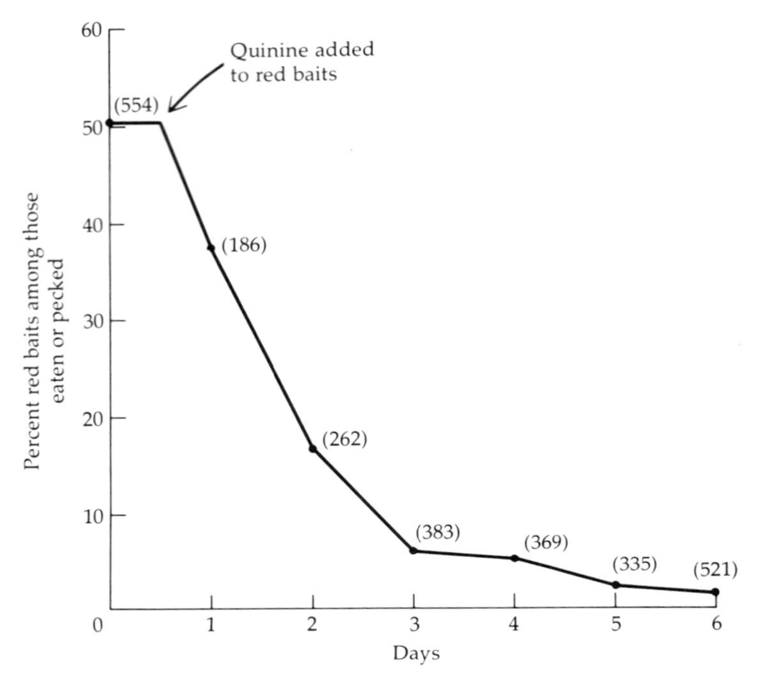

FIGURE 21. Wild birds quickly learn to avoid red baits that have been rendered distasteful to them by the addition of quinine. The ordinate gives the percentage of red (as opposed to blue) baits eaten or pecked, plotted against days. Numbers by the points are total number of baits eaten or pecked on that day. (Data from Ford, 1971.)

chippus), which is the mimic (Figure 22). In these experiments, butterflies were fed to caged Florida scrub jays (Brower, 1958). Birds fed the monarch (the model) reacted violently to its bad taste and quickly learned to avoid it; thereafter they also avoided the viceroy (the mimic). On the other hand, birds never fed the monarch would eat the viceroy. (Such birds did not feed on the viceroy very happily, though; when given a choice between viceroys and unrelated, highly palatable butterflies, the birds spurned the viceroys; this shows that ''palatability'' is really a matter of degree and also that in the monarch–viceroy example, the distinction between Batesian and Müllerian mimicry is slightly blurred.)

In any case, the sort of evidence just described shows that predators do behave as Bates had originally surmised. There is, however, another important test of the theory of Batesian mimicry; if mimics really do mimic a model because of the selective advantage it confers, then butterflies that live in

FIGURE 22. The monarch butterfly (*Danaus plexippus*), the model, and the viceroy butterfly (*Limenitis archippus*), its Batesian mimic. (Courtesy of T. Eisner.)

habitats beyond the geographical range of the model ought not to be mimetic. A splendid illustration of this expected geographical correlation is provided by the highly distasteful model, *Battus philenor,* and its palatable Batesian mimic, *Limenitis arthemis.* The map in Figure 23 shows the essential situation, in which the solid line depicts the approximate northern limit of the model, whereas the dashed lines pertain to *L. arthemis.* Below the lower dashed line at about 42 degrees latitude, *L. arthemis* mimics *B. philenor* in having extreme blue-green iridescence of the dorsal hind wings;

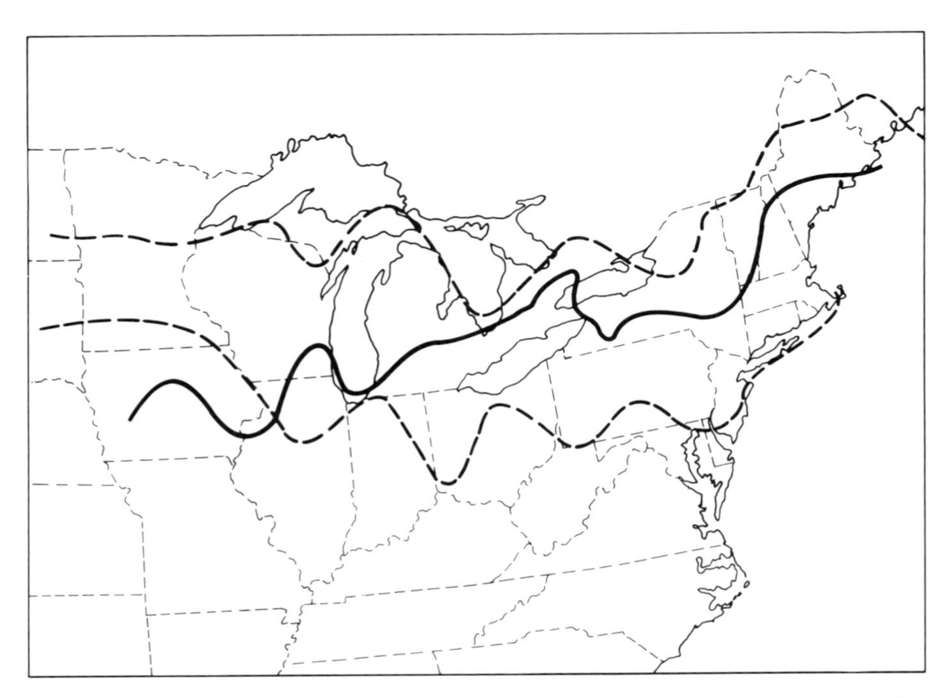

FIGURE 23. The northern limit of the extremely distasteful butterfly model *Battus philenor* (solid line) in the United States coincides with Batesian mimicry by its mimic *Limenitis arthemis.* Below the lower dashed line *L. arthemis* is an excellent mimic; above the upper dashed line there is no mimicry of *B. philenor.* In the intermediate zone between the dashed lines, coinciding with the geographical limit of *B. philenor,* the quality of mimicry is extremely variable due to hybridization between mimetic and nonmimetic forms. (After Platt and Brower, 1968.)

above the upper dashed line at about 43 degrees latitude, *L. arthemis* is nonmimetic, lacking the iridescence but having white bands and extensive red-orange spotting on the wings. Between these two limits the mimetic and nonmimetic forms hybridize, and the butterflies found there are extremely variable, some being rather good mimics while others are very poor. The important point here is that the excellent mimicry found in *L. arthemis* is strikingly correlated with the presence of its model.

Evolution of mimicry. Mimicry is a complex adaptation involving many aspects of morphology and behavior (behavior because mimics often imitate such behaviors of models as flight or resting postures). Such an adaptation could not be expected to evolve gradually, since early steps in the process, when mimicry is poor, would serve only to make the incipient mimic conspicuous to predators without conferring an equal protection due to good mimicry. On the other hand, it is also absurd to imagine that such perfect mimicry as is observed could arise in a single mutational step. Thus, we have the dilemma that mimicry could evolve neither gradually nor all at once. Fortunately, the evolution of mimicry has been greatly clarified by genetic studies of Batesian mimics in the genus *Papilio.* The evidence is too voluminous to quote in detail, but the overall situation is that the principal part of the mimicry is due to a single allele, usually dominant, that by itself promotes a tolerable resemblance to the model. This single allele is said to be at the MAJOR MIMICRY LOCUS. Details of the mimicry phenotype are due to other genes, called MODIFIER GENES, that further improve the resemblance (see Turner, 1977).

The situation, then, seems to be that mutation of a single major mimicry locus, which promotes perhaps a crude resemblance to the model, is the first step in the evolution of mimicry. Details of the mimetic pattern are left to modifiers of relatively small effect. This situation is well illustrated by *Papilio dardanus,* a highly polymorphic butterfly found in Madagascar and the southern half of Africa. The Malagasy race is nonmimetic, but in continental Africa many geographically distinct races occur that are Batesian mimics of different models. (Only the females in this species are mimetic, by the way. Butterflies make great use of vision during courtship, and females choose their mates based largely on color pattern; mimicry in males does not evolve because a mimetic pattern in males too greatly diminishes the chance of being chosen for mating.) When various African mimics are crossed to the Malagasy race, the distinctive details of the mimicry are lost in the hybrids, but the major genes, which are dominant and apparently are alleles, are expressed and produce almost identical patterns.

The initial mutation in the evolution of mimicry need not produce a detailed resemblance to the model. Often a crude resemblance will confer some advantage. In Trinidad, for instance, wing margins of male *Hyalophora promethea* moths were painted black to "mimic" the female of the "model" *Parides anchises.* The moths were released along with unpainted control moths and recaptured, and from the recapture rates a selective advantage due to the artificial mimicry could be calculated. In one such experiment, the selective advantage was around 50 percent, although the estimate is a very approximate one and the advantage depends markedly on the number of mimics relative to the number of models (Brower et al., 1967). Indeed, with too many mimics relative to the number of models, the predators may taste a few and learn that animals with warning coloration are largely palatable; this consideration sets an upper limit to the number of mimics that can be maintained without a breakdown in protection due to mimicry (Charlesworth and Charlesworth, 1975).

Modifiers and Supergenes. Two types of modifiers must be distinguished: SPECIFIC MODIFIERS are modifiers expressed only in the presence of the major mimicry gene and improve the resemblance to the model; NONSPECIFIC MODIFIERS are expressed in all genotypes, and they lead to improved mimicry (and hence increased fitness) in the presence of the major gene but merely promote conspicuousness (and hence decreased fitness) in the absence of the major gene. The distinction between specific modifiers and nonspecific modifiers is important because specific modifiers in natural populations typically exhibit no genetic linkage with the major gene, where nonspecific modifiers are usually tightly linked. Moreover, Charlesworth and Charlesworth (1976a, 1976b) have shown that this relationship between type of modifier and genetic linkage is exactly what is expected on theoretical grounds when the major mimicry gene is polymorphic.

The action of specific modifiers can be illustrated by comparison of the African *Papilio dardanus* with a relative in the Far East, *P. polytes.* In both species, populations that are polymorphic for major mimicry genes also contain specific modifiers that alter the lengths of the tails on the hind wings of the mimetic forms. In *P. dardanus,* the modifiers shorten the tails to improve resemblance to the model; in *P. polytes,* the modifiers lengthen the tails. These modifiers of tail length are unlinked to the major mimicry locus, and the occurrence of shortening or lengthening modifiers is limited to populations in which the mimic exhibits either shorter or longer tails, respectively. That is to say, modification of tail length in the direction of the model does occur when mimics in either *P. dardanus* or *P. polytes* are crossed to other members of the same population; when outcrossed to other populations in these species that lack the appropriate mimetic forms, however, modification of tail length does not occur because the nonmimetic populations do not contain the appropriate modifiers.

Specific modifiers need not be linked to the major mimicry locus because they are expressed only in mimetic forms; in nonmimetic genotypes, they cause no decrease in fitness. For nonspecific modifiers, however, which are expressed in all genotypes and do cause a reduction in fitness in nonmimetic forms, the situation is quite different. Unless the major mimicry gene is fixed or nearly fixed in the population, nonspecific modifiers that are unlinked to the major locus will tend to be eliminated by natural selection. Only if the nonspecific modifiers are tightly linked to the polymorphic mimicry locus will the modifiers increase in frequency. Polymorphism for a major mimicry gene therefore creates a sort of evolutionary "sieve," which sifts through mutations of nonspecific modifier loci as they occur and preserves only those that are tightly linked. In time, there evolves a SUPERGENE consisting of the major mimicry locus and around it a cluster of tightly linked nonspecific modifiers. Within such a supergene, recombination is rare and linkage disequilibrium is extreme; so the block of loci causing mimicry is usually inherited as a single genetic unit, which in genetic crosses creates the illusion that nearly perfect mimicry is due to a single locus (Charlesworth and Charlesworth, 1976a).

Mimicry supergenes have been well analyzed in the Oriental species *Papilio memnon,* where rare recombinants found in natural populations permit the genetic structure of the supergene to be inferred. Six loci have been identified, all within a recombination distance of 0.6 percent (see Turner, 1977). The component loci in the supergene, given in linear order, control the presence or absence of tails, blue marks on the hindwings, white marks on the hindwings, color pattern of the forewing, color of the basal triangle on the forewing, and color of the abdomen. The situation in *P. memnon* is by no means unique, as the multiple mimetic forms in *P. dardanus* and *P. polytes* also seem to be controlled by supergenes of the same sort.

In summary, Batesian mimicry is a complex adaptation involving many aspects of morphology and behavior. It is quite clear that the adaptation evolves in two major stages, first the selection of a major mutation that provides a sudden, often crude, but satisfactory sort of mimicry, and second the selection of specific and nonspecific modifiers that in small increments progressively improve and finally perfect the resemblance. What is not so clear — and this may ultimately be the most interesting question raised by studies of Batesian mimicry — is how many other major adaptations evolve by means of an analogous two-step process.

Population Genetics and Speciation

As noted in Chapter 1, the focus of population genetics is on microevolution — evolutionary changes that occur in local populations. By means of migration and interdeme selection, such changes can be spread throughout an entire species. From the standpoint of population genetics, therefore, the species is perhaps the largest unit of concern. On the other hand, species do split and give rise to new species, and it is the process of SPECIATION (species formation) that ultimately accounts for the obvious and wide-ranging diversity we see among living organisms. It is generally assumed that the study of population genetics is relevant to understanding the genetic changes that occur during speciation, the argument being that the sorts of genetic change involved in the formation of new species are, at least in some cases, not very different from those that occur in the evolution of a single species. Although the relevance of population genetics to understanding speciation may indeed be profound, the sad truth is that we know almost nothing about the genetics of species formation.

THE SPECIES CONCEPT

To discuss the genetic basis of species formation we must be more precise in defining the word "species." A precise but not rigid definition is what is needed, for there is so much diversity among living organisms that an excessively rigid definition would be bound to create difficulty when applied to ambiguous cases.

The first thing to note is that a single, all-inclusive definition of "species" applicable to all situations does not (and perhaps cannot) exist. In many cases a less-than-ideal but pragmatic approach must be followed. For example, a biologist involved in the study of a rare, obscure insect in which genetic studies are impossible or impractical cannot apply a genetic definition of species; the biologist must settle for some other definition, one that can actually be applied to the case in hand. Before the advent of electrophoresis, taxonomists often had to base decisions on morphology, and two populations were considered distinct species if they were sufficiently different morphologically. Such a criterion involved some risks — in certain instances, insect larvae and adults were assigned to different species, in other cases freely interbreeding mimetic and nonmimetic forms of Batesian mimics were assigned to different species, and in still other cases sibling species were not distinguished. (SIBLING SPECIES are species that are genetically distinct but almost identical morphologically.) Paleontologists, too, must often make distinctions between species based largely on morphology, even in almost continuous fossil lineages.

Although there is no single satisfactory definition of "species," two rather general definitions accommodate most cases of interest. Simpson (1961) provides the following definitions.

GENETICAL SPECIES: Genetical species are groups of actual or potentially interbreeding natural populations that are reproductively isolated from other such groups. (This definition is usually called the BIOLOGICAL SPECIES concept, but the term genetical species is more precise.)

EVOLUTIONARY SPECIES: An evolutionary species is a lineage (an ancestral-descendant sequence of populations) evolving separately from others and forming a single unit in its evolutionary role and tendencies.

These definitions warrant a brief discussion, and it is convenient to consider the evolutionary species first.

The Evolutionary Species. This definition is most useful when genetic studies are impossible or impractical, as is the case in paleontological studies of extinct forms, or when dealing with primarily asexual organisms such as bacteria, or when dealing with preserved museum or herbarium specimens. The word "role" as used in the definition is equivalent to the naturalist's word NICHE, meaning a population's whole way of life and relationship to its environment. (Whether two groups of organisms actually share the same niche is often a difficult question of judgment, of course, even for specialists, but much can be learned from, for example, morphology.)

The Genetical Species. This concept is the key one in population genetics, and the word "species" will hereafter mean "genetical species" unless otherwise specified. What the above definition states, in essence, is that two populations belong to the same species if they are actually or potentially capable of exchanging genes — that is to say, if they share a common GENE POOL. Populations that belong to different species do not share a common gene pool by definition and are said to be REPRODUCTIVELY ISOLATED.

The provision for "potential" interbreeding in the genetical definition of species is necessary because many populations do not interbreed due to geographical separation. A hypothetical population on an island may not actually interbreed with the mainland population, for example, but only because of geographical separation; if brought together, the populations may interbreed freely and produce fertile offspring, so they do share a common gene pool in the definition used here. By the same token, reproductive isolation is often a matter of degree. Individuals of two distinct species may hybridize on occasion and even produce fertile offspring, yet reproductive isolation is unbroken as long as gene exchange is sufficiently infrequent that each gene pool evolves independently of the other.

Although the concept of evolutionary species is more general than that of genetical species, evolutionary species almost always coincide with genetical species in sexually reproducing organisms. The fundamental reason for this correspondence is known to ecologists as the PRINCIPLE OF COMPETITIVE EXCLUSION, which states that two distinct species cannot occupy the same niche indefinitely: one species must inevitably change its niche or be driven to extinction by the other (Gause, 1934; Hardin, 1960). Thus, two coexisting populations that share a common niche will almost always interbreed, and two populations with distinct niches will almost always be reproductively isolated.

MODES OF REPRODUCTIVE ISOLATION

The establishment of reproductive isolation is obviously a crucial step in the origin of species of sexually reproducing organisms, but, unfortunately, almost nothing is known about the details of the process. It is known that existing species are reproductively isolated by any of a variety of mechanisms (usually by more than one), and these can be classified broadly according to whether they act prior to mating (PREMATING ISOLATION) or afterward (POSTMATING ISOLATION). (Table IV provides a typical sort of classification of mechanisms resulting in reproductive isolation.) Mechanisms of reproductive isolation found in existing species are, however, themselves products of evolution, and they say little about the processes by which they evolved, especially about the early stages.

A number of experiments have been carried out to determine whether artificial selection can create or amplify behavioral differences between populations and lead to increased premating reproductive isolation. An example involving the sibling species *Drosophila pseudoobscura* and *D. persimilis* is provided by Kessler (1966), who mixed males of one species with females of the other and artificially selected a "high-isolation" line by mating animals that did not mate interspecifically to members of their own species and a "low-isolation" line by remating animals that did mate interspecifically to members of their own species. Artificial selection was successful. After 16 generations the low-isolation line produced 22 percent interspecific matings in an experimental test, and the high-isolation line produced only 2 percent interspecific matings; both of these values are significantly

TABLE IV. Classification of modes of reproductive isolation.

I. Premating isolating mechanisms (prevention of interspecific crosses)
   A. Seasonal or habitat isolation
   B. Behavioral isolation
   C. Mechanical isolation (incompatibility of reproductive structures)

II. Postmating isolating mechanisms (limitation of success of interspecific crosses)
   A. Gametic mortality
   B. Zygotic mortality
   C. Hybrid inviability
   D. Hybrid sterility (partial or complete)

Based on Mayr, 1970.

different from the 10 percent interspecific matings found in the unselected control populations.

Knight and collaborators (1956) have demonstrated that artificial selection for premating reproductive isolation can be effective even within a single species. In experiments with *D. melanogaster,* flies homozygous for either the recessive *ebony*-body (*e*) mutation or the recessive *vestigial*-wing (*vg*) mutation were mixed, and in each generation the offspring of the *e* × *vg* matings (which are heterozygous for both *e* and *vg* and therefore phenotypically normal) were discarded, thus preserving only the offspring of *e* × *e* and *vg* × *vg* matings. After 20 to 30 generations, significant premating isolation had occurred, the selected populations yielding about 39 percent *e* × *vg* matings as compared to 45 percent in the unselected controls.

Not only is direct selection for premating reproductive isolation successful, but Powell (1978) has shown that it rather frequently arises by chance in populations subject to periodic bottlenecks. Powell studied eight populations of *D. pseudoobscura,* each descended from a single pair of flies. Each population was allowed to expand in size (or "flush") in its own large population cage, and after the flush each population was reestablished from a single pair of flies isolated from each cage. After four such founder-flush cycles for each population, Powell found no evidence of postmating reproductive isolation, but three of the eight populations gave evidence of some premating reproductive isolation.

These experiments show that genetic variation affecting reproductive isolation does occur in populations and that at least partial reproductive isolation can evolve in response to artificial selection or, as in the experiments of Powell (1978), by pure chance (see also Thoday and Gibson, 1962, 1970; and Scharloo, 1971). We do not know, however, whether artificial selection can push reproductive isolation to completion (that is, to the formation of new genetical species), nor do we know the genetic basis of the response to selection. The issues are, nevertheless, amenable to experimental analysis and deserve much more attention than they have so far received.

CHAPTER FIVE

THE GEOGRAPHY OF SPECIATION

The key element in speciation among sexual organisms is the evolution of reproductive isolation, that is to say, the evolution of mechanisms that restrict or prevent gene flow (migration) between populations. Geographical barriers, such as rivers or mountains, often suffice to prevent gene flow between populations, and such geographical effects are the basis of ALLOPATRIC SPECIATION, a mode of speciation mediated by physical separation of the populations undergoing speciation (Mayr, 1963, 1970). Alternative modes of speciation involving no pronounced geographical separation are also known. In PARAPATRIC SPECIATION (White, 1968, 1974), populations that are contiguous undergo the evolution of reproductive isolation; in SYMPATRIC SPECIATION (Maynard Smith, 1966), a new species arises entirely within the geographical range of its parental form (for recent reviews, see Bush, 1975; Carson, 1975, 1978; White, 1978; Templeton, 1980). Allopatric, parapatric, and sympatric speciation are the principal modes of species formation; although more detailed classifications are useful for some purposes (White, 1978), we will focus on these three broad categories.

Allopatric Speciation. There are actually two types of allopatric speciation, differing somewhat in their genetic concomitants. The types are distinguished by whether or not a marked founder effect or population bottleneck is involved.

Type Ia allopatric speciation (in the terminology of Bush, 1975) can occur when a geographical barrier (a river, desert, mountain range, glacial tongue — any number of things) splits a species with a wide geographical range into two or more large subpopulations between which gene flow is severely restricted (Figure 24A). In such a situation, each geographically isolated subpopulation is free to evolve in its own way, adapting to its own local conditions. Although the initial large sizes of the subpopulations insure that each will carry essentially all the genetic variation present in the original population, the subpopulations will diverge genet-

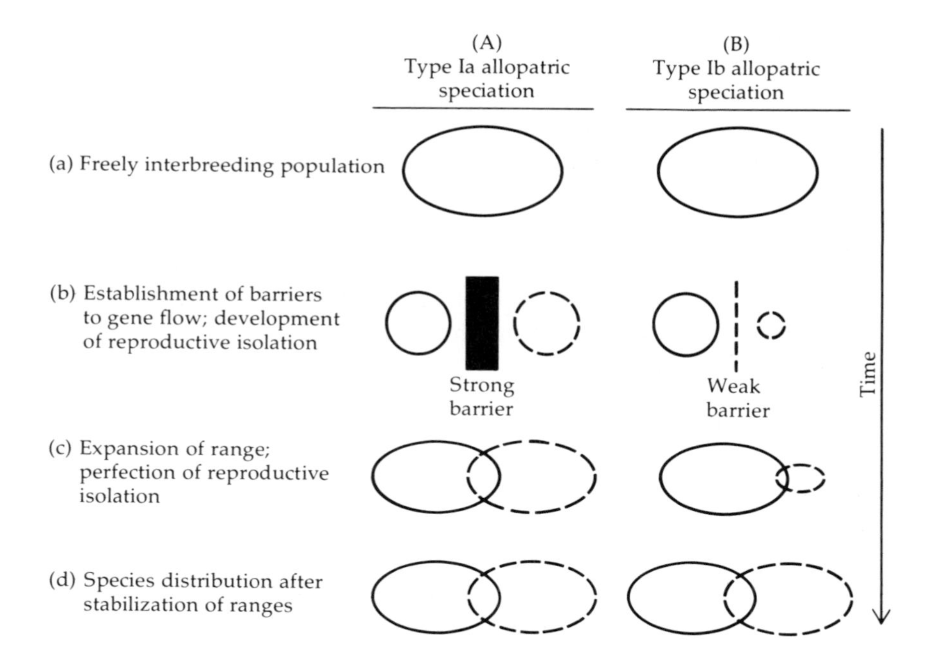

FIGURE 24. (A) Type Ia allopatric speciation, in which a strong geographical barrier to gene flow splits a population (a) into two large subpopulations (b) that evolve independently and develop incipient reproductive isolation. Reproductive isolation is perfected should the barrier then disappear or become surmounted (c), leading to overlapping species ranges (d). (B) Type Ib allopatric speciation, in which a small subpopulation (with consequent large founder effects) splits off from a large population (a) and (b) and, by random genetic drift, develops incipient reproductive isolation; the populations thereafter follow patterns (c) and (d) of Type Ia allopatric speciation. (After Bush, 1975.)

ically, each evolving toward an adaptive peak appropriate to its local conditions. While the subpopulations are geographically isolated, there is no direct selection for reproductive isolation. Reproductive isolation could evolve as a correlated response to natural selection for other characteristics, however, or genetic differentiation could become so pronounced that hybrid genotypes would be selectively inferior. In the latter case, should the subpopulations again come into contact, there would be selection in favor of premating reproductive barriers (this is called the WALLACE EFFECT), because those organisms that hybridize produce inferior offspring. Type Ia allopatric speciation is apparently a slow process, and the sorts of genetic changes that occur in populations undergoing speciation are the same sorts of microevolutionary changes that occur in any large evolving population.

Type Ib allopatric speciation (Figure 24B) is geographically similar to Type Ia, but Type Ib includes a pronounced founder effect due to a small number of propagules fortuitously founding a new colony geographically isolated from the parental population. The initial colony will be small, perhaps consisting of a single fertilized female, and the resulting population will therefore experience extensive inbreeding and random genetic drift (Carson, 1968, 1971; Templeton, 1980). Indeed, because of the change in allele frequencies, the population may evolve to a new selective peak even in the absence of an environmental change. As with Type Ia allopatric speciation, initial reproductive barriers are assumed to arise by chance and to be perfected first in the area of overlap should the populations regain contact.

Allopatric speciation is generally agreed to be the principal mode of speciation in most groups of plants and animals. Type Ia appears to be characteristic of most outbreeding animals of high vagility (VAGILITY refers to the ability of organisms to disperse) that typically have low reproductive rates and exist at population densities near the carrying capacity of the environment. Included here are large mammals such as felids (cats) and canids (dogs), most birds, certain reptiles and amphibians, and most large fish. Social organization into family groups, harems, or herds promotes Type Ib speciation, especially in organisms of low vagility such as primates. High vagility in organisms that exist in patchy environments also seems to promote speciation according to the Type Ib mode.

Allopatric speciation has been particularly well-studied in the Hawaiian Drosophilidae, whose 300 known endemic species (the total is estimated at 700 or more) have evolved explosively in the few million years or less since the islands were formed by volcanic action. (The age of Kauai, the oldest island, is about 5.6 million years; that of Hawaii, the youngest island, is about one million years.) The Hawaiian drosophilids are extraordinarily diverse in morphology and sexual behavior, and since the Hawaiian Islands lack native fleshy fruits like those that provide food and shelter for most continental species, the Hawaiian species have had to acquire highly specialized and sometimes bizarre feeding habits, such as feeding on slime fluxes from trees, decaying bark, or even on spider eggs. Speciation in the Hawaiian drosophilids has been studied extensively by Carson and colleagues (Carson, 1971, 1975, 1978; Carson and Kaneshiro, 1976; Carson and Bryant, 1979), who have argued that such rapid speciation may be promoted by repeated cycles of population expansion (population flush) under favorable environmental conditions when selection is relaxed (Carson, 1968); such conditions may permit more genotypes to survive and reproduce than would at other times, and, since emigration pressure will be greatest at the peak of the flush when the population is most dense, there will be an enhanced likelihood that founders of new colonies will be genotypically quite distinct from the parental population. Speciation by this sort of FOUNDER-FLUSH model is a special case of Type Ib speciation, and the finding of Powell (1978) that founder-flush cycles often lead to incipient reproductive isolation lends credence to the hypothesis.

Speciation by the Type Ib mechanism can be extremely rapid, in contrast to Type Ia, which tends to be slow. Among the cichlid fish of the great lakes of East Africa, for example, five species are endemic to Lake Nabugabo, though the lake is only 3500 years old, and more than 150 species are endemic to 750,000-year-old Lake Victoria (Fryer and Iles, 1972, discussed in Bush, 1975).

Parapatric speciation. Contiguous populations sometimes undergo speciation, especially those of organisms having very low vagility that exist in numerous small subpopulations; such parapatric speciation has been studied in the flightless morabine grasshoppers of Australia (White, 1969), snails of the genus *Partula* (Clarke and Murray, 1969), the burrowing mole rats of Israel (Nevo and Cleve, 1978), and plants that have limited pollen dispersal, such as those cited earlier in connection with heavy metal tolerance.

In the case of parapatric speciation, low vagility combined with small subpopulations serves to accentuate the effects of random genetic drift and promotes genetic differentiation. Postmating reproductive isolation due to hybrid inviability or infertility can arise rapidly, especially if chromosomal rearrangements, which may become fixed in small populations due to chance, are involved. In the approximately 200 species of flightless grasshoppers of the subfamily Morabinae, for example, 39 chromosomal fusions (due to translocations) and 22 dissociations (due to chromosome breakage) are known to have occurred, and most species are chromosomally unique. Zones of overlap between morabine species are invariably narrow; hybridization does occur in these zones, but gene flow between the species is essentially eliminated by the low fertility of the chromosomally abnormal hybrids (White, 1978).

Sympatric speciation. In certain cases, a new species can evolve well within the dispersal range of the parental species. This mode of speciation is called sympatric speciation, and it is apparently limited to certain special types of organisms such as insect parasites, especially those that are host-specific and in which courtship and mating occur on or in the host. In a population of this sort, any individuals having an altered host specificity will court each other and mate while associated with their special host, and they will in this manner be reproductively isolated from the parental population, the premating reproductive barrier being brought about by the distinct host preferences. Under these circumstances there is no strong selection for premating reproductive isolation, and in at least one case, which involves European cherry fruit flies of the genus *Rhagoletis* (Boller and Bush, 1973), species can be hybridized and yield fully fertile offspring.

The number of genetic loci involved in host specificity may be few; for example, those concerned with chemoreception. A single locus controls recognition of host plants in the gall-forming fruit fly *Procecidochares,* for example (Huettel and Bush, 1972). Moreover, survival on a new host need not be very good at first; subsequent evolution of a conventional sort will progressively adapt the new species to life on its new host.

With sympatric speciation, therefore, a new species can arise almost instantaneously and may differ from the parental species at only a few loci. There is another virtually instantaneous mode of speciation that often occurs sympatrically or parapatrically, and this mode of speciation is of fundamental importance in the evolution of plants, especially higher plants. The mode of speciation is by means of the formation of fertile autopolyploids or allopolyploids, and its importance in plant evolution was detailed in Chapter 1. Polyploidy leads to instantaneous postmating reproductive isolation because the interspecific hybrids are chromosomally unbalanced and highly infertile. (There may, of course, be subsequent selection for premating reproductive isolation.) Although autopolyploids would seem to be genetically much the same as their diploid ancestors, the alteration of gene dosage in the autopolyploids seems to affect gene regulation radically in such a way that the biochemical milieu in the autopolyploids is unique (Levy, 1976).

GENETICS OF SPECIES DIFFERENCES

The amount of genetic change involved in speciation is a difficult question to investigate experimentally because species are almost never caught in the act of speciation (or, if they are, the situation is unrecognized). Extant species will typically have undergone a long period of evolutionary divergence since their speciation occurred, so genetic differences between extant species are certainly greater, and perhaps much greater, than those involved in the speciation process itself. Furthermore, the amount of genetic change involved in speciation is likely to depend on the type of organism and the mode of speciation; Type Ia allopatric speciation probably involves the greatest change because it requires the longest time (though much of the genetic differentiation in this case would be attributable to evolutionary adaptation to local environments while the populations were geographically isolated), whereas sympatric speciation, when it occurs, may involve fewer genetic changes. A further point involves the types of genetic changes that occur during speciation, specifically whether they primarily involve visible changes in chromosome structure, changes in structural genes (that is, genes that code for polypeptides), or changes in regulatory genes (that is, genes that modify the expression of other genes).

Chromosomal Differences. The KARYOTYPES (that is, chromosomal complements) of literally thousands of species have been examined, though often at rather low levels of resolution not involving, for example, special procedures for producing chromosome bands or identifying chromosomal regions rich in some particular kinds of DNA such as highly repetitive DNA or ribosomal DNA. For those taxa that have been examined at relatively high resolution (and such species as the drosophilids are exceptionally suited for comparative cytogenetics because of their giant polytene salivary gland chromosomes), the situation is clear: most species, but not all, differ in their karyotypes (see Wilson et al., 1974, and White, 1978, for reviews). Of the several hundred described Hawaiian drosophilids, for instance, only a small fraction are HOMOSEQUENTIAL (having identical banding patterns on their polytene chromosomes); most of the karyotypic differences are inversions rather than the more complex rearrangements that result in chromosome fusion or dissociation. Such complex rearrangements are common in the morabine grasshoppers, however; and in some morabine genera, each species is karyotypically unique. Nor is extensive karyotype evolution restricted to insects. In the evolutionary divergence of humans and chimpanzees, there have been at least five recognizable pericentric inversions and one chromosome fusion (resulting in human chromosome number 2; Miller, 1977; de Grouchy et al., 1978). The situation is essentially the same in plants, though polyploidy represents still another level of karyotypic diversity (Grant, 1971; Stebbins, 1976; Levin and Wilson, 1976).

Although karyotype evolution occurs in virtually all lineages, the rate of karyotypic change can vary a hundredfold in different groups; this is apparent in Table V, which provides estimates of the number of karyotypic changes per lineage per million years in various groups of vertebrates. (The term "total karyotypic changes" in the table refers to changes in the number of chromosomes in the karyotype plus changes in the number of chromosome arms.) As seen in Table V, the average rate of total karyotypic change in mammals is about ten times greater than that in other ver-

TABLE V. Rates of karyotypic change in vertebrate genera.

Group	Number of genera examined	Average age of genera × 10^6 years	Karyotypic changes per lineage per 10^6 years		
			Chromosome number	Arm number	Total
Mammals					
Horses	1	3.5	0.609	0.786	1.395
Primates	13	3.8	0.333	0.413	.746
Lagomorphs	3	5.0	0.230	0.403	.633
Rodents	50	6.0	0.178	0.253	.431
Artiodactyls	15	4.2	0.364	0.197	.561
Insectivores	7	8.1	0.074	0.113	.187
Marsupials	15	5.6	0.052	0.124	.176
Carnivores	10	12.9	0.042	0.036	.078
Bats	15	9.0	0.028	0.031	.059
Whales	2	6.5	0.000	0.025	.025
Average for mammals		6.5	0.129	0.166	.295
Other vertebrates					
Lizards	16	20.1	0.027	0.031	.058
Snakes	14	12.1	0.007	0.041	.048
Turtles and crocodiles	14	45.2	0.0006	0.0016	.002
Frogs	15	26.4	0.011	0.012	.023
Salamanders	11	23.4	0.006	0.008	.014
Teleosts	12	5.7	0.003	0.026	.029
Average for nonmammals		22.1	0.009	0.020	.029

From Bush et al., 1977.

tebrate groups; and even among mammals, the rates vary considerably. The highest known rate of karyotype evolution in mammals is found in horses, primates are second, and the lowest rate is in whales. In general, high rates of karyotype evolution are associated with organisms that have low vagility (e.g., rodents) or that typically exist in relatively small interbreeding groups (e.g., herds in horses or primate troops); both of these characteristics would promote fixation of chromosome rearrangements by random genetic drift (Wilson et al., 1975; Bush et al., 1977).

The role of chromosome rearrangements in the speciation process itself is still unclear, however. In groups that undergo speciation predominantly according to the Type Ib allopatric model or according to the parapatric model, chromosomal differentiation owing to founder effects might be common and indeed crucial to the origin of postmating reproductive isolation. In such groups as frogs and whales, which are thought to undergo speciation primarily according to the Type Ia allopatric model, karyotypic changes that arise during speciation would be expected to be less frequent.

Structural Genes. The opening sections of Chapter 3 reviewed how comparisons of the amino acid sequence of proteins from various species can be used to infer evolutionary relationships among those species. Indeed, because of the MOLECULAR CLOCK — that is, the roughly constant rate of amino acid substitution that seems to occur in many proteins — such comparisons also yield estimates of the length of time since two present-day species underwent evolutionary divergence. Crucial as such observations are for studying relationships between higher taxonomic categories (genera, families, orders, and so on), their use in studying the speciation process itself is limited because rates of amino acid substitution are generally so slow that recently diverged species will usually have few or no differences in amino acid sequence (see Figures 18, 19, and 20 in Chapter 3 for examples).

Genetic differences between closely related species at the level of structural genes can be studied by electrophoresis

(Chapter 2). With regard to electrophoretic variants (allozymes), most closely related species differ only in allele frequencies rather than in the presence or absence of particular alleles (see Ayala, 1975; Avise, 1976; Gottlieb, 1976). Comparison of populations, therefore, calls for some appropriate measure of genetic distance. A number of measures have been discussed earlier (such as F_{ST} in Chapter 3 and the chord-length measure in Box C of Chapter 3; see also Box F of this chapter for an important method of analysis of population structure).

Hierarchical Structure of Populations

F

Populations are usually structured as groups within groups. A total population may, for instance, be divided into subpopulations, and these further divided into demes within subpopulations. As a real example, a species of freshwater fish in some geographical area is divided into populations within lakes, and populations within lakes may be further divided into schools. In such cases, it is of interest to apportion the total genetic differentiation according to the different levels of population structure.

The table here presents hypothetical frequencies of an allele in seven randomly mating demes grouped into three subpopulations. Below the data are calculations to extract various quantities of interest, such as the mean allele frequency in each subpopulation ($\bar{p}_{i.}$) and the overall mean ($\bar{p}_{..}$, which we will also symbolize as $\bar{p}$). From Box A of Chapter 4 we know that the total sum of squares

	Subpopulation 1	**Subpopulation 2**	**Subpopulation 3**	Grand Totals
Deme 1	.9 (.18)	.6 (.48)	.4 (.48)	
Deme 2	.8 (.32)	.5 (.50)	.3 (.42)	
Deme 3	.7 (.42)			
$\Sigma_j p_{ij}$	2.4	1.1	.7	4.2
n_i	3.0	2.0	2.0	7.0
$\bar{p}_{i.}$	.8 (.32)	.55 (.495)	.35 (.455)	[$\bar{p}_{..}$ = .6]
$n_i\bar{p}^2_{i.}$	1.92	.605	.245	2.77
$\Sigma_j p^2_{ij}$	1.94	.61	.25	2.80

F is $TSS = \Sigma\Sigma p_{ij}^2 - (\Sigma n_i)\bar{p}^2 = 2.8 - (7)(.6)^2 = .280$ and that the between-subpopulation sum of squares is $BSS = \Sigma n_i \bar{p}_{i.}^2 - (\Sigma n_i)\bar{p}^2 = 2.77 - (7)(.6)^2 = .250$. The within-subpopulation sum of squares is then $WSS = TSS - BSS = .03$.

Define $\sigma_{DS}^2 = WSS/\Sigma n_i = .004286$ (in this case), $\sigma_{ST}^2 = BSS/\Sigma n_i = .035714$, and $\sigma_{DT}^2 = TSS/\Sigma n_i = .04$. (Since $TSS = BSS + WSS$, of course, $\sigma_{DT}^2 = \sigma_{DS}^2 + \sigma_{ST}^2$.)

a. Calculate the variance in allele frequency for each of the subpopulations in the table; let σ_i^2 represent the variance in the ith subpopulation. Show that the weighted average variance among demes within subpopulations, which is $(\Sigma n_i \sigma_i^2)/\Sigma n_i$, is numerically equal to σ_{DS}^2.

b. The weighted variance between population means is defined as $\Sigma n_i(\bar{p}_{i.} - \bar{p})^2/\Sigma n_i$. Calculate this variance and show that it numerically equals σ_{ST}^2.

c. The total variance in allele frequency is $\Sigma\Sigma(p_{ij} - \bar{p})^2/\Sigma n_i$. Show that this numerically equals σ_{DT}^2.

Since the demes are randomly mating, the heterozygosity for the jth deme in the ith subpopulation is $H_{ij} = 2p_{ij}(1 - p_{ij})$. (The values of H_{ij} are tabulated in parentheses alongside the corresponding allele frequencies.) The heterozygosity in the ith subpopulation, were it a single random mating unit, would be $H_i = 2\bar{p}_{i.}(1 - \bar{p}_{i.})$. (These values are tabulated alongside the corresponding values of $\bar{p}_{i.}$.) The heterozygosity in the total population, were it a single random mating unit, would be $H_T = 2\bar{p}(1 - \bar{p}) = .48$.

d. Define $H_D = (\Sigma\Sigma H_{ij})/\Sigma n_i$ as the average heterozygosity within demes, and define $H_S = (\Sigma n_i H_i)/\Sigma n_i$ as the weighted average heterozygosity within subpopulations. Calculate H_D and H_S and show that (numerically) $H_S - H_D = 2\sigma_{DS}^2$, $H_T - H_S = 2\sigma_{ST}^2$, and $H_T - H_D = 2\sigma_{DT}^2$.

$F_{DS} = (H_S - H_D)/H_S$ measures the reduction in heterozygosity associated with the division of subpopulations into demes.

$F_{ST} = (H_T - H_S)/H_T$ measures the reduction in heterozygosity associated with the division of the total population into subpopulations.

$F_{DT} = (H_T - H_D)/H_T$ measures the reduction in heterozygosity associated with the aggregate effect of all levels of population structure.

e. Calculate F_{DS}, F_{ST}, and F_{DT}, and show numerically that $(1 - F_{DS})(1 - F_{ST}) = 1 - F_{DT}$.

f. The weighted average of $\bar{p}_{i.}(1 - \bar{p}_{i.})$ over all subpopulations is $[\Sigma n_i \bar{p}_{i.}(1 - \bar{p}_{i.})]/\Sigma n_i$, which we will symbolize as $\tilde{p}(1 -$

$\tilde{p}$). (Note that $\tilde{p}(1 - \tilde{p}) = H_S/2$.) Calculate $\sigma^2_{DS}/\tilde{p}(1 - \tilde{p})$, $\sigma^2_{ST}/\bar{p}(1 - \bar{p})$, and $\sigma^2_{DT}/\bar{p}(1 - \bar{p})$, and show that these numerically equal F_{DS}, F_{ST}, and F_{DT}, respectively. F

You will have noted by now that F_{DS} for the data in the table is much smaller than F_{ST}. This finding means that most of the genetic differentiation among demes is associated with differentiation between subpopulations; little is associated with differentiation between demes within subpopulations.

g. Selander et al. (1969) studied the *Esterase*-5 locus in the house mouse (*Mus musculus*) in 23 regions (*D*) grouped within five geographical subdivisions (*S*) of Texas (*T*). Thus $\Sigma n_i = 23$, and for their data $\Sigma\Sigma p^2_{ij} = 15.794698$, $\Sigma n_i \bar{p}^2_{i.} = 15.349228$, and $\bar{p} = .816$. Calculate σ^2_{DS}, σ^2_{ST}, σ^2_{DT}, H_D, H_S, H_T, F_{DS}, F_{ST}, and F_{DT}. Interpret the amount of genetic differentiation found at the levels of subdivision and region based on the relative values of F_{DS} and F_{ST}.

(Note: Estimation of variance components from actual data is somewhat more complex than indicated above, as the calculated variances must be corrected for sampling errors in the mean as well as for other accidents of sampling; the procedure is spelled out in Wright [1978, p. 86 and following]; if the sample sizes are reasonably large the corrections are usually minor, however; see also Cockerham, 1973.)

A widely used measure of genetic distance is due to Nei (1975). The basis of the measure is a so-called NORMALIZED IDENTITY, which expresses the probability that a randomly chosen allele from each of two different populations will be identical, relative to the probability that two randomly chosen alleles from the same population will be identical. ("Identical" in this context means "indistinguishable," not "identical by descent.") The situation is illustrated in Table VI, which shows the alleles found at a locus and the allele frequencies that occur in two hypothetical populations, denoted X and Y. Define J_{XX} as the probability that two alleles chosen at random from population X are identical; obviously $J_{XX} = \Sigma p_i^2 = p_1^2 + p_2^2 + p_3^2 + p_4^2 + p_5^2 = (.6)^2 + (.3)^2 + (.05)^2 + (.05)^2 + (0)^2 = .455$. (Note that with random mating, J_{XX} equals the homozygosity in population X.) Likewise J_{YY} is defined as the probability that two alleles chosen at random

TABLE VI. Frequencies of five alleles at a locus in each of two hypothetical populations.

Alleles	Population X	Population Y
1	$p_1 = .60$	$q_1 = .30$
2	$p_2 = .30$	$q_2 = .60$
3	$p_3 = .05$	$q_3 = 0$
4	$p_4 = .05$	$q_4 = 0$
5	$p_5 = 0$	$q_5 = .10$

See the text for calculation of the normalized identity (I) and the standard genetic distance (D) for this locus.

from population Y are identical, and $J_{YY} = \Sigma q_i^2 = (.3)^2 + (.6)^2 + (0)^2 + (0)^2 + (.1)^2 = .460$. Now, define J_{XY} as the probability that two alleles are identical when one is chosen from population X and the other from population Y; in this case $J_{XY} = \Sigma p_i q_i = (.6)(.3) + (.3)(.6) + (.05)(0) + (.05)(0) + (0)(.10) = .360$. Nei defines the normalized identity, I, for this locus as

$$I = \frac{J_{XY}}{\sqrt{J_{XX} J_{YY}}} = \frac{.360}{\sqrt{(.455)(.460)}} = .7869$$

and the STANDARD GENETIC DISTANCE, D, as

$$D = -\log_e I = -\log_e(.7869) = .2396$$

Of course, the genetic distance between two populations should be based on as many loci as possible. For many loci, J_{XX}, J_{YY}, and J_{XY} are defined as the arithmetic means of the corresponding quantities, where the averages are taken over all loci (including monomorphic loci). Quantities I and D are then calculated as above in terms of the average J's.

Relationships between standard genetic distances observed for various taxonomic categories are about as expected. Higher categories are on the average more different than lower ones. Genetic distances are highly variable, however, so some taxonomic categories overlap. Nei (1976) estimates that for local races of a species, D ranges from nearly 0 to 0.05; for subspecies $D = 0.02$ to 0.2. (The distinction between local races and subspecies is somewhat arbitrary; a RACE is usually defined as a group of populations of a species that are geographically separated from other such groups and that differ in allele frequencies. If races differ enough in easily observable characteristics, they are often given Latin names and formally recognized as subspecies.) For full species, $D = 0.10$ to 2.0; sibling species, which are morphologically nearly identical, generally have D values two to three times smaller than the D's between morphologically distinct species. For genera, D is usually greater than 1.

If nucleotide substitutions in a gene occur randomly and independently, then D can be interpreted as the average number of electrophoretically detectable nucleotide substitutions that have occurred between the populations since they underwent speciation. For example, the minimum value of D known between species is $D = 0.05$, which occurs between the sibling species *D. pseudoobscura* and *D. persimilis*. *Drosophila* is estimated to have roughly 5000 loci coding for structural genes, so $D = 0.05$ translates into $5000 \times 0.05 = 250$ electrophoretically detectable nucleotide substitutions separating the sibling species; if we make the assumption that only about a quarter of all nucleotide substitutions are detected by routine electrophoresis (the validity of this assumption is discussed in Chapter 2), then the species would differ in about $250 \times 4 = 1000$ nucleotide substitutions (an average of 500 substitutions in each lineage) since they underwent speciation. Unfortunately, we do not know how many of these substitutions were directly involved in the speciation process itself.

Regulatory Genes. One of the interesting findings from studies of genetic distance is that humans and chimpanzees, while taxonomically assigned to different families, have $D = 0.62$, a value typical of that between species and even exceeded by some sibling species in *Drosophiia* (King and

Wilson, 1975; see also Wilson et al., 1974). Such a small value of genetic distance between humans and chimpanzees does not adequately reflect their extensive morphological divergence, but comparisons of the amino acid sequences of various proteins, as well as studies of protein differences by immunological methods, fully support the remarkable similarity of structural genes between the species. King and Wilson (1975) have therefore suggested that evolution at the level of morphology may occur by means of changes in regulatory genes rather than by changes in structural genes; indeed, they suggest that major evolutionary adaptations of all types may involve regulatory genes.

The role of regulatory-gene changes in adaptive evolution is difficult to assess at present, partly because appropriate techniques for the large-scale study of regulatory genes have not yet been developed and partly because the concept of "regulatory gene" is still somewhat fuzzy and not operationally defined. Polymorphisms of regulatory genes (as of structural genes) do exist in natural populations, so the genetic variation on which natural selection can act is available (see, for example, Mukai and Cockerham, 1977; McDonald et al., 1977; Barnes and Birley, 1978; Lewis and Gibson, 1978; McDonald and Ayala, 1978; Johnson et al., 1980; and Laurie-Ahlberg et al., 1980). Another line of evidence in favor of the regulatory gene hypothesis is that levels of enzyme activity, which are often under the control of regulatory genes, can vary enormously among species. For example, the amount of pancreatic ribonuclease per gram of pancreas is a thousand times higher in ruminants than it is in humans, presumably in order to digest the RNA of cellulose-degrading microbial symbionts that pass from the rumen into the intestine. The evolution of microbe-dependent ruminant nutrition has been accompanied by major changes in the activity of several other digestive enzymes as well (for a review, see Wilson et al., 1977).

Studies of adaptive evolution by means of changes in gene regulation are only beginning, however, and they have yet to be carried out on a large scale in natural populations and systematized. It nevertheless seems likely that regulatory genes indeed are of major importance in microevolution and perhaps also in many aspects of speciation.

Problems

1. Taking into account both maximum long-term evolutionary response and the rapidity of immediate evolutionary response, why is the population structure depicted in Figure 6A most suited to the sort of fitness model typified by Table I and the population structure in Figure 6B most suited to the fitness model in Table II?
2. For a neighborhood size of N and a rate of migration of m per generation, under which conditions would the shifting balance process be most favored: (a) $N = 50$, $Nm = .25$; (b) $N = 100$, $Nm = .50$; (c) $N = 50$, $Nm = .10$; (d) $N = 100$, $Nm = .20$?
3. How could one determine experimentally whether the phenotypic "surface" (see Figure 2) of a quantitative trait had multiple peaks?
4. How can group selection in the strict sense occur when members of groups are genetically related? (Recall that the term "group selection in the strict sense" is used in the text to mean behavior that is detrimental to the individual but beneficial to other members of the group.)
5. "Cheap" altruism is altruism with no fitness cost to the altruist ($x = 0$). For what values of the coefficient of relationship would cheap altruism be favored?
6. For a population in which there is no inbreeding, the coefficient of relationship between half sibs is 1/4. If an individual undergoes a loss of .4 units in fitness by performing an altruistic act toward a group of n half siblings, each of whom gains .2 units in fitness as a result of the altruism, how large must n be in order for the altruistic behavior to be favored by kin selection?
7. Why is routine addition of one or more antibiotics to cattle feed as a preventive against bacterial infection a dangerous procedure in the long run?

8. Why might evolution under natural conditions favor genes that produce enzymes that chemically modify streptomycin rather than mutations that so alter the membrane as to make it impermeable to the antibiotic or that produce altered ribosomes insensitive to the antibiotic?

9. Consider a moth population with a dominant allele for melanism at a frequency of .01 and with a recessive allele for melanism at a frequency of .01 at an unlinked locus. With a change in environment that favors the melanistic forms, which allele frequency will initially increase the fastest and why?

10. Why should there be selection in favor of self-fertilization of heavy metal tolerant plants on mine sites? What phenomenon associated with inbreeding would militate against self-fertilization?

11. From the data in Figure 19B, calculate the relative survival probabilities due to bird predation for melanic and nonmelanic moths on (a) dark backgrounds, (b) light backgrounds, assuming the absence of habitat selection of the sort shown to occur in Figure 19A. (Note: the height of the bars in Figure 19B, from left to right, is 26%, 74%, 86%, 14%.)

12. Why are Batesian mimics more highly favored when the number of mimics is small relative to the number of models?

13. Assign each of the following modes of speciation into either of two groups. Group I: "random genetic drift an important element in speciation," Group II: "random genetic drift of little or no importance in speciation." Modes of speciation to be assigned: (a) allopatric type Ia, (b) allopatric type Ib, (c) parapatric, (d) sympatric, (e) polyploidy.

14. A locus with four alleles A_1, A_2, A_3, A_4 has allele frequencies .12, .80, .06, .02, respectively, in one population, and frequencies .30, .56, .10, .04, respectively, in a related population. Calculate the normalized identity and the standard genetic distance for this locus.

15. For a locus with two alleles, A and a, the allele frequency of A in one population is .3 and in another population it is .5; for another locus with two alleles, B and b, the allele frequency of B is .2 in the first population and .3 in the second.

Calculate the normalized identity and the standard genetic distance for the two populations based on these loci. (Assume linkage equilibrium.)

Further Readings

Antonovics, J., A. D. Bradshaw and R. G. Turner. 1971. Heavy metal tolerance in plants. *Adv. Ecol. Res.* 7:1-85.

Bush, G. L. 1975. Modes of animal speciation. *Ann. Rev. Ecol. Syst.* 6:339-364.

Carson, H. L. and K. Y. Kaneshiro. 1976. *Drosophila* of Hawaii: Systematics and ecological genetics. *Ann. Rev. Ecol. Syst.* 7:311-346.

Cook, L. M. 1971. *Coefficients of Natural Selection.* Hutchinson, London.

Creed, R. (ed.). 1971. *Ecological Genetics and Evolution.* Blackwell, Oxford.

Emlen, J. M. 1973. *Ecology: An Evolutionary Approach.* Addison-Wesley, Reading, Massachusetts.

Ford, E. B. 1975. *Ecological Genetics.* 4th ed. Chapman and Hall, London.

Gould, S. J. 1977. *Ontogeny and Phylogeny.* Harvard Univ. Press, Cambridge, Massachusetts.

Grant, V. 1971. *Plant Speciation.* Columbia Univ. Press, New York.

Jain, S. K. and A. D. Bradshaw. 1966. Evolutionary divergence among adjacent plant populations. I. Evidence and its theoretical analysis. *Heredity* 21:407-442.

Kettlewell, H. B. D. 1973. *The Evolution of Melanism.* Clarendon Press, Oxford.

Mayr, E. 1970. *Populations, Species, and Evolution.* Harvard Univ. Press, Cambridge, Massachusetts.

Nei, M. 1975. *Molecular Population Genetics and Evolution.* American Elsevier, New York.

Parsons, P. A. 1973. *Behavioural and Ecological Genetics.* Clarendon, Oxford.

Pianka, E. R. 1978. *Evolutionary Ecology.* 2nd ed. Harper & Row, New York.

Roughgarden, J. 1979. *Theory of Population Genetics and Evolutionary Ecology: An Introduction.* Macmillan, New York.

Sheppard, P. M. 1967. *Natural Selection and Heredity.* Hutchinson, London.

Stebbins, G. L. 1971. *Chromosomal Evolution in Higher Plants.* Addison-Wesley, Reading, Massachusetts.

Turner, J. R. G. 1977. Butterfly mimicry: The genetical evolution of an adaptation. *Evol. Biol.* 10:163–206.

White, M. J. D. 1973. *Animal Cytology and Evolution.* Cambridge Univ. Press, Cambridge.

White, M. J. D. (ed.). 1974. *Genetic Mechanisms of Speciation in Insects.* Australia and New Zealand Book Co., Sydney.

White, M. J. D. 1978. *Modes of Speciation.* W. H. Freeman, San Francisco.

Wilson, E. O. 1975. *Sociobiology.* Harvard Univ. Press, Cambridge, Massachusetts.

Wright, S. 1977. *Evolution and the Genetics of Populations,* Vol. 3: *Experimental Results and Evolutionary Deductions.* Univ. of Chicago Press, Chicago.

Wright, S. 1978. *Evolution and the Genetics of Populations,* Vol. 4: *Variability Within and Among Natural Populations.* Univ. of Chicago Press, Chicago.

Answers to Chapter-End Problems

CHAPTER 1.

1. 1/2; 2/3
2. 2/3: 1/3
3. $[6!/(4!2!)](3/4)^4(1/4)^2 + [6!/(5!1!)](3/4)^5(1/4) + [6!/6!](3/4)^6 = .83057$
4. 4/12; 1/4; 1/12; because Pr{FC = J of H} = Pr{FC = H|FC = J} × Pr{FC = J} from the multiplication rule involving conditional probabilities.
5. *Ab*/*ab*, *AB*/*ab*, *ab*/*ab*, *aB*/*ab* in proportions .44, .06, .06, .44.
6. $\chi_1^2 = 4$ and $p \approx .048$, so there is evidence of linkage (just barely significant at the 5 percent level). Estimate $r = .40$ with standard error $= \sqrt{(.4)(.6)/100} = .049$
7. Since one child is known to be a girl, there are only three possible sex distributions, namely BG, GB, and GG, and each has probability 1/3. The probability that the other child is a boy is therefore 2/3. (The intuitive answer to this question is 1/2, which is wrong because the fact that one child is a girl is very significant information.)
8. The woman's mother must be heterozygous. The probability that the woman herself is heterozygous is therefore 1/2. If she is heterozygous, the probability that she has an affected son is 1/2. The overall probability of her having an affected son is calculated by use of the multiplication rule involving conditional probabilities, i.e., Pr{affected son} = Pr{affected son|woman is heterozygous} × Pr{woman is heterozygous} = 1/2 × 1/2 = 1/4.
9. Because the deficiency cannot produce the normal gene product to compensate for the mutant allele on the homologue. Note the similarity here with recessive genes on the X chromosome in males.
10. eleven (cf. Figure 11)
11. Of the chromosome strands not involved in the crossover, one carries the inversion and one is structurally normal. Of

those involved in the crossover, one carries a duplication of chromosomal material to the left of the leftmost inversion breakpoint and a deficiency of the chromosomal material to the right of the rightmost breakpoint, the other carries a deficiency of the leftmost material and a duplication of the rightmost material, but both are monocentric. With pericentric inversions in *Drosophila,* there is no dicentric chromosome bridge to consign abnormal gametes to nonfunctional products of meiosis in the female.

12. The original messenger RNA has the nucleotide sequence AUGACACUUUCUAAAGGCUGA, translated as the amino acid sequence M-T-L-S-K-G (see Table I). A substitution of A for the first G in the DNA template leads to a substitution of U for the first C in the messenger RNA, and therefore a substitution of isoleucine (I) for threnonine (T) in the polypeptide. Insertion of A after the first C in the DNA template leads to the messenger RNA sequence AUGUACACUUUCUAAAGGCUAA, which is translated as M-Y-T-F (because UAA is an ochre terminator). Note that the frameshift mutation causes a shift in reading frame during translation.

13. Population size decreases every generation until ultimate extinction. If $N_0 > K$ in logistic growth with $r > 0$, population size decreases toward K.

14. Expected in F_2 is 93.75:31.25 with Mendelian segregation. $\chi_1^2 = 1.41$ with $p \approx .2$, so there is no reason to reject hypothesis of Mendelian segregation. The normal allele is dominant.

15. Letting A, B, C represent the dominant phenotype at each of the loci and a, b, c represent the recessive phenotype, the expectations are obtained by multiplying as follows: [(3/4)A + (1/4)a][(3/4)B + (1/4)b][(3/4)C + (1/4)c] = (27/64)ABC + (9/64)ABc + (9/64)AbC + (9/64)aBC + (3/64)Abc + (3/64)aBc + (3/64)abC + (1/64)abc. For 639 offspring the expected ratio is 269.578:89.859:89.859:89.859:29.953:29.953:29.953:9.984. For Mendel's data, $\chi_7^2 = 2.67$ (the degrees of freedom equal 7 because there are 8 classes of data), and the associated $p \approx .92$; Mendel's results are very, very good indeed.

CHAPTER 2

1. five, depending on the number of fast (F) and slow (S) polypeptides in the molecule, namely, *FFFF, FFFS, FFSS, FSSS,* and *SSSS.*

2. expected aa = $(.7807)^2 \times 57$ = 34.7; ab = 2(.7807)(.2193) × 57 = 19.5; bb = $(.2193)^2 \times 57$ = 2.7
3. frequency M = (2202 + 1496)/6200 = .5965; of N = .4035; expected MM, MN, NN phenotypes = 1103, 1492.3, 504.7, respectively. χ^2 = .00363 + .00917 + .00573 = .01853, $p \approx .96$. This is an unusually good fit to the random-mating expectations, and there is no reason to reject the hypothesis.
4. frequency $d = \sqrt{170/400}$ = .65192; frequency of D = 1 − .65192 = .34808. Let frequency of $D = p$, that of $d = q$. There will be $p^2 + 2pq$ Rh$^+$, of which $2pq$ are heterozygous, thus frequency of heterozygotes among Rh$^+$ is $2pq/(p^2 + 2pq) = 2q/(1 + q)$ = .789288. Therefore .789288 × 230 = 181.5 would be expected to be heterozygous. (Another approach is to calculate the expected number as $2pq \times 400$ = 181.5.)
5. $q = \sqrt{10^{-4}}$ = .01; Frequency of heterozygotes = $2pq = 2(1 - q)q$ = .0198, or about one in 50.5 individuals.
6. 3/4 and 1/4; $(3/4)^2$ = 9/16; 2(3/4)(1/4) = 6/16; $(1/4)^2$ = 1/16
7. q = .05 (because the locus is sex-linked); frequency of affected females = q^2 = .0025; frequency of heterozygous females = $2pq = 2(1 - q)q$ = .095
8. frequency $A_1A_1 = (.1)^2$ = .01; A_1A_2 = 2(.1)(.2) = .04; A_2A_2 = .04; A_2A_3 = .12; A_3A_3 = .09; A_3A_4 = .24; A_4A_4 = .16; A_1A_3 = .06; A_1A_4 = .08; A_2A_4 = .16
9. frequency A_1B_1 = (.3)(.2) = .06; A_1B_2 = .09; A_1B_3 = .15; A_2B_1 = .14; A_2B_2 = .21; A_2B_3 = .35; because $p_2 = 1 - p_1$ = .7 and $q_3 = 1 - q_1 - q_2$ = .5
10. P = 3/5 = 60 percent. The polymorphic loci are numbers 1, 2, and 5. Let H_i be the heterozygosity for locus i. H_1 = 2(.37)(.63) = .4662; H_2 = .1128; H_3 = .00995; H_4 = 0; H_5 = .3720. Average heterozygosity = $(\Sigma H_i)/5$ = .19219, or roughly 19 percent.
11. parent–offspring F = 1/4; brother–sister F = 1/4
12. Use the formula $c(1 + 15q)/(c + 16q - cq)$ with c = .015 and q = .02. The formula works out to be .05826. That is, almost 6 percent of the affected individuals arise from first cousin matings.
13. Among nonrelative matings, frequencies of AA, Aa, and aa are $(.995)^2$ = .990025; 2(.995)(.005) = .009950; and $(.005)^2$ = .000025, respectively. Among offspring of second cousins use formula $p^2(1 - F) + pF$; $2pq(1 - F)$; $q^2(1 - F) + qF$ with F = 1/64; the frequencies of AA, Aa, and aa are .990336; .009328; .000336, respectively. Note that the frequency of aa in off-

spring of second cousin matings is .000336/.000025 = 13.44 times what it is in matings of nonrelatives.

14. Use α symbols as in text to denote identity by descent. For two-way cross ($\alpha_1\alpha_1 \times \alpha_2\alpha_2$), hybrid is $\alpha_1\alpha_2$, gametes are $(1/2)\alpha_1 + (1/2)\alpha_2$, and offspring (after open pollination) are $(1/4)\alpha_1\alpha_1$, $(1/2)\alpha_1\alpha_2$, $(1/4)\alpha_2\alpha_2$, so $F = 1/4 + 1/4 = 1/2$. For three-way cross ($\alpha_1\alpha_1 \times \alpha_2\alpha_2$) $\times$ $\alpha_3\alpha_3$, hybrid population is $(1/2)\alpha_1\alpha_3$ and $(1/2)\alpha_2\alpha_3$; gametic frequencies are $(1/4)\alpha_1$, $(1/4)\alpha_2$, $(1/2)\alpha_3$. After open pollination, frequency $\alpha_1\alpha_1 = 1/16$, of $\alpha_2\alpha_2 = 1/16$, of $\alpha_3\alpha_3 = 1/4$, all other gentoypes having alleles that are not identical by descent; thus $F = 1/16 + 1/16 + 1/4 = 3/8$.

15. Expand $(pA + qa)^4 = p^4AAAA + 4p^3qAAAa + 6p^2q^2AAaa + 4pq^3Aaaa + q^4aaaa$.

CHAPTER 3

1. Probability of fixation of neutral mutation = initial frequency $= 1/2N_a$, where N_a is actual size (assuming equal numbers of males and females). Thus $1/2N_a = .05$, so $N_a = 10$.

2. $N_e = \{(1/5)[(1/500) + (1/1500) + (1/10) + (1/50) + (1/1000)]\}^{-1} = 40.431$

3. Effective number of alleles = reciprocal of homozygosity. The numbers for these examples are 2.80 and 8.

4. $(.5)(.5)/2N = .01707$, so $N = 7.32$.

5. Use $F_t = 1 - (1 - 1/2N)^t$ for $N = 50$ and $t = 200$; $F = .866$.

6. $(7 \times 10^{-7})/(3 \times 10^{-5} + 7 \times 10^{-7}) = .0228$.

7. $H_S = (.48 + .18)/2 = .33$; $H_T = 2(.25)(.75) = .375$; $\bar{p} = .25$ and $\overline{p^2} = .085$, so $\sigma^2 = .0225$. $(H_T - H_S)/H_T = .12$; $\sigma^2/\bar{p}\bar{q} = .12$.

8. Average frequency in separated populations = $(.36 + .81)/2 = .585 = \overline{q^2}$; average frequency in fused population = $(.75)^2 = .5625 = \bar{q}^2$; $\sigma^2 = \overline{q^2} - \bar{q}^2 = .0225$.

9. Use $p_t = p_{t-1}(1 - m) + \bar{p}m$ with $m = .10$ and $\bar{p} = .25$. In one population, p goes from .4 to .385; other goes from .1 to .115. Eventual frequency is .25 in both populations.

10. Use Table VIII with $p = .16 + .48/2 = .4$. Here $\bar{w} = .76$, $w_{11} - w_{12} = .2$, $w_{12} - w_{22} = .2$, so in gametes for next generation $p = .46316$ and $q = .53684$. Zygotes in next generation have frequencies .2145, .4973, and .2882 for AA, Aa, and aa, respectively.

11. Use $\hat{p} = (w_{12} - w_{22})/(2w_{12} - w_{11} - w_{22})$ with $\hat{p} = .8$, $w_{11} = .98$, and $w_{12} = 1$. Solution for w_{22} is $w_{22} = .92$.

12. $\hat{q} = \sqrt{\mu/s} = \sqrt{5 \times 10^{-6}/.8} = .0025$; $\hat{q} = \mu/hs = 5 \times 10^{-6}/(.035)(.8) = .000179$.

13. Use $F_t = 1 - (1 - 1/2N)^t$ with $N = 30$ and $t = 50$; $F = .56844$ after 50 generations. Average frequencies of *AA*, *Aa*, and *aa* are $p^2(1 - F) + pF$, $2pq(1 - F)$, and $q^2(1 - F) + qF$, where $p = .4$ and $q = .6$; these turn out to be .2964, .2072, and .4964, respectively. In the population with $p = .2$, the frequencies of *AA*, *Aa*, and *aa* are p^2, $2pq$, $q^2 = .04, .32, .64$, respectively, because of random mating within the subpopulations.

14. Use $F = \sigma^2/\bar{p}\bar{q}$ for $\bar{p} = .4$, $\bar{q} = .6$, and F calculated as in Problem 13 for $t = 10, 50$, and 100. For $t = 10$, $F = .15471$, so $\sigma^2 = .03713$; for $t = 50$, $F = .56844$, so $\sigma^2 = .13643$; for $t = 100$, $F = .81376$, so $\sigma^2 = .19530$.

15. The protein has $120 \times 3 = 360$ nucleotide sites, which change at rate $.5 \times 10^{-9}$ amino acid-changing substitutions per nucleotide site per year. We thus expect $360 \times .5 \times 10^{-9} = 1.8 \times 10^{-7}$ amino acid changes per year. The species have each evolved from the common ancestor for 180×10^6 years, for a total divergence time of $2 \times 180 \times 10^6 = 3.6 \times 10^8$ years. Thus expected number of amino acid differences is $3.6 \times 10^8 \times 1.8 \times 10^{-7} = 64.8$, or about 65.

CHAPTER 4.

1. 33.3 percent.

2. 150; when the females have a mean of 165 and the males have a mean of 140, the overall parental mean is 152.5 and the selection differential is therefore 12.5; therefore, the mean of the offspring is expected to be 145. (Note: selection practiced in only one sex has the effect of halving the selection differential.)

3. 58.56.

4. Let the unknown mean $= x$. Then $R = 25 - x$ and $S = 35 - x$, and from $R = h^2S$ we have $25 - x = .4(35 - x)$, or $x = 18.333$.

5. Use the formula (Total response) $=$ h^2 $\times$ (Cumulative selection differential), so the answer is 130.

6. From Table IX, paternal half-sib covariance is $\sigma_a^2/4$, so $\sigma_a^2 = 60$. σ^2 is given as 240, so $h^2 = 25$ percent. $R = 8$, so the mean of the next generation is expected to be 108.

7. $\sigma_g^2 + \sigma_e^2 = 240$ and $\sigma_e^2 = 190$, so $\sigma_g^2 = 50$ and broad sense heritability is $50/240 = 20.8$ percent.

8. Maximum = 70 percent, minimum = 0 percent.

9. For Jones, $.24 = \sigma_a^2/4$, so $\sigma_a^2 = .96$; for Smith, $\sigma_a^2 = .48$.

10. For $x = 21, 15, 9$, $\mu^* = 15$, $a = 6$, $d = 0$; for $y = \ln(x)$ the numbers are $y = 3.04452, 2.70805, 2.19722$, respectively, and $\mu^* = 2.62087$, $a = .42365$, $d = .08718$. The latter values should be used because the variate y satisfies the assumption of normality in the theory; x does not.

11. For the A locus, $\sigma_a^2 = 2pq[a + (q - p)d]^2 = .18$; for the B locus, $\sigma_a^2 = 2pq[a + (q - p)d]^2 = .0672$. Since there is no linkage disequilibrium or epistasis, the contributions of the loci to the total σ_a^2 are additive; the total $\sigma_a^2 = .18 + .0672 = .2472$, and $h^2 = .2472/\sigma^2$.

12. Because $\Delta p = (Z/B)pq[a + (q - p)d]$ and $\sigma_a^2 = 2pq[a + (q - p)d]^2$. The locus is at an equilibrium of allele frequency ($\Delta p = 0$) if and only if either $p = 0$ or $p = 1$ or $a + (q - p)d = 0$. In all three possible cases, $\sigma_a^2 = 0$. For overdominance, of course, the case of interest is the equilibrium at which $a + (q - p)d = 0$, or $p = (a + d)/2d$.

13. $\sigma_a^2 = 2pq[a + (q - p)d]^2 = .007318$; $\sigma_d^2 = (2pqd)^2 = .001129$; $\sigma_g^2 = \sigma_a^2 + \sigma_d^2 = .008447$.

14. $d > a$ implies overdominance, so there will be an equilibrium at which $\sigma_a^2 = 0$, which implies $a + (q - p)d = 0$, or $p = (a + d)/2d$ (see the answer to Problem 12). Thus, if we call the alleles A and A' so that the phenotypic values of AA, AA', $A'A'$ are $\mu^* + a$, $\mu^* + d$, $\mu^* - a$, respectively, then the equilibrium frequency of A is $p = .91667$.

15. Since the phenotypic values of AA, AA', $A'A'$ are $\mu^* + a$, $\mu^* + d$, $\mu^* - a$, respectively, and these occur in the proportions 1/4, 1/2, 1/4, with $d = 0$ the mean of the F_2 is $(1/4)(\mu^* + a) + (1/2)(\mu^*) + (1/4)(\mu^* - a) = \mu^*$. The genetic variance, which is the average of the squared deviations from the mean, is $(1/4)(\mu^* + a - \mu^*)^2 + (1/2)(\mu^* + d - \mu^*)^2 + (1/4)(\mu^* - a - \mu^*)^2 = a^2/2$. This is the genetic variance for one locus. For n completely equivalent, unlinked, and additive loci, all contribute equally to the genetic variance, so the genetic variance in the F_2 is $\sigma_g^2 = na^2/2$. The environmental variance in the F_2 is presumed to be the same as in the inbred lines, namely $\sigma_e^2 = \sigma^2$. Since $\sigma_{\text{total}}^2 = \sigma_g^2 + \sigma_e^2$, we have, for the F_2, $\sigma_2^2 = na^2/2 + \sigma^2$.

For the second part, use $R = 2na$ to obtain $R^2 = 4n^2a^2$. Use the expression above for σ_2^2 to obtain $\sigma_2^2 - \sigma^2 = na^2/2$. Now eliminate a^2 from the two equations (divide the equation for

R^2 by that for $\sigma_2^2 - \sigma^2$) and obtain $n = R^2/[8(\sigma_2^2 - \sigma^2)]$. Note that R, σ_2^2, and σ^2 are all measurable quantities.

CHAPTER 5

1. Table I implies multiple fitness peaks, and small deme size with limited migration allows exploration of fitness surface by random genetic drift. Table II implies a simple, one-peak fitness surface, and small deme size with limited migration tends to impede the short-term rate of evolution toward this peak because of random genetic drift; so the most appropriate population structure for Table II is the mass selection one of Figure 6B.
2. The shifting balance process is favored the smaller the neighborhood size and the smaller the amount of migration, so the best answer is (c).
3. Establish several large populations with different initial allele frequencies and practice directional selection. If the selection limit is the same in all, one peak is suggested; different selection limits suggest more than one peak. The populations should be large to minimize random genetic drift. Different initial allele frequencies can be established by crossing inbreds to produce an F_1, then backcrossing the F_1 to each inbred for one or more generations.
4. If the members of the group are related, then the decrease in fitness to the altruist can be compensated for by an increase in survival probability of the altruist's kin, so the mechanism is one of kin selection.
5. Since $x = 0$, any $r \geq 0$ will do.
6. We would need $r \geq x/ny$, where $r = 1/4$, $x = .4$, $y = .2$. Thus $1/4 \geq .4/.2n$, or $n \geq 8$.
7. Because it encourages evolution and spread of antibiotic resistance in the bacterial populations.
8. Because mutations affecting the structure of membranes or ribosomes have detrimental fitness effects in the absence of the antibiotic.
9. From Chapter 3, $\Delta p = pq^2s/\overline{w}$ for a favored dominant and $\Delta p = p^2qs/\overline{w}$ for a favored recessive, s being the selection coefficient. The ratio of Δp for dominant to Δp for a recessive, assuming the $\overline{w}$'s are approximately equal, is q/p; in this case, 99. So the dominant will initially increase in frequency most rapidly.
10. Most pollen blown in from surrounding areas is from non-

tolerant plants, so more alleles for tolerance occur in pollen from mine site. Inbreeding depression militates against selfing.

11. Survival probability = 1 − probability of being eaten. In dark background, (a) survival ratio for melanic:nonmelanic is 74:26, which is 1:.35 or 2.85:1; in light background, (b) survival ratio for melanic:nonmelanic is 14:86, which is .16:1 or 1:6.14.

12. Because dupes learn more easily that the mimics are unpalatable due to repeated unpleasant experiences with the more numerous model.

13. Group I: b and c; Group II: a, d, and e.

14. Let a subscript 1 refer to first population, a subscript 2 refer to the second. $J_{11} = .6584$, $J_{22} = .4152$, $J_{12} = .4908$. $I = .938709$, $D = .06325$.

15. Let subscript 1 refer to population 1, 2 refer to 2. For the A locus, $J_{11} = .58$, $J_{22} = .50$, $J_{12} = .50$; for the B locus, $J_{11} = .68$, $J_{22} = .58$, $J_{12} = .62$. For the overall J's, one calculates the arithmetic average (including monomorphic loci, though in this case, there are none); the averages are $J_{11} = .63$, $J_{22} = .54$, $J_{12} = .56$. Then $I = .96011$ and $D = .040708$.

Answers to Box Problems

CHAPTER 1

A

a. $r = 147/879 = .16724$, s.e. $= \sqrt{r(1-r)/879} = .01259$. With 96 percent confidence assert that a second estimate of ρ would be in the interval from .14206 to .19242; with 99.7 percent confidence assert that a second estimate would lie in the interval .12947 to .20501.

b. Let p = frequency of *Adh*-1 allele $= 138/248 = .55645$, s.e. $= .03155$. 96 percent confidence interval is .49335 to .61955; 99.7 percent confidence interval is .46180 to .65110. Interpret confidence intervals in terms of second estimates as in (a) above.

B

a. $16!/(9!3!3!1!) \times (9/16)^9(3/16)^3(3/16)^3(1/16) = .024521$

b. $(1-r)/2$, $(1-r)/2$, $r/2$, $r/2$, for the four types, which are .4, .4, .1, .1; probability of obtaining 4:4:1:1 in 10 offspring is $10!/(4!4!1!1!)(.4)^4(.4)^4(.1)^1(.1)^1 = .04129$

C

a. $N_{50} = 269.159$; $N_{100} = 724.465$; $N_{150} = 1949.960$; $N_{200} = 5248.490$; $N_{250} = 14126.772$

b.

	$r_0 = \ln(1 + r)$	$r_0 = r$	$r_0 = r - r^2/2$
$N(50)$	269.159	271.828	269.123
$N(100)$	724.465	738.906	724.274
$N(150)$	1949.960	2008.554	1949.192
$N(200)$	5248.490	5459.815	5245.732
$N(250)$	14126.772	14841.316	14117.496

c. $\int_{N_0}^{N_t}[1/N(1 + aN)]dN = \ln[N_t/(1 + aN_t)] - \ln[N_0/(1 + aN_0)]$, where $a = 1/K$. This equals $\int_0^t r dt = r(t - 0) = rt$, so $e^{rt} = [N_t/(1 - N_t/K)]/[N_0/(1 - N_0/K)] = \{[KN_t/(K - N_t)]\}/\{[KN_0/(K - N_0)]\}$, or $N_0e^{rt}/(K - N_0) = N_t/(K - N_t)$. Let $\theta = N_0e^{rt}/(K - N_0)$. Then $N_t = K\theta - N_t\theta$, or $N_t = K\theta/(1 + \theta) = K/(1 + 1/\theta) = K/\{1 + [(K - N_0)/N_0]e^{-rt}\}$. Substitute $C = (K - N_0)/N_0$ to obtain the desired result.

d.

Generation	Exact	Approximate
N_1	1050	1049.958
N_2	1099.875	1099.668
N_3	1149.376	1148.885
N_4	1198.261	1197.375
N_5	1246.295	1244.919

D

Quantity	a. Unweighted regression	b. Weighted regression
(1)	4	42
(2)	5.0000	6.80952
(3)	30.0000	48.09524
(4)	21.5000	26.47619
(5)	123.5000	187.28571
(6)	16.0000	6.99556
(7)	5.0000	1.72568
(8)	3.2000	4.05380
(9)	5.5000	−1.12824

E **a.** From $dN_1/dt = 0$, $N_1 + \alpha_{21}N_2 = K_1$, or $N_1 = K_1 - \alpha_{21}N_2$. From $dN_2/dt = 0$, $N_2 + \alpha_{12}N_1 = K_2$, or (substituting for N_1), $N_2 + \alpha_{12}(K_1 - \alpha_{21}N_2) = K_2$, from which $N_2 = (K_2 - \alpha_{12}K_1)/(1 - \alpha_{21}\alpha_{12})$. Thus $N_1 = K_1 - \alpha_{21}[(K_2 - \alpha_{12}K_1)/(1 - \alpha_{21}\alpha_{12})]$, from which $N_1 = (K_1 - \alpha_{21}K_2)/(1 - \alpha_{21}\alpha_{12})$.

b. $\hat{N}_1 = 1000$; $\hat{N}_2 = 2000$. Values of r_1 and r_2 affect the rate at which N_1 and N_2 approach $\hat{N}_1$ and $\hat{N}_2$, higher values of r being associated with faster rates of approach.

F In a large but nongrowing population, each allele in the parental generation must give rise, on the average, to one allele in the offspring generation. If the mutant allele is rare, it will always be heterozygous with a normal allele; with blending inheritance, both alleles become "contaminated" before transmission to the next generation, so a single rare mutant allele in the parental generation leads to two "blended" alleles in the offspring generation. The number of mutant alleles therefore increases as 1, 2, 4, 8, 16, ..., 2^n.

With Mendelian inheritance there is no blending of alleles, so the effect of a mutant allele remains constant (assuming no change in the environment, of course). Moreover, each individual in a nongrowing population contributes an average of two gametes to the next generation. A rare mutant allele will always be heterozygous, and among the two gametes contributed to the next generation by a heterozygote, one gamete, on the average, will carry the mutant allele (because of segregation). Therefore, the average number of mutant alleles is expected to remain constant with Mendelian inheritance in a very large population.

CHAPTER 2

A **a.** frequency of *a* allele = (2 × 35 + 19)/114 = .7807; frequency of *b* allele = 1 − .7807 = .2193

b. frequency of F allele = (726 + 90 + 111 + 1)/1158 = .8014; of S = .1986

c. frequency of F allele = (726 + 90 + 172 + 32)/1158 = .8808; of S = .1192

d. frequency of NS-bearing = (90 + 1 + 32 + 0)/1158 = .1062; of non-NS = .8938

B

a. monomers: A, B, E; dimer: D; no information: C

b. A: 2 alleles (*Fast* and *Slow*); B: 3 alleles (*Fast, Intermediate,* and *Slow*); C: 1 allele; D: 2 alleles (*Fast* and *Slow*); E: 2 alleles (*Fast* and *Slow*)

c. A: A^FA^F = 32 individuals; A^FA^S = 16; A^SA^S = 2; frequency A^F = (64 + 16)/100 = .80; frequency A^S = .20

B: B^FB^F = 7 individuals; B^FB^I = 13; B^IB^I = 5; B^FB^S = 12; B^IB^S = 9; B^SB^S = 4; frequency B^F = (14 + 13 + 12)/100 = .39; frequency B^I = (13 + 10 + 9)/100 = .32; frequency B^S = 1 − .39 − .32 = .29

C: frequency of the one allele found = 1.0

D: D^FD^F = 8 individuals; D^FD^S = 24; D^SD^S = 18; frequency D^F = (16 + 24)/100 = .4; frequency D^S = .6

E: E^FE^F = 0 individuals; E^FE^S = 1; E^SE^S = 49; frequency E^F = (0 + 1)/100 = .01; frequency E^S = .99

d. polymorphic: A, B, D

monomorphic: C, E (because frequency of most common allele equals .99)

proportion of polymorphic loci: P = 3/5 = 60 percent.

e. average heterozygosity: A = 16 individuals out of 50, so H_A = 16/50 = .32; H_B = (13 + 12 + 9)/50 = .68; H_C = 0; H_D = 24/50 = .48; H_E = 1/50 = .02; average heterozygosity: H = (.32 + .68 + 0 + .48 + .02)/5 = .30

C

a. Characteristic equation is $\lambda^5 - (1/2)\lambda^3 - (3/8)\lambda^2 - (1/8)\lambda = 0$. $\lambda = 1$ is a solution because $1 - 1/2 - 3/8 - 1/8 = 0$. At stable age structure, age classes 0, 1, 2, 3, and 4 are found in the ratio 1:1/2:1/4:1/8:1/16, respectively. Expressed as proportions, age classes 0, 1, 2, 3, and 4 are in frequencies 16/31:8/31:4/31:2/31:1/31, or in percentage units 51.6:25.8:12.9:6.5:3.2, respectively.

b. $N_t = N_{t-1}\lambda = (N_{t-2}\lambda)\lambda = (N_{t-3}\lambda)\lambda\lambda = \ldots = N_{t-t}\lambda^t = N_0\lambda^t$. Set $N_{t*} = 2N_0$, so $N_{t*} = N_0\lambda^{t*} = 2N_0$, so $\lambda^{t*} = 2$, or $t^*\ln\lambda = \ln 2$. Thus $\ln\lambda = (\ln 2)/t^*$, or $\lambda = \exp[(\ln 2)/t^*]$. For Sweden, t^* = 173 years, so λ = 1.00401. For Mexico, t^* = 19.8 years, so λ = 1.03563. Sweden's population increases at .401 percent per year; Mexico's increases at 3.563 percent per year.

D

a.
$$\begin{aligned} Q' &= PQ + 2PR + Q^2/2 + QR \\ &= 2[(P)(Q/2) + PR + Q^2/4 + (Q/2)(R)] \\ &= 2(P + Q/2)(Q/2 + R) = 2pq \end{aligned}$$

b. $R' = Q^2/4 + QR + R^2 = (Q/2 + R)^2 = q^2$
c. frequency of heterozygotes $= 2pq = 2\,(1 - q)\,q = 2q - 2q^2 \approx 2q$ because q^2 is much smaller than q
d. for two alleles, the Punnett square is

		A (p)	a (q)
heterozygous parent	A (1/2)	AA ($p/2$)	Aa ($q/2$)
	a (1/2)	Aa ($p/2$)	aa ($q/2$)

Thus the frequency of heterozygous offspring is $p/2 + q/2 = (p + q)/2 = 1/2$
e. $(2pq)^2 = 4(p^2)(q^2)$

E **a.** preliminary estimates: $r = .58376082$, $p = .28030481$, $q = .12866579$, and $\theta = .00726858$.
b. final estimates: $\hat{r} = .5895$, $\hat{p} = .2813$, $\hat{q} = .1291$.
c. expected numbers: 0 = 715.9, A = 846.3, B = 348.0, AB = 149.7.
d. $\chi^2 = 3.7275$, p $\approx$ .06, which is not significant. There is no reason to reject the hypothesis of Hardy–Weinberg proportions. The χ^2 has 1 degree of freedom because there are 4 (to start with) − 1 (for using the sample size) − 1 (for estimating $\hat{r}$) − 1 (for estimating $\hat{p}$) = 1. A degree of freedom is not deducted for estimating $\hat{q}$ because $\hat{q} = 1 - \hat{p} - \hat{r}$.

F **a.** $f_n - m_n = (1/2)(m_{n-1} + f_{n-1}) - f_{n-1} = (1/2)m_{n-1} - (1/2)f_{n-1}$
$= -(1/2)(f_{n-1} - m_{n-1})$
b. Calculate $(2f_n + m_n)/3$. It equals $(m_{n-1} + f_{n-1} + f_{n-1})/3 = (2f_{n-1} + m_{n-1})/3$. Thus $(2f_n + m_n)/3 = (2f_0 + m_0)/3$ for all values of n, so $(2f_n + m_n)/3$ is a constant. To show that the allele frequencies do not change when the allele frequency in both sexes is $\hat{f} = \hat{m} = (2f_n + m_n)/3$, just substitute: $f_n = (1/2)(\hat{m} + \hat{f}) = (1/2)[(2f_n + m_n)/3 + (2f_n + m_n)/3] = (2f_n + m_n)/3 = \hat{f}$; $m_n = \hat{f} = (2f_n + m_n)/3 = \hat{m}$. Thus, when $f = \hat{f}$ and $m = \hat{m}$, f and m remain constant thereafter. Note that $\hat{f} = \hat{m} = (2f_0 + m_0)/3$.
c. $m_0 = .2$, $f_0 = .8$; $m_1 = .8$, $f_1 = (1/2)(.2 + .8) = .5$; $m_2 = .5$, $f_2 = (1/2)(.5 + .8) = .65$; $m_3 = .65$, $f_3 = (1/2)(.65 + .5) = .575$; $m_4 = .575$, $f_4 = (1/2)(.575 + .65) = .6125$; $m_5 = .6125$, $f_5 = (1/2)(.6125 + .575) = .59375$; $m_6 = .59375$; $f_6 = (1/2)(.59375 + .6125) = .60313$. Ultimate frequencies in both sexes $= \hat{m} = \hat{f} = (2f_0 + m_0)/3 = .60$. Genotype frequencies among females: AA (frequency = .36), Aa (.48), aa (.16); among males: A (frequency .6), a (.4).

G **a.** $P_{11} = (P_{11} + P_{12})(P_{11} + P_{21}) + (P_{11}P_{22} - P_{12}P_{21}) = P_{11}(P_{11} +$

$P_{12} + P_{21} + P_{22}) + P_{12}P_{21} - P_{12}P_{21} = P_{11}$. The remaining three expressions are solved similarly.

b. $D_{max} = P^*_{11} - p_1q_1$, where P^*_{11} is the maximum possible value of P_{11}. $P_{11} \leq \min(p_1, q_1)$. Now $D_{max} = p_1 - p_1q_1 = p_1(1 - q_1) = p_1q_2$ when $p_1 < q_1$, and $D_{max} = q_1(1 - p_1) = q_1p_2$ when $p_1 > q_1$. However, $p_1q_2 < p_2q_1$ if and only if $p_1(1 - q_1) < (1 - p_1)q_1$, or $p_1 < q_1$. Thus $D_{max} = \min(p_1q_2, q_1p_2)$. $D_{min} = P^*_{11} - p_1q_1$, where now P^*_{11} is the minimum possible value of P_{11}. $P_{11} \geq \max(p_1 + q_1 - 1, 0)$. Now $D_{min} = 0 - p_1q_1 = -p_1q_1$ if $p_1 + q_1 - 1 < 0$, and $D_{min} = p_1 + q_1 - 1 - p_1q_1 = p_1(1 - q_1) - (1 - q_1) = -(1 - q_1)(1 - p_1) = -p_2q_2$ if $p_1 + q_1 - 1 > 0$. On the other hand, $-p_1q_1 < -p_2q_2$ if and only if $-p_1q_1 < -(1 - p_1)(1 - q_1)$, or $p_1 + q_1 - 1 > 0$. Thus $D_{min} = \max(-p_1q_1, - p_2q_2)$.

Gamete	Frequency	Value of x (A locus)	Value of y (B locus)	xy
A_1B_1	P_{11}	1	1	1
A_1B_2	P_{12}	1	0	0
A_2B_1	P_{21}	0	1	0
A_2B_2	P_{22}	0	0	0

$\bar{x} = P_{11} + P_{12} = p_1$ and $\overline{x^2} = P_{11} + P_{12} = p_1$, so $\sigma^2_x = p_1 - p_1^2 = p_1p_2$. $\bar{y} = P_{11} + P_{21} = q_1$ and $\overline{y^2} = P_{11} + P_{21} = q_1$, so $\sigma^2_y = q_1 - q_1^2 = q_1q_2$. $\overline{xy} = P_{11}$, so $Cov(x,y) = P_{11} - p_1q_1 = D$ (from part a). $\rho = Cov(x,y)/\sigma_x\sigma_y = D/\sqrt{p_1p_2q_1q_2}$.

d. For Mukai et al. data in the text, $D = -.011794500$, $\chi^2 = 16.16505358$, $N = 1158$, $\rho = -.118150155$. $\rho^2N = 16.16505367$ (the small discrepancy is due to round-off error). For the Clegg et al. data in the text, $D = .025101865$, $\chi^2 = 40.83137838$, $N = 3049$, $\rho = .115722635$. $\rho^2N = 40.83137943$; again there is a small round-off error.

e. Let $A_1 = \alpha Gpdh - 1^F$ and B_1 = non-*NS*. Observed numbers of A_1B_1, A_1B_2, A_2B_1, and A_2B_2 are 837, 91, 198, and 32, respectively, with $N = 1158$. $D = (837 \times 32 - 91 \times 198)/N^2 = .006537$. Using allele frequencies from Box A (b and d), $\rho = .053184$, and $\chi^2 = \rho^2N = 3.275$, $p \approx .07$. Thus there is no reason to reject the hypothesis of linkage equilibrium.

H

Paths are (1) E<u>B</u>DG, (2) E<u>A</u>DG, and (3) E<u>A</u>CG.

a. For an autosomal locus, $F_I = (1/2)^4(1 + F_B) + (1/2)^4(1 + F_A) + (1/2)^4(1 + F_A) = (1/2)^4(5/4) + (1/2)^4(2) + (1/2)^4(2) = 21/64$.

b. For a sex-linked locus, paths (2) and (3) are excluded because they pass through two or more males in succession, and path (1) has three females in it. Thus $F_I = (1/2)^3(1 + F_B) = (1/2)^3(5/4) = 5/32$.

I **a.** $H_0 - H_t = 2pqF_t$, so $(H_0 - H_t)/H_0 = F_t$
b. $(H_0 - H_t)/H_0 = (1/2)(H_0 - H_{t-1})/H_0 + (1/4)(H_0 - H_{t-2})/H_0 + 1/4$, or $H_0 - H_t = (1/2)(H_0 - H_{t-1}) + (1/4)(H_0 - H_{t-2}) + (1/4)H_0$, or $H_t = (1/2)H_{t-1} + (1/4)H_{t-2}$. Assume $H_0 = H_1 = 1/2$. Then $H_2 = (1/2)(1/2) + (1/4)(1/2) = 3/8$; $H_3 = (1/2)(3/8) + (1/4)(1/2) = 5/16$; $H_4 = (1/2)(5/16) + (1/4)(3/8) = 8/32$; $H_5 = (1/2)(8/32) + (1/4)(5/16) = 13/64$.
c. From (b), $(H_t/H_{t-1}) = (1/2) + (1/4)(H_{t-2}/H_{t-1})$, so $\lambda = (1/2) + (1/4)(1/\lambda)$, or $4\lambda^2 - 2\lambda - 1 = 0$. Use the quadratic formula $x = (-b \pm \sqrt{b^2 - 4ac})/2a$ for the solutions of $ax^2 + bx + c = 0$ to conclude $\lambda = (2 \pm \sqrt{4 + 16})/8 = (1 \pm \sqrt{5})/4$. The positive root is .80902, which is the root we want; the negative root is $-.30902$.

CHAPTER 3

A **a.** $(e^{-x})^2 = e^{-2x} = 1 - 2x$ (approximately, for x small); also $e^{-x_1} = 1 - x_1$, $e^{-x_2} = 1 - x_2$, so $1 - 2x = (1 - x_1)(1 - x_2) = 1 - x_1 - x_2 + x_1x_2 = 1 - x_1 - x_2$, for x_1x_2 small. Thus $2x = x_1 + x_2$. For the hypothetical insects $1/N = (10^{-1} + 10^{-2} + 10^{-3} + 10^{-4})/4 = .027775$, or $N = 36.0$.
b. 4 bulls, 100 cows, $N = 15.385$; 4 bulls, 200 cows, $N = 15.686$.

B **a.** Let H_i be $\bar{H}_S$ for subpopulations in group i; $H_1 = .3625$, $H_2 = .3425$; $H_3 = .3125$; $H_4 = .2774$; $H_5 = 0$. For all groups, $H_T = .375$. F_{ST} for groups 1, 2, 3, 4, 5 is .03333, .08667, .16667, .26027, 1.00000, respectively. Let σ_i^2 be the variance in allele frequency for the ith group; for groups 1, 2, 3, 4, 5, $\sigma_i^2 = .00625$, .01625, .03125, .0488, .1875, respectively. For each group, $\bar{p}(1 - \bar{p}) = .1875$, and $\sigma_i^2/\bar{p}(1 - \bar{p})$ gives the same values for F_{ST} as above.
b. $H_S^{(i)} = .021758$, .280878, .318798, .037278 for alleles $i = 1, 2, 3, 4$, respectively, in subpopulation 1, and .255, .355278, .148878, 0 in subpopulation 2. $\bar{H}_S^{(i)} = .138379$, .318078, .233838, .018639 are the respective averages. $\bar{p}_i = .0805$, .469, .441, .0095 are the average allele frequencies for alleles $i = 1, 2, 3, 4$. Thus $H_T^{(i)} = .1480395$, .498078, .493038, .0188195. $F_{ST}^{(i)} = .065256232$, .361389180, .525720127, .009591116 for $i = 1, 2, 3, 4$. Weighted average $F_{ST} = .387781256$. $H_S = 1 - \Sigma p_i^2 = .329356$ for subpopulation 1 and .379578 for subpopulation 2, average = .354467. $H_T = .5789875$. $F_{ST} = .387781256$. Also, $\sigma_i^2 = .00483025$, .09, .1296, .00009025 for alleles 1, 2, 3, 4, and $F_{ST}^{(i)} = \sigma_i^2/\bar{p}_i(1 - \bar{p}_i)$ are same as above. $F_{ST} = \Sigma\sigma_i^2/\Sigma\bar{p}_i(1 - \bar{p}_i)$ can also be verified.

C **a.** For the first part, $\bar{p} = 1/2$, $\bar{q} = 1/2$, $\sigma^2 = x^2$, so $F_{ST} = 4x^2$. Also, $d^2 = 1 - \sqrt{(.5 - x)(.5 + x)} - \sqrt{(.5 - x)(.5 + x)} = 1 - 2\sqrt{.25 - x^2} = 1 - 2(.5 - x^2) = 2x^2$, approximately, so $2d^2 = 4x^2 = F_{ST}$.

x	F_{ST}	Exact $2d^2$
.1	.04	.04041
.2	.16	.16697
.3	.36	.40000
.35	.49	.57171
.4	.64	.80000
.49	.9604	1.602005

For the optional part, from elementary trigonometry, $\sin\theta_1 = \sqrt{p_1}$, $\sin\theta_2 = \sqrt{p_2}$, $\cos\theta_1 = \sqrt{q_1}$, $\cos\theta_2 = \sqrt{q_2}$, $\theta = \theta_1 - \theta_2$. Cos $\theta = \cos\theta_1\cos\theta_2 + \sin\theta_1\sin\theta_2 = \sqrt{q_1q_2} + \sqrt{p_1p_2}$, $d^2 = 1 - \sqrt{p_1p_2} - \sqrt{q_1q_2}$ (by definition) $= 1 - \cos\theta$. Length of chord $= 2\sqrt{(1-\cos\theta)/2} = \sqrt{2}\sqrt{(1-\cos\theta)} = \sqrt{2}\sqrt{d^2} = d\sqrt{2}$.

b. $d_{12} = .08897$, $d_{13} = .12847$, $d_{14} = .20226$, $d_{23} = .03962$, $d_{24} = .11381$, $d_{34} = .07427$. Therefore populations 2 and 3 are to be pooled. Call this pooled population number 5; it has allele frequencies $p = (.25 + .30)/2 = .275$ and $q = .725$. For populations 1, 4, and 5, $d_{15} = .10901$, $d_{45} = .09377$, and $d_{14} = .20226$, respectively (d_{14} is the same as above, of course). Therefore populations 4 and 5 should be pooled. The final tree is:

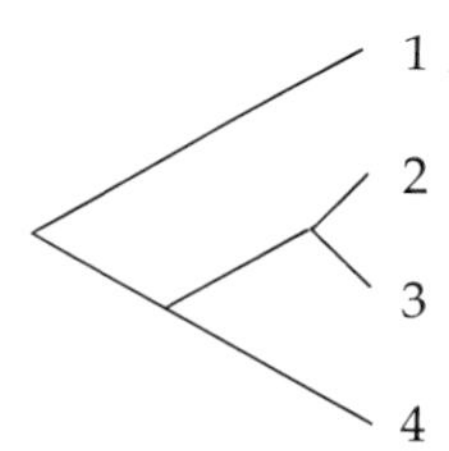

c. D_m for the populations in Table III (from top of table to bottom) = .0135, .0031, .0147, .0281, .0275, .0067, .0267.

D

a. $\hat{F} = xy + (1-x)y\hat{F}$, where $x = 1/2N$ and $y = (1-\mu)^2$. Therefore, $\hat{F} = xy/[1-(1-x)y] = xy/(1-y+xy) = 1/[(1/xy)-(1/x)+1] = 1/\{(1/x)[(1/y)-1]+1\}$. Then substitute $1 - 2\mu$ for y.

b. Because the equation for $\hat{F}$ has the same form, with m replacing μ.

c. $F_t = x + (1-x)F_{t-1}$, where $x = 1/2N$ is the probability of autozygosity provided neither allele is a migrant or a mutant, and $(1 - m - \mu)^2$ is the probability that neither is a migrant or a mutant. The equation for $\hat{F}$ again has the same form as in (a), with $m + \mu$ replacing μ.

E

a. Probability of fixation $= 10^{-4}$, probability of loss $= .9999$; average time to fixation = 16,000 generations; average time to loss = 14.74 generations.

b. .016

c. 8.402 generations

F **a.** Stepping stone model, $\hat{F} = .97249$; island model, $\hat{F} = .33331$; increase in $\hat{F}$ of 191.77 percent.

b. $4\delta\sigma\sqrt{2\mu} = 4N\sqrt{m}\sqrt{2\mu} = 4N\sqrt{2m\mu}$; $\hat{F} = .97249$ for $\delta = 50$, $\sigma = .10$

c. $\hat{F} = .24691$

d. $N = 17.725$ in part (b), and $N = 20.012$ in part (c). The neighborhood sizes are tolerably the same.

G **a.** Subtract p from both sides of expression for p', giving

$$\begin{aligned}\Delta p &= (1/\bar{w})[p^2w_{11} + pqw_{12} - p(p^2w_{11} + 2pqw_{12} + q^2w_{22}]\\ &= (1/\bar{w})[p^2(1 - p)w_{11} + pq(1 - 2p)w_{12} - pq^2w_{22}]\\ &= (pq/\bar{w})[pw_{11} + (q - p)w_{12} - qw_{22}]\\ &= (pq/\bar{w})[p(w_{11} - w_{12}) + q(w_{12} - w_{22})].\end{aligned}$$

b. Recall from basic calculus that if $\int F(x)dx = G(x)$ then $\int_a^b F(x)dx = G(b) - G(a)$. All of the problems in this part involve the formula. For ease of representation, let $I(p)$ represent $\int_{p_0}^{p_t} f(p)dp$, where $f(p) = 1/pq^2$ in part (1), $1/pq$ in part (2), $1/p^2q$ in part (3), and $1/pq$ in part (4). Furthermore, let $J = \int_0^t Cdt$, where $C = s$ in parts (1), (3), and (4), and $s/2$ in part (2). (1) $I(p) = 1/q_t - \ln(q_t/p_t) - 1/q_0 + \ln(q_0/p_0) = \ln(p_tq_0/p_0q_t) + (1/q_t) - (1/q_0)$ and $J = st$. (2) $I(p) = -\ln(q_t/p_t) + \ln(q_0/p_0) = \ln(p_tq_0/p_0q_t)$ and $J = (s/2)t$. (3) $I(p) = -(1/p_t) - \ln(q_t/p_t) + (1/p_0) + \ln(q_0/p_0) = \ln(p_tq_0/p_0q_t) - (1/p_t) + (1/p_0)$ and $J = st$. (4) $I(p)$ is the same as in (2), but here $J = st$.

c. Times for cases 1 through 4 are 10,818.01, 1838.05, 10,818.01, and 919.02 generations, respectively.

H **a.** Subtract $\nu/(\mu + \nu)$ from both sides of the equation for p_t, then factor $(1 - \mu - \nu)$ from the right-hand side. At equilibrium for given numbers, $p = .02$. When $p_0 = 1$, p_t for halfway to equilibrium is $(1 + .02)/2 = .51$. Also $1 - \mu - \nu = .9999$. Solve $(.51 - .02) = (1 - .02)(.9999)^t$ for t, or $.5 = (.9999)^t$; $t = \ln(.5)/\ln(.9999) = 6931.12$ generations. For p to go halfway to equilibrium the equation for t is always $.5 = (1 - \mu - \nu)^t$ for this model.

b. $p_0/\bar{p} = 2$ implies $p_0 = 2\bar{p}$ or $p_0 - \bar{p} = \bar{p}$. $p_t - \bar{p} = \bar{p}(.99)^{69}$, or $p_t = 1.4998\bar{p} \approx 1.5\bar{p}$; 69 generations is thus halfway to equilibrium.

c. $100 = 25 + t$, so $t = 75$ generations required to change allele frequency to one-fourth its initial value; $1/\sqrt{.0004} = 25 + t$, or $50 = 25 + t$, so $t = 25$ generations required to change genotype frequency.

d. Let $p_0w^t = x$. Then $p_t/(1 - p_t) = x/q_0$, or $p_t = (x/q_0)/[1 + (x/q_0)]$, thus $p_t = x/(q_0 + x)$. In the numerical example, $w = .98$. For $p_0 = .99$, $p_{50} = .973$; for $p_0 = .60$, $p_{50} = .353$.

I Let x represent time, y the natural logarithm of p/q; then $x_0 = 5$, $y_0 = .24522$, and $x_1 = 24$, $y_1 = 2.86718$. The slope $b = .1380 = \ln w$, so $w = 1.14798$; intercept $= -.44478$, which is the initial value of $\ln(p/q)$, so $p/q = .64097$, or $p = .3906$ initially. Strain A increased in fitness by $.14798/120 = .00123$ per generation, or about .12 percent per generation.

J Mutation: $\hat{q} = .01961$, $L = \mu = 2 \times 10^{-6}$.
Segregation: $\hat{q} = .01961$, $L = 1.02 \times 10^{-4}$.
The segregation load is so much larger (despite equality of allele frequency) because of the small reduction in fitness in the AA homozygotes. In fact, the segregation load $= \hat{p}^2 s + \hat{q}^2 t = 1 \times 10^{-4} + 2 \times 10^{-6}$.

K **a.** $p_i(1 - p_i) = p_j(1 - p_j)$ implies $p_i - p_j = p_i^2 - p_j^2 = (p_i - p_j)(p_i + p_j)$. Assume $p_i - p_j \neq 0$. Then $p_i + p_j$ must equal 1, but $p_i + p_j$ cannot equal 1 because there are more than two alleles. Thus it must be that $p_i - p_j = 0$ for all i and j, or $p_i = p_j$ for all i and j. Let $p_i = p$ for all i. We must have $\Sigma p_i = 1$, so $np = 1$, or $p = 1/n$. Now, at equilibrium, $\theta = 1 - \Sigma p_i^2 = 1 - \Sigma(1/n)^2 = 1 - (1/n) = (n - 1)/n$. $\bar{w} = \theta - 2p(1 - p)$, and $p = 1/n$, so $\bar{w} = (n - 2)(n - 1)/n^2 = (n^2 - 3n + 2)/n^2$. For $n = 3$, 30, and 300, $\bar{w} = .2222$, .9022, and .9900, respectively.
b. $\bar{w} = (n^2 - 3n + 2)/n^2$, so $1 - \bar{w} = (3n - 2)/n^2$. For $n = 3$, 30, and 300, this equals .7777, .0977, .0098, respectively. (The small discrepancies between these numbers and $1 - \bar{w}$ as calculated in (a) are due to round-off error.) This helps explain why there are so many alleles, because initial selective advantage to a new mutation is very large.

L **a.** For *Arrowhead* chromosome:

Frequency	Legit	Altitude	2N
.25	.517	259	1146
.35	.294	914	760
.37	.253	1402	1450
.44	.115	1890	478
.45	.096	2438	44
.55	−.096	2621	82
.50	0	3018	10

Weighted regression of legit on altitude:
slope $= -.000232336$
intercept $= .561342172$
$a = |\text{slope}|^{-1} = 4304.102814$. Note: $(.000232336)^{-1} = 4304.111287$; the value for a given has more significant digits because it was calculated prior to rounding off.

$g = 5.09188 \times 10^{-6}$ per meter, assuming $\sigma^2 = 101{,}500$ m^2. $2759g = .014048496$, or a selection across the gradient of about 1.4 percent.

b. Legit (.24) = .542, legit (.30) = .401. Let x corresponding to $p = .24$ be 0, x for $p = .30$ is then 2000 km. Thus slope = $(.401 - .542)/2000 = -7.05 \times 10^{-5}$/km. Then $a = |\text{slope}|^{-1} = 14184.397$ km, and $g = 1.401610 \times 10^{-11}$ assuming $\sigma^2 = 10$ km^2. Finally, $2000g = 2.80 \times 10^{-8}$ is the change in selection across Japan required to account for the cline — an incredibly small amount of selection, almost surely too small not to be swamped by drift or other factors. Cavalli-Sforza and Bodmer (1971) calculate how long it would take for such a cline to be established by selection (20,000 generations!) and they conclude that the cline is very likely due to founder effects and migration rather than selection.

CHAPTER 4

A **a.** $\bar{x} = 12.17$; $\bar{y} = 4.67$; $\sigma_x^2 = 7.72$; $\sigma_y^2 = 6.52$; $\sigma_{xy} = -7.00$; $r = -.987$; $b = -.907$

b. $TSS = 23.00000$; $WSS = 18.21667$; $BSS = 4.78333$; $BMS = 2.392$; $WMS = 2.024$; $\tilde{n} = 3.917$; $V_B = .093957$; $V_W = 2.024$; $t = .04436$

B **a.** A proportion p of the time the substitution of A' for A changes AA to AA' and a proportion q of the time changes AA' to $A'A'$. Thus the average effect of the substitution is $p(\mu + d - \mu - a) + q(\mu - a - \mu - d) = d(p - q) - a = -[a + (q - p)d] = -\alpha$. As shown in the text, $\sigma_a^2 = 2pq[a + (q - p)d]^2$, so $\sigma_a^2 = 2pq\alpha^2$.

b. When the favored allele is dominant, $d = a$ and $\alpha = a + (q - p)a = 2qa$, so $\Delta p = pq^2(2ia/\sigma)$. For a favored dominant, $w_{11} = 1$, $w_{12} = 1$, $w_{22} = 1 - s$, so $\Delta p = pq^2s$, assuming $\bar{w} \approx 1$. Therefore $s = 2ia/\sigma$. When the favored allele is recessive, $d = -a$ and $\alpha = 2pa$ so $\Delta p = p^2q(2ia/\sigma)$, but then $w_{11} = 1$, $w_{12} = 1 - s$, $w_{22} = 1 - s$, so $\Delta p = p^2qs$ (for $\bar{w} \approx 1$), and $s = 2ia/\sigma$.

c. For $h^2 = .09$ and $i = .80$, $s = .24, .16, .069$, respectively, for $n = 8, 18, 98$; for $h^2 = .09$ and $i = 2.06$, $s = .618, .412, .176$ for $n = 8, 18, 98$; for $h^2 = .16$ and $i = .8$, $s = .32, .213, .091$ for $n = 8, 18, 98$; and for $h^2 = .16$ and $i = 2.06$, $s = .824, .549, .235$ for $n = 8, 18, 98$.

C **a.** $\mu_S = (1/B)\{(1/\sqrt{2\pi}\sigma)\int_T^\infty x\exp(X)\,dx\}$, where $X = -(x - \mu)^2/2\sigma^2$. So $\mu_S = (1/B)\{(1/\sqrt{2\pi}\sigma)[-\sigma^2\exp(X)|_T^\infty + \mu\int_T^\infty \exp(X)dx]\} = (1/B)\{(1/\sqrt{2\pi}\sigma)\{0 + \sigma^2\exp[-(T - \mu)^2/2\sigma^2]\} + (\mu/\sqrt{2\pi}\sigma)\int_T^\infty \exp[-(x - \mu)^2/2\sigma^2]dx\} = (1/B)(\sigma^2Z + \mu B) = (\sigma^2Z/B) + \mu$.

b. i for $B = .10$ is 1.76; i for $B = .01$ is 2.66, so the factor of increase in i is $2.66/1.76 = 1.5$.

c. For $B = .70$, $1 - B = .30$, so $t = -.52$ and $z = .34769$, thus $i = .50$; for $B = .10$, $i = 1.76$; and for $B = .01$, $i = 2.66$. For cattle, 84 percent is the proportion of total intensity of selection due to sire, 16 percent due to dam; for swine and chickens, 60 percent is due to selection of sires, 40 percent due to selection of dams.

D

a. Thorax length: $n = 60$, $2a/\sigma = .20$. Six-week weight: $n = 32$, $2a/\sigma = .25$.

b. $(U - D)^2 = 8n\sigma_a^2 = 19{,}040$, so $U - D = 137.99$. Thus $(U - D)/\sigma = 16.87$ — i.e., a response of about 8σ in each direction.

E

a. For $p = .4$, $\overline{w} = .7520$; $p' = .489361702$, $\overline{w}' = .795654142$; thus $\Delta\overline{w} = .043654142$. On the other hand, $a + (q - p)d = .28$ for $p = .4$, and $\sigma_a^2 = .037632$, so $\sigma_a^2/\overline{w} = .050042553$, which is 14.63 percent larger than $\Delta\overline{w}$ above. Since $1 + (pq/2\overline{w})(w_{11} - 2w_{12} + w_{22}) = .872340426$, $\Delta\overline{w} = (\sigma_a^2/\overline{w})(.872340426) = .043654142$, which is exactly equal to the $\Delta\overline{w}$ given above.

b. $\overline{w} = .97520$, $p' = .406890894$, $\overline{w}' = .975582091$, $\Delta\overline{w} = .000382091$; also $a + (q - p)d = .028$ so $\sigma_a^2 = .000376320$ and $\sigma_a^2/\overline{w} = .000385890$, which is only 1 percent larger than $\Delta\overline{w}$ calculated directly.

c. At equilibrium, $\hat{p} = t/(s + t)$. $\sigma_a^2 = 0$ when $a + (q - p)d = 0$ or $p = (a + d)/2d$. Substitute $a = (t - s)/2$ and $d = (t + s)/2$ to obtain $p = \hat{p}$. For $s = .2$, $t = .6$, $\hat{p} = .75$ and $d = .4$, so at equilibrium $\sigma_a^2 = 2.25 \times 10^{-2}$; for $s = .02$, $t = .06$, $\hat{p} = .75$, and $d = .04$, so at equilibrium $\sigma_a^2 = 2.25 \times 10^{-4}$.

d. $d\overline{w}/dp = -2ps + 2(1 - p)t = 0$ when $p = t/(s + t)$, which is $\hat{p}$. $d^2\overline{w}/dp^2 = -2s - 2t < 0$, so $\hat{p}$ represents a maximum of $\overline{w}$.

F

a. $n = 3$, $B = .10$ $(i = 1.76)$; $\sqrt[3]{B} \approx .45$ $(i' = .88)$; gains in the ratio $1.76{:}3 \times .88{:}1.76\sqrt{3}$, or 1.76:2.64:3.05. In the case $n = 3$, $B = .01$ $(i = 2.66)$; $\sqrt[3]{B} \approx .20$ $(i' = 1.40)$; gains in the ratio $2.66{:}3 \times 1.40{:}2.66\sqrt{3}$, or 2.66:4.20:4.61.

b. For uncorrelated traits, $I = a_1h_1^2x_1 + a_2h_2^2x_2 = (81.5)(.47)x_1 + (20.1)(.47)x_2 = 38.3x_1 + 9.4x_2$.

G

a. $\mu' = (p + \Delta p)^2(\mu^* + a) + 2(p + \Delta p)(q - \Delta p)(\mu^* + d) + (q - \Delta p)^2(\mu^* - a) = [p^2 + 2p\Delta p + (\Delta p)^2](\mu^* + a) + 2[pq - p\Delta p + q\Delta p - (\Delta p)^2](\mu^* + d) + [q^2 - 2q\Delta p + (\Delta p)^2](\mu^* - a) = (p^2 + 2pq + q^2)\mu^* + (p^2 - q^2)a + 2pqd + \Delta p\mu^*(2p - 2p + 2q - 2q) + \Delta pa(2p + 2q) + 2(q - p)\Delta pd = \mu^* + (p - q)(p + q)a + 2pqd + 2\Delta pa(p + q) + 2(q - p)\Delta pd = \mu + 2[a + (q - p)d]\Delta p$.

b. 1. $\mu^* + a - \mu = \mu^* + a - \mu^* - (p - q)a - 2pqd = a(1 - p + q) - 2pqd = 2qa - 2pqd = 2q(a - pd)$.

2. $\mu^* + d - \mu = \mu^* + d - \mu^* - (p - q)a - 2pqd = (q - p)a + (1 - 2pq)d$.

3. $\mu^* - a - \mu = \mu^* - a - \mu^* - (p - q)a - 2pqd = -a(1 + p - q) - 2pqd = -2pa - 2pqd = -2p(a + qd)$.
4. $2q[a + (q - p)d] - 2q^2d = 2qa + d[2q(q - p) - 2q^2] = 2qa + d[2q^2 - 2pq - 2q^2] = 2q(a - pd)$.
5. $(q - p)[a + (q - p)d] + 2pqd = a(q - p) + d[(q - p)^2 + 2pq] = a(q - p) + d[q^2 - 2pq + p^2 + 2pq] = a(q - p) + d(1 - 2pq)$ [this last step because $p^2 + 2pq + q^2 = 1$, so $1 - 2pq = p^2 + q^2$] $= (q - p)a + (1 - 2pq)d$.
6. $-2p[a + (q - p)d] - 2p^2d = -2pa - d[2p(q - p) + 2p^2] = -2pa - d[2pq - 2p^2 + 2p^2] = -2p(a + qd)$.

H Carcass grade, $h^2 = .162$; thickness of fat, $h^2 = .341$.

I **a.** $M = (1 - F_{IT})(2\bar{p}^2 + 2\bar{p}\bar{q}) + F_{IT}(2\bar{p}) = 2\bar{p}(1 - F_{IT}) + 2\bar{p}F_{IT} = 2\bar{p}$. Mean square $= (1 - F_{IT})(4\bar{p}^2 + 2\bar{p}\bar{q}) + F_{IT}(4\bar{p})$. Variance $= V_{IT}$ = mean square $- M^2 = (1 - F_{IT})(4\bar{p}^2 + 2\bar{p}\bar{q}) + F_{IT}(4\bar{p}) - 4\bar{p}^2 = 4\bar{p}^2 - 4\bar{p}^2F_{IT} + 2\bar{p}\bar{q} - 2\bar{p}\bar{q}F_{IT} + 4\bar{p}F_{IT} - 4\bar{p}^2 = 4\bar{p}(1 - \bar{p})F_{IT} + 2\bar{p}\bar{q} - 2\bar{p}\bar{q}F_{IT} = 2\bar{p}\bar{q}F_{IT} + 2\bar{p}\bar{q} = 2\bar{p}\bar{q}(1 + F_{IT})$.

b. $V_{IS} = V_{IT} - V_{ST} = (1 + F_{IT})V_0 - 2F_{ST}V_0 = (1 + F_{IT} - 2F_{ST})V_0$.

c. $1 - F_{IS} = H_I/H_S$; $1 - F_{ST} = H_S/H_T$; $1 - F_{IT} = H_I/H_T$. Now multiply the first two and cancel.

d. $1 - F_{ST} = (1 - 1/2N)^t$ where $N = 20$ and $t = 10$ (because there are 10 generations of random genetic drift with $N = 20$, the first nine with random mating within flocks and the last one with inbreeding), so $1 - F_{ST} = .7763$. Also $1 - F_{IS} = 1 - .25 = .75$. Now use the product formula in part (c): $1 - F_{IT} = (.7763)(.75) = .5822$, so $F_{IT} = .4178$.

J **a.** Gains (in multiples of $h^2\sigma i$) equal $\sqrt{K}$; for $k = 1, 2, 3, 4$, and 5, these are 1.000, 1.319, 1.519, 1.661, and 1.768.

b. Note $1/K = (1/k) + (k - 1)t/k$; as k becomes large, $1/k$ goes to 0, and $(k - 1)/k$ goes to 1, so $1/K$ goes to t, or K to $1/t$. Maximum gain for large k is then $h^2\sigma ri\sqrt{1/t}$.

c. $h_1 = \sqrt{.2} = .447$, $h_2 = \sqrt{.5} = .707$; $R_1'/\sigma_1 = ih_1h_2r_A = (2)(.447)(.707)(.29) = .183$. On the other hand, $R_1/\sigma_1 = h_1^2i^* = (.2)i^*$, and for R_1/σ_1 to be .183 we need $i^* = .915$. Negative genetic correlation implies that direct selection for high milk production will lead to decrease in butterfat content as correlated response, unless some selection index is used that is based on both traits.

K **a.** $A = r\hat{h}^2 = (.28)(.79) = .221$. Then (from the exact formulas) $h^2 = .746$, $\hat{\sigma}_a^2 = \sigma_a^2(1.284)$, $\hat{\sigma}_p^2 = \sigma_p^2(1.212)$, thus there is a 28.4 percent increase in σ_a^2, a 21.2 percent increase in σ_p^2, and a $(.79 - .746)/.79 = 5.9$ percent increase in h^2. Using the approximations, $h^2 = .753$, $\hat{\sigma}_a^2 = \sigma_a^2(1.221)$, $\hat{\sigma}_p^2 = \sigma_p^2(1.175)$, for a 22.1 percent increase in σ_a^2, a 17.5 percent increase in σ_p^2, and a 4.9 percent increase in h^2.

b. Let $H = h^2$. Substitute A/r for $\hat{h}^2$ in the formula to obtain $H = (A/r)\{(1 - A)/[1 - (A/r)A]\} = (A - A^2)/(r - A^2)$, from which $A^2(1 - H) - A + rH = 0$ follows with some rearrangement.
c. $\hat{F} = .024$. For this trait, $h^2 = 15/60 = .25$ and $1 - h^2 = .75$, so A satisfies $(.75)A^2 - A + (.125) = 0$. Recall the solutions of $ax^2 + bx + c = 0$ are $x = (-b \pm \sqrt{b^2 - 4ac})/2a$; here x represents A, and $a = .75$, $b = -1$, $c = .125$. The solutions for A are 1.194 (not the one we want because it's larger than 1) and .139 (the one we want). So $A = .139$. Now $\hat{h}^2 = A/r = .278$, and, from the exact formulas, $\hat{\sigma}_a^2 = 17.42$ and $\hat{\sigma}_p^2 = 62.43$. Note, as a check, that $\hat{\sigma}_a^2/\hat{\sigma}_p^2 = .279$, in agreement with the .278 above. (The small discrepancy is the result of round-off error.)

L

a. For $B = .01$, $t = 2.33$ and $z = .02665$, so $\mu_S = z/B = 2.66$. For $B = .10$, $t' = 1.28$, so $\mu' = t - t' = 1.05$. Thus $h^2 = 2\mu'/\mu_S = 79$ percent.
b. For $F_p = .001$, $t = 3.09$, and $z = .00337$, so $z/F_p = \mu_S = 3.37$. Now $h^2 = 2\mu'/\mu_S = .860$, so $\mu' = (.860)(3.37)/2 = 1.45$. Then $t' = t - \mu' = 3.09 - 1.45 = 1.64$, and from the table in Box C the value of B corresponding to $t = 1.64$ is $B = .05$. B corresponds to F_o, so the probability that the child is affected is 5 percent.

CHAPTER 5

A

a. $p_1^2(1 - 2\alpha + 2\epsilon) + 2p_1q_1(1 - \alpha + \epsilon) + q_1^2 = 1 - (2p_1^2 + 2p_1q_1)(\alpha - \epsilon) = 1 - 2p_1(\alpha - \epsilon)$
b. $p_1^2(1 - \alpha + 2\epsilon) + 2p_1q_1(1 + \epsilon) + q_1^2(1 - \alpha) = 1 - (p_1^2 + q_1^2)\alpha + 2p_1(p_1 + q_1)\epsilon$
c. $p_1^2(1 + 2\epsilon) + 2p_1q_1(1 - \alpha + \epsilon) + q_1^2(1 - 2\alpha) = 1 + 2p_1(p_1 + q_1)\epsilon - 2q_1(p_1 + q_1)\alpha$
d. $p_2^2(1 - 2\alpha + 2\epsilon) + 2p_2q_2(1 - \alpha + 2\epsilon) + q_2^2(1 + 2\epsilon) = 1 - 2p_2(p_2 + q_2)\alpha + 2\epsilon$
e. $p_2^2(1 - \alpha + \epsilon) + 2p_2q_2(1 + \epsilon) + q_2^2(1 - \alpha + \epsilon) = 1 - \alpha(p_2^2 + q_2^2) + \epsilon$
f. $p_2^2 + 2p_2q_2(1 - \alpha) + q_2^2(1 - 2\alpha) = 1 - 2q_2(p_2 + q_2)\alpha$
g. $p_2^2[1 - 2p_1(\alpha - \epsilon)] + 2p_2q_2[1 - (p_1^2 + q_1^2)\alpha + 2p_1\epsilon] + q_2^2[1 - 2q_1\alpha + 2p_1\epsilon] = 1 - 2\alpha[p_1p_2^2 + p_1^2p_2q_2 + q_1^2p_2q_2 + q_1q_2^2] + 2p_1\epsilon$. The term in brackets equals $p_1p_2(p_2 + p_1q_2) + q_1q_2(q_1p_2 + q_2) = p_1p_2(p_2 + q_2 - q_1q_2) + q_1q_2(p_2 - p_1p_2 + q_2) = p_1p_2(1 - q_1q_2) + q_1q_2(1 - p_1p_2) = p_1p_2 + q_1q_2 - 2p_1p_2q_1q_2 = p_1p_2 + 1 - p_1 - p_2 + p_1p_2 - 2p_1p_2 + 2p_1^2p_2 + 2p_1p_2^2 - 2p_1^2p_2^2 = 1 - p_1 - p_2 + 2p_1p_2(p_1 + p_2 - p_1p_2)$. Also $\overline{w} = p_1^2[1 + 2\epsilon - 2p_2\alpha] + 2p_1q_1[1 + \epsilon - (p_2^2 + q_2^2)\alpha] + q_1^2(1 - 2q_2\alpha) = 1 + 2p_1\epsilon - 2\alpha[p_1^2p_2 + p_1q_1p_2^2 + p_1q_1q_2^2 + q_1^2q_2]$, and the term in brackets can again be reduced to $p_1p_2 + q_1q_2 - 2p_1p_2q_1q_2$.

B $K = 1500$, $\alpha = .25$, $r = .05$, $\hat{N}_1 = 1000$, $\hat{N}_2 = 2000$. $R_r/\sigma^2_{ar} = 0$, $R_K/\sigma^2_{aK} = 3.333 \times 10^{-5}$, $R_\alpha/\sigma^2_{a\alpha} = -.067$.

C **a.** For identical twins, parent-offspring, full sibs, half sibs, first cousins, and uncle–nephew (in that order), $F_{IR} = 1/2, 1/4, 1/4, 1/8, 1/16, 1/8$, so $r = 1, 1/2, 1/2, 1/4, 1/8, 1/4$.

b. Since we are dealing in life and death, $x = y = 1$. The *total* gain in the probability of transmission of an allele from saving n relatives from death is $nry = nr$. For 2 sibs, $nr = 1$; for 4 nephews, $nr = 1$; for 8 cousins, $nr = 1$, In all cases, the loss in probability of transmission of an allele when Haldane lays down his life is $x = 1$. Thus, in all cases, Haldane breaks exactly even in terms of kin selection.

D $s = .224$

E **a.** 5 for *B. betularia*, 4.7 for *B. cognataria*

b. $C = \int_{p_0}^{p_t}(2/p)dp = 2\ln p_t - 2\ln p_0$

c. $C = \int_{p_0}^{p_t}[(1 + p)/p^2]dp = -1/p_t + \ln p_t + 1/p_0 - \ln p_0$

d. For $p_0 = .01$, C is 4.6 for dominant, 103.6 for recessive, 9.2 for additive; when $p_0 = .001$, the corresponding numbers are 6.9, 1005.9, and 13.8.

e. For $n = 10$, 30, 100, the number of substitutions per generation is .1388, .2404, and .4389, respectively.

f. For $n = 10$, 30, 100, the cost per substitution is .80, .46, and .25, respectively.

F **a.** $\sigma_1^2 = .006667$; $\sigma_2^2 = .002500$; $\sigma_3^2 = .002500$. Weighted average $= (3\sigma_1^2 + 2\sigma_2^2 + 2\sigma_3^2)/7 = .004286 = \sigma^2_{DS}$.

b. Weighted average variance between population means is $[3(.8 - .6)^2 + 2(.55 - .6)^2 + 2(.35 - .6)^2]/7 = .035714 = \sigma^2_{ST}$.

c. Total variance in allele frequency is $[(.3)^2 + (.2)^2 + (.1)^2 + 0^2 + (-.1)^2 + (-.2)^2 + (-.3)^2]/7 = .040000 = \sigma^2_{DT}$.

d. $H_D = 2.8/7 = .4$; $H_S = [3(.32) + 2(.495) + 2(.455)]/7 = .408571$; $H_T = .48$; $H_S - H_D = 2(.004286) = 2\sigma^2_{DS}$; $H_T - H_S = 2(.035714) = 2\sigma^2_{ST}$; $H_T - H_D = 2(.04) = 2\sigma^2_{DT}$.

e. $F_{DS} = .020978$; $F_{ST} = .148810$; $F_{DT} = .166667$, and $(1 - F_{DS})(1 - F_{ST}) = (1 - F_{DT})$.

f. $\sigma^2_{DS}/\bar{p}(1 - \bar{p}) = .020980$; $\sigma^2_{ST}/\bar{p}(1 - \bar{p}) = .148808$; $\sigma^2_{DT}/\bar{p}(1 - \bar{p}) = .166667$. The small discrepancies with the answers in (e) are due to round-off error.

g. $TSS = 15.794698 - 23(.816)^2 = .480010$; $BSS = 15.349228 - 23(.816)^2 = .034546$; $WSS = TSS - BSS = .445464$. $\sigma^2_{DS} = WSS/23 = .019368$; $\sigma^2_{ST} = BSS/23 = .001502$; $\sigma^2_{DT} = TSS/23 = .020870$; $H_T = 2\bar{p}(1 - \bar{p}) = .300288$; $H_S = H_T - 2\sigma^2_{ST} = .297284$; $H_D = H_T - 2\sigma^2_{DT} = .258548$; $F_{DS} = (H_S - H_D)/H_S = .1303$; $F_{ST} = (H_T - H_S)/H_T =$

.0100; $F_{DT} = (H_T - H_D)/H_T = .1390$, or $F_{DS} = 2\sigma^2_{DS}/H_S = .1303$; $F_{ST} = \sigma^2_{ST}/\bar{p}(1 - \bar{p}) = .0100$, $F_{DT} = \sigma^2_{DT}/\bar{p}(1 - \bar{p}) = .1390$. Since $F_{ST} \ll F_{DS}$, most of the genetic differentiation is between regions within subdivisions; little differentiation is between subdivisions relative to the total.

Bibliography

Alexander, R. D. 1974. The evolution of social behavior. *Ann. Rev. Ecol. Syst.* 5:325–384.

Allard, R. W. 1960. *Principles of Plant Breeding.* John Wiley and Sons, New York.

Allard, R. W. 1975. The mating system and microevolution. *Genetics* 79:115–126.

Allard, R. W., G. R. Babbel, M. T. Clegg and A. L. Kahler. 1972. Evidence for coadaptation in *Avena barbata. Proc. Natl. Acad. Sci. U.S.A.* 69:3043–3048.

Allard, R. W., A. L. Kahler and B. S. Weir. 1972. The effect of selection on esterase allozymes in a barley population. *Genetics* 72:489–503.

Allison, A. C. 1964. Polymorphism and natural selection in human populations. *Cold Spring Harbor Symp. Quant. Biol.* 29:139–149.

Anderson, E. S. 1969. Ecology and epidemiology of transferable drug resistance. In G. E. W. Wolstenholme and M. O'Connor (eds.). *Bacterial Episomes and Plasmids.* Little, Brown and Co., Boston, pp. 102–115.

Anderson, W. W. 1969. Polymorphism resulting from the mating advantage of rare male genotypes. *Proc. Natl. Acad. Sci. U.S.A.* 64:190–197.

Antonovics, J. 1968. Evolution in closely adjacent plant populations. V. Evolution of self-fertility. *Heredity* 23:219–238.

Antonovics, J. and A. D. Bradshaw. 1970. Evolution in closely adjacent plant populations. VII. Clinal patterns at a mine boundary. *Heredity* 25:349–362.

Antonovics, J., A. D. Bradshaw and R. G. Turner. 1971. Heavy metal tolerance in plants. *Adv. Ecol. Res.* 7:1–85.

Atwood, S. S. 1947. Cytogenetics and breeding of forage crops. *Adv. Genet.* 1:1–67.

Avery, P. J. and W. G. Hill. 1979. Distribution of linkage disequilibrium with selection and finite population size. *Genet. Res., Camb.* 33:29–48.

BIBLIOGRAPHY

Avise, J. C. 1976. Genetic differentiation during speciation. In F. J. Ayala (ed.). *Molecular Evolution.* Sinauer Associates, Sunderland, Massachusetts, pp. 106–122.

Ayala, F. J. 1975. Genetic differentiation during the speciation process. *Evol. Biol.* 8:1–78.

Ayala, F. J. (ed.). 1976. *Molecular Evolution.* Sinauer Associates, Sunderland, Massachusetts.

Ayala, F. J. and C. A. Campbell. 1974. Frequency-dependent selection. *Ann. Rev. Ecol. Syst.* 5:115–138.

Ayala, F. J., J. R. Powell and Th. Dobzhansky. 1971. Polymorphism in continental and island populations of *Drosophila willistoni. Proc. Natl. Acad. Sci. U.S.A.* 68:2480–2483.

Ayala, F. J., J. R. Powell and M. L. Tracey. 1972. Enzyme variability in the *Drosophila willistoni* group: V. Genetic variation in natural populations of *Drosophila equinoxialis. Genet. Res., Camb.* 20:19–112.

Baker, W. K. 1975. Linkage disequilibrium over space and time in natural populations of *Drosophila montana. Proc. Natl. Acad. Sci. U.S.A.* 72:4095–4099.

Barker, W. C., L. K. Ketcham and M. O. Dayhoff. 1978. A comprehensive examination of protein sequences for evidence of internal gene duplication. *J. Mol. Evol.* 10:265–281.

Barnes, B. W. and A. J. Birley. 1978. Genetical variation for enzyme activity in a population of *Drosophila melanogaster.* IV. Analysis of alcohol dehydrogenase activity in chromosome substitution lines. *Heredity* 40:51–57.

Begon, M. 1977. The effective size of a natural *Drosophila subobscura* population. *Heredity* 38:13–18.

Bell, A. E. 1974. Genetic modeling with laboratory animals. In *First World Congress on Genetics Applied to Livestock Production,* Madrid. Gráficas Orbe, Madrid, pp. 415–424.

Bell, A. E. 1977. Heritability in retrospect. *J. Hered.* 68:297–300.

Bennett, J. H. 1954. Panmixia with tetrasomic and hexasomic inheritance. *Genetics* 39:150–158.

Bennett, J. H. and C. R. Oertel. 1965. The approach to a random association of genotypes with random mating. *J. Theor. Biol.* 9:67–76.

Benveniste, R. and J. Davies. 1973. Aminoglycoside antibiotic-inactivating enzymes in Actinomycetes similar to those present in clinical isolates of antibiotic-resistant bacteria. *Proc. Natl. Acad. Sci. U.S.A.* 70:2276–2280.

Berg, D. E., J. Davies, B. Allet and J.-D. Rochaix. 1975. Transposition of R factor genes to bacteriophage λ. *Proc. Natl. Acad. Sci. U.S.A.* 72:3628-3632.

Berger, P. J. 1977. Multiple-trait selection experiments: Current status, problem areas and experimental approaches. In E. Pollak, O. Kempthorne and T. B. Bailey, Jr. (eds.). *International Conference on Quantitative Genetics.* Iowa State Univ. Press, Ames, Iowa, pp. 191-203.

Bernstein, S. C., L. H. Throckmorton and J. L. Hubby. 1973. Still more genetic variability in natural populations. *Proc. Natl. Acad. Sci. U.S.A.* 70:3928-3931.

Bishop, J. A. and L. M. Cook. 1975. Moths, melanism and clean air. *Sci. Am.* 232:90-99.

Blixt, S. 1974. The pea. In R. C. King (ed.). *Handbook of Genetics,* Vol. 2: *Plants, Plant Viruses,* and *Protists.* Plenum, New York, pp. 181-221.

Bodmer, W. F. and J. G. Bodmer. 1974. The HL-A histocompatibility antigens and disease. In *Tenth Symp. in Advanced Medicine.* Pitman Medical Press, London, pp. 157-174.

Boller, E. F. and G. L. Bush. 1973. Evidence for genetic variation in populations of the European cherry fruit fly *Rhagoletis cerasi* (Diptera: Tephritidae) based on physiological parameters and hybridization experiments. *Entomol. Exp. Appl.* 17:279-293.

Bonhomme, F. and R. K. Selander. 1978. Estimating total genic diversity in the house mouse. *Biochem. Genet.* 16:287-298.

Brncic, D. 1970. Studies on the evolutionary biology of Chilean species of Drosophila. *Evol. Biol.* 4, Suppl:401-437.

Brower, J. V. Z. 1958. Experimental studies of mimicry in some North American butterflies. Part I. The Monarch, *Danaus plexippus,* and Viceroy, *Limenitis archippus archippus. Evolution* 12:32-47.

Brower, L. P., L. M. Cook and H. J. Croze. 1967. Predator response to artificial Batesian mimics released in a neotropical environment. *Evolution* 21:11-23.

Brown, A. H. D. 1979. Enzyme polymorphism in plant populations. *Theor. Pop. Biol.* 15:1-42.

Brown, J. L. 1966. Types of group selection. *Nature* 211:870.

Brues, A. M. 1969. Genetic load and its varieties. *Science* 164:1130-1136.

Bryant, E. H. 1976. A comment on the role of environmental variation in maintaining polymorphisms in natural populations. *Evolution* 30:188-190.

Bryson, V. and H. J. Vogel (eds.). 1965. *Evolving Genes and Proteins.*

Academic Press, New York.

Bukhari, A., J. Shapiro and S. Adhya (eds.). 1977. *DNA Insertion Elements, Plasmids, and Episomes.* Cold Spring Harbor Laboratory, Cold Spring Harbor, New York.

Bundgaard, J. and F. B. Christiansen. 1972. Dynamics of polymorphisms: I. Selection components in an experimental population of *Drosophila melanogaster. Genetics* 71:439–460.

Buri, P. 1956. Gene frequency in small populations of mutant *Drosophila. Evolution* 10:367–402.

Burla, H., A. B. da Cunha, A. R. Cordeiro, Th. Dobzhansky, C. Malagolowkin and C. Pavan. 1949. The *willistoni* group of sibling species of *Drosophila. Evolution* 3:300–314.

Bush, G. L., S. M. Case, A. C. Wilson and J. L. Patton. 1977. Rapid speciation and chromosomal evolution in mammals. *Proc. Natl. Acad. Sci. U.S.A.* 74:3942–3946.

Bush, G. L. 1975. Modes of animal speciation. *Ann. Rev. Ecol. Syst.* 6:339–364.

Caisse, M. and J. Antonovics. 1978. Evolution in closely adjacent plant populations. IX. Evolution of reproductive isolation in clinal populations. *Heredity* 40:371–384.

Calhoon, R. E. and B. B. Bohren. 1974. Genetic gains from reciprocal recurrent and within-line selection for egg production in the fowl. *Theor. Appl. Genet.* 44:364–372.

Campbell, A. M. 1969. *Episomes.* Harper & Row, New York.

Carson, H. L. 1967. Permanent heterozygosity. *Evol. Biol.* 1:143–168.

Carson, H. L. 1968. The population flush and its genetic consequences. *In* R. C. Lewontin (ed.). *Population Biology and Evolution.* Syracuse Univ. Press, Syracuse, New York, pp. 123–137.

Carson, H. L. 1971. Speciation and the founder principle. *Stadler Genetic Symp.* 3:51–70.

Carson, H. L. 1975. The genetics of speciation at the diploid level. *Am. Nat.* 109:83–92.

Carson, H. L. 1978. Speciation and sexual selection in Hawaiian *Drosophila.* In P. F. Brussard (ed.). *Ecological Genetics: The Interface.* Springer-Verlag, New York, pp. 93–107.

Carson, H. L. and P. J. Bryant. 1979. Change in a secondary sexual character as evidence of incipient speciation in *Drosophila silvestris. Proc. Natl. Acad. Sci. U.S.A.* 76:1929–1932.

Carson, H. L., D. E. Hardy, H. T. Spieth and W. S. Stone. 1970. The evolutionary biology of the Hawaiian Drosophilidae. *Evol. Biol.* 4, Suppl:437–544.

BIBLIOGRAPHY

Carson, H. L. and K. Y. Kaneshiro. 1976. *Drosophila* of Hawaii: Systematics and ecological genetics. *Ann. Rev. Ecol. Syst.* 7:311-346.

Carter, L. O. 1961. The inheritance of congenital pyloric stenosis. *Brit. Med. Bull.* 17:251-254.

Cavalli-Sforza, L. L. 1966. Population structure and human evolution. *Proc. Roy. Soc., Ser. B* 164:362-379.

Cavalli-Sforza, L. L. 1974. The genetics of human populations. *Sci. Am.* 231:81-89.

Cavalli-Sforza, L. L. and W. F. Bodmer. 1971. *The Genetics of Human Populations.* W. H. Freeman, San Francisco.

Cavalli-Sforza, L. L. and A. W. F. Edwards. 1967. Phylogenetic analysis. Models and estimation procedures. *Am. J. Hum. Genet.* 19:233-257.

Cavalli-Sforza, L. L. and M. Feldman. 1976. Evolution of continuous variation: Direct approach through joint distribution of genotypes and phenotypes. *Proc. Natl. Acad. Sci. U.S.A.* 73:1689-1692.

Cavalli-Sforza, L. L. and M. W. Feldman. 1978. The evolution of continuous variation: III. Joint transmission of genotype, phenotype and environment. *Genetics* 90:391-425.

Chakraborty, R., P. A. Fuerst and M. Nei. 1978. Statistical studies on protein polymorphism in natural populations. II. Gene differentiation between populations. *Genetics* 88:367-390.

Chakraborty, R. and M. Nei. 1977. Bottleneck effects on average heterozygosity and genetic distance with the stepwise mutation model. *Evolution* 31:347-356.

Charlesworth, B. 1970. Selection in populations with overlapping generations. I. The use of Malthusian parameters in population genetics. *Theor. Pop. Biol.* 1:352-370.

Charlesworth, D. and B. Charlesworth. 1975. Theoretical genetics of Batesian mimicry. I. Single-locus models. *J. Theor. Biol.* 55:283-303.

Charlesworth, D. and B. Charlesworth. 1976a. Theoretical genetics of Batesian mimicry. II. Evolution of supergenes. *J. Theor. Biol.* 55:305-324.

Charlesworth, D. and B. Charlesworth. 1976b. Theoretical genetics of Batesian mimicry. III. Evolution of dominance. *J. Theor. Biol.* 55:325-337.

Charlesworth, B. and D. L. Hartl. 1978. Population dynamics of the segregation distorter polymorphism in *Drosophila melanogaster. Genetics* 89:171-192.

Choi, Y. 1978. Genetic load and viability variation in Korean natural populations of *Drosophila melanogaster. Theor. Appl. Gen.* 53:65–70.

Choy, S. C. and B. S. Weir. 1978. Exact inbreeding coefficients in populations with overlapping generations. *Genetics* 89:591–614.

Christiansen, F. B., J. Bundgaard and J. S. F. Barker. 1977a. On the structure of fitness estimates under post-observational selection. *Evolution* 31:843–853.

Christiansen, F. B. and O. Frydenberg. 1973. Selection component analysis of natural polymorphisms using population samples including mother-offspring combinations. *Theor. Pop. Biol.* 4:425–445.

Christiansen, F. B. and O. Frydenberg. 1974. Geographical patterns of four polymorphisms in *Zoarces viviparus* as evidence of selection. *Genetics* 77:765–770.

Christiansen, F. B., O. Frydenberg and V. Simonsen. 1977b. Genetics of *Zoarces* populations. X. Selection component analysis of the Est III polymorphism using samples of successive cohorts. *Hereditas* 87:129–150.

Clarke, B. C. 1966. The evolution of morph-ratio clines. *Am. Nat.* 100:389–402.

Clarke, B. 1970. Darwinian evolution of proteins. *Science* 168:1009–1011.

Clarke, B. 1972. Density-dependent selection. *Am. Nat.* 106:1–13.

Clarke, B. and J. Murray. 1969. Ecological genetics and speciation in land snails of the genus *Partula. Biol. J. Linn. Soc.* 1:31–42.

Clarke, B. and P. O'Donald. 1964. Frequency-dependent selection. *Heredity* 19:201–206.

Clayton, G. A., G. R. Knight, J. A. Morris and A. Robertson. 1956b. An experimental check on quantitative genetical theory. III. Correlated responses. *J. Genet.* 55:171–180.

Clayton, G. A., J. A. Morris and A. Robertson. 1953. Selection for abdominal chaetae in a large population of *Drosophila melanogaster.* In *Symp. Genetics of Population Structure.* Premiata Tipografia Successori Filli Fusi, Pavia, Italy, pp. 7–15.

Clayton, G. A., J. A. Morris and A. Robertson. 1956a. An experimental check on quantitative genetical theory. I. Short-term responses to selection. *J. Genet.* 55:131–151.

Clayton, G. A. and A. Robertson. 1957. An experimental check on quantitative genetical theory. II. The long-term effects of selection. *J. Genet.* 55:152–170.

Clegg, M. T. and R. W. Allard. 1972. Patterns of genetic differentiation in the slender wild oat species *Avena barbata*. *Proc. Natl. Acad. Sci. U.S.A.* 69:1820–1824.

Clegg, M. T., R. W. Allard and A. L. Kahler. 1972. Is the gene the unit of selection? Evidence from two experimental plant populations. *Proc. Natl. Acad. Sci. U.S.A.* 69:2474–2478.

Clowes, R. C. 1972. Molecular structure of bacterial plasmids. *Bacteriol. Rev.* 36:361–405.

Cockerham, C. C. 1954. An extension of the concept of partitioning hereditary variance for analysis of covariances among relatives when epistasis is present. *Genetics* 39:859–882.

Cockerham, C. C. 1956. Effects of linkage on the covariances between relatives. *Genetics* 41:138–141.

Cockerham, C. C. 1959. Partitions of hereditary variance for various genetic models. *Genetics* 44:1141–1148.

Cockerham, C. C. 1973. Analyses of gene frequencies. *Genetics* 74:679–700.

Cockerham, C. C. and T. Mukai. 1978. Effects of marker chromosomes on relative viability. *Genetics* 90:827–849.

Comstock, R. E. 1977. Quantitative genetics and the design of breeding programs. In E. Pollak, O. Kempthorne and T. B. Bailey, Jr. (eds.). *International Conference on Quantitative Genetics.* Iowa State Univ. Press, Ames, Iowa, pp. 705–718.

Comstock, R. E. 1978. Quantitative genetics in maize breeding. In D. B. Walden (ed.). *Maize Breeding and Genetics.* John Wiley and Sons, New York, pp. 191–206.

Comstock, R. E., H. F. Robinson and P. H. Harvey. 1949. A breeding procedure designed to make maximum use of both general and specific combining ability. *Agron. J.* 41:360–367.

Cook, S. C. A., C. Lefébvre and T. McNeilly. 1972. Competition between metal tolerant and normal plant populations on normal soil. *Evolution* 26:366–372.

Cotterman, C. W. 1940. A Calculus for Statistico-genetics. Ph.D. Thesis, Ohio State Univ., Columbus, Ohio.

Cox, E. C. and T. C. Gibson. 1974. Selection for high mutation rates in chemostats. *Genetics* 77:169–184.

Coyne, J. A., A. A. Felton and R. C. Lewontin. 1978. Extent of genetic variation at a highly polymorphic esterase locus in *Drosophila pseudoobscura*. *Proc. Natl. Acad. Sci. U.S.A.* 75:5090–5093.

Crick, F. 1979. Split genes and RNA splicing. *Science* 204:264–271.

Crow, J. F. 1945. A chart of the χ^2 and t distributions. *J. Amer. Stat. Assn.* 40:376.

Crow, J. F. 1976. *Genetics Notes,* 7th ed. Burgess, Minneapolis.

Crow, J. F. and C. Denniston (eds.). 1974. *Genetic Distance.* Plenum Press, New York.

Crow, J. F. and J. Felsenstein. 1968. The effect of assortative mating on the genetic composition of a population. *Eugen. Quart.* 15:85–97.

Crow, J. F. and M. Kimura. 1970. *An Introduction to Population Genetics Theory.* Harper & Row, New York.

Crow, J. F. and R. G. Temin. 1964. Evidence for the partial dominance of recessive lethal genes in natural populations of *Drosophila. Am. Nat.* 98:21–33.

Crozier, R. H. 1977. Evolutionary genetics of the Hymenoptera. *Ann. Rev. Entomol.* 22:263–288.

Crumpacker, D. W. 1967. Genetic loads in maize (*Zea mays* L.) and other cross-fertilized plants and animals. *Evol. Biol.* 1:306–424.

Crumpacker, D. W., J. Pyati and L. Ehrman. 1977. Ecological genetics and chromosomal polymorphism in Colorado populations of *Drosophila pseudoobscura. Evol. Biol.* 10:437–470.

Crumpacker, D. W. and J. S. Williams. 1973. Density, dispersion, and population structure in *Drosophila pseudoobscura. Ecol. Monographs* 43:499–538.

Darlington, P. J., Jr. 1978. Altruism: Its characteristics and evolution. *Proc. Natl. Acad. Sci. U.S.A.* 75:385–389.

Davidson, E. H. and R. J. Britten. 1979. Regulation of gene expression: Possible role of repetitive sequences. *Science* 204:1052–1059.

Davies, J. and D. I. Smith. 1978. Plasmid-determined resistance to anti-microbial agents. *Ann. Rev. Microbiol.* 32:469–518.

Day, T. H., P. C. Hillier and B. Clarke. 1974a. Properties of genetically polymorphic isozymes of alcohol dehydrogenase in *Drosophila melanogaster. Biochem. Genet.* 11:141–153.

Day, T. H., P. C. Hillier and B. Clarke. 1974b. The relative quantities and catalytic activities of enzymes produced by alleles at the alcohol dehydrogenase locus in *Drosophila melanogaster. Biochem. Genet.* 11:155–165.

de Grouchy, J., C. Turleau and C. Finaz. 1978. Chromosomal phylogeny of the primates. *Ann. Rev. Genet.* 12:289–328.

De Jong, G. and W. Scharloo. 1976. Environmental determination of selective significance or neutrality of amylase variants in *Drosophila melanogaster. Genetics* 84:77–94.

de Magalhães, L. E. and M. A. Q. Rodrigues Pereira. 1976. Frequency-dependent mating success among mutant ebony of *Drosophila melanogaster. Experientia* 32:309–310.

BIBLIOGRAPHY

Dempster, E. R. 1955. Maintenance of genetic heterogeneity. *Cold Spring Harbor Symp. Quant. Biol.* 20:25–32.

Dempster, E. R. and I. M. Lerner. 1950. Heritability of threshold characters. *Genetics* 35:212–236.

Dice, L. R. and W. B. Howard. 1951. Distance of dispersal by prairie deer mice from birthplaces to breeding sites. *Contr. Lab. Vert. Biol. Univ. Mich.* 50:1–15.

Dickerson, R. E. 1971. The structure of cytochrome c and the rates of molecular evolution. *J. Molec. Evol.* 1:26–45.

Dobzhansky, Th. 1948. Genetics of natural populations: XVI. Altitudinal and seasonal changes produced by natural selection in certain populations of *Drosophila pseudoobscura* and *Drosophila persimilis*. *Genetics* 33:158–176.

Dobzhansky, Th. 1970. *Genetics of the Evolutionary Process*. Columbia University Press, New York.

Dobzhansky, Th. and J. R. Powell. 1974. Rates of dispersal of *Drosophila psuedoobscura* and its relatives. *Proc. Roy. Soc., Ser. B.* 187:281–298.

Dobzhansky, Th. and B. Spassky. 1963. Genetics of natural populations: XXXIV. Adaptive norm, genetic load and genetic elite in *Drosophila pseudoobscura. Genetics* 48:1467–1485.

Dobzhansky, Th., B. Spassky and T. Tidwell. 1963. Genetics of natural populations: XXXII. Inbreeding and the mutational and balanced loads in natural populations of *D. pseudoobscura. Genetics* 48:361–373.

Dudley, J. W. 1977. Seventy-six generations of selection for oil and protein percentage in maize. In E. Pollak, O. Kempthorne and T. B. Bailey, Jr. (eds.). *International Conference on Quantitative Genetics*. Iowa State Univ. Press, Ames, Iowa, pp. 459–473.

Dunn, L. C. 1965. *A Short History of Genetics*. McGraw-Hill, New York.

Dykhuizen, D. 1978. Selection for tryptophan auxotrophs of *Escherichia coli* in glucose-limited chemostats as a test of the energy conservation hypothesis of evolution. *Evolution* 32:125–150.

Edwards, A. W. F. 1967. Fundamental theorem of natural selection. *Nature* 215:537–538.

Edwards, A. W. F. 1971. Distance between populations on the basis of gene frequencies. *Biometrics* 27:873–881.

Ehrman, L. 1970. The mating advantage of rare males in *Drosophila. Proc. Natl. Acad. Sci. U.S.A.* 515:345–348.

Ehrman, L. and P. A. Parsons. 1976. *The Genetics of Behavior*. Sinauer Associates, Sunderland, Massachusetts.

Ehrman, L. and J. Probber. 1978. Rare *Drosophila* males: The mysterious matter of choice. *Am. Scientist* 66:216–222.

Elton, C. and M. Nicholson. 1942. The ten-year cycle of the lynx in Canada. *J. Animal Ecol.* 11:215–244.

Emlen, J. M. 1973. *Ecology: An Evolutionary Approach.* Addison-Wesley, Reading, Massachusetts.

Endler, J. A. 1977. *Geographic Variation, Speciation, and Clines.* Princeton University Press, Princeton.

Endrizzi, J. E. 1974. Alternate-1 and alternate-2 disjunctions in heterozygous reciprocal translocations. *Genetics* 77:55–60.

Eshel, I. and D. Cohen. 1976. Altruism, competition, and kin selection in populations. In S. Karlin and E. Nevo (eds.). *Population Genetics and Ecology.* Academic Press, New York, pp. 537–546.

Evans, F. C. and F. E. Smith. 1952. The intrinsic rate of natural increase for the human louse, *Pediculus humanus* L. *Am. Nat.* 86:18–29.

Ewens, W. J. 1972. The sampling theory of selectively neutral alleles. *Theor. Pop. Biol.* 3:87–112.

Ewens, W. J. 1977. Selection and neutrality. In F. B. Christiansen and T. M. Fenchel (eds.). *Measuring Selection in Natural Populations.* Springer-Verlag, New York, pp. 159–176.

Ewens, W. J. 1979. Testing the generalized neutrality hypothesis. *Theor. Pop. Biol.* 15:205–216.

Ewens, W. J. and M. W. Feldman. 1976. The theoretical assessment of selective neutrality. In S. Karlin and E. Nevo (eds.). *Population Genetics and Ecology.* Academic Press, New York, pp. 303–338.

Falconer, D. S. 1955. Patterns of response in selection experiments with mice. *Cold Spring Harbor Symp. Quant. Biol.* 20:178–196.

Falconer, D. S. 1960. *Introduction to Quantitative Genetics.* Ronald Press, New York.

Falconer, D. S. 1965. The inheritance of liability to certain diseases estimated from the incidence among relatives. *Ann. Hum. Genet.* 29:51–76.

Falconer, D. S. 1971. Improvement of litter size in a strain of mice at a selection limit. *Genet. Res., Camb.* 17:215–235.

Falconer, D. S. 1977. Some results of the Edinburgh selection experiments with mice. In E. Pollak, O. Kempthorne and T. J. Bailey, Jr. (eds.). *International Conference on Quantitative Genetics.* Iowa State Univ. Press, Ames, Iowa, pp. 101–115.

Fasman, G. D. (ed.). 1976. *Handbook of Biochemistry and Molecular*

Biology, 3rd ed., *Vol 2: Nucleic Acids.* CRC Press, Cleveland.

Feldman, M. and L. L. Cavalli-Sforza. 1976. Cultural and biological processes: Selection for a trait under complex transmission. *Theor. Pop. Biol.* 9:238–259.

Feldman, M. W. and F. B. Christiansen. 1975. The effect of population subdivision on two loci without selection. *Genet. Res., Camb.* 24:151–162.

Feldman, M. W., I. Franklin and G. J. Thomson. 1974. Selection in complex genetic systems. I. The symmetric equilibria of the three-locus symmetric viability model. *Genetics* 76:135–162.

Feldman, M. W. and R. C. Lewontin. 1975. The heritability hang-up. *Science* 190:1163–1168.

Feldman, M. W., R. C. Lewontin, I. R. Franklin and F. B. Christiansen. 1975. Selection in complex genetic systems. III. An effect of allele multiplicity with two loci. *Genetics* 79:333–347.

Felsenstein, J. 1965. On the biological significance of the cost of gene substitution. *Am. Nat.* 105:1–11.

Felsenstein, J. 1971. Inbreeding and variance effective numbers in populations with overlapping generations. *Genetics* 68:581–597.

Felsenstein, J. 1976. The theoretical population genetics of variable selection and migration. *Ann. Rev. Genet.* 10:253–280.

Felsenstein, J. 1979. A symmetry principle in the evolution of cooperative relationships. Unpublished manuscript.

Finnerty, V. and G. Johnson. 1979. Post-translational modification as a potential explanation of high levels of enzyme polymorphism: Xanthine dehydrogenase and aldehyde oxidase in *Drosophila melanogaster. Genetics* 91:695–722.

Fisher, R. A. 1930. *The Genetical Theory of Natural Selection.* Clarendon, Oxford.

Fisher, R. A. 1936. Has Mendel's work been rediscovered? *Ann. Sci.* 1:115–137.

Fisher, R. A. 1949. *The Theory of Inbreeding,* 2nd ed. Oliver and Boyd, London.

Fisher, R. A. 1950. Gene frequencies in a cline determined by selection and diffusion. *Biometrics* 6:353–361.

Fitch, W. M. 1973. Aspects of molecular evolution. *Ann. Rev. Genet.* 7:343–380.

Fitch, W. M. 1976. Molecular evolutionary clocks. In F. J. Ayala (ed.). *Molecular Evolution.* Sinauer Associates, Sunderland, Massachusetts, pp. 160–178.

Fitch, W. M. 1977. Phylogenies constrained by the crossover process as illustrated by human hemoglobins and a thirteen cycle, eleven-amino-acid repeat in human apolipoprotein A-1. *Genetics* 86:623–644.

Fitch, W. M. and C. H. Langley. 1976. Protein evolution and the molecular clock. *Fed. Proc.* 35:2092–2097.

Fitch, W. M. and E. Margoliash. 1967. Construction of phylogenetic trees. *Science* 155:279–284.

Fitch, W. M. and E. Margoliash. 1970. The usefulness of amino acid and nucleotide sequences in evolutionary studies. *Evol. Biol.* 4:67–110.

Flavell, R. B. and D. B. Smith. 1976. Nucleotide sequence organization in the wheat genome. *Heredity* 37:231–252.

Ford, H. A. 1971. The degree of mimetic protection gained by new partial mimics. *Heredity* 27:227–236.

Forster, G. G., M. J. Whitten, T. Prout and R. Gill. 1972. Chromosome rearrangements for the control of insect pests. *Science* 176:875–880.

Franklin, I. R. 1977. The distribution of the proportion of the genome which is homozygous by descent in inbred individuals. *Theor. Pop. Biol.* 11:60–80.

Franklin, I. and R. C. Lewontin. 1970. Is the gene the unit of selection? *Genetics* 65:707–734.

Fryer, G. and T. D. Iles. 1972. *The Cichlid Fishes of the Great Lakes of Africa: Their Biology and Evolution.* TFH, Hong Kong.

Gardner, C. O. 1978. Population improvement in maize. In D. B. Walden (ed.). *Maize Breeding and Genetics.* John Wiley and Sons, New York, pp. 207–228.

Gardner, P., D. H. Smith, H. Beer and R. C. Moellering, Jr. 1969. Recovery of resistance (R) factors from a drug-free community. *Lancet* (Oct. 11, 1969):774–776.

Garrison, R. J., V. E. Anderson and S. C. Reed. 1968. Assortative marriage. *Eugen. Quart.* 15:113–127.

Gause, G. F. 1934. *The Struggle for Existence.* Williams and Wilkins, Baltimore.

Giblett, E. R. 1977. Genetic polymorphisms in human blood. *Ann. Rev. Genet.* 11:13–28.

Gibson, J. 1972. Differences in the number of molecules produced by two allelic electrophoretic variants in *D. melanogaster*. *Experientia* 28:975–976.

Gillespie, J. H. 1972. The effects of stochastic environments on allele frequencies in natural populations. *Theor. Pop. Biol.* 3:241–248.

Gillespie, J. H. 1978. A general model to account for enzyme variation in natural populations: V. The SAS-CFF model. *Theor. Pop. Biol.* 14:1-45.

Gillespie, J. and K. Kojima. 1968. The degree of polymorphism in enzymes involved in energy production compared to that in nonspecific enzymes in two *D. ananassae* populations. *Proc. Natl. Acad. Sci. U.S.A.* 61:582-585.

Gillespie, J. H. and C. H. Langley. 1974. A general model to account for enzyme variation in natural populations. *Genetics* 76:837-848.

Gillois, M. 1966. Le concept d'identité et son importance en génétique. *Annales de Génétique* 9:58-65.

Goldberg, R. B. 1978. DNA sequence organization in the Soybean plant. *Biochem. Genet.* 16:45-68.

Goodenough, U. 1978. *Genetics,* 2nd ed. Holt, Rinehart and Winston, New York.

Goodman, M. 1976. Protein sequences in phylogeny. In F. J. Ayala (ed.). *Molecular Evolution,* Sinauer Associates, Sunderland, Massachusetts, pp. 141-160.

Gottesman, I. I. and J. Shields. 1967. A polygenic theory of schizophrenia. *Proc. Natl. Acad. Sci. U.S.A.* 58:199-205.

Gottlieb, L. D. 1976. Biochemical consequences of speciation in plants. In F. J. Ayala (ed.). *Molecular Evolution.* Sinauer Associates, Sunderland, Massachusetts, pp. 123-140.

Gould, S. J. 1977. *Ontogeny and Phylogeny.* Belknap, Cambridge, Massachusetts.

Grant, V. 1971. *Plant Speciation.* Columbia Univ. Press, New York.

Griffing, B. 1960. Theoretical consequences of truncation selection based on the individual phenotype. *Aust. J. Biol. Sci.* 13:309-343.

Griffiths, D. J. 1950. The liability of seed crops of perennial ryegrass (*Lolium perenne*) to contamination by wind-borne pollen. *J. Agric. Res.* 40:19-38.

Gromko, M. H. 1977. What is frequency-dependent selection? *Evolution* 31:438-442.

Haldane, J. B. S. 1948. The theory of a cline. *J. Genet.* 48:277-284.

Haldane, J. B. S. 1956. The theory of selection for melanism in Lepidoptera. *Proc. Roy. Soc., Ser. B* 145:303-308.

Haldane, J. B. S. 1957. The cost of natural selection. *J. Genet.* 55:511-524.

Haldane, J. B. S. 1960. More precise expressions for the cost of natural selection. *J. Genet.* 57:351-360.

Haldane, J. B. S. and S. D. Jayakar. 1963. Polymorphism due to selection of varying direction. *J. Genet.* 58:237–242.

Hamilton, W. D. 1972. Altruism and related phenomena, mainly in social insects. *Ann. Rev. Ecol. Syst.* 3:193–232.

Hamrick, J. L. and R. W. Allard. 1972. Microgeographical variation in allozyme frequencies in *Avena barbata. Proc. Natl. Acad. Sci. U.S.A.* 69:2100–2104.

Hardin, G. 1960. The competitive exclusion principle. *Science* 131:1292–1298.

Hardy, G. H. 1908. Mendelian proportions in a mixed population. *Science* 28:41–50. (Reprinted in J. H. Peters (ed.). 1959. *Classic Papers in Genetics.* Prentice-Hall, Englewood Cliffs, New Jersey.)

Harper, R. A. and F. B. Armstrong. 1973. Alkaline phosphatase of *Drosophila melanogaster.* II. Biochemical comparison among four allelic forms. *Biochem. Genet.* 10:29–38.

Harris, D. L. 1977. Past, present and potential contributions of quantitative genetics to applied animal breeding. In E. Pollak, O. Kempthorne, and T. B. Bailey, Jr. (eds.). *International Conference on Quantitative Genetics.* Iowa State Univ. Press, Ames, Iowa, pp. 587–614.

Harris, E. L., J. C. Christian and W. E. Nance. 1979. Genetic variance in nonverbal intelligence: Data from kinships of identical twins. *Science* 205:1153–1154.

Harris, H. 1966. Enzyme polymorphisms in man. *Proc. Roy. Soc., Ser. B* 164:298–310.

Harris, H. 1969. Enzyme and protein polymorphism in human populations. *Brit. Med. Bull.* 25:5–13.

Harris, H. and D. A. Hopkinson. 1972. Average heterozygosity in man. *Ann. Hum. Genet., Lond.* 36:9–20.

Harris, H. and D. A. Hopkinson. 1976. *Handbook of Enzyme Electrophoresis in Human Genetics.* American Elsevier, New York.

Harris, H., D. A. Hopkinson and Y. H. Edwards. 1977. Polymorphism and the subunit structure of enzymes: A contribution to the neutralist-selectionist controversy. *Proc. Natl. Acad. Sci. U.S.A.* 74:698–701.

Harrison, G. A., J. S. Wiener, J. M. Tanner and N. A. Barnicot. 1964. *Human Biology: An Introduction to Human Evolution, Variation and Growth.* Oxford, London.

Hartl, D. L. 1972. A fundamental theorem of natural selection for sex linkage or arrhenotoky. *Am. Nat.* 106:516–524.

Hartl, D. L. 1977. *Our Uncertain Heritage: Genetics and Human Diversity.* J. B. Lippincott, Philadelphia.

Hartl, D. L. and S. W. Brown. 1970. The origin of male haploid genetic systems and their expected sex ratio. *Theor. Pop. Biol.* 1:165–190.

Hartl, D. L. and R. D. Cook. 1973. Balanced polymorphisms of quasineutral alleles. *Theor. Pop. Biol.* 4:163–172.

Hartl, D. L. and D. Dykhuizen. 1979. A selectively driven molecular clock. *Nature* 281:230-231.

Hayes, H. K. 1963. *A Professor's Story of Hybrid Corn.* Burgess, Minneapolis.

Hayman, B. I. 1954. The theory and analysis of diallel crosses. *Genetics* 39:789-809.

Hazel, L. N. and J. L. Lush. 1942. The efficiency of three methods of selection. *J. Hered.* 33:393-399.

Hedrick, P. W. 1955. Genetic similarity and distance: Comments and comparisons. *Evolution* 29:362-366.

Hedrick, P. W., M. E. Ginevan and E. P. Ewing. 1976. Genetic polymorphism in heterogeneous environments. *Ann. Rev. Ecol. Syst.* 7:1-32.

Hill, W. G. 1972. Effective size of populations with overlapping generations. *Theor. Pop. Biol.* 3:278-289.

Hill, W. G. 1976. Non-random association of neutral linked genes in finite populations. In S. Karlin and E. Nevo (eds.). *Population Genetics and Ecology.* Academic Press, New York, pp. 339-376.

Hill, W. G. 1977. Correlation of gene frequencies between neutral linked genes in finite populations. *Theor. Pop. Biol.* 11:239-248.

Hill, W. G. 1978. Estimation of heritability by regression using collateral relatives; linear heritability estimation. *Genet. Res., Camb.* 32:265-274.

Hill, W. G. and P. J. Avery. 1978. Estimating number of genes by genotype assay. *Heredity* 40:397-403.

Hiraizumi, Y. 1964. Prezygotic selection as a factor in the maintenance of variability. *Cold Spring Harbor Symp. Quant. Biol.* 29:51-60.

Hiraizumi, Y. and J. F. Crow. 1960. Heterozygous effects on viability, fertility, rate of development and longevity of *Drosophila* chromosomes that are lethal when homozygous. *Genetics* 45:1071-1083.

Hoenigsberg, H. F., J. J. Palomino, M. J. Hayes, I. Z. Zandstra and G. G. Rojas. 1977. Population genetics in the American tropics. X. Genetic load differences in *Drosophila willistoni* from Columbia. *Evolution* 31:805-811.

Holgate, P. 1966. A mathematical study of the founder principle of evolutionary genetics. *J. Appl. Prob.* 3:115–128.

Hollingsworth, T. H. 1969. *Historical Demography.* Cornell Univ. Press, Ithaca.

Hood, L., J. H. Campbell and S. C. R. Elgin. 1975. The organization, expression, and evolution of antibody genes and other multigene families. *Ann. Rev. Genet.* 9:305–354.

Hubby, J. L. and R. C. Lewontin. 1966. A molecular approach to the study of genic heterozygosity in natural populations. I. The number of alleles at different loci in *Drosophila pseudoobscura. Genetics* 54:577–594.

Huettel, M. D. and G. L. Bush. 1972. The genetics of host selection and its bearing on sympatric speciation in *Procecidochares* (Diptera: Tephritidae). *Entomol. Exp. Appl.* 15:465–480.

Hull, F. H. 1945. Recurrent selection for specific combining ability in corn. *J. Am. Soc. Agron.* 37:134–145.

Iltis, H. 1932. *Life of Mendel* (trans. by E. and C. Paul). Norton, New York.

Ives, P. T. 1945. The genetic structure of American populations of *Drosophila melanogaster. Genetics* 30:167–196.

Jacob, F. 1977. Evolution and tinkering. *Science* 196:1161–1166.

Jacquard, A. 1974. *The Genetic Structure of Populations* (trans. by D. and B. Charlesworth). Springer-Verlag, New York.

Jain, S. K. and A. D. Bradshaw. 1966. Evolutionary divergence among adjacent plant populations. I. The evidence and its theoretical analysis. *Heredity* 21:407–441.

Jenkins, M. T. 1978. Maize breeding during the development and early years of hybrid maize. In D. B. Walden (ed.). *Maize Breeding and Genetics.* John Wiley and Sons, New York, pp. 13–28.

Jinks, J. L. and P. Towey. 1976. Estimating the number of genes in a polygenic system by genotypic assay. *Heredity* 37:69–82.

Johannsen, W. 1903. Über Erblichkeit in Populationen und in reinen Linien. Gustav Fisher: Jena. [Translated in part in Peters, J. A. (ed.). 1959. *Classic Papers in Genetics.* Prentice-Hall, Englewood Cliffs, New Jersey, pp. 21–26.]

Johansson, I. and J. Rendel. 1968. *Genetics and Animal Breeding.* W. H. Freeman, San Francisco.

Johnson, G. B. 1977. Hidden heterogeneity among electrophoretic alleles. In F. B. Christiansen and T. M. Fenchel (eds.). *Measuring Selection in Natural Populations.* Springer-Verlag, New York, pp. 223–244.

Johnson, G. B., V. Finnerty and D. L. Hartl. 1980. To be published.

Jones, J. S., B. H. Leith and P. Rawlings. 1977. Polymorphism in *Cepaea*. *Ann. Rev. Ecol. Syst.* 8:109–144.

Jukes, T. H. 1966. *Molecules and Evolution*. Columbia Univ. Press, New York.

Jungen, H. and D. L. Hartl. 1979. Average fitness of populations of *Drosophila melanogaster* as estimated using compound-autosome strains. *Evolution* 33:359–370.

Kahler, A. L., M. T. Clegg and R. W. Allard. 1975. Evolutionary changes in the mating system of an experimental population of barley (*Hordeum vulgare* L.). *Proc. Natl. Acad. Sci. U.S.A.* 72:943–946.

Karlin, S. 1969. *Equilibrium Behavior of Population Genetic Models with Non-Random Mating*. Gordon and Breach, New York.

Karlin, S. 1975. General two-locus selection models: Some objectives, results and interpretations. *Theor. Pop. Biol.* 7:364–398.

Karlin, S. 1979a. Principles of polymorphism and epistasis for multilocus systems. *Proc. Natl. Acad. Sci. U.S.A.* 76:541–545.

Karlin, S. 1979b. Models of multifactorial inheritance. *Theor. Pop. Biol.* 15:308–438.

Karlin, S. and M. W. Feldman. 1970. Linkage and selection: Two locus symmetric viability models. *Theor. Pop. Biol.* 1:39–71.

Karlin, S. and B. Levikson. 1974. Temporal fluctuations in selection intensities: Case of small population size. *Theor. Pop. Biol.* 6:383–412.

Karlin, S. and U. Lieberman. 1974. Random temporal variation in selection intensities: Case of large population size. *Theor. Pop. Biol.* 6:355–382.

Karlin, S. and J. Raper. 1979. Sexual selection encounter models. *Theor. Pop. Biol.* 15:246–256.

Kastritsis, C. D. and D. W. Crumpacker. 1966. Gene arrangements in the third chromosome of *Drosophila pseudoobscura*. I. Configurations with tester chromosomes. *J. Hered.* 57:151–158.

Katz, A. J. and R. A. Cardellino. 1978. Estimation of fitness components in *Drosophila melanogaster*. I. Heterozygote viability indices. *Genetics* 88:139–148.

Katz, A. J. and F. D. Enfield. 1977. Response to selection for increased pupa weight in *Tribolium castaneum* as related to population structure. *Genet. Res., Camb.* 30:237–246.

Kaufman, P. K., F. D. Enfield and R. E. Comstock. 1977. Stabilizing selection for pupa weight in *Tribolium castaneum*. *Genetics* 87:327–341.

Kearsey, M. J. and J. L. Jinks. 1968. A general method of detecting additive, dominance and epistatic variation for metrical traits. I. Theory. *Heredity* 23:403–409.

Kempthorne, O. 1969. *An Introduction to Genetic Statistics.* Iowa State Univ. Press, Ames, Iowa.

Kempthorne, O. 1978. Logical, epistemological and statistical aspects of nature-nurture data interpretation. *Biometrics* 34:1–23.

Kerr, W. E. 1969. Some aspects of the evolution of social bees. *Evol. Biol.* 3:119–176.

Kerr, W. E. 1974. Advances in cytology and genetics of bees. *Ann. Rev. Entomol.* 19:253–268.

Kerr, W. E. and S. Wright. 1954. Experimental studies of the distribution of gene frequencies in very small populations of *Drosophila melanogaster.* III. Aristapedia and spineless. *Evolution* 8:293–301.

Kessler, S. 1966. Selection for and against ethological isolation between *Drosophila pseudoobscura* and *Drosophila persimilis. Evolution* 20:634–645.

Kettlewell, H. B. D. 1973. *The Evolution of Melanism: The Study of a Recurring Necessity.* Clarendon, Oxford.

Kidd, K. K. and L. L. Cavalli-Sforza. 1973. An analysis of the genetics of schizophrenia. *Social Biol.* 20:254–265.

Kidwell, J. F., M. T. Clegg, F. M. Stewart and T. Prout. 1977. Regions of stable equilibria for models of differential selection in the two sexes under random mating. *Genetics* 85:171–183.

King, J. L. and T. H. Jukes. 1969. Non-Darwinian evolution: Random fixation of selectively neutral mutations. *Science* 164:788–798.

King, M. C. and A. C. Wilson. 1975. Evolution at two levels: Molecular similarities and biological differences between humans and chimpanzees. *Science* 188:107–116.

Kingman, J. F. C. 1977. The population structure associated with Ewens' sampling formula. *Theor. Pop. Biol.* 11:274–283.

Kimura, M. 1955. Solution of a process of random genetic drift with a continuous model. *Proc. Natl. Acad. Sci. U.S.A.* 41:144–150.

Kimura, M. 1958. On the change of population fitness by natural selection. *Heredity* 12:145–167.

Kimura, M. 1964. *Diffusion Models in Population Genetics.* Methuen, London.

Kimura, M. 1968. Evolutionary rate at the molecular level. *Nature* 217:624–626.

Kimura, M. 1976. Population genetics and molecular evolution.

Johns Hopkins Medical J. 138:253-261.

Kimura, M. 1977. The neutral theory of molecular evolution and polymorphism. *Scientia* 112:687-707.

Kimura, M. and J. F. Crow. 1964. The number of alleles that can be maintained in a finite population. *Genetics* 49:725-738.

Kimura, M. and T. Ohta. 1971. *Theoretical Aspects of Population Genetics.* Princeton Univ. Press, Princeton.

Kimura, M. and T. Ohta. 1974. On some principles governing molecular evolution. *Proc. Natl. Acad. Sci. U.S.A.* 71:2848-2852.

Kimura, M. and T. Ohta. 1978. Stepwise mutation model and distribution of allelic frequencies in a finite population. *Proc. Natl. Acad. Sci. U.S.A.* 75:2868-2872.

Kimura, M. and G. H. Weiss. 1964. The stepping stone model of population structure and the decrease of genetic correlation with distance. *Genetics* 49:561-576.

Kleckner, N. 1977. Translocatable elements in procaryotes. *Cell* 11:11-23.

Knight, G. R., A. Robertson and C. H. Waddington. 1956. Selection for sexual isolation within a species. *Evolution* 10:14-22.

Koehn, R. K. 1978. Physiology and biochemistry of enzyme variation: The interface of ecology and population genetics. In *Ecological Genetics: The Interface.* P. Brussard (ed.). Springer-Verlag, New York, pp. 51-72.

Kojima, K. 1971. Is there a constant fitness for a given genotype? No! *Evolution* 25:281-285.

Lakovaara, S. and A. Saura. 1971. Genetic variation in marginal populations of *Drosophila subobscura. Hereditas* 69:77-82.

Lalouel, J. M. 1977. A conceptual framework of Malécot's model of isolation by distance. *Ann. Hum. Genet., Lond.* 40:355-360.

Lande, R. 1976. The maintenance of genetic variability by mutation in a polygenic character with linked loci. *Genet. Res., Camb.* 26:221-235.

Lande, R. 1977. Statistical tests for natural selection on quantitative characters. *Evolution* 31:442-444.

Langley, C. H., K. Ito and R. A. Voelker. 1977. Linkage disequilibrium in natural populations of *Drosophila melanogaster.* Seasonal variation. *Genetics* 86:447-454.

Langley, C. H., D. B. Smith and F. M. Johnson. 1978. Analysis of linkage disequilibria between allozyme loci in natural populations of *Drosophila melanogaster. Genet. Res., Camb.* 32:215-230.

Latter, B. D. H. 1975. Enzyme polymorphisms: Gene frequency

distributions with mutation and selection for optimal activity. *Genetics* 79:325–331.

Laurie-Ahlberg, C. C., G. P. Maroni, G. C. Bewley, J. C. Lucchesi and B. S. Weir. 1980. Quantitative genetic variation of enzyme activities in natural populations of *Drosophila melanogaster. Proc. Natl. Acad. Sci. U.S.A.* (in press).

Lees, D. R. and E. R. Creed. 1977. The genetics of the *insularia* forms of the peppered moth, *Biston betularia. Heredity* 39:67–73.

Leigh, E. G., Jr. 1977. How does selection reconcile individual advantage with the good of the group? *Proc. Natl. Acad. Sci. U.S.A.* 74:4542–4546.

Leigh Brown, A. J. and C. H. Langley. 1979. Reevaluation of level of genic heterozygosity in natural populations of *Drosophila melanogaster* by two-dimensional electrophoresis. *Proc. Natl. Acad. Sci. U.S.A.* 76:2381–2384.

Lerner, I. M. 1958. *The Genetic Basis of Selection.* John Wiley and Sons, New York.

Levene, H. 1953. Genetic equilibrium when more than one ecological niche is available. *Am. Nat.* 87:331–333.

Levin, B. R. and W. L. Kilmer. 1975. Interdemic selection and the evolution of altruism: A computer simulation study. *Evolution* 28:527–545.

Levin, D. A. 1978. Genetic variation in annual Phlox: Self-compatible versus self-incompatible species. *Evolution* 32:245–263.

Levin, D. A. and H. W. Kerster. 1968. Local gene dispersal in *Phlox. Evolution* 22:130–139.

Levin, D. A. and H. W. Kerster. 1974. Gene flow in seed plants. *Evol. Biol.* 7:139–220.

Levin, D. A. and A. C. Wilson. 1976. Rate of evolution in seed plants: Net increase in diversity of chromosome numbers and species numbers through time. *Proc. Natl. Acad. Sci. U.S.A.* 73:2086–2090.

Levins, R. 1966. The strategy of model building in population biology. *Amer. Sci.* 54:421–431.

Levy, M. 1976. Altered glycoflavone expression in induced autotetraploids of *Phlox drummondii. Biochem. Syst. and Ecol.* 4:249–254.

Levy, M. and D. A. Levin. 1975. Genetic heterozygosity and variation in permanent translocation heterozygotes of the *Oenothera biennis* complex. *Genetics* 79:493–512.

Lewin, B. 1977. *Gene Expression,* Vol. 3. *Plasmids and Phages.* John

Wiley and Sons, New York.

Lewis, N. and J. Gibson. 1978. Variation in amount of enzyme protein in natural populations. *Biochem. Genet.* 16:159-170.

Lewontin, R. C. 1970. The units of selection. *Ann. Rev. Ecol. Syst.* 1:1-18.

Lewontin, R. C. 1972. The apportionment of human diversity. *Evol. Biol.* 6:381-398.

Lewontin, R. C. 1974a. *The Genetic Basis of Evolutionary Change.* Columbia Univ. Press, New York.

Lewontin, R. C. 1974b. The analysis of variance and the analysis of causes. *Am. J. Hum. Genet.* 26:400-411.

Lewontin, R. C. 1975. Genetic aspects of intelligence. *Ann. Rev. Genet.* 9:387-406.

Lewontin, R. C. and C. C. Cockerham. 1959. The goodness-of-fit test for detecting natural selection in random mating populations. *Evolution* 13:561-564.

Lewontin, R. C., L. R. Ginzburg and S. D. Tuljapurkar. 1978. Heterosis as an explanation for large amounts of genetic polymorphism. *Genetics* 88:149-169.

Lewontin, R. C. and J. L. Hubby. 1966. A molecular approach to the study of genic heterozygosity in natural populations. II. Amount of variation and degree of heterozygosity in natural populations of *Drosophila pseudoobscura. Genetics* 54:595-609.

Lewontin, R. C., D. Kirk and J. Crow. 1968. Selective mating, assortative mating, and inbreeding: Definitions and implications. *Eugen. Quart.* 15:141-143.

Lewontin, R. C. and M. J. D. White. 1960. Interaction between inversion polymorphisms of two chromosome pairs in the grasshopper, *Moraba scurra. Evolution* 14:116-129.

Li, C. C. 1967a. Castle's early work on selection and equilibrium. *Am. J. Hum. Gen.* 19:70-74.

Li, C. C. 1967b. Fundamental theorem of natural selection. *Nature* 214:505-506.

Li, W.-H. 1978. Maintenance of genetic variability under the joint effect of mutation, selection and random drift. *Genetics* 90:349-382.

Loehlin, J. C., G. Lindzey and J. N. Spuhler. 1975. *Race Differences in Intelligence.* W. H. Freeman, San Francisco.

Lonnquist, J. H. 1964. A modification of the ear-to-row procedure for the improvement of maize populations. *Crop Sci.* 4:227-228.

Lonnquist, J. H. and N. E. Williams. 1967. Development of maize hybrids through selection among full-sib families. *Crop Sci.* 7:369–370.

Lotka, A. J. 1922. The stability of the normal age distribution. *Proc. Natl. Acad. Sci. U.S.A.* 8:339–345.

Lotka, A. J. 1925. *Elements of Physical Biology.* Williams and Wilkins, Baltimore.

Lush, J. L. 1937. *Animal Breeding Plans.* Iowa State Univ. Press, Ames, Iowa.

Lush, J. L. 1945. *Animal Breeding Plans,* 3rd ed. Iowa State Univ. Press, Ames, Iowa.

Lush, J. L. 1947a. Family merit and individual merit as bases for selection. Part I. *Am. Nat.* 81:241–261.

Lush, J. L. 1947b. Family merit and individual merit as bases for selection. Part II. *Am. Nat.* 81:362–379.

Lyttle, T. W. 1979. Experimental population genetics of meiotic drive systems. II. Accumulation of genetic modifiers of Segregation Distorter (SD) in laboratory populations. *Genetics* 91:339–357.

MacIntyre, R. J. 1976. Evolution and ecological value of duplicate genes. *Ann. Rev. Ecol. Syst.* 7:421–468.

MacNair, M. 1979. The genetics of copper tolerance in the yellow monkey flower, *Mimulus guttatus.* I. Crosses to nontolerants. *Genetics* 91:553–563.

Malécot, G. 1950. Quelques schémas probabilistes sur la variabilité des populations naturelles. *Ann. Univ. Lyon Sci. A* 13:37–60.

Malécot, G. 1967. Identical loci and relationship. *Proc. Fifth Berkeley Symp. Math. Statist. Prob.* 4:317–332.

Malécot, G. 1969. *The Mathematics of Heredity* (trans. by D. M. Yermanos). W. H. Freeman, San Francisco.

Malécot, G. 1972. Structure géographique et variabilité d'une grande population. *Proc. 4th Int. Cong. Hum. Genet.* pp. 138–154.

Mandel, S. P. H. 1971. Owen's model of a genetical system with differential viability between the sexes. *Heredity* 26:49–63.

Marinković, D., F. J. Ayala and M. Andjelković. 1978. Genetic polymorphism and phylogeny of *Drosophila subobscura. Evolution* 32:164–173.

Markert, C. L., J. B. Shaklee and G. S. Whitt. 1975. Evolution of a gene. *Science* 189:102–114.

Markow, T. A. 1978. A test for the rare male mating advantage in coisogenic strains of *Drosophila melanogaster. Genet. Res., Camb.* 32:123–127.

Maruyama, T. and M. Kimura. 1978. Theoretical study of genetic variability, assuming stepwise production of neutral and very slightly deleterious mutations. *Proc. Natl. Acad. Sci. U.S.A.* 75:919–922.

Matessi, C. and S. D. Jayakar. 1976. Conditions for the evolution of altruism under Darwinian selection. *Theor. Pop. Biol.* 9:360–387.

May, R. M., J. A. Endler and R. E. McMurtie. 1975. Gene frequency clines in the presence of selection opposed by gene flow. *Am. Nat.* 109:659–676.

Maynard Smith, J. 1966. Sympatric speciation. *Am. Nat.* 100:637–650.

Maynard Smith, J. 1968. "Haldane's dilemma" and the rate of evolution. *Nature* 29:1114–1116.

Mayr, E. 1963. *Animal Species and Evolution.* Harvard Univ. Press, Cambridge, Massachusetts.

Mayr, E. 1970. *Populations, Species, and Evolution.* Harvard Univ. Press, Cambridge, Massachusetts.

McClearn, G. E. and J. C. DeFries. 1973. *Introduction to Behavioral Genetics.* W. H. Freeman, San Francisco.

McDonald, J. F. and F. J. Ayala. 1978. Genetic and biochemical basis of enzyme activity variation in natural populations. I. Alcohol dehydrogenase in *Drosophila melanogaster. Genetics* 89:371–388.

McDonald, J. F., G. K. Chambers, J. David and F. J. Ayala. 1977. Adaptive response due to changes in gene regulation: A study with *Drosophila. Proc. Natl. Acad. Sci. U.S.A.* 74:4562–4566.

McMillan, I. and A. Robertson. 1974. The power of methods for the detection of major genes affecting quantitative characters. *Heredity* 32:349–356.

McNeilly, T. 1968. Evolution in closely adjacent plant populations. III. *Agrostis tenuis* on a small copper mine. *Heredity* 23:99–108.

Mettler, L. E., A. A. Voelker and T. Mukai. 1977. Inversion clines in populations of *Drosophila melanogaster. Genetics* 87:169–176.

Meyer, H. H. and F. D. Enfield. 1975. Experimental evidence on limitations of the heritability parameter. *Theor. Appl. Genet.* 45:268–273.

Meynell, G. G. 1972. *Bacterial Plasmids.* M.I.T. Press, Cambridge, Massachusetts.

Milkman, R. 1970. The genetic basis of natural variation in *Drosophila melanogaster. Adv. Genet.* 15:55–114.

Milkman, R. 1973. Electrophoretic variation in *Escherichia coli* from natural sources. *Science* 182:1024–1026.

Milkman, R. 1978. Selection differentials and selection coefficients. *Genetics* 88:391–403.

Miller, D. A. 1977. Evolution of primate chromosomes. *Science* 198:1116–1124.

Molin, C. D. 1979. An external scent as the basis for a rare-male mating advantage in *Drosophila melanogaster*. *Am. Nat.* 113:951–954.

Moran, P. A. P. 1964. On the non-existence of adaptive topographies. *Ann. Hum. Genet., Lond.* 27:383–393.

Morton, N. E. 1961. Morbidity of children from consanguineous marriages. In A. G. Steinberg (ed.). *Progress in Medical Genetics,* Vol. 1. Grune and Stratton, New York, pp. 261–291.

Morton, N. E. 1969. Human population structure. *Ann. Rev. Genet.* 3:53–74.

Morton, N. E. 1977. Isolation by distance in human populations. *Ann. Hum. Genet., Lond.* 40:361–365.

Morton, N. E. 1978. Effect of inbreeding on IQ and mental retardation. *Proc. Natl. Acad. Sci. U.S.A.* 75:3906–3908.

Morton, N. E., J. F. Crow and H. J. Muller. 1956. An estimate of mutational damage in man from data on consanguineous marriages. *Proc. Natl. Acad. Sci. U.S.A.* 42:855–863.

Morton, N. E., H. M. Dick, N. C. Allan, M. M. Izatt, R. Hill and S. Yee. 1977. Bioassay of kinship in northwestern Europe. *Ann. Hum. Genet., Lond.* 41:249–255.

Morton, N. E. and C. J. MacLean. 1974. Analysis of family resemblance. III. Complex segregation of quantitative traits. *Am. J. Hum. Genet.* 26:489–503.

Morton, N. E., N. Yasuda, C. Miki and S. Yee. 1968. Bioassay of population structure under isolation by distance. *Am. J. Hum. Genet.* 20:411–419.

Mourant, A. E., A. C. Kopeć and K. Domaniewska-Sobczak. 1976. *The Distribution of the Human Blood Groups and Other Polymorphisms,* 2nd ed. Oxford Univ. Press, New York.

Mukai, T. 1977. Genetic variance for viability and linkage disequilibrium in natural populations of *Drosophila melanogaster*. In F. B. Christiansen and T. M. Fenchel (eds.). *Measuring Selection in Natural Populations.* Springer-Verlag, New York, pp. 97–112.

Mukai, T. and C. C. Cockerham. 1977. Spontaneous mutation rates at enzyme loci in *Drosophila melanogaster*. *Proc. Natl. Acad. Sci. U.S.A.* 74:2514–2517.

Mukai, T., L. E. Mettler and S. Chigusa. 1971. Linkage disequilibrium in a local population of *Drosophila melanogaster*. *Proc. Natl. Acad. Sci. U.S.A.* 68:1065-1069.

Mukai, T., T. K. Watanabe and O. Yamaguchi. 1974. The genetic structure of natural populations of *Drosophila melanogaster*. XII. Linkage disequilibrium in a large local population. *Genetics* 77:771-793.

Mukai, T. and O. Yamaguchi. 1974. The genetic structure of natural populations of *Drosophila melanogaster*. XI. Genetic variability of local populations. *Genetics* 76:339-366.

Mulcahy, D. L. and S. M. Kaplan. 1979. Mendelian ratios despite nonrandom fertilization? *Am. Nat.* 113:419-425.

Muller, H. J. 1950. Our load of mutations. *Am. J. Hum. Genet.* 2:111-176.

Nagylaki, T. 1978. Random genetic drift in a cline. *Proc. Natl. Acad. Sci. U.S.A.* 75:423-426.

Nassar, R. F. 1972. Further evidence on multiple peak epistasis in *Drosophila melanogaster*. *Aust. J. Biol. Sci.* 25:565-572.

Neal, N. P. 1935. The decrease in yielding capacity in advanced generations of hybrid corn. *J. Amer. Soc. Agron.* 27:666-670.

Neel, J. V. 1978. The population structure of an Amerindian tribe, the Yanomama. *Ann. Rev. Genet.* 12:365-414.

Neel, J. V. and W. J. Schull. 1962. The effect of inbreeding on mortality and morbidity in two Japanese cities. *Proc. Natl. Acad. Sci. U.S.A.* 48:573-582.

Neel, J. V. and E. A. Thompson. 1978. Founder effect and the number of private polymorphisms observed in Amerindian tribes. *Proc. Natl. Acad. Sci. U.S.A.* 75:1904-1908.

Nei, M. 1975. *Molecular Population Genetics and Evolution.* American Elsevier, New York.

Nei, M. 1976. Mathematical models of speciation and genetic distance. In S. Karlin and E. Nevo (eds.). *Population Genetics and Ecology.* Academic Press, New York, pp. 723-765.

Nei, M. 1977. F-statistics and analysis of gene diversity in subdivided populations. *Ann. Hum. Genet., Lond.* 41:225-233.

Nei, M. 1978. The theory of genetic distance and evolution of human races. *Jap. J. Human. Genet.* 23:341-369.

Nei, M., R. Chakraborty and P. A. Fuerst. 1976. Infinite allele model with varying mutation rate. *Proc. Natl. Acad. Sci. U.S.A.* 73:4164-4168.

Nei, M., P. A. Fuerst and R. Chakraborty. 1976. Testing the neutral mutation hypothesis by distribution of single locus heterozygosity. *Nature* 262:491–493.

Nei, M. and Y. Imaizumi. 1966. Genetic structure of human populations. *Heredity* 21:183–190.

Nei, M., T. Maruyama and R. Chakraborty. 1975. The bottleneck effect and genetic variability in populations. *Evolution* 29:1–10.

Nevo, E. 1978. Genetic variation in natural populations: Patterns and theory. *Theor. Pop. Biol.* 13:121–177.

Nevo, E. and H. Cleve. 1978. Genetic differentiation during speciation. *Nature* 275:125.

Newcombe, H. B. 1964. In M. Fishbein (ed.), Papers and Discussions of the *Second International Conference on Congenital Malformations.* International Medical Congress, New York.

Nordskog, A. W. and F. G. Giesbrecht. 1964. Regression in egg production in the domestic fowl when selection is relaxed. *Genetics* 50:407–416.

Novick, A. 1955. Mutagens and antimutagens. *Brookhaven Symp. Biol.* 8:201–215.

O'Donald, P. 1969. "Haldane's dilemma" and the rate of natural selection. *Nature* 221:815–816.

O'Donald, P. 1977. Theoretical aspects of sexual selection. *Theor. Pop. Biol.* 12:298–334.

O'Donald, P. 1978. Theoretical aspects of sexual selection: A generalized model of mating behavior. *Theor. Pop. Biol.* 13:226–243.

O'Donald, P. 1979. Theoretical aspects of sexual selection: Variation in threshold of female mating response. *Theor. Pop. Biol.* 15:191–204.

Ohno, S. 1970. *Evolution by Gene Duplication.* Springer-Verlag, New York.

Ohta, T. 1978. Theoretical population genetics of repeated genes forming a multigene family. *Genetics* 88:845–861.

Olby, R. 1974. *The Path to the Double Helix.* Univ. of Washington Press, Seattle.

Orozco, F. 1976. A dynamic study of genotype-environment interaction with egg laying of *Tribolium castaneum. Heredity* 37:157–172.

Oster, G., I. Eshel and D. Cohen. 1977. Worker-queen conflict and the evolution of social insects. *Theor. Pop. Biol.* 12:49–85.

Parsons, P. A. 1977. Genes, behavior, and evolutionary processes: The genus *Drosophila. Adv. Genet.* 19:1–32.

Pearl, R. 1927. The growth of populations. *Quart. Rev. Biol.* 2:532–548.

Pearson, K. and A. Lee. 1903. On the laws of inheritance in man. I. Inheritance of physical characteristics. *Biometrika* 2:357–462.

Petit, C. and L. Ehrman. 1969. Sexual selection in *Drosophila*. *Evol. Biol.* 3:177–223.

Pinsker, W., P. Lankinen and D. Sperlich. 1978. Allozyme and inversion polymorphism in a central European population of *Drosophila subobscura*. *Genetica* 48:207–214.

Pirchner, F. 1969. *Population Genetics in Animal Breeding*. W. H. Freeman, San Francisco.

Platt, A. P. and L. P. Brower. 1968. Mimetic versus disruptive coloration in intergrading populations of *Limenitis arthemis* and *Astyanax* butterflies. *Evolution* 22:699–718.

Pollak, E. 1978. With selection for fecundity the mean fitness does not necessarily increase. *Genetics* 90:383–389.

Ponzoni, R. W. and J. W. James. 1978. Possible biases in heritability estimates from intraclass correlation. *Theor. Appl. Genet.* 53:25–27.

Poole, R. W. 1978. The statistical prediction of population fluctuations. *Ann. Rev. Ecol. Syst.* 9:427–448.

Powell, J. R. 1978. The founder-flush speciation theory: An experimental approach. *Evolution* 32:465–474.

Powell, J. R., Th. Dobzhansky, J. E. Hook and H. E. Wistrand. 1976. Genetics of natural populations. XLIII. Further studies of rates of dispersal of *Drosophila pseudoobscura* and its relatives. *Genetics* 82:493–506.

Prakash, S. 1977a. Gene polymorphism in natural populations of *Drosophila persimilis*. *Genetics* 85:513–520.

Prakash, S. 1977b. Further studies on gene polymorphism in the mainbody and geographically isolated populations of *Drosophila pseudoobscura*. *Genetics* 85:713–719.

Prakash, S. and R. C. Lewontin. 1968. A molecular approach to the study of genic heterozygosity. III. Direct evidence of coadaptation in gene arrangements of *Drosophila*. *Proc. Natl. Acad. Sci. U.S.A.* 59:398–405.

Prevosti, A. 1966. Chromosomal polymorphisms in western Mediterranean populations of *Drosophila subobscura*. *Genet. Res., Camb.* 7:149–158.

Prout, T. 1962. The effects of stabilizing selection on the time of development in *Drosophila melanogaster*. *Genet. Res., Camb.* 3:364–382.

Prout, T. 1965. The estimation of fitness from genotypic frequencies. *Evolution* 19:546–551.

Prout, T. 1969. The estimation of fitness from population data. *Genetics* 63:949–967.

Prout, T. 1971a. The relation between fitness components and population prediction in *Drosophila*. I. The estimation of fitness components. *Genetics* 68:127–149.

Prout, T. 1971b. The relation between fitness components and population prediction in *Drosophila*. II. Population prediction. *Genetics* 68:151–167.

Race, R. R. and R. Sanger. 1975. *Blood Groups in Man,* 6th ed. J. B. Lippincott, Philadelphia.

Radinsky, L. 1978. Do albumin clocks run on time? *Science* 200:1182–1183.

Ramshaw, J. A. M., J. A. Coyne and R. C. Lewontin. 1980. The sensitivity of gel electrophoresis as a detector of genetic variation. *Genetics* (in press).

Ramshaw, J. A. M. and W. F. Eanes. 1978. Study of the charge-state model for electrophoretic variation using isoelectric focusing of esterase-5 from *Drosophila pseudoobscura. Nature* 275:68–70.

Reeck, G. R., E. Swanson and D. C. Teller. 1978. The evolution of histones. *J. Mol. Evol.* 10:309–317.

Reed, T. E. and J. V. Neel. 1959. Huntington's chorea in Michigan. *Am. J. Hum. Genet.* 11:107–136.

Reich, T., J. Rice, C. R. Cloninger, R. Wette and J. James. 1979. The use of multiple thresholds and segregation analysis in analyzing the phenotypic heterogeneity of multifactorial traits. *Ann. Hum. Genet., Lond.* 42:371–390.

Richardson, R. H. 1974. Effects of dispersal, habitat selection and competition on a speciation pattern of *Drosophila* endemic to Hawaii. In. M. J. D. White (ed.). *Genetic Mechanisms of Speciation in Insects.* Australia and New Zealand Book Co., Sydney, pp. 140–164.

Richmond, R. C. 1972. Enzyme variability in the *Drosophila willistoni* group. III. Amounts of variability in the superspecies *D. paulistorum. Genetics* 71:87–112.

Ritossa, F. 1976. The bobbed locus. In M. Ashburner and E. Novitski (eds.). *The Genetics and Biology of Drosophila,* Vol. 1b. Academic Press, London, pp. 801–846.

Roberts, R. C. 1966. The limits to artificial selection for body weight in the mouse. II. The genetic nature of the limits. *Genet. Res., Camb.* 8:361–375.

Roberts, R. C. 1967. The limits to artificial selection for body weight in the mouse. III. Selection from crosses between previously selected lines. *Genet. Res., Camb.* 9:73–85.

Robertson, A. 1955. Prediction equations in quantitative genetics. *Biometrics* 11:95–98.

Robertson, A. 1960. A theory of limits in artificial selection. *Proc. Roy. Soc., Ser. B* 153:234–249.

Robertson, A. 1962. Selection for heterozygotes in small populations. *Genetics* 47:1291–1300.

Robertson, A. 1967. Animal breeding. *Ann. Rev. Genet.* 1:295–312.

Robertson, A. 1970. A theory of limits in artificial selection with many linked loci. In K. Kojima (ed.). *Mathematical Topics in Population Genetics.* Springer-Verlag, New York, pp. 246–288.

Robertson, F. W. 1955. Selection response and the properties of genetic variation. *Cold Spring Harbor Symp. Quant. Biol.* 20:166–177.

Robertson, F. W. 1957. Studies in quantitative inheritance. XI. Genetic and environmental correlation between body size and egg production in *Drosophila melanogaster. J. Genet.* 55:428–443.

Robinson, H. F., R. E. Comstock and P. H. Harvey. 1949. Estimates of heritability and degree of dominance in corn. *Agron. J.* 41:353–359.

Ruddle, F. H., T. H. Roderick, T. B. Shows, P. G. Weigl, R. K. Chipman and P. K. Anderson. 1969. Measurement of genetic heterogeneity by means of enzyme polymorphism. *J. Hered.* 60:321–322.

Russell, W. A. and S. A. Eberhart. 1975. Hybrid performance of selected maize lines from reciprocal recurrent and testcross selection programs. *Crop. Sci.* 15:1–4.

Salceda, V. M. 1977. Carga genetica en siete poblaciones naturales de *Drosophila melanogaster* (Meigen) de diferentes localidades de Mexico. *Sobretiro de Agrociencia* 28:47–52.

Sandler, L. and E. Novitski. 1957. Meiotic drive as an evolutionary force. *Am. Nat.* 91:105–110.

Sayre, R. 1975. *Rosalind Franklin and DNA.* Norton, New York.

Schaffer, H. E., D. G. Yardley and W. W. Anderson. 1977. Drift or selection: A statistical test of gene frequency over generations. *Genetics* 87:371–379.

Scharloo, W. 1971. Reproductive isolation by disruptive selection: Did it occur? *Am. Nat.* 105:83–86.

Schull, W. J. and J. V. Neel. 1965. *The Effects of Inbreeding on Japanese Children.* Harper & Row, New York.

Scudo, F. M. 1967. Selection on both haplo and diplophase. *Genetics* 56:693–704.

Selander, R. K. 1976. Genetic variation in natural populations. In

F. J. Ayala (ed.). *Molecular Evolution.* Sinauer Associates, Sunderland, Massachusetts, pp. 21–45.

Selander, R. K., W. G. Hunt and S. Y. Yang. 1969. Protein polymorphism and genetic heterozygosity in two European subspecies of the house mouse. *Evolution* 23:379–390.

Selander, R. K., S. Y. Yang and W. G. Hunt. 1969. Polymorphism in esterases and hemoglobins in wild populations of the house mouse. In M. R. Wheeler (ed.). *Studies in Genetics: V.* Publication 6918, Univ. of Texas, Austin, pp. 271–328.

Semeonoff, R. 1977. Can polymorphism be maintained by selection favoring an intermediate optimum phenotype? *Heredity* 39:373–381.

Seyffert, W. and G. Forkmann. 1976. Simulation of quantitative characters by genes with biochemically definable action. VII. Observation and discussion of nonlinear relationships. In S. Karlin and E. Nevo (eds.). *Population Genetics and Ecology.* Academic Press, New York, pp. 431–440.

Shelby, C. E., R. T. Clark and R. R. Woodward. 1955. The heritability of some economic characteristics of beef cattle. *J. Animal Sci.* 14:372–385.

Sheppard, P. M. 1959. The evolution of mimicry: A problem in ecology and genetics. *Cold Spring Harbor Symp. Quant. Biol.* 24:131–140.

Shields, J. 1962. *Monozygotic Twins Brought Up Apart and Brought Up Together.* Oxford, London.

Shull, G. H. 1909. A pure line method of corn breeding. *Am. Breeders Mag.* 1:98–107.

Simmonds, N. W. 1977. Approximations for i, intensity of selection. *Heredity* 38:413–414.

Simmons, M. J. and J. F. Crow. 1977. Mutations affecting fitness in *Drosophila* populations. *Ann. Rev. Genet.* 11:49–78.

Simpson, G. G. 1961. *Principles of Animal Taxonomy.* Columbia Univ. Press, New York.

Singh, R. S., R. C. Lewontin and A. A. Felton. 1976. Genetic heterogeneity within electrophoretic "alleles" of xanthine dehydrogenase in *Drosophila pseudoobscura. Genetics* 84:609–629.

Slatkin, M. 1972. On treating the chromosome as the unit of selection. *Genetics* 72:157–168.

Slatkin, M. 1977. Gene flow and genetic drift in a species subject to frequent local extinctions. *Theor. Pop. Biol.* 12:253–262.

Slatkin, M. and T. Maruyama. 1975. Genetic drift in a cline. *Genetics* 81:209–222.

Smith, C. 1975. Quantitative inheritance. In G. Fraser and O. Mayo (eds.). *Textbook of Human Genetics.* Blackwell, Oxford, pp. 382–441.

Smith, C. A. B. 1977. A note on genetic distance. *Ann. Hum. Genet., Lond.* 40:463–479.

Smith, R. L. 1974. *Ecology and Field Biology,* 2nd ed. Harper & Row, New York.

Smouse, P. E. 1974. Likelihood analysis of recombinational disequilibrium in multiple-locus gametic frequencies. *Genetics* 76:557–565.

Somero, G. N. 1978. Temperature adaptation of enzymes. *Ann. Rev. Ecol. Syst.* 9:1–30.

Spencer, W. P. 1947. Mutations in wild populations of *Drosophila. Advan. Genet.* 1:359–402.

Sperlich, D., H. Feuerbach-Mravlag, P. Lang, A. Michaelidis and A. Pentzos-Daponte. 1977. Genetic load and viability distribution in central and marginal populations of *Drosophila subobscura. Genetics* 86:835–848.

Spielman, R. S., J. V. Neel and F. H. F. Li. 1977. Inbreeding estimation from population data: Models, procedures and implications. *Genetics* 85:355–371.

Spiess, E. B. 1970. Mating propensity and its genetic basis in *Drosophila. Evol. Biol.* 4, suppl. 315–380.

Spiess, E. B. and R. J. Schuellein. 1956. Chromosomal adaptive polymorphism in *Drosophila persimilis.* I. Life cycle components under near optimal conditions. *Genetics* 41:501–516.

Spieth, H. T. 1968. Evolutionary implications of sexual behavior in *Drosophila. Evol. Biol.* 2:157–193.

Spofford, J. B. 1969. Heterosis and the evolution of duplications. *Am. Nat.* 103:407–432.

Sprague, G. F. 1967. Plant breeding. *Ann. Rev. Genet.* 1:269–294.

Sprague, G. F. 1978. Introductory remarks to the session on the history of hybrid corn. In D. B. Walden (ed.). *Maize Breeding and Genetics.* John Wiley and Sons, New York, pp. 11–12.

Spuhler, J. N. 1968. Assortative mating with respect to physical characteristics. *Eugen. Quart.* 15:128–140.

Stalker, H. D. 1976. Chromosome studies in wild populations of *D. melanogaster. Genetics* 82:323–347.

Stebbins, G. L. 1950. *Variation and Evolution in Plants.* Columbia Univ. Press, New York.

Stebbins, G. L. 1976. Chromosomes, DNA and plant evolution. *Evol. Biol.* 9:1–34.

Stebbins, G. L. 1977. *Processes of Organic Evolution,* 3rd ed. Prentice-Hall, Englewood Cliffs, New Jersey.

Stern, C. 1973. *Principles of Human Genetics,* 3rd ed. W. H. Freeman, San Francisco.

Stern, C. and E. R. Sherwood (eds.). 1966. *The Origin of Genetics: A Mendel Source Book.* W. H. Freeman, San Francisco.

Sturtevant, A. H. 1965. *A History of Genetics.* Harper & Row, New York.

Sved, J. A. 1968. The stability of linked systems of loci with small population size. *Genetics* 59:543–563.

Sved, J. A. 1971. An estimate of heterosis in *Drosophila melanogaster. Genet. Res., Camb.* 18:97–105.

Sved, J. A. and F. J. Ayala. 1970. A population cage test for heterosis in *Drosophila pseudoobscura. Genetics* 66:97–113.

Tartof, K. D. 1975. Redundant genes. *Ann. Rev. Genetics* 9:355–385.

Tashian, R. E., M. Goodman, R. E. Ferrell and R. J. Tanis. 1976. Evolution of carbonic anhydrase in primates and other mammals. In M. Goodman and R. E. Tashian (eds.). *Molecular Anthropology.* Plenum, New York.

Taylor, C. E. and J. R. Powell. 1977. Microgeographic differentiation of chromosomal and enzyme polymorphisms in *Drosophila persimilis. Genetics* 85:681–695.

Teissier, G. 1942. Persistence d'un gène léthal dans une population de *Drosophiles. Compt. Rend. Acad. Sci.* 214:327–330.

Templeton, A. R. 1974. Density dependent selection in parthenogenetic and self-mating populations. *Theor. Pop. Biol.* 5:229–250.

Templeton, A. R. 1979. Once again, why 300 species of Hawaiian *Drosophila? Evolution* 33:513–517.

Templeton, A. 1980. The theory of speciation via the founder principle. *Genetics* (in press).

Thoday, J. M. 1972. Disruptive selection. *Proc. Roy. Soc., Ser. B* 182:109–143.

Thoday, J. M. and T. B. Boam. 1959. Effects of disruptive selection. II. Polymorphism and divergence without isolation. *Heredity* 13:205–218.

Thoday, J. M. and J. B. Gibson. 1962. Isolation by disruptive selection. *Nature* 193:1164–1166.

Thoday, J. M. and J. B. Gibson. 1970. The probability of isolation by disruptive selection. *Am. Nat.* 104:219–230.

Thomson, G. 1977. The effect of a selected locus on linked neutral loci. *Genetics* 85:753–788.

Thomson, G. and W. F. Bodmer. 1977. The genetics of HLA and disease associations. In F. B. Christiansen and T. M. Fenchel (eds.). *Measuring Selection in Natural Populations.* Springer-Verlag, New York, pp. 545–564.

Trippa, G., A. Loverre and A. Catamo. 1976. Thermostability studies for investigating non-electrophoretic polymorphic alleles in *Drosophila melanogaster. Nature* 260:42.

Trivers, R. L. and H. Hare. 1976. Haplodiploidy and the evolution of the social insects. *Science* 191:249–263.

Turelli, M. 1977. Random environments and stochastic calculus. *Theor. Pop. Biol.* 12:140–178.

Turner, H. N. and S. S. Y. Young. 1969. *Quantitative Genetics in Sheep Breeding.* Cornell Univ. Press, Ithaca.

Turner, J. R. G. 1970. Changes in mean fitness under natural selection. In K. Kojima (ed.) *Mathematical Topics in Population Genetics.* Springer-Verlag, New York, pp. 32–78.

Turner, J. R. G. 1972. Selection and stability in the complex polymorphism of *Moraba scurra. Evolution* 26:334–343.

Turner, J. R. G. 1977. Butterfly mimicry: The genetical evolution of an adaptation. *Evol. Biol.* 10:163–206.

Turner, J. R. G., M. S. Johnson and W. F. Eanes. 1979. Contrasted modes of evolution in the same genome: Allozymes and adaptive change in *Heliconius. Proc. Natl. Acad. Sci. U.S.A.* 76:1924–1928.

United Nations Demographic Yearbook. 1973. U. N. Publishing Service, New York.

van Delden, W., A. C. Boerma and A. Kamping. 1978. The alcohol dehydrogenase polymorphism in populations of *Drosophila melanogaster.* I. Selection in different environments. *Genetics* 90:161–191.

van Delden, W., A. Kamping and H. van Dijk. 1975. Selection at the alcohol dehydrogenase locus in *Drosophila melanogaster. Experientia* 31:418–420.

Van Valen, L. 1963. Haldane's dilemma, evolutionary rates and heterosis. *Am. Nat.* 97:185–190.

Van Valen, L. 1973. A new evolutionary law. *Evolutionary Theory* 1:1–30.

Voelker, R. A., C. C. Cockerham, F. M. Johnson, H. E. Schaffer, T. Mukai and L. E. Mettler. 1978. Inversions fail to account for allozyme clines. *Genetics* 88:515–527.

Volterra, V. 1926. *La Lutte Pour La Vie.* Gauthier, Paris.

Wade, M. J. 1976. Group selection among laboratory populations of *Tribolium. Proc. Natl. Acad. Sci. U.S.A.* 73:4604–4607.

Wade, M. J. 1977. An experimental study of group selection. *Evolution* 31:134–153.

Wade, M. J. 1978. A critical review of the models of group selection. *Quart. Rev. Biol.* 53:101–114.

Wahlund, S. 1928. Zuzammensetzung von Populationen und Korrelationserscheinungen vom Standpunkt der Vererbungslehre aus betrachtet. *Hereditas* 11:65–106.

Wallace, B. 1968. *Topics in Population Genetics.* W. W. Norton, New York.

Wallace, B. 1970. *Genetic Load.* Prentice-Hall, Englewood Cliffs, New Jersey.

Wallace, B. 1975. Hard and soft selection revisited. *Evolution* 29:465–473.

Wangersky, P. J. 1978. Lotka-Volterra population models. *Ann. Rev. Ecol. Syst.* 9:189–218.

Watanabe, T. 1967. Infectious drug resistance. *Sci. Am.* 217:19–27.

Watanabe, T. K. and T. Watanabe. 1977. Enzyme and chromosome polymorphisms in Japanese natural populations of *Drosophila melanogaster. Genetics* 85:319–329.

Watson, J. D. 1968. *The Double Helix.* Atheneum, New York.

Watson, J. D. and F. H. C. Crick. 1953. Molecular structure of nucleic acids. *Nature* 171:737–738.

Watterson, G. A. 1977. Heterosis or neutrality? *Genetics* 85:789–814.

Wehrhahn, C. and R. W. Allard. 1965. The detection and measurement of the effects of individual genes involved in the inheritance of a quantitative character in wheat. *Genetics* 51:109–119.

Weinberg, W. 1908 (trans. by S. H. Boyer). On the demonstration of heredity in man. 1963. *Papers on Human Genetics.* Prentice-Hall, Englewood Cliffs, New Jersey.

Weir, B. A., R. W. Allard and A. L. Kahler. 1972. Analysis of complex alloyzme polymorphisms in a barley population. *Genetics* 72:505–523.

Weir, B. S., A. H. D. Brown and D. R. Marshall. 1976. Testing for selective neutrality of electrophoretically detectable protein polymorphisms. *Genetics* 84:639–659.

Weir, B. S. and C. C. Cockerham. 1978. Testing hypotheses about linkage disequilibrium with multiple alleles. *Genetics* 88:633–642.

Westoll, T. S. 1949. On the evolution of the Dipnoi. In G. L. Jepson, E. Mayr and G. G. Simpson (eds.). *Genetics, Paleon-*

tology, and Evolution. Princeton Univ. Press, Princeton, New Jersey.
White, M. J. D. 1968. Models of speciation. *Science* 159:1065–1070.
White, M. J. D. 1969. Chromosomal rearrangements and speciation in insects. *Ann. Rev. Genet.* 3:75–98.
White, M. J. D. (ed.). 1974. *Genetic Mechanisms of Speciation in Insects.* Australia and New Zealand Book Co., Sydney.
White, M. J. D. 1978. *Modes of Speciation.* W. H. Freeman, San Francisco.
Williams, G. C., R. K. Koehn and J. B. Mitton. 1973. Genetic differentiation without isolation in the American eel, *Anguilla rostrata. Evolution* 27:192–204.
Wills, C. 1973. In defense of naive pan-selectionism. *Am. Nat.* 107:23–34.
Wilson, A. C., G. L. Bush, S. M. Case and M.-C. King. 1975. Social structuring of mammalian populations and rate of chromosomal evolution. *Proc. Natl. Acad. Sci. U.S.A.* 72:5061–5065.
Wilson, A., S. S. Carlson and T. J. White. 1977. Biochemical evolution. *Ann. Rev. Biochem.* 46:573–639.
Wilson, A. C., L. R. Maxson and V. M. Sarich. 1974. Two types of molecular evolution. Evidence from studies of interspecific hybridization. *Proc. Natl. Acad. Sci. U.S.A.* 71:2843–2847.
Wilson, A. C., V. M. Sarich and L. R. Maxson. 1974. The importance of gene rearrangement in evolution: evidence from studies on rates of chromosomal, protein, and anatomical evolution. *Proc. Natl. Acad. Sci. U.S.A.* 71:3028–3030.
Wilson, D. S. 1975. A theory of group selection. *Proc. Natl. Acad. Sci. U.S.A.* 72:143–146.
Wilson, E. O. 1971. *The Insect Societies.* Harvard Univ. Press, Cambridge, Massachusetts.
Wilson, E. O. 1975. *Sociobiology: The New Synthesis.* Harvard Univ. Press, Cambridge, Massachusetts.
Wilson, E. O. and W. H. Bossert. 1971. *A Primer of Population Biology.* Sinauer Associates, Sunderland, Massachusetts.
Wilson, S. P., H. D. Goodale, W. H. Kyle and E. F. Godfrey. 1971. Long term selection for body weight in mice. *J. Hered.* 62:228–234.
Wolstenholme, G. E. W. and M. O'Connor (eds.). 1969. *Bacterial Episomes and Plasmids.* Little, Brown and Co., Boston.
Wright, S. 1958. *Systems of Mating and Other Papers.* Iowa State College Press, Ames.
Wright, S. 1964. The distribution of self-incompatibility alleles in

populations. *Evolution* 18:609–619.

Wright, S. 1967. "Surfaces" of selective value. *Proc. Natl. Acad. Sci. U.S.A.* 58:165–172.

Wright, S. 1969. *Evolution and the Genetics of Populations.* Vol. 2. *The Theory of Gene Frequencies.* Univ. of Chicago Press, Chicago.

Wright, S. 1970. Random drift and the shifting balance theory of evolution. In K. Kojima (ed.). *Mathematical Topics in Population Genetics.* Springer-Verlag, New York, pp. 1–31.

Wright, S. 1977. *Evolution and the Genetics of Populations.* Vol. 3. *Experimental Results and Evolutionary Deductions.* Univ. of Chicago Press, Chicago.

Wright, S. 1978. *Evolution and the Genetics of Populations.* Vol. 4. *Variability Within and Among Natural Populations.* Univ. of Chicago Press, Chicago.

Wynne-Edwards, V. C. 1962. *Animal Dispersion in Relation to Social Behavior.* Oliver and Boyd, Edinburgh.

Yamazaki, T. 1977. Enzyme polymorphism and functional difference: Mean, variance, and distribution of heterozygosity. In M. Kimura (ed.). *Proc. 2nd Taniguchi Int. Symp. on Biophysics: Mol. Evol. and Polymorphism,* pp. 127–147.

Yardley, D. G. 1978. Selection at the amylase locus of *D. melanogaster*: A word of caution. *Evolution* 32:920–921.

Yasuda, N. 1968. Distribution of matrimonial distance in the Mishima District. *Proc. XIIth Int. Cong. Genet.* 2:178–179.

Yokoyama, S. and J. Felsenstein. 1978. A model of kin selection for an altruistic trait considered as a quantitative character. *Proc. Natl. Acad. Sci. U.S.A.* 75:420–422.

Yokoyama, S. and M. Nei. 1979. Population dynamics of sex-determining alleles in honey bees and self-incompatibility alleles in plants. *Genetics* 91:609–626.

Zouros, E. 1976. The distribution of enzyme and inversion polymorphism over the genome of *Drosophila*: Evidence against balancing selection. *Genetics* 83:169–179.

Zuckerkandl, E. 1976. Evolutionary processes and evolutionary noise at the molecular level. II. A selectionist model for random fixations in proteins. *J. Mol. Evol.* 7:269–311.

Author Index

AUTHOR INDEX

AUTHOR INDEX

AUTHOR INDEX

AUTHOR INDEX

AUTHOR INDEX

Subject Index

SUBJECT INDEX

SUBJECT INDEX

SUBJECT INDEX